The Lisheen Mine Archaeological Project
1996–8

The Lisheen Mine Archaeological Project
1996–8

Managing Editor

Margaret Gowen

Contributing Editors

Michael Phillips
John Ó Néill
Irish Archaeological Wetland Unit

Wordwell

2005

First published 2005
Wordwell Ltd
(on behalf of Margaret Gowen & Co. Ltd)
PO Box 69
Bray
Co. Wicklow
www.wordwellbooks.com

A CIP catalogue record for this book is available from the British Library..

ISBN 1 869857 83 6

Production and layout: Irish Archaeological Wetland Unit.

This publication has received support from the Heritage Council

Printed by Cromwell Press Trowbridge, Wiltshire, England

Contents

Foreword

In 1983 a major conference on wetland archaeology took place in London. Papers concerning wetland archaeology in most areas of Europe were presented but, conspicuously, Ireland was absent. It is not true to say that at the time there was no wetland archaeology in Ireland. Indeed, writing a year later, John Coles was able to state "… destruction of Irish peatlands has been going on for decades, and there has been no national response".

This all changed, however, in 1985 as the archaeological importance of the wetlands came to be recognised. Soon systematic survey and selective excavation began, along with an element of scientific analysis of the evidence.

It must be admitted, however, as more and more trackways and other archaeological features were exposed, the scientific and archaeological exploitation of this material has not been satisfactory. Proper wetland excavation is complex, time-consuming and exceptionally expensive and is, indeed, only possible in the context of a major national initiative.

In fact, only a single wetland excavation project in Ireland has involved the sort of scale, detailed analysis and scientific rigour which should be the norm. This is the Lisheen Mine Archaeological Project.

Generous funding by Minorco/Lisheen enabled all aspects of the excavations at Lisheen, over two long seasons, to be carried out to the highest level of archaeological excellence. In addition, as is evident from this volume, an extensive and intensive range of detailed environmental studies was carried out, both by Irish and foreign specialists, on a scale of intensity which had never before taken place in this country. Indeed, it may well be that such a detailed campaign of specialist study will not happen again in any Irish bog.

The present volume is a fitting monument to the exceptional quality of the excavations and the research which followed. This book, indeed, is a masterly synopsis of what wetland archaeology can be. It is thus a matter of sadness that in all probability future generations, viewing the wastelands of former Irish bogs, will regard Lisheen with regret as a lone example of what should have been.

Professor Barry Raftery

Acknowledgements

This very large project involved a great number of people, many of whom made a very significant contribution. The project was designed to be integrated yet to allow each specialist total discretion with the scope and content of their own projects. The approach focused on the integration of the final results and was facilitated by a high level of communication and structured dialogue, on and off site, between the specialists and archaeologists. The generosity of all involved and the open and enthusiastic exchange of information at an academic and technical level underpinned the project design in a way that was almost unimaginable at the outset; and the quality of the outcome speaks for itself.

At its earliest stage, the project was very fortunate to have the input of the Irish Archaeological Wetland Unit and of Aonghus Moloney and Prof. Barry Raftery in particular. Prior to the commencement of fieldwork, Dr Ann Lynch of the National Monuments Section, Department of the Environment, Heritage and Local Government (formerly *Dúchas*) provided much needed advice and assistance on a number of aspects of the project design. The project was very fortunate in being able to resource the academic relationships between Prof. Barry Raftery, Dr Wil Casparie and Dr Chris Caseldine. The scope of the project design was guided and assisted by each of them.

The astonishing contribution of Wil Casparie in particular must be acknowledged. The great weight of his extensive knowledge, experience and generosity permeated all aspects of the fieldwork and reached a great number of the individuals involved, all of whom learned from him. Wil's work was supported by the huge efforts of Bernie Owens, Geodetic Surveyor. Undaunted by the scale of the undertaking, and driven by a project manager who felt instinctively that a contour survey of the peat basin would assist Wil's work, Bernie spent weeks and weeks over a 2-year period out on the bog in all sorts of conditions, probing, measuring and recording. The project must acknowledge this extremely important contribution. In many respects, everything else fell in behind this baseline survey.

Both Margaret Stokes and Mike Dane of Minorco Lisheen Ltd (now Lisheen Mine Ltd) oversaw the development of the project design and its operational practicalities, together with the re-design of the Tailings Management Facility and the early, difficult servicing of the large field staff in 1996. Margaret Stokes remained with the project and has to be credited with securing all aspects of its resourcing from its earliest stages.

The site required a very large staff, all of whom worked in very varied and often extremely inhospitable conditions. The teams under the supervision of the excavation directors have to be credited with tenacity and a focus on doing an excellent job.

Given the long history of the project, compiling, editing and producing the finished report became a mammoth task that required the input of a large number of people. The final production and editing of the report was overseen and co-ordinated by Michael Phillips. Text editing was undertaken by Margaret Gowen, John Ó Néill, Eileen Reilly, Sarah Cross May and Lindsay Rafter. Maps and images were prepared by Kieron Goucher, Penny Iremonger, Shirley Markley, Sadbh McIlveen, Bernie Owens and Mario Sughi. Other assistance was provided by Maria Pullen, Nessa Walsh and Keay Burridge. Final thanks must be given to Conor McDermott, Nathalie Rynne and Michael Stanley of the Irish Archaeological Wetland Unit for bringing all the graphics and text to publication.

Wil Casparie would like to acknowledge the help and assistance of the following individuals:

Many aspects, if not all, of the peat-stratigraphical study have been discussed in the most pleasant and enthusiastic way with the archaeological directors of the Lisheen Archaeological Project—Sarah Cross May, Cara Murray, John Ó Néill, Paul Stevens, Tim Coughlan and Malachy Conway (Operations Manager); as well as Simon Dick (excavation supervisor) and the specialists—Bernie Owens, Eileen Reilly, Ingelise Stuijts, Chris Caseldine and Ben Gearey. Also thanks to Margaret Gowen, who designed the project brief and managed the Lisheen Archaeological Project, and Prof. Barry Raftery, the academic consultant of the project, who were excellent and inspiring partners in both the work and extensive scientific discussions. For all the archaeological, geographic, palaeobotanical and other helpful and very valuable information and comments, and for the great hospitality over many years, I am indebted to all the above mentioned colleagues. Without their expertise and help I would not have been able to complete the study in this way.

I was also involved in the archaeological survey of Derryville Bog, conducted by the Irish Archaeological Wetland Unit, and would like to thank the staff members, both current and former, Conor McDermott, Aonghus Moloney and Nóra Bermingham for the excellent and fruitful co-operation, nice discussions and warm friendship over the years.

I discussed all the main hydrological aspects of the study with my friend Jan Streefkerk, Senior Hydrologist of the National Forestry Service of the Netherlands (Staatsbosbeheer), and expert in bog hydrology, and would like to thank him very much for his large interest in the study and for his very valuable advice.

Eileen Reilly would like to acknowledge the help and assistance of the following individuals:

Thank you to Margaret Gowen for the opportunity to work on this material and to all the field directors during the 1996/7 seasons—Sarah Cross May, Cara Murray, Paul Stevens, John Ó Néill, Tim Coughlan, and also Eoin Sullivan (supervisor).

Sincere thanks to Dr David, Smith Institute of Archaeology and Antiquity, University of Birmingham, for his help with the identifications, useful discussions on the findings and for the use of the Gorham and Girling Collections. Thanks also to Dr Nicki Whitehouse, School of Archaeology and Palaeoecology, Queen's University Belfast, for help in identifying certain species. Thanks to Mike Morris (retired), Institute of Terrestrial Ecology, Furzebrook, for informal comments on some of the insects found at Derryville, and thank you to Mark Brennan, Berkeley Library, Trinity College Dublin, for providing access to much of the reference material.

Special thanks to:

Department of the Environment, Heritage and Local Government (formerly Dúchas *the Heritage Service)*
Dr Ann Lynch, Ed Bourke and Ciara O'Donnell (licensing section), Victor Buckley, Hugh Carey, Tom Condit and Barry O'Reilly

National Museum of Ireland
Mary Cahill, Raghnall Ó Floinn and Andy Halpin

Lisheen Mine Ltd
Margaret Stokes, Mike Dane, Willie Stapleton, Rita King, Eileen Kennedy, Ann Haslett, Simon Archer, Bernard Hyde and Maurice Mackarel

Bord na Móna
Paddy Behan, Mike Reddin, Dónal Ryan, Liam Dempsey and Ray Yates

Mulcair Civil Engineering
Pat McCarthy and Emer O'Brien

Kvaerner-Cementation (Ireland)
Eric Copeland, Mark Lloyd and Steve Barber

ESB
John Connolly, Nicholas Dunne and Tadgh O'Brien

Irish Archaeological Wetland Unit
Aonghus Moloney, Conor McDermott and Nóra Bermingham

Beta Analytic Inc.

Palaeoecology Centre, Queen's University Belfast
Dave Brown and Gerry McCormac

AOC Scotland
John Barber and Clare Ellis

GeoArc Ltd.

Support
Michael King, Tom Murphy and Sean Spain

Focus Films
Ger Cantwell

Landowners
Andrew Cantwell, Richard Little, Jim O'Grady and Denis Sheehy

Report editing and production
Michael Phillips, Margaret Gowen, John Ó Néill, Eileen Reilly, Sarah Cross May, Lindsay Rafter, Kieron Goucher, Penny Iremonger, Shirley Markley, Sadbh McIlveen, Bernie Owens, Mario Sughi, Maria Pullen, Nessa Walsh and Keay Burridge
Conor McDermott, Nathalie Rynne and Michael Stanley of the Irish Archaeological Wetland Unit

Key personnel and excavation team 1996, 1997 and 1998

Project management
Margaret Gowen (project design and management)

Academic consultant
Prof. Barry Raftery

Specialists
Dr Wil Casparie (peat development, morphology and hydrology)
Dr Chris Caseldine (palynology and related studies)
Dr Ben Gearey (palynology, testate amobae and related studies)
Dr Jackie Hatton (palynology and related studies)
Dr Ingelise Stuijts (wood species and growth analysis)
Eileen Reilly (palaeoentomology)
Bernie Owens (geodetic survey)

Archaeological directors
Cara Murray 96E202, 96E066
Paul Stevens 96E298, 97E372, 97E036, 97E051, 97E168
John Ó Néill 97E158, 97E439
Sarah Cross May 97E160

Operations and dryland archaeological assessment 1997
Malachy Conway 96E237

Directors of early excavation work
Irish Archaeological Wetland Unit 94E106
Tim Coughlan 96E203

Supervisors
Simon Dick, Helen Farrell, Steve Jones, Alice Mayers, Eoin Sullivan and Chris Swain

Illustrators
Sarah Ryan-Stevens, Nina Koeberl and Elspeth Logan

Archaeologists
Peter Barker, Karl Brady, Jerome Cameron, Robert Chapple, Sinéad Cafferkey, Zoë Clarke, Chris Conway, John Conway, Matthew Lancelot Crook, Siobhán Deery, Rupert Detheridge, Bernie Doherty, Jo Dullaghan, Stuart Elder, Declan Enright, Niall Fennelly, John Fletcher, Aideen Foley, Lucy Ford-Hutchinson, Anke Halmschlag, Joanne Hamilton, Penny Johnston, Doreen Keating, Mark Keegan, Eoghan Kieran, Bríd Kirby, Chris Long, Susan Lyons, Meriel McClatchie, Melanie McQuade, Gwynfor Maurice, Bernice Molloy, Cathy Moore, Declan Moore, Danny O'Brien, Brenda O'Meara, Andrew Orr, Sinéad Phelan, Natasha Powers, Francis Scully, Ted Smyth, Michael Stanley, Katherina Stephens and Angela Wallace

Local personnel
Regina Cleary, Eleanor Cleary, Mary Doherty, Evelyn Dunne, John Dunne, Dan Egan, Oliver Esmonde, Zane Everard, Denis Kennedy, James Kiely, John Kiely, Mark Kiely, Mark King, Mary Rose Maher, Brigena Moore, Sarah Murphy, Sinéad O'Dwyer, Michael Purdue, Claire Stapleton, Michael Stiepel and Denis Taylor

Photography/video
Mark Moraghan, John Sunderland and Martin Gavin

Survey
Bernie Owens, Kieron Goucher and Derek Copeland. Also *Minorco Lisheen surveyors* Paul Kelly, Paul Moran and Harry Twomey

Dublin-based operations
Nessa Walsh, Joyce Hickey and Dolores Cotter

1. Introduction

Margaret Gowen and John Ó Néill

Background

The Lisheen Archaeological Project was conducted as part of the planning requirement for the development of the Lisheen Mine, a large lead/zinc mine located near Moyne, Thurles, County Tipperary (see Figs 1.1 and 1.2). The project arose as a result of the findings of the Environmental Impact Study (EIS) prepared for planning purposes in advance of the development. The EIS examined a wide study area, focusing on nine townlands (Figs 1.3 and 1.4). Part of the study area included a 'peninsula' of bog extending southwards from a large area of Bord na Móna-worked raised peat bog called Derryville Bog, which is part of the southern extent of the Littleton raised bog complex. The basin in which this portion of the bog formed incorporates portions of the townlands of Derryfadda, Killoran and Cooleeny. It became the focus of intense archaeological interest when it was chosen as the site of the Tailings Management Facility (TMF) for the mine, an embanked reservoir some 72 hectares in extent.

The development

The mine development covered an area of over 200 hectares (Fig. 1.5). The TMF reservoir formed the easternmost element of the mine complex, with the mine 'decline' and above-ground plant site located adjacent to the bog on marginal agricultural land to the west of the bog and the TMF site. Several 'borrow area' quarry sites for the TMF embankment material were located further to the west.

Archaeological results of the EIS

The EIS desk study searches, aerial photography and fieldwork surveys revealed a great number of surviving recorded archaeological sites and monuments in the study area generally, but the great majority of these sites were located to the west and south of the proposed development area and no known recorded monuments on dryland were disturbed. A feature of the recorded sites was that they were generally located at slightly higher elevations than the lands on which the mine's surface development occurred, and most of the recorded sites were of Early Christian and later date. There were no entries in the Record of Monuments and Places (RMP) for the low-lying marginal land close to the peat basin and just two trackway sites were recorded within the bog. However, a preliminary field inspection of the development area for the purposes of the preparation of an impact assessment revealed a great number and range of potential archaeological sites within the bog. As design progressed and the bog became the chosen location for the TMF, it was decided that the area should be the subject of an intensive field survey appraisal. This first detailed appraisal was carried out by the Irish Archaeological Wetland Unit (IAWU) of University College Dublin.

Preliminary evaluation of the TMF site

The preliminary evaluation and survey by the IAWU in 1994 and 1995 indicated that the area was rich in archaeological remains. The survey formed an extension to ongoing archaeological survey and evaluation of the Littleton raised bog complex as a whole, which was in progress at the time. The results of this first survey in the Lisheen study area occasioned a substantial change in the design layout of the TMF in order to avoid a particularly rich archaeological area in the southwest corner of the study area (subsequently named the Cooleeny Complex). After preliminary investigation, this area was eventually cordoned off from the TMF development land in 1996 by a membrane of agricultural-grade plastic (see fold-out map), in an approach similar to that used for conservation purposes at Corlea, County Longford.

The 1995 IAWU survey formed the basis for the following intensive two-year wetland field project (1996–7), the results of which are presented in this book, along with the results from the dryland excavations of 1997 and 1998.

The project design

The framework for the archaeological project design was based on the planning requirement to mitigate for the impact of the development by creating a full archaeological record of the area in advance of construction. The project design sought to achieve more than the creation of a record of archaeological sites and features on a piecemeal, site-by-site basis but to use the opportunity presented by the scale of the construction project and the

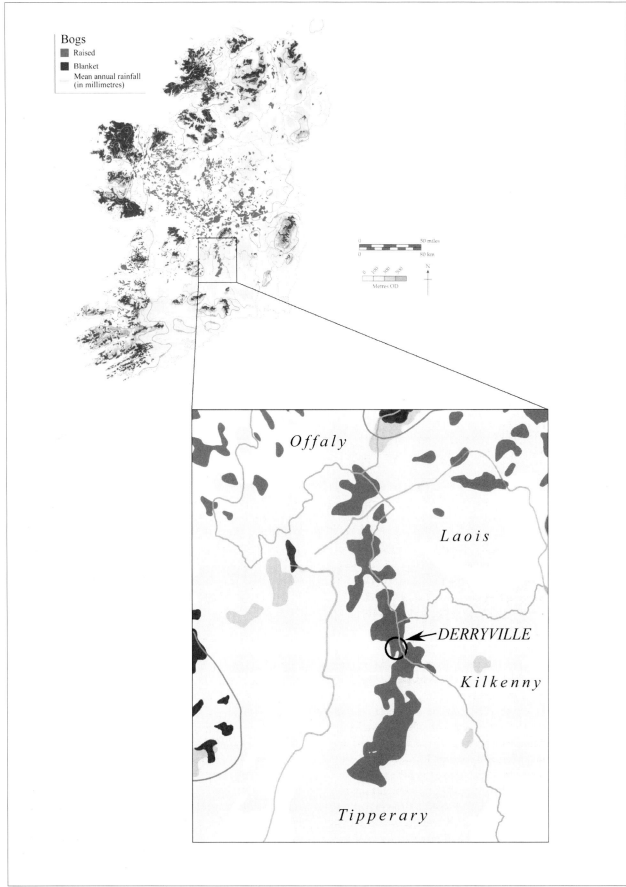

Fig. 1.1 *Location of Derryville Bog and surrounding peatlands (after Aalen* et al. *1997).*

landscape of the area to integrate a study of the palaeoenvironmental developent of the bog basin and adjoining lands with the results of detailed excavation, study and recording of its archaeological remains. The project design, using an interdisciplinary approach, sought to reveal the progression and dynamics of the bog development on the site in its general palaeoenvironmental context and to understand the links between this progression and the nature of the archaeological activity associated with it.

The project was undertaken in two linked field operations. The first was the integrated wetland palaeoenvironmental and archaeological study of the peat basin in which the TMF was to be located, conducted as a pre-development project. The second was the later recording and study of sites excavated as a consequnce of identification during the monitoring of soil-stripping on the dryland to the west of the TMF during the site preparation work for the plant site, in advance of bulk excavation for the mine 'decline'.

The first phase of work in the 72ha TMF development area was concentrated within the well-defined peat basin. The task set by the planning requirement was to resolve and provide a pre-intervention record of the area in advance of the mine development. This requirement was met by a project that sought to record the basin and surrounding topography, the growth of the raised bog system, in this particular location, in detail, and to record the full range, variety, chronology and relational context of the archaeological activity within its particular evolving palaeoecological framework. Although comparisons with this project design can be made with similar projects at the Somerset Levels, Flag Fen and the Mountdillon Bogs, the pre-defined limits of the Lisheen Mine complex created the need for a well-defined, fully realised project design from the outset, as the opportunity to revisit sites later would not exist. The project was also funded in a manner that ensured the security and realisation of the overall project design concept.

The chosen interdisciplinary approach to the wetland project was essential to its academic design. In the first instance, it sought to provide a highly detailed and comprehensive survey of the underlying topography of the peat basin and the dryland relief of its surroundings. This singular approach, using intensive geodetic survey through a 2-season probing record of the peat basin as the other elements of fieldwork were underway, served to facilitate an extremely accurate study of the raised bog peat development and its hydrology. It also supported the highly accurate inclusion and incorporation of an analysis of the palynology of the peat basin and a range of related palaeoenvironmental studies. Apart from the baseline pollen analysis and the highly detailed study of the peat stratigraphy in the basin, the palaeoenvironmental studies included a large-scale and intensive analysis of the wood and charcoal from the archaeological sites, Coleopteran studies and a highly specialised study of testate amoebae as hydrological indicators.

While addressing the project objectives these varied analyses were conducted in a highly integrated project management and communications framework, lending the project a dynamic character that facilitated the exploitation of productive avenues of research as the project evolved and trends became evident from different sets of data. The project design aimed to achieve an extension of the archaeological excavation results using these specialist analyses, and it was hoped the results would place the archaeological activity within the palaeoenvironmental and chronological context of the development of the peat basin in a way that would illustrate how the bog development influenced archaeological activity. The 2-year timetable of the project and its intensive fieldwork programme, coupled with mid-season and end of season reviews by the project teams, significantly helped to maximise the results.

What was not anticipated at the outset was the degree to which the underlying topography of the basin would inform the peat development study and the understanding of the particular hydrology of the bog at the various phases of its development. Also surprising and rewarding for the team was the extent to which that study, in turn, would inform a very high level of definition in relation to the impact of human activity on phases of peat development. The testate amobae study was ground-breaking in this context in that its measurement of hydrological change could be readily and sucessfully compared with significant features and trends noted in the peat stratigraphy. Finally, the incorporation of a significant ^{14}C and dendrochronological dating programme provided a particularly high level of definition and linkage for all the results.

Monitoring of preparatory earthmoving on dryland during the construction phase of the project gave rise to the discovery of a variety of archaeological sites, none of which had any surface expression and all of which could be later linked to the sequence of archaeological activity defined by the findings of the study in the peat basin.

The findings

Sixty-six sites and features of various types, dating from the late Neolithic to the medieval period, were excavated within the bog. Archaeological monitoring of the dryland construction operations to the west of the bog revealed over twenty additional, largely prehistoric sites, many of them burnt mounds (*fulachta fiadh*). The project was fortunate that two of the sites were habitation sites, one with the remains of two well-preserved round houses. A flat cemetery of over thirty unmarked cremated pit burials was found to be broadly contemporary with the habitation sites. It was noteworthy that the discovery of the prehistoric sites occurred at lower topographic contour levels than the Early Christian and medieval sites in the vicinity, including those listed in the Record of Monuments and Places (Fig. 1.3).

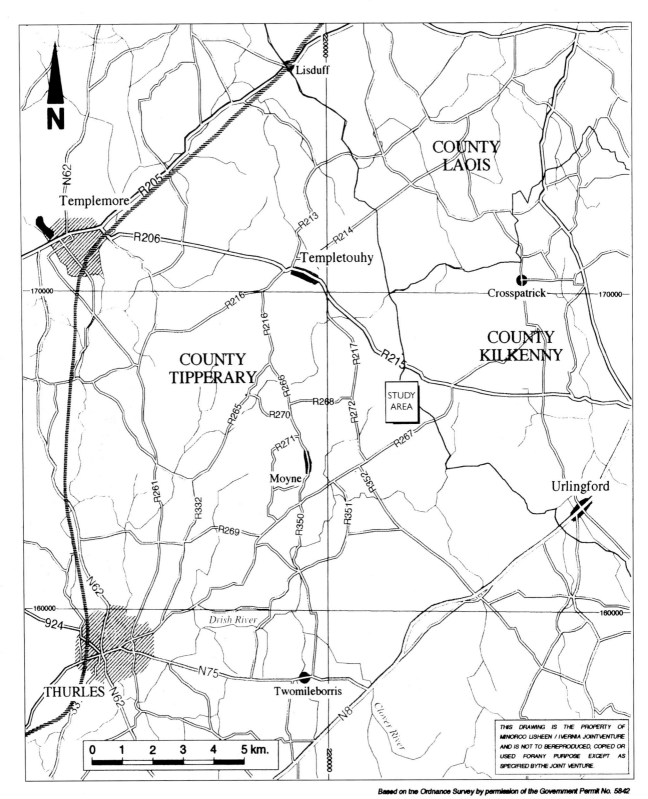

Fig. 1.2 Regional location map of the study area (inset).

As anticipated, and hoped for, the use of a multidisciplinary approach established both an overall palaeoenvironmental framework for the archaeological sequence and a microenvironmental context for individual sites. Moreover, the extensive palaeoenvironmental data gathered provided a context for the dryland excavations, which allowed them to be placed within an unusually well-defined, location-specific palaeoenvironmental sequence. The end product is a unique synthesis of cultural and environmental history, as seen through the surviving micro and macrofossil record of Derryville Bog and its hinterland.

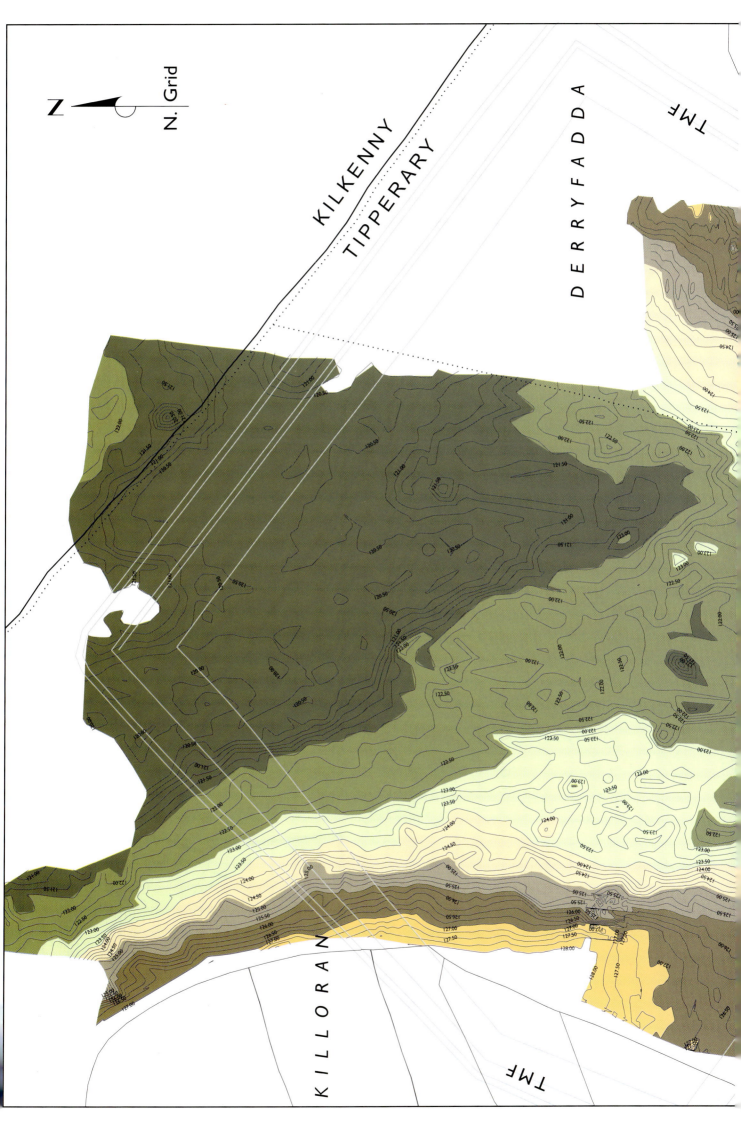

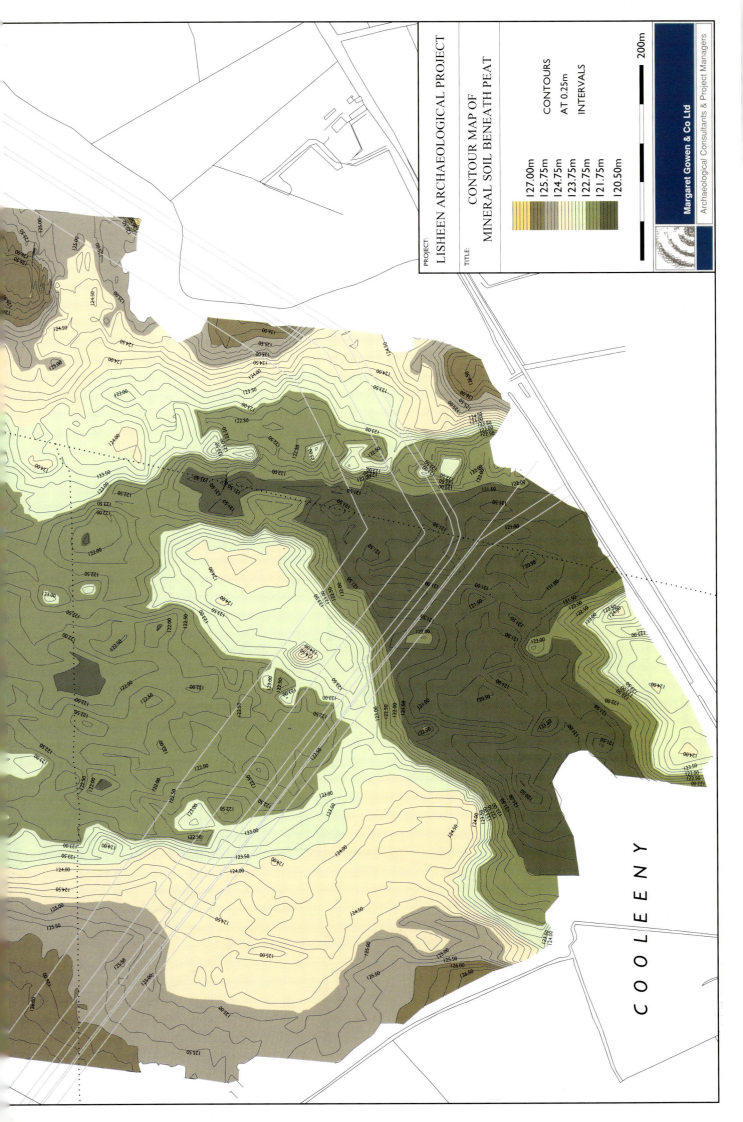

PROJECT: LISHEEN ARCHAEOLOGICAL PROJECT

TITLE: CONTOUR MAP OF
MINERAL SOIL BENEATH PEAT

CONTOURS
AT 0.25m
INTERVALS

127.00m
125.75m
124.75m
123.75m
122.75m
121.75m
120.50m

200m

Margaret Gowen & Co Ltd
Archaeological Consultants & Project Managers

C O O L E E N Y

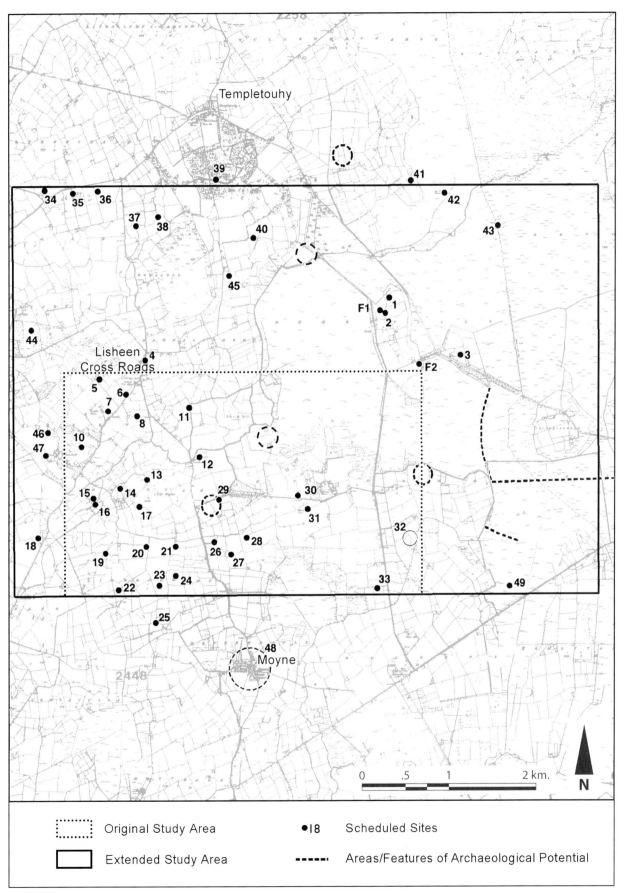

Templetouhy

Lisheen
Cross Roads

Moyne

0   .5   1   2 km.

N

⬚ Original Study Area   ●18   Scheduled Sites

▭ Extended Study Area   -----   Areas/Features of Archaeological Potential

Fig. 1.3 Overview of the original and extended study areas including recorded monuments.

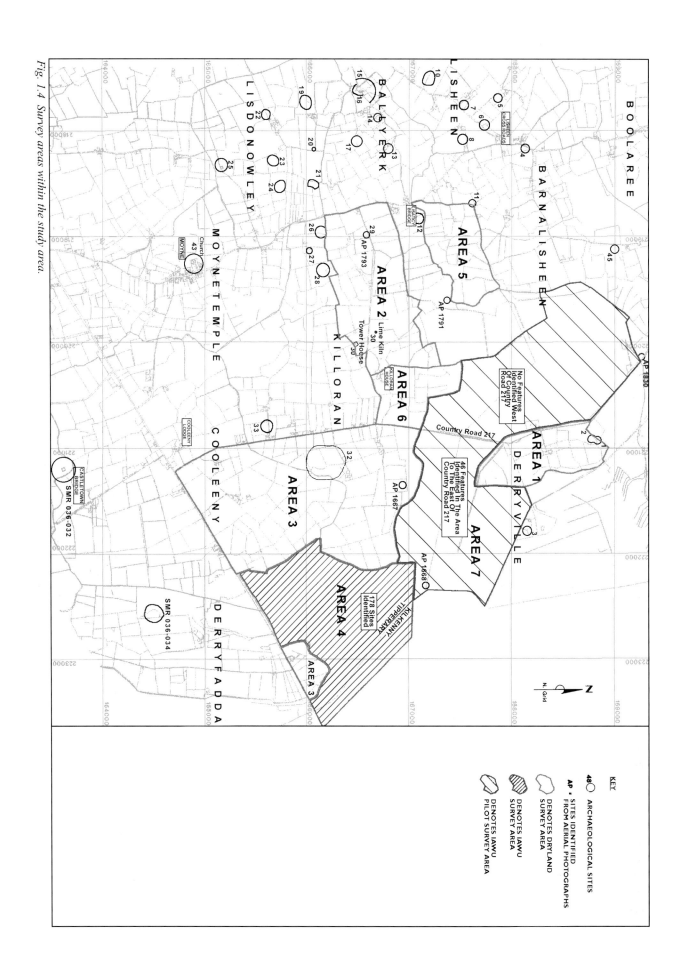

Fig. 1.4 Survey areas within the study area.

The project results suggest that palaeoenvironmental studies in a wetland context can be linked to archaeological findings in an extremely focused, useful and detailed manner. On completion, the project was able to illustrate not just archaeology in context, but a full description of the development of the peat basin, its surrounding landscape and, more particularly, events in the bog that both precipitated archaeological activity or were the result of archaeological activity.

In late 1999, the final results of the Lisheen Archaeological Project were published on a very limited basis as *The Lisheen Archaeological Project, 1996–98* (Gowen 1999). In addition to copies lodged with the Department of the Environment, Heritage and Local Government and the National Museum of Ireland, further copies were presented to a number of other institutions to facilitate access to the research results. These included the Royal Irish Academy, the Royal Society of Antiquaries of Ireland and the Department of Archaeology, University College Dublin.

This current volume presents an abridged version of the final report and, therefore, owes much to the original report structure. The research for the project was completed in 1999 and, with the exception of some additional bibliographic details, this volume reflects the state of knowledge at that time. Although many of the studies appear as stand alone chapters, the project design approach that was adopted meant that there was regular integration of study themes and analytical focus between the individuals involved. Thus, while each specialist or archaeologist was primarily concerned with his or her own discipline, the collective application of an integrated study was consciously fashioned as an interactive operation over several years.

Firstly, the results cover the history of Derryville Bog, describing the underlying geology and topography, the peat morphology and development of the bog, and the archaeological chronology. Following this are the palaeoenvironmental results, describing the findings of the palynology, testate amoebae, wood and charcoal and Coleoptera analyses. Finally, the archaeological excavations are described, including discussions on woodworking, terrain sensitivity and landscape context. Lists of radiocarbon and dendrochronology dates are provided at the end of this volume for ease of reference. A number of additional files can be downloaded from the Margaret Gowen & Co. Ltd website (www.mglarc.com), including an unabridged version of Dr Wil Casparie's peat morphology analysis, which was condensed for the printed book; the IAWU survey catalogue; Dr Ingelise Stuijt's site-specific wood species catalogue; Eileen Reilly's Coleoptera species lists in taxonomic order, and miscellaneous maps, files, drawings and photographs that could not be accommodated in the printed book.

Funding

All aspects of the archaeological fieldwork, environmental study, analysis and preliminary report preparation for this project were funded in full by Lisheen Mine Ltd. The project was supervised and administered for Lisheen Mine Ltd by Margaret Stokes, Environmental Manager, as part of the planning requirement for the project, which received its full planning permission in July 1997.

Site codes and field numbers

Three separate coding systems were used throughout the project and are represented within this book. During the initial IAWU survey, 178 individually identified sites were listed with codes issued by year, bog and site number—95DER0001 to 95DER0178 (DER representing an abbreviation of Derryville Bog). The Lisheen Archaeological Project adopted a shorter version of this coding system in the field over subsequent years, expressing the site codes as DER1 to DER178. As new wetland sites came to light during the excavations they were assigned new site codes using this system (i.e. site codes DER201 to DER400, although not all codes were assigned in sequence and so the code numbers do not reflect the total number of sites identified).

Following excavation, all of the wetland sites were allocated final site names according to the townland in which they were located, but retained their assigned site code number: thus DER22 became Cooleeny 22; DER2 became Derryfadda 2 and DER18 became Killoran 18. Sites are referred to in the text by the final site name, while the header for each catalogue entry (see Chapter 10) includes the site code in the DER form for archival purposes. References to sites with an IAWU site code refer exclusively to unexcavated sites recorded during the initial survey (IAWU 1996a).

From the outset, sites excavated on the dryland were assigned site codes and names according to the townland in which they were located. Thus, the three sites in Barnalisheen townland were assigned Barnalisheen 1, Barnalisheen 2 and Barnalisheen 3, and the sole site in Cooleeny townland was assigned Cooleeny 1. The twenty-nine sites in Killoran townland were assigned Killoran 1 to Killoran 31, although the site names Killoran 18 and Killoran 20 were reserved for two of the wetland sites. The header for each catalogue entry (see Chapter 11) includes the site codes in the following form: BAR1 to BAR 3, COO1 and KIL01 to KIL31.

During the excavations each drain and the adjacent field to the east were allocated a unique number for reference purposes. A total of fifty-eight fields were numbered from west to east across the bog. This field numbering system was used to indicate the location of palaeoenvironmental sampling sites and excavation cuttings.

Fig. 1.5 Location map of excavated dryland sites in Killoran and Barnalisheen townlands.

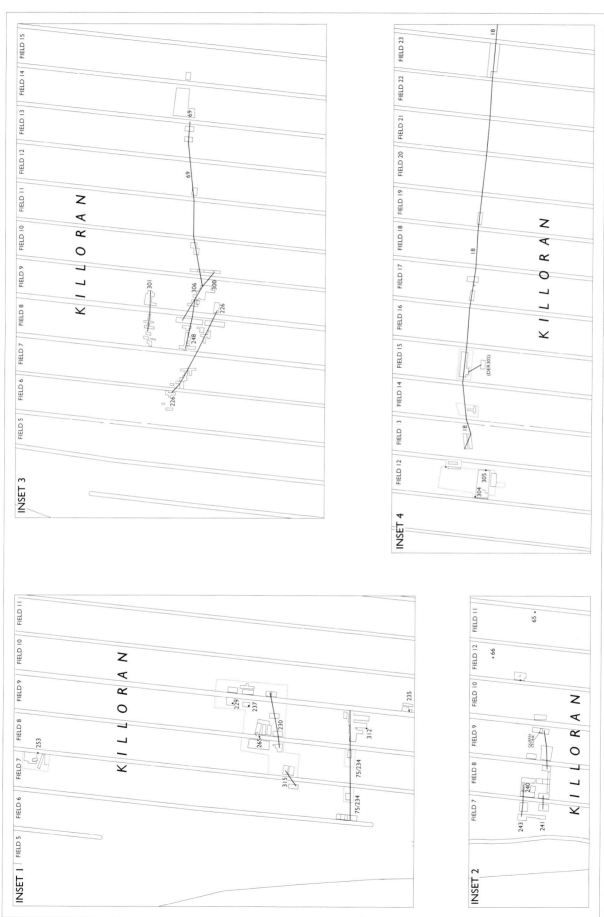

Fig. 1.6 Insets to fold-out plan of overall study area.

2. Geology and topography

Wil Casparie and Margaret Gowen

Geology

The Lisheen lead/zinc deposit occurs within the 'Rathdowney trend,' which runs for 40km in a northeast–southwest direction between Abbeyleix, County Laois and Thurles, County Tipperary. Lower Carboniferous Limestone underlies the area. To the northwest lie the Devil's Bit Mountains, a Silurian inlier overlain by sandstone and conglomerates of Devonian–Carboniferous age. To the southwest lie the Upper Carboniferous sandstone and coal-bearing sequences of the Slieve Ardagh Hills area.

In the Lisheen area, the Lower Carboniferous sequence records a marine transgression over Old Red Sandstone (Devonian) alluvial plane deposits (represented by the Mellom House to Ballysteen formations) and the establishment of continuous marine deposits, predominantly carbonate sedimentation (represented by the Ballysteen, Waulsortian and Crosspatrick formations (RPS Cairns Ltd 1995, fig. 7.1.1)). The ore-bearing deposit of sulphide mineralisation typically lies at depths of 170–210m, associated with the Killoran, Derryville and other faults in several large, fault-bounded blocks.

Topography and relief

The Lisheen Mine area is located between Templemore, Urlingford and Thurles, close to the villages of Templetouhy and Moyne.

The topography of the area is characterised by a generally flat plain with gentle north–south undulations, sloping gradually towards the south. The plain measures approximately 25km east–west, and its elevation is 130m Ordnance Datum (OD); there are some low hills within the area, not exceeding 196m OD.

Two rivers, the Drish River and the River Suir, drain it. The latter flows north–south on the western side of the area and joins the Drish south of Thurles. Within the mine area, raised bog developed in a system of glacial valleys or a glacial tongue-basin lying between 140m OD in the south and about 100m OD near Roscrea. Derryville Bog lies between 128m and 120m OD.

Soils

The conclusions of a study conducted by Pete Coxon as part of the 1995 EIS suggested that the limestone bedrock lies close to the surface throughout the EIS study area (*ibid.*, fig. 8.1.1). It is suggested that much of the undulating nature of the landscape can be generally attributed to karst development in the bedrock. Outcropping rock occurs in some locations, and the presence of a well-preserved lime kiln just west of Killoran House provides evidence for limestone-processing.

Thin till composed of a sandy diamictite containing predominantly local limestone clasts covers much of the higher ground. Sand and gravel deposits occur locally, including a deposit at Carrick Hill, and, in one low-lying area studied, silty soils that appeared to have been derived from alluvium were found.

The majority of the farmland in the area lies to the south and west, extending unbroken by bog towards Moyne and Thurles. On the eastern side, the study area is dominated by Bord na Móna-worked raised bog, which forms part of what is known as the Littleton raised bog complex. About two-thirds of the farmland consists of brown earth or grey-brown podzolic soils developed from the calcareous till on undulating topography. The remainder of the farmland is composed of gley, peaty gley and peat formed in depressions from calcareous till or alluvium.

The farmland is mainly in dairy farming and is made up of small, irregular, improved pasture fields enclosed by hedgerows. The pattern of fields has remained virtually unchanged since the 1903 Ordnance Survey (OS). The land usage has been kind to the field monuments up to quite recently, but land improvement has seen the levelling of ringfort-type enclosures in the locality.

The surface mine development is located at the southwestern edge of the Derryville raised bog complex, incorporating one of the characteristic series of 'peninsulas' of bog that extend southwards from it. In this case, the spit of bog crosses the Crosspatrick road and extends in a north–south basin towards the Drish River. The TMF has been constructed within this bog.

Geographical context of Derryville Bog

Derryville Bog belongs to a string of raised bogs east of a line from Thurles through Templemore to Roscrea in County Tipperary. The total length of this string is over 40km, running from the northwest flank of the

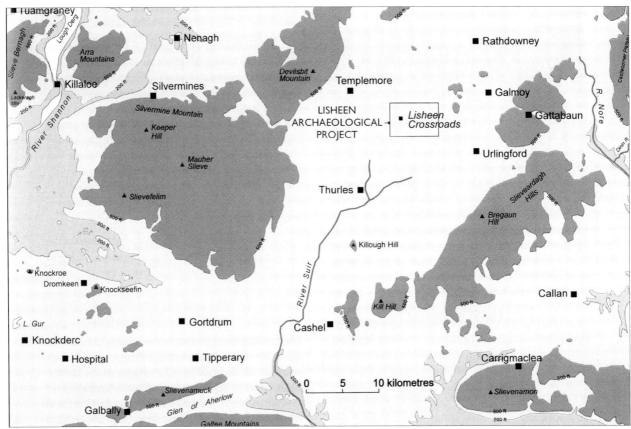

Fig. 2.1 The topography of central Tipperary indicating the location of the extended study area of the Lisheen Archaeological Project as shown in Fig. 1.3. (after Archer et al. 1996. Reproduced with permission of the Geological Survey of Ireland).

Slieveardagh Hills, between Horse and Jockey and New Birmingham to the south, to the valley of the River Nore to the north, between Roscrea and Knock. Most of these raised bogs are situated just on and west of the border between County Tipperary on the western side and Counties Kilkenny and Laois on the eastern side. The bogs developed in a system of glacial valleys or a glacial tongue-basin between 140m OD in the south to about 100m OD near Roscrea. Derryville Bog lies between 128m and 120m OD.

The width of these bogs generally varies between 2km and 5km; the portion of Derryville Bog in which this

project has been conducted is about 1km wide. Within this area, the main discharge direction is to the north.

In both the western and eastern margins of the bog, a complex system of elevations of dryland and peat-filled depressions occurs owing to the presence of a number of underlying ridges in the till, with a general northwest–southeast orientation. The height variation amounts to about 3m. This area was situated just inside the southern extent of the ice sheet of the last glacial period, so it can be assumed that these ridges are mostly moraines or drumlins. Localised boulder clay is present in the subsoil of the eastern fringe of Derryville Bog.

3. Peat morphology and bog development

Wil Casparie

Introduction

Aims

In this chapter the development of Derryville Bog (Fig. 3.1) will be discussed. The research is based on peat-stratigraphical survey of the bog, the geodetic survey and detailed levelling of the peat basin by Bernie Owens, the palaeoenvironmental information of the archaeological excavations, documented in close co-operation with the archaeological directors, and the research by the specialists of the Lisheen Archaeological Project, as delivered in their particular chapters.

The main aims of the peat-stratigraphical study were:

1. To describe the overall conditions of peat growth necessary for understanding the development of this bog, in which so many trackways, platforms and stake rows were found.

2. To analyse in as much detail as possible the peat sequence, in order to reconstruct environmental conditions of the bog as spatial occurrence for thousands of years.

3. To understand the contexts in which so many trackways and platforms were constructed in this small bog.

An outline of different aims in peat research

Raised bogs, their peat sequence and their vegetation have been subjects for scientific investigation for many decades. The aims of research were their occurrence, stratigraphy, growing mechanisms and spatial situation, but also the impact of climate, the vegetation dynamics, the typical hydrology, the occurrence of disasters and the accessibility in earlier times as can be recorded from the many publications in this field of science. Next to this, the commercial aspects of peat deposits in relation to the use of peatland as arable soils, and peat extraction, for instance, have been important aims for peat investigations. A complete or even a representative overview of all the peat research cannot be presented. Only a few examples will be mentioned here.

Overbeck (1975) and Feehan and O'Donovan (1996) can be seen as authors of extensive monographs, generally focused on many aspects of peat growth. The peat sequence and peat types are among others described by Grosse-Brauckmann (1980) and, specifically for one Irish bog, by Bloetjes and van der Meer (1992) and Hill

(1992). Studies of the peat stratigraphy and analyses of the peat growth mechanisms have been made by Casparie (1972; 1993) and Casparie and Streefkerk (1992). For the climatic impact a larger number of papers can be mentioned, such as those by Barber (1981), Dupont (1986; 1987), van Geel (1976) and van der Molen (1992). The hydrology and palaeoecological development of mires, and more specifically of raised bogs, have been studied a number of times by, among others, Streefkerk and Casparie (1989), Casparie and Moloney (1994) and Petzelberger *et al.* (1999). The vegetation development and the impact of deforestation of uplands surrounding raised bogs have been studied many times based on palaeoecological information from peat deposits. A few examples are mentioned here: Caseldine *et al.* 1997, van Geel *et al.* 1989 and Casparie 1992. By far the studies of most importance for understanding the prehistoric and historic access to raised bog surfaces are peat-stratigraphical investigations combined with wetland archaeological research. The best examples of such studies are those of the Somerset Levels by John Coles and others (1975–89), detailed in the *Somerset Levels Papers*, and of the German raised bogs by Hajo Hayen (1985; 1987; 1991). Further mention can be made of Casparie (1982; 1984; 1986; 1987), Casparie and Molema (1990), Casparie and Moloney (1996) and Casparie *et al.* (1992). Basic information on modern Irish trackway research as has been conducted by the Lisheen Archaeological Project is presented by Raftery (1990; 1996) and by the IAWU (1996a). The latter report, in particular, was the basis for the Lisheen Archaeological Project.

The peat-stratigraphical study of Derryville Bog, as presented in this chapter, has been undertaken in close co-operation with the large number of investigations presented in this book, such as those by Gowen, Caseldine, Gearey, Hatton, Cross May, Murray, Ó Néill, Reilly, Stevens and Stuijts. The Lisheen Archaeological Project adopted a multidisciplinary approach. This chapter on bog development, and focused on the problems of access in earlier periods, is a central part of that approach.

Methods

During the fieldwork campaigns all the peat faces of the sixty Bord na Móna (BnM) drains were analysed and a large number of them were investigated and documented, based on detailed observations. The spread, distribution and extent of a large number of features were established

Fig. 3.1 Derryville Bog in the Lisheen area, Counties Tipperary, Laois and Kilkenny. The uplands are indicated with their (modern) parcelling structure, the peatlands (mostly Bord na Móna properties) with the general mire symbol. The rectangle with broken lines indicates the plan of Figs 3.2, 3.3 and 3.13–3.19. The study area is situated between Killoran, Cooleeny and Derryfadda.

and documented. The botanical composition of many peat samples was identified and described in the field. All of this information resulted in the presentation of over sixty drawings of peat sections, profiles and diagrams, the design of six distribution and extent maps, and the description of peat types, distribution patterns, peat-growing processes and bog hydrological dynamics. A representative selection of nine of these peat sections and peat profiles are presented in this chapter. Based on the extensive survey and documentation of the peat types eight distribution maps have been designed, covering the bog development from about 5000 BC on. As previously mentioned, all aspects of the peat-stratigraphical survey have been discussed with the members of the Lisheen Archaeological Project.

The topographical and geographical situation

Derryville Bog belongs to a string of raised bogs, east of a line from Thurles through Templemore to Roscrea in County Tipperary. To this string belongs the 'Littleton Bog Complex' of BnM. Its total length is over 40km, running from the northwest flank of the Slieveardagh Hills on the south to the valley of the River Nore to the north. The bogs developed in a system of glacial valleys or a glacial tongue-basin on levels between 140m OD in the south to about 100m OD near Roscrea. Derryville Bog lies between 128m and 120m OD and is situated close to the middle of this string of bogs. The width of the bogs varies between 2km and 5km; the portion of Derryville Bog in which this project has been conducted measures about 1km. Within this area the main discharge direction is to the north, but substantial discharge in a southerly direction also occurred. In general, the subsoil slopes about 1 *promille* from south to north.

Most of the subsoil in the basin consists of plastic clays. The western margin (Cooleeny and Killoran upland) and the eastern margin (Derryfadda) are probably drumlins of the Last Glacial, with their levels 10–15m above the Derryville basin, as the simplified contour map (1m intervals) indicates (Fig. 3.2). In the basin, with its rather irregular relief, and on the slopes of the uplands a number of ridges occur (above 124.0m OD). These are probably moraines. The most impressive moraine separates two gullies in the subsoil of the basin, of which a southern gully (indicated as 'S' in Fig. 3.3) and a northern gully (indicated as 'N' in Fig. 3.3) are easily detectable in the contour map (Fig. 3.2). All of the gully soils are below 122.0m OD.

Overview of the bog development

Presentation of the information

Three main peat types were identified: fen peat, wood peat (which was mostly bog marginal forest peat) and raised bog peat. This is demonstrated in the selected peat sections through the bog (Figs 3.4–3.10). Figs 3.4–3.6 concern the north–south sections through the western marginal, central and eastern marginal zones of the bog, respectively. Fig. 3.7 is the cross-section through the eastern half of the bog. In Fig. 3.8a a simplified schematic cross-section through the western half of the bog is presented. In Fig. 3.8b two detailed peat profiles are presented. Fig. 3.9 presents the cross-section through the peats of the Cooleeny area. Fig. 3.10 depicts a peat-stratigraphical detail in the western marginal area of Derryville Bog. The location of the peat sections and profiles and of the gullies designated S and N are given in Fig. 3.3. The data for Figs 3.5–3.7 are partly derived from the archaeological survey conducted by the IAWU in 1995.

In Fig. 3.11 a very schematic overview of the peat growth is presented. The development is plotted against time, between 6500 BC and AD 1000. Part of this is the rise of the bog water table, depicted in Fig. 3.12.

Next to this, the development of the main discharge systems for the peat types in question was studied and documented. Excluded from the scheme of Fig. 3.11 are aspects of the spatial, vegetational and hydrological development of a large number of features in this bog or in regions or parts of it. Such information would have complicated the scheme too much. For that reason the maps (Figs 3.13–3.20) were composed depicting the bog surfaces during succeeding stages of bog development, between 5000 BC and AD 500. Based on the presented peat types and their hydrology, these maps indicate very clearly the accessibility of Derryville Bog in prehistoric times. In these maps, the location of the trackways and platforms dated to the time window covering the bog surface are depicted, except for the period 100 BC–AD 300 (Fig. 3.19). In this period trackways and platforms were absent from the peat deposits. Indicated in this map are archaeological features of the preceding period: 400–100 BC. These maps (Figs 3.13–3.20) will be discussed below.

Bog development

Fen peat growth commenced between 7000 BC and 6500 BC (Figs 3.11 and 3.12) in the two gully-like depressions (S and N of Fig. 3.3) of glacial origin, in the basin between Cooleeny, the Killoran upland to the west and the Derryfadda upland to the east. In the minerotrophic fen peat environment with its quick-rising water table, *Phragmites australis* (reed) and *Menyanthes trifoliata* (bog bean) were important dwellers, next to a number of sedge species, rushes, aquatic species and mosses. Locally, wood horizons were present. Between 5200 BC and 5000 BC the two bog systems made hydrological contact when the intervening moraine ridge became overgrown. Initially, discharge was in a northern direction up to 4500 BC, when a discharging water flow in a southern direction came into existence, up to 3000 BC, on only a limited scale.

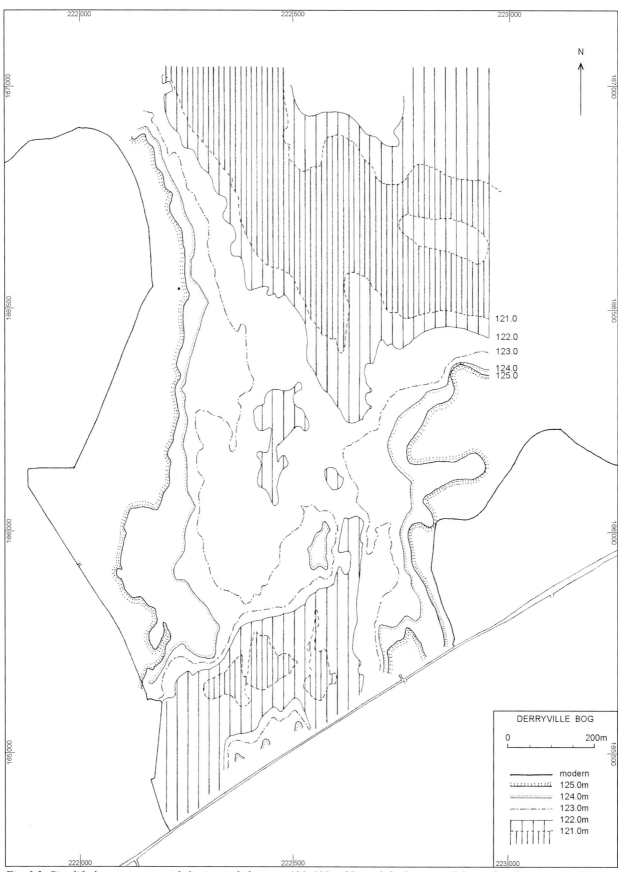

Fig. 3.2 Simplified contour map with 1m intervals between 121–125m OD, and the location of the modern bog margins. The two depressions in the subsoil (levels below 122m) are mentioned as the northern gully N and the southern gully S, respectively, in Fig. 3.3.

From *c*. 2500 BC onwards, the bog margins became covered with forests, where *Alnus glutinosa* (alder) was the predominant tree but where many *Quercus petraea* (sessile oaks) were encountered, mostly rooting in the mineral slopes of the bog margins. These forests continued growing until the beginning of the first millennium AD. They were the main domains of most of the prehistoric activity in Derryville Bog. In the meantime the discharge systems shifted from the central area to the western zone of the bog, resulting in the development of the two main discharge channels: one to the south (Cooleeny system) and one to the north (Derryville system). The location of the latter is largely determined by extra water supplied by Killoran Bog on the western upland, from *c*. 1600 BC (Killoran system). For the location of these systems, see Fig. 3.17: the Cooleeny system south of the watershed, the Derryville system north of the watershed and the Killoran system on the western upland.

By 1800 BC, when about 3m of fen peat had formed, raised bog growth started. The standard ombrotrophic development from highly humified *Sphagnum* peat (mostly *Sphagnum* Section Acutifolia) over moderately humified (since *c*. 900 BC) to poorly humified *Sphagnum* peat accumulation (from *c*. 400 BC) can be identified. In both latter types *Sphagnum imbricatum* is the main peat-former. Extensions of *Sphagnum cuspidatum* (water moss) peat, bog bean and also *Scheuchzeria palustris* (rannock rush) are important indicators of specific environmental conditions, spatial structures, such as hummocks and hollows, pools and erosion systems, and major shifts in bog hydrology. The thickness of the raised bog peats reached 4m in non-compacted conditions. Most peat that grew after *c*. AD 500 has vanished owing to the peat-extraction activities of BnM.

In the discussion of the bog development special attention will be given to the hydrological dynamics of the peat-growing systems; the hydrological approach will follow the discussion of the fen peat deposits and will precede that of the bog marginal peats and the raised bog peats.

Dating framework

The Lisheen Archaeological Project produced large numbers of dates, both dendrochronological and radiocarbon dates. Many of them were used for the time-scale as designed by Caseldine *et al.* (see Chapter 6). The following picture of the bog development is based on many of these dates. For the sake of legibility, the dendrochronological dates will mostly be given without their ± probability and the radiocarbon dates will mostly be presented as the middle value of their calibrated range as *circa* dates. In both cases, the laboratory indication and dating number will be omitted. All dates are given in BC or AD. Archaeological dates will not be used, except for the distinguished archaeological zones/periods. For the total list of dates, see Chapters 17 and 18.

The recorded peat types

The peat types recorded during the peat-stratigraphical survey and discussed in this paper will be presented here as symbols for the peat sections (Figs 3.4–3.11). For the maps (Figs 3.13–3.20), the specific symbols will be presented and explained per map.

Subsoil

1. Mineral subsoil, mostly glacial clay and loam-till, occasionally boulder clay and sand

Fen peat and related deposits

2. Fen peat
3. Oxidised top of fen peat
4. Sandy peat
5. *Phragmites australis* (reed) remains
6. *Menyanthes trifoliata* (bog bean) remains
7. Wood-rich peat, wood layers, mostly *Alnus glutinosa* (alder)
8. Wood-rich peat of Killoran Bog
9. *Quercus* (oak) stumps

Raised bog peat and related deposits or features

10. Highly humified *Sphagnum* peat
11. Moderately humified *Sphagnum* peat
12. Poorly humified *Sphagnum* peat
13. *Sphagnum cuspidatum* (water moss) peat, mostly poorly humified
14. *Scheuchzeria palustris* (rannock rush) remains
15. *Eriophorum vaginatum* (hare's tail grass) remains
16. *Calluna vulgaris* (heather, ling) remains
17. Lake-bottom sediment or pool fill
18. Amorphous peat, desiccation layer or forest soil
19. Dopplerite (humic colloids)
20. Erosion gully, filled with poorly humified *Sphagnum* peat
21. Series of water channels in the peat; probably remains of bog lakes

Miscellaneous symbols

22. Platform or track remains (cross section)
23. Level of causeway Cooleeny 31 in length section

The bog development

Fen peat

General aspects of the peat growth

Fen peat growth had already commenced *c*. 7000 BC at a level of about 120.50m OD in two separate nuclei (Fig. 3.3, S and N), the gullies S (Fig. 3.5, 50–310m) and N (Fig. 3.5, 840–1330m, and Fig. 3.6, 750–1100m). It continued up to 1800–1400 BC, with the bog surface at about 124.55–124.90m OD, respectively. The fen peat

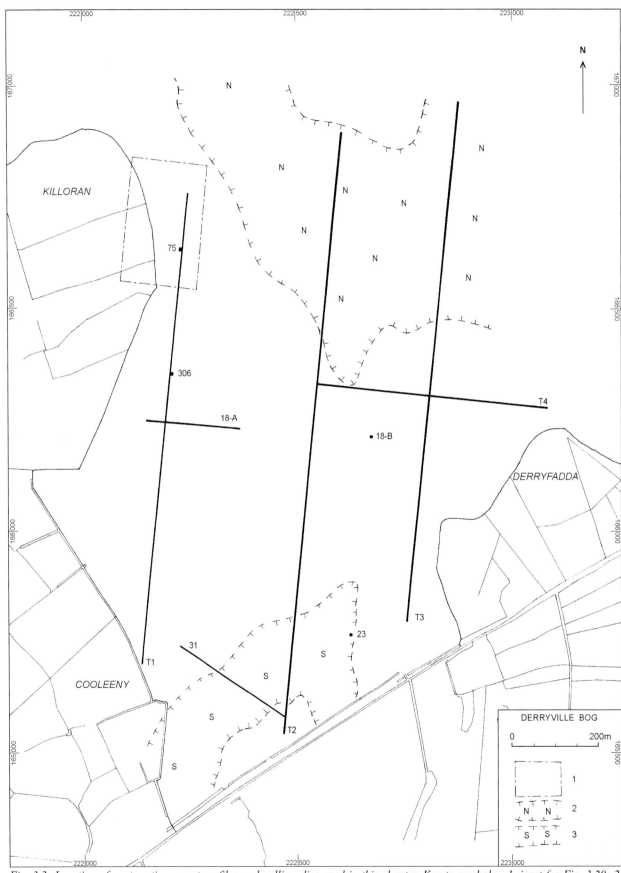

Fig. 3.3 Location of peat sections, peat profiles and gullies, discussed in this chapter. Key to symbols: 1. inset for Fig. 3.20; 2. northern gully; 3. southern gully; T1–T4: peat sections Trenches 1–4 (Figs 3.4–3.7). The numbers of the other sections and profiles (Figs 3.8–3.10) refer to the archaeological sites Killoran 18, Derryfadda 23, Cooleeny 31, Killoran 75 and Killoran 306.

deposits are on average 3m deep. The width of the fen peat deposits in the basin is about 600m.

The main constituents of the peat-growing vegetation are many monocotylous plant species such as grasses, reeds, sedges and rushes, bog bean, a number of moss species and aquatic plant species. Wood, generally alder, is present in minor quantities. Mostly, the peat accumulation took place in wet conditions, verging on an open-water environment.

Before 6000 BC the Boreal pine (*Pinus*) forests in this basin, present to levels of about 122.10m OD on the mineral subsoil, were already drowned (Fig. 3.5, 290m). From *c.* 5200–5000 BC the two peat-growing nuclei became hydrologically connected (Fig. 3.13) when the moraine ridge between the two centres of fen peat growth became partly overgrown at a level of 122.50m OD (Fig. 3.5, 300–820m, also see below). The discharge developed in a northerly direction. From *c.* 4500–4200 BC on, at a level of about 123.0m OD, the first discharge in a south to southwest direction (Cooleeny area) came into existence, increasing in importance from 3300–3000 BC onwards (Fig. 3.14). This relates to the fen peat deposits in Fig. 3.9, between 40m and 210m (below 122.5m OD). After increasing discharge it can be assumed that forest peat growth increased over the bog surface in just the narrowest stretch of the basin, as Fig. 3.14 suggests. This indicates some sort of watershed in this zone.

The rise in the rate of peat growth was considerable (Fig. 3.12). The growth rate of the fen peat between 6000 BC and 1800 BC is on average 6.09cm per 100 years, slowly decreasing from 6.67cm per 100 years in the beginning to 5.08cm per 100 years between 2200 BC and 1800 BC. Fluctuations in the rise of the bog water table are documented, mostly dated after 3300 BC. The slower rise of the bog water table between 3300 BC and 2700 BC, as Fig. 3.12 shows, can be related to increasing discharge via the Cooleeny outlet (see below). Wood peat layers also indicate hampered rise or even standstill situations or superficial desiccation. Such layers, mostly present above a slope in the subsoil as Figs 3.4–3.7 indicate, can be approximately dated to 3500–2700 BC, 3400–2900 BC, 2700–2200 BC and 2500–2000 BC.

The lake or pool at the top of the fen peat in Fig. 3.11 existed in the Cooleeny area (the lake-bottom sediment in Fig. 3.9) and dated to between *c.* 2300–2200 BC and 1800 BC. This most probably developed after a bog burst, when the Cooleeny Bog surface lowered by compaction of the peat and the depression became subsequently flooded (Fig. 3.15).

In the southern half of Derryville Bog (Fig. 3.5, 0–400m), the transition to raised bog peat, dated to 1800 BC, occurred here and there via an alder wood peat layer. This is indicative of stagnation in the rise of the ground water table but perhaps not for local superficial desiccation, for in the northern half of Derryville Bog (Fig. 3.5, 400–1360m) no wood layers are recorded. In general, the transition from minerotrophic to ombrotrophic peat growth can be ascribed to the increasing influence of the direct rainfall on the water chemistry in the basin.

Hydrological aspects: water supply and discharge systems

The most important factor in the growth of the fen peat can be identified as the rising water table in the basin, commencing before *c.* 7000 BC at a level of 120.50m OD, and continuing up to 1800 BC (bog surface at 124.55m OD) and 1400 BC (bog surface at 124.90m OD, see Fig. 3.11), respectively. Notable changes in water supply and discharge influenced bog development.

Up to *c.* 5200 BC two different fen peat systems developed independently of each other (Fig. 3.13), in the nuclei identified as S and N in Fig. 3.3. From 5200–5000 BC onwards, with the bog surface at a level of about 122.70m OD, the moraine ridge between the two nuclei became overgrown. Between 4500 and 4300 BC *c.* 90% of the ridge in question was covered with fen peat. At about 3500 BC only one isolated summit remained from this ridge (Fig. 3.14). The discharge of the water surplus of both fen peat systems in a northern direction from *c.* 5000 BC onwards was located in the central zone of the basin, being a 150m- to 200m-wide wet channel-like system, bordered by large reed fields (Figs 3.14 and 3.15). The nature of the discharge of the two fen peat systems preceding 5000 BC is unclear.

After about 4500–4200 BC discharge in a southwestern to southern direction, via the Cooleeny gully (the southern gully of Fig. 3.3), came into existence over an outlet at a level of *c.* 123.0m OD, flowing to the upper course of a tributary of the Drish River. Up to *c.* 3300 BC this discharge was only moderate but with the rising bog surface the amount of discharge rose considerably (Fig. 3.14). After *c.* 2500–2200 BC the central channel-like discharge system running in a northern direction (Fig. 3.15) shifted to the west. This was most probably caused by the increasing southwesterly discharge via the Cooleeny gully (Fig. 3.15). A bog burst dated to 2200 BC (designated bog burst A), can also be seen as a consequence of the rise of the discharge and the westward shift of the discharge system.

From *c.* 3300 BC onwards, the rise of the bog water table in the southern stretches of Derryville Bog appears to have fluctuated to a larger degree, giving way to different types of peat to develop as wood levels from 3300–2000 BC (see below). Other examples are inclusions of raised bog peat in the northern stretches of the bog (Fig. 3.5, 700–750m and 800–860m), dated to between 3000 and 2000 BC, and the development of the bog lake in the Cooleeny area (Fig. 3.9, 110–230m), dated to between 2300–2200 BC and 1800 BC. The rise of the fen peat bog surface slowed down somewhat in the period 3300–2700 BC (see below).

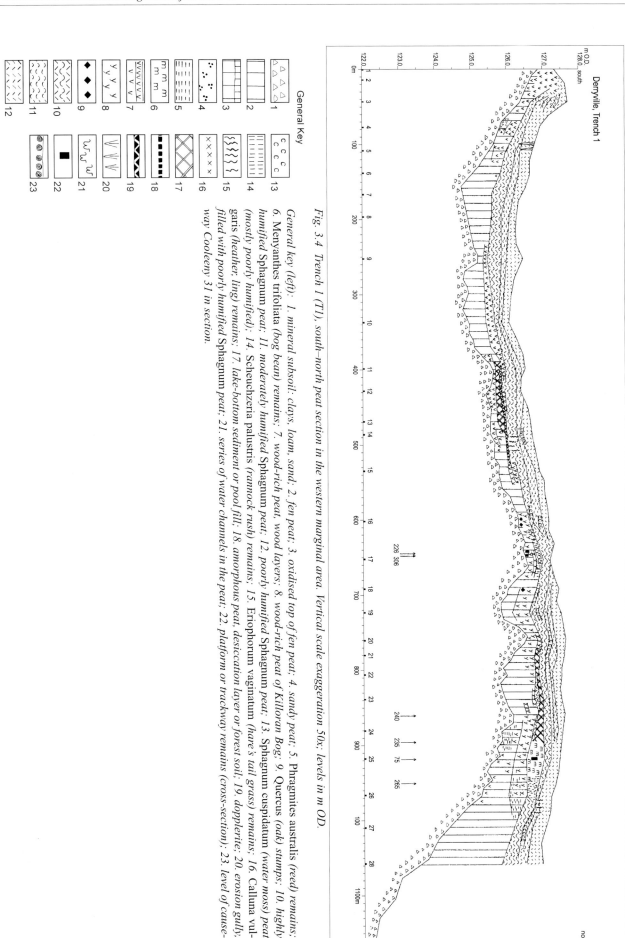

General Key

Fig. 3.4 Trench 1 (T1), south–north peat section in the western marginal area. Vertical scale exaggeration 50x; levels in m OD.

General key (left): 1. mineral subsoil: clays, loam, sand; 2. fen peat; 3. oxidised top of fen peat; 4. sandy peat; 5. Phragmites australis (reed) remains; 6. Menyanthes trifoliata (bog bean) remains; 7. wood-rich peat, wood layers; 8. wood-rich peat of Killoran Bog; 9. Quercus (oak) stumps; 10. highly humified Sphagnum peat; 11. moderately humified Sphagnum peat; 12. poorly humified Sphagnum peat; 13. Sphagnum cuspidatum (water moss) peat (mostly poorly humified); 14. Scheuchzeria palustris (rannock rush) remains; 15. Eriophorum vaginatum (hare's tail grass) remains; 16. Calluna vulgaris (heather, ling) remains; 17. lake-bottom sediment or pool fill; 18. amorphous peat, desiccation layer or forest soil; 19. dopplerite; 20. erosion gully, filled with poorly humified Sphagnum peat; 21. series of water channels in the peat; 22. platform or trackway remains (cross-section); 23. level of causeway Cooleeny 31 in section.

The other effect of the increasing discharge via the Cooleeny area was the dividing of the bog into two different hydrological systems: the northern half of Derryville Bog corresponding to the N peat-growing nucleus in Fig. 3.5 (from 720m to the north) and the southern half corresponding to the S nucleus in Fig. 3.5 (0–600m), although see also Figs 3.3 and 3.15. The intervening west–east watershed (Fig. 3.5, 600–700m; see also Fig. 3.16) will be discussed below.

Specific features

Raised bog inclusions

A number of inclusions of raised bog are documented in the fen peat deposits of peat sections not presented here. In Fig. 3.5, between 700m and 1000m raised bog accumulations are present on top of the fen peat or in shallow depressions of the fen peat surface, at remarkably low levels in the peat stratigraphy. These inclusions and other accumulations concern nearly all highly humified *Sphagnum* peat and they are all situated in the northern half of Derryville Bog. In Fig. 3.15, four of these occurrences are depicted. Given their levels in the fen peat deposits in a number of peat faces (between 123.70m and 124.20m OD) it is acceptable to say that the inclusions can be dated to between 3000 and 2500 BC. Most likely their presence can be attributed to the slower rise of the minerotrophic water table in the period mentioned. The balance between minerotrophic and ombrotrophic water-type conditions would have shifted into conditions favourable for raised bog peat growth in a number of locations.

As well as this, a number of locations with raised bog peat in fen peat environments concern the more recent fill or redeposited peat remains in erosion gullies of the Cooleeny system. The best example is in Fig. 3.5 between 110m and 240m. Between 110m and 145m the cross-section through the main erosion gully of bog burst D (see below) is cut (lowest levels about 122.0m OD). These features cannot be interpreted as raised bog inclusions.

Water channels

As Fig. 3.5 indicates, between 980m and 1260m, the fen peat and the basic raised bog peat layers contained large amounts of water in the northern stretches of Derryville Bog, probably not present in the peat deposits but as water channels in the peat. Although information concerning spatial and stratigraphical aspects is not known in detail, it demonstrates the overwhelming availability of water during the development of the bog. The supply of water never seems to have been the limiting factor for peat growth. This was not as soaks, originating as seepage with the western upland (Killoran upland) acting as an infiltration area. More likely, it relates to pools or lakes located, as indicated in Fig. 3.13, just above the lowest areas of the mineral subsoil and probably dating

from about 5000 BC on. The fen peat deposits, very rich in reed remains, documented in Fig. 3.6 between 830m and 1120m can be assumed to have been the eastern shore area of these pools or lakes, perhaps from about 5000 BC onwards. Based on stratigraphic considerations, the lakes already existed at *c*. 3300 BC, although a later date, based on renewed development (e.g. after erosion), can not be excluded. This could have occurred between about 3000 BC and 2000 BC. It is unlikely that a permanent lake was present here as the occurrence of gyttja (lake bottom sediment) was not recorded during the survey. For this reason, Fig. 3.15 does not show the presence of pools or lakes, but only the identified water channels. This is possibly related to renewed lake development linked to the specific subsoil relief of the area (the N gully). During raised bog growth after 1800 BC the area continued to be extremely wet (Fig. 3.16).

Wood peat

In the central zone of Derryville Bog, the fen peat growth became replaced by raised bog growth at about 1800 BC, while near the western and eastern uplands the wood-rich fen peat evolved into wood peat (Figs 3.4, 3.6 and 3.7). In Killoran Bog, similar wood peat deposits also developed. In the western upland, this wood peat is, in fact, bog marginal forest peat. These bog marginal forests could maintain themselves for a very long time, growing upwards with the increasing bog water table. They could extend temporarily after bog bursts, as indicated in Fig. 3.9 between 180–220m and shown at a level of 124.0–124.50m OD (see below). Eventually they became overgrown by raised bog peat, mainly after AD 500.

The expansion of wood peat (alder and pine) growth over fen peat in the central zone after the temporary lowering of the bog water table is well illustrated in Figs 3.14 and 3.15, and dated between 3500 BC and 1800 BC. Their spread indicates the development of the watershed between the southern and the northern systems (in the S and the N gullies, respectively, in Fig. 3.3).

The Cooleeny area

The southern gully in the Derryville basin (S in Fig. 3.3), with its base at about 120.50m OD, seems to be a depression without an outlet. This can explain the early commencement of fen peat growth mentioned above. Its spread is indicated in Fig. 3.13. Only when the bog water table had risen to about 122.80–123.0m OD and considerable parts of the moraine ridge to the north of the southern gully had become overgrown (shortly after the period as depicted in Fig. 3.13) could discharge of the fen peat deposits in this gully occur in a northern direction. This can be dated to about 5000–4800 BC or shortly after.

With the rising bog water table, discharge in a south-southwestern direction came into existence, via an outlet outside the study area. Based on peat-stratigraphical data it is not unlikely that this outlet had a level of about

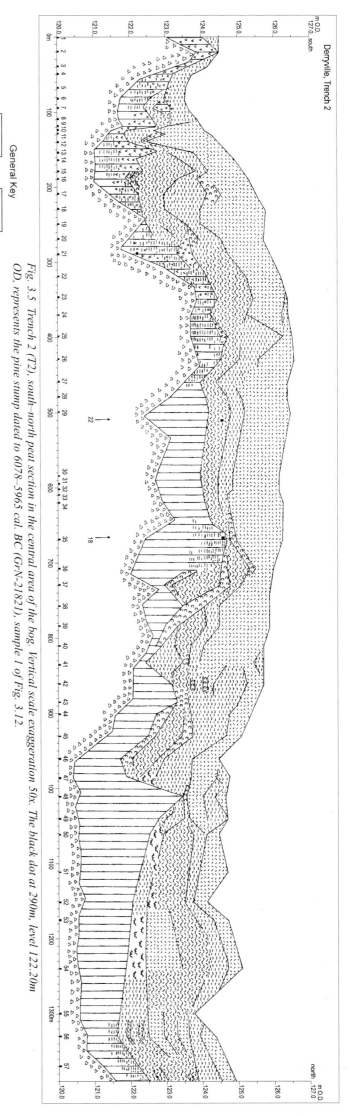

Fig. 3.5 Trench 2 (T2), south–north peat section in the central area of the bog. Vertical scale exaggeration 50x. The black dot at 290m, level 122.20m OD, represents the pine stump dated to 6078–5965 cal. BC (GrN-21821), sample 1 of Fig. 3.12.

General Key (left): 1. mineral subsoil: clays, loam, sand; 2. fen peat; 3. oxidised top of fen peat; 4. sandy peat; 5. Phragmites australis (reed) remains; 6. Menyanthes trifoliata (bog bean) remains; 7. wood-rich peat, wood layers; 8. wood-rich peat of Killoran Bog; 9. Quercus (oak) stumps; 10. highly humified Sphagnum peat; 11. moderately humified Sphagnum peat; 12. poorly humified Sphagnum peat; 13. Sphagnum cuspidatum (water moss) peat (mostly poorly humified); 14. Scheuchzeria palustris (rannoch rush) remains; 15. Eriophorum vaginatum (hare's tail grass) remains; 16. Calluna vulgaris (heather; ling) remains; 17. lake-bottom sediment or pool fill; 18. amorphous peat, desiccation layer or forest soil; 19. doplerite; 20. erosion gully, filled with poorly humified Sphagnum peat; 21. series of water channels in the peat; 22. platform or trackway remains (cross-section); 23. level of causeway Cooleeny 31 in section.

122.80m OD. If this is correct southwestern discharge could have occurred since about 4800 BC. Discharge in a northern direction maintained itself until about 3300–3000 BC (Fig. 3.14), as can be concluded from the dates obtained from the centrally situated discharge zone of the fen peat, as depicted in Fig. 3.15. The level of the bog surface can be estimated at about 123.70–123.90m OD, not more than one metre above the supposed south-southwestern outlet level. Although the exact location of the outlet (perhaps a threshold or spillway) is not known, owing to the significant modern reclamation activities in this area, it can most likely be found 800–1000m to the southwest of the Cooleeny peat section of Fig. 3.9 (see also Fig. 3.3 for the location of this section). The main discharge could not have started before about 3000 BC, when the slope of the bog surface in a south-southwest direction was about 1 *promille*.

Fig. 3.9 (between 60m and 180m) indicates the fill of a bog lake at the top of the fen peat, as previously mentioned. The inundation of this area of fen peat, situated just above the lowest depression in the mineral subsoil of the southern gully, can be roughly dated to *c.* 2300 BC or earlier. The presence of this lake-bottom sediment can be seen in relation to the beginning of superficial discharge of the bog surface in a southwestern direction. The lake just above the lowest subsoil levels of the southern gully can be understood as a water accumulation. It formed after a (superficial) desiccation of the very wet fen peat deposits, resulting in a local depression at the former bog surface, and was flooded for some time. The lake existed to about 1800 BC. The presence of a wood layer at the base of the lake-bottom sediment, as depicted between 60m and 120m in Fig. 3.9, supports the suggestion of desiccation, which can be dated to shortly after 3000 BC. This may have coincided with the aforementioned slower rise of the bog water table in Fig. 3.12, between 3300 and 2700 BC.

The location of this bog lake is depicted in Fig. 3.15 as 'pool'. A corresponding development is documented below trackway Cooleeny 22. Here, a pool fill deposit in the upper fen peat levels was surveyed and most likely dates from between *c.* 2200 and 1800 BC. The location of this pool is also indicated in Fig. 3.15.

After *c.* 2200 BC (bog burst A, see below) peat growth developed in the Cooleeny gully with numerous erosion phenomena, much more than in the northern half of Derryville Bog. This indicates an increasing slope of the bog surface to over 3 *promille*, the value that stands for the repeated occurrence of eroding water flows on raised bog surfaces. This must also have played a role in the westward shift of the north–south discharge zone of the fen peat deposits after *c.* 3000 BC.

Killoran Bog—extra water supply

On the Killoran upland, west of Derryville Bog, a raised bog or blanket bog had developed, of which the exact location, size and age are not well known. Its level is about 5–8m above the surface of Derryville Bog.

The eastern stretch of the Killoran Bog is indicated in Figs 3.16–3.19. This area of bog consists of forest peat on the eastern slopes of the Killoran upland. The bog in question supplied its discharge water into the western zone of the Derryville basin. From *c.* 1600 BC onwards, the resulting discharge channel of this system and the Derryville discharge system occurred as the 'northern discharge channel' (see below). This was an extremely wet area, nearly or completely impassable in prehistory. Its course is indicated in Figs 3.16–3.19. It can be accepted that this extra water supply started some centuries earlier, about 2200–2000 BC. The gully in question appears in Fig. 3.4, Trench 1, between 760m and 1000m, at the top of the fen peat.

Early human influence on the fen peat growing environment c. 1440 BC

From about 1800 BC ombrotrophic conditions were already prevalent in the two raised bog complexes, as depicted in Fig. 3.16. The intermediate fen peat zone can be seen as the watershed between these complexes. In this zone, a stone-built causeway, Killoran 18, dated to 1440 BC, was constructed (see catalogue entry in Chapter 10). The level of the bog surface was 124.90m OD. The causeway was intended to cross the entire bog and was about 530m long, mostly marked by the use of stones, except in the westernmost 50m where the walking surface was constructed of wood. This is somewhat schematically depicted in Fig. 3.8a. Its walking surface was about 0.70–0.90m wide. The stretch of fen peat between the raised bog complexes became suddenly overgrown by raised bog peat, as is depicted in Fig. 3.8b. This shift can be dated to 1420 BC. It can be assumed that the construction of Killoran 18 influenced the bog surface to such a degree that it was the causative factor in the transition from fen peat to raised bog peat. This will be explained as follows.

Usually, growing fen peat of the types encountered in Derryville Bog has a water content of about 90%. Such surfaces are very soft and for that reason are hardly or not passable. We can exclude that the stone causeway was constructed under these conditions. In conditions with lower water content, such peats would develop into wood-rich or even forest peat deposits. This did not occur here. However, constructing a causeway of stones and pebbles, such as Killoran 18, would not result in a usable crossing of the bog with such a soft bog surface. The stones would have pressed down too much in the soft peat.

Considering the fact that the builders succeeded in constructing the road in a reasonable way, it can be concluded that the water content of the upper fen peat layers must have been lowered to about 80%, most probably after a serious drop in the water table. This will have caused the bog surface to shrink by *c.* 20cm. A date of *c.* 1450 BC is likely for this event.

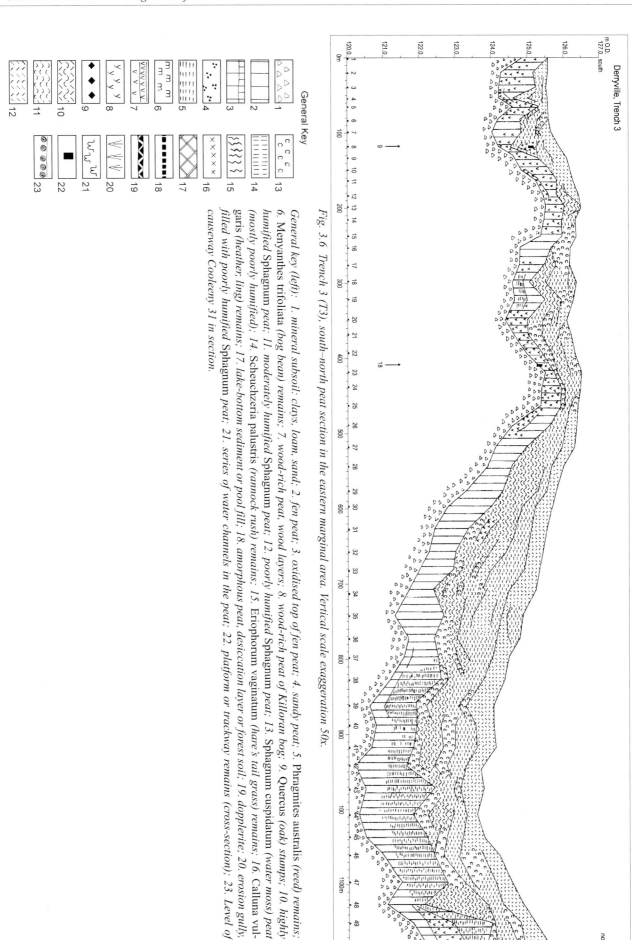

Fig. 3.6 Trench 3 (T3), south–north peat section in the eastern marginal area. Vertical scale exaggeration 50x.

General Key

General key (left): 1. mineral subsoil: clays, loam, sand; 2. fen peat; 3. oxidised top of fen peat; 4. sandy peat; 5. Phragmites australis (reed) remains; 6. Menyanthes trifoliata (bog bean) remains; 7. wood-rich peat, wood layers; 8. wood-rich peat of Killoran bog; 9. Quercus (oak) stumps; 10. highly humified Sphagnum peat; 11. moderately humified Sphagnum peat; 12. poorly humified Sphagnum peat; 13. Sphagnum cuspidatum (water moss) peat (mostly poorly humified); 14. Scheuchzeria palustris (rannoch rush) remains; 15. Eriophorum vaginatum (hare's tail grass) remains; 16. Calluna vulgaris (heather, ling) remains; 17. lake-bottom sediment or pool fill; 18. amorphous peat, desiccation layer or forest soil; 19. dopplerite; 20. erosion gully, filled with poorly humified Sphagnum peat; 21. series of water channels in the peat; 22. platform or trackway remains (cross-section); 23. Level of causeway Cooleeny 31 in section.

The heavy stones used to provide a walking surface must have compressed the fen peat surface to an extra degree, to about 10–15cm more than a wooden walking surface, amounting to perhaps 25–30cm. This allowed the water surplus of the southern raised bog complex to discharge over the entire stone-built stretch of the causeway in a northern direction. The very acidic ombrotrophic water induced the quick shift of the fen peat zone into raised bog peat growth. Judging from the growing characteristics of a number of alder scrubs, germinated during the construction of the causeway, this overgrowth can be dated to about 1420 BC (Casparie and Stevens 2001).

The western stretch of the causeway, documented as Killoran 305 (dated to 1415 BC), with its wooden walking surface, was pressed down into the underlying peats by 20–40cm (Fig. 3.8a, between 120m and 170m), indicating that this concerns a zone of very soft and thus wetter peat. This softer peat can be interpreted as a deposit of the 'northern discharge zone' (see below and Figs 3.16–3.18), draining part of the southern raised bog complex. This zone was really impassable.

A complicating factor stimulating extra flooding was the blocking of the discharge channel by the wood of trackway Killoran 305. This resulted in wood peat growth in the discharge zone, as indicated by symbol 11 in Fig. 3.17.

Judging from these data it can be concluded that the sudden shift from fen peat growth to raised bog peat growth (Fig. 3.8b) in this watershed zone was caused by human interference in the bog environment dating from 1440 BC.

The hydrological approach

The rising bog surface

The time/depth curve of Fig. 3.12 indicates the rising bog water table. The curve is based on radiocarbon and dendrochronological dates of numerous archaeological and palaeobotanical 'landfall' samples. These wood and peat samples were situated on the mineral subsoil, so in reconstructing the rise of the bog water table, and thus the peat-forming surface, there are no problems relating to compaction and shrinkage of the peat deposits.

The drawn line is the best estimate of the general rise of the bog water table between *c.* 6000 BC and AD 500, interrupted by five steep drops caused by bog bursts. The bend in the curve at a level of 124.50m OD represents the shift from minerotrophic to ombrotrophic peat growth, dated to 1800 BC in Derryville Bog.

Some of the data deviated from this curve, providing evidence for the bog bursts on the one hand and for water supply from nearby Killoran Bog on the other hand. Both are discussed below. The existence and the extent of the three hydrological systems, namely the Derryville system, the Killoran system and the Cooleeny system, will be discussed below. The rather complex occurrence of discharge channels in the study area and the watersheds between the hydrological systems is also discussed below.

Bog bursts

Evidence for bog bursts can be demonstrated by the presence of samples with largely different ages in close proximity. For example, in Fig. 3.12, sample 3 is a bog oak overlying platform Derryfadda 218, from which sample 4 was taken. Sample 5 is a bog oak from underneath platform Derryfadda 213. Sample 8 is bog oak E8 and sample 10 is bog oak E9 from the same location. The remarkable difference in ages between two such features indicates serious drops in the bog water table suddenly occurring and collapsing the surface of the bog. The sudden drops in the water table can be calculated for each event as up to one metre or slightly more. The peat-stratigraphical survey indicated important erosion phenomena in relation to these features, especially in the Cooleeny system (see Fig. 3.9, 0–80m). In general, the level of the water table recovered after one or two centuries (see also Chapter 6). It can be stated that all the dates below the curve are related to bog bursts, five of which are well documented. It is likely that at least four more bog bursts occurred in the study area. These are only documented and dated in the peat stratigraphy. The bog bursts are indicated in Fig. 3.11 with arrows. Bog burst A can be dated to 2200 BC; bog burst B to 1250 BC; bog burst C to 820 BC; bog burst D to 595 BC with a second drop in the water table at about 450–400 BC; and bog burst E is dated to about 100 BC. The arrows with broken lines in Fig. 3.11 also indicate likely bog bursts whereby the absence of archaeological finds meant that only the peat-stratigraphical features could be dated.

It can be assumed that bog bursts are quite normal phenomena in Irish raised bog development. Feehan and O'Donovan (1996, 319–412) have listed and discussed a large number of historically documented bog bursts. In Tumbeagh Bog, County Offaly, the present author has studied the peat stratigraphy in relation to the find of a late medieval bog body. At least four bog bursts could be established between about 1000 BC and AD 1400. Bog bursts are also known from raised bogs from the Continent (Casparie 1972, 221–6; Casparie *et al.* 1983, 149–52; van der Sanden 1996, 23–4). Bog bursts can be seen as important self-cleaning mechanisms for raised bogs; together with the sudden discharge of masses of water, masses of nutrients were removed that had previously frustrated ombrotrophic raised bog growth. This resulted in extreme acidic and oligotrophic environmental growing conditions, favorable for poorly humified *Sphagnum* peat growth.

Killoran Bog and its water discharge

In Fig. 3.12, the points 19–33, plotted clearly above the curve, all belong to samples (trackways and bog oaks) from the western bog margin. This indicates that the water supply to the western marginal area may have originated from several sources at levels higher than 127.58m OD (as indicated by sample 21, Killoran 241). The survey of

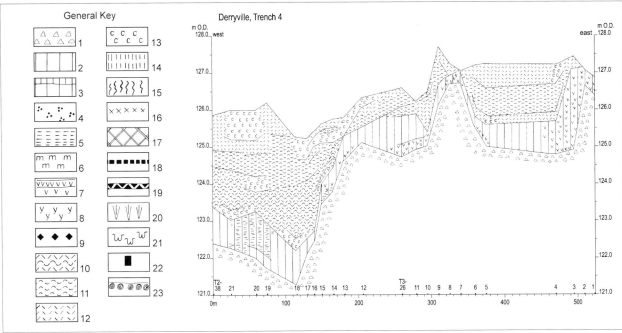

Fig. 3.7 Trench 4 (T4), west–east cross-section through the eastern half of the bog. Vertical scale exaggeration 50x.
General key (above): 1. mineral subsoil: clays, loam, sand; 2. fen peat; 3. oxidised top of fen peat; 4. sandy peat; 5. Phragmites australis (reed) remains; 6. Menyanthes trifiolata (bog bean) remains; 7. wood-rich peat, wood layers; 8. wood-rich peat of Killoran Bog; 9. Quercus (oak) stumps; 10. highly humified Sphagnum peat; 11. moderately humified Sphagnum peat; 12. poorly humified Sphagnum peat; 13. Sphagnum cuspidatum (water moss) peat (mostly poorly humified); 14. Scheuchzeria palustris (rannock rush) remains; 15. Eriophorum vaginatum (hare's tail grass) remains; 16. Calluna vulgaris (heather, ling) remains; 17. lake-bottom sediment or pool fill; 18. amorphous peat, desiccation layer or forest soil; 19. dopplerite; 20. erosion gully, filled with poorly humified Sphagnum peat; 21. series of water channels in the peat; 22. platform or trackway remains (cross-section); 23. level of causeway Cooleeny 31 in section.

the Killoran upland revealed that an ombrotrophic bog developed on this upland with *Sphagnum* as the main peat-former. It was probably a kind of very shallow blanket bog of at least six hectares over mineral soil with a slope of about 1%. Its level was about 129.0–133.0m OD. Peat growth commenced some time after 2700–2500 BC. Its discharge in an eastern direction was downslope via a number of gullies into the Derryville basin, at least since 1600 BC, or perhaps since *c.* 2000 BC. These discharge flows facilitated forest peat growth on the slopes in question, in which trackways were encountered. The spread of these forest peat deposits is given in Figs 3.16–3.19. The trackways occur in three clusters, so at least three gullies existed, supplying water to the discharge channels of Derryville Bog (see below). The location of some recorded gullies is given in Figs 3.16–3.19 with open arrows in the Killoran forest peat deposits.

Hydrological systems

In the Derryville area three different hydrological regimes were distinguished. As previously mentioned, after *c.* 3300 BC the originally south to north directed discharge system in the central zone of fen peat growing in Derryville Bog (Figs 3.14 and 3.15) shifted in a western direction. The increasing discharge in a south-southwest-

ern direction played a role in this. This developed into a regime of its own in the Cooleeny area. Next to these systems the increasing water supply from Killoran Bog, as discussed above, must have had its influence in this shift. From about 1600 BC onwards, three different hydrological systems were active in the area. In Figs 3.16–3.18, the spatial relationship of the hydrological systems is illustrated from around 1600 BC to about 400 BC, respectively, based on the recorded discharge channels and watersheds.

Derryville system

The superficial discharge in a northerly direction from the central zone developed from about 5000 BC (Fig. 3.13) and continued to about 3000 BC (Fig. 3.14). Then it decreased until 1800–1700 BC (Fig. 3.15), probably owing to the commencement of raised bog growth in the two raised bog complexes (Fig. 3.16). Two lightly constructed oak plank footpaths, Derryfadda 23 (1590 BC), and Cooleeny 22 (1517 BC), in the southern raised bog complex indicate that this shift was already completed before 1600 BC. A large number of trackways, platforms and *fulachta fiadh* in the western zone of Derryville Bog, nearly all dated to between 1600 BC and 200 BC (Killoran 237, Killoran 18, Killoran 305, Killoran 253, Killoran 230, Killoran 235, Killoran 240, Killoran 241, Killoran 243, Killoran 69, Cooleeny 64,

Killoran 234, Killoran 75, Killoran 312, Killoran 314, Killoran 301, Killoran 248, Killoran 306 and Killoran 226), enables us to indicate exactly the location and even some shifts of the western boundary of this system in the succeeding periods (Figs 3.16–3.19). These archaeological structures mark the position of the northern discharge channel being the western boundary of this system (see below). In general, erosion phenomena occur only very superficially in this system. It can be assumed that up to about 100 BC the relief of the bog surface was minimal, perhaps about 1 *promille*.

Cooleeny system

This system corresponds, roughly speaking, to the southerly gully S (Figs 3.2, 3.3 and 3.13). It had its predecessors in the fen peat deposits from perhaps *c.* 4500 BC onwards, when minimal amounts of superficial water started to discharge in a south-southwestern direction (Fig. 3.14), with increasing discharge from about 3000 BC on (Fig. 3.15). The ombrotrophic Cooleeny system commenced about 1800 BC with the growth of raised bog peat (Fig. 3.16). In this system, extensive erosion phenomena, related to its discharge channel, have been surveyed. The heaviest bog bursts in Derryville Bog can be located in this system. It can be accepted that its surface was relatively highly sloped, e.g. 3–5 *promille*. Its northern margin bordered the Derryville system over the watershed zones as depicted in Figs 3.17 and 3.18.

Killoran system

The Killoran Bog itself developed independently from the systems in the Derryville basin. Its surface was about 5–8m higher than Derryville Bog. From *c.* 1600 BC on, or even a few centuries earlier, it discharged its surplus water via three or four gullies sloping down into the northern discharge channel of the Derryville system (see Figs 3.16–3.19). These gullies, and their surroundings, were covered with forest peat (Figs 3.16–3.19), in which a number of trackways and platforms were found; the oldest (Killoran 241) is dated to 1547 BC. This forest peat growth on the eastern slopes of Killoran upland occurred about 2500 BC.

The hydrology of Killoran Bog is incompletely known. Nevertheless, it could be stated that there are important differences in the water chemistry of the identified gullies, which are indicated with arrows in Figs 3.16–3.19. The water flow of the northernmost gully was very rich in minerals (Fig. 3.20), whereas the other gullies seemed to have discharged predominantly acidic bog water.

Since *c.* 1300 BC, the Killoran water supply resulted in the accumulation of extra forest peat and a rise of the bog surface of about 0.8–0.9m above that of Derryville Bog. In Fig. 3.8a this is very schematically depicted, based on data recorded in relation to the excavation of the causeway Killoran 18 and trackway Killoran 305.

About 100 BC–AD 200 this difference in level was reduced to low values, caused by the continuing rise in the level of the bog surface of Derryville Bog to 127.50m OD. Around AD 500, large stretches of the Killoran Bog forest peat deposits eventually became overgrown by ombrotrophic peat, as is depicted in Fig. 3.10, stages 11–12, at a level of about 127.80m OD in an uncompacted state. This can be accepted as the end of the water supply from the Killoran system.

The southern stretch of Killoran Bog was affected *c.* 820 BC by heavy erosion, as samples 25 (bog oak from Killoran 306/69), 26 (Killoran 69) and 29 (Killoran 306) indicate on the curve of Fig. 3.12. This can be related to bog burst C, documented in the southern zone of Derryville Bog. This bog burst flowed in a northern direction.

Discharge channels

Two main discharge channels have been identified, the Northern discharge channel, discharging the northern parts of Derryville Bog and the Killoran Bog systems, and the Southern discharge channel, discharging the Cooleeny Bog system, the southern half of Derryville Bog. The course of both channels and their fluctuations in the succeeding periods are given in Figs 3.16–3.19.

Northern discharge channel

The gully developed between the Killoran and the Derryville systems. Both are systems with an off-centre-discharge gully (gully B in Figs 3.16–3.19).

Its location is well documented by the many trackways and platforms in the western bog marginal area, as already indicated above. In general, the channel bordered on the eastern margin of the Killoran forest peat deposits and the western margin of the Derryville raised bog deposits, as depicted in Figs 3.16–3.19.

During its existence, between *c.* 1600 BC and AD 200, the gully shifted about 50–70m to the west as a consequence of the rising Derryville Bog surface from 124.60m to 127.0m OD (see Figs 3.11 and 3.12). The gully could also be identified by its peat types, deviating from the surrounding peat deposits. In Fig. 3.20 a detail of this is depicted. Four phases can be distinguished in its development. In the first phase, up to *c.* 1200 BC, a *c.* 50m-wide zone with very humid fen peat formed here (with the western terminus of causeway Killoran 18, dated to 1440 BC, and trackway Cooleeny 64, dated to *c.* 1220 BC), where to the west the Killoran forest peat (with trackway Killoran 305, dated to *c.* 1415 BC) and to the east raised bog peat developed (with trackways Derryfadda 23, dated to 1590 BC, and Cooleeny 22, dated to 1517 BC). This was a soak-like mesotrophic and extra damp to wet area with soft peats and slow water flow in a northerly direction (see above and Fig. 3.16).

Between 1440 and 1200 BC this channel was blocked by the construction of the wooden stretch of causeway Killoran 18 (Fig. 3.16, symbol 11).

Secondly, from *c.* 1200 BC to between 820 and 600 BC, a gully with water flow still in a northerly direction developed here over a length of about 250m (Fig. 3.17, gully stretches B and D, including stretch C). The gully was identifiable by its amorphous organic pool-fill sediment, as recorded in Fig. 3.4, 750–900m. Its width was 20–25m. Dating from about 820 BC (bog burst C), severe erosion in a northerly direction took place in the region of Killoran 69 (Fig. 3.17); the gully in this region (Fig. 3.17, gully stretch C) was emptied by this erosion.

Thirdly, inversion of the water flow of the gully stretch D occured around 600 BC, south of Killoran 69, Killoran 226, Killoran 246 and Killoran 306 (located in Fig. 3.18, in gully area C). The top of the amorphous lake-bottom deposit became desiccated (Fig. 3.18, gully area D). It can be accepted that area D of the gully lost its function (discharging in a northern direction) and transformed the watershed (Fig. 3.18, stretch C) between the two systems (Fig. 3.18, areas B and D, respectively). Most probably this event can be exactly dated to 595 BC, in relation to bog burst D.

The fourth phase, from *c.* 600 BC to about 100 BC, is characterised by peat types of open-water conditions in the remaining gully area B of Figs 3.18 and 3.19. In the middle stretch (area C of Fig. 3.18), marking the starting point of the inverted southerly direction of the water flow, erosion phenomena are recognisable. In the peat section of Fig. 3.4, the gully remains from the location of Killoran 306/69 in a northern direction could hardly be recognised for a length of about 100m (Fig. 3.4, 650–750m). Further to the north, to Killoran 240 (Fig. 3.4, 750–880m), an amorphous pool-fill sediment remained of the gully, only 10m wide (Fig. 3.18, the narrow stretch of gully D). In a northern direction, this fill shifted to a mixture of bog bean and rannock rush peats in a well-developed 10–15m-wide gully which opened in the pool near Killoran 75 (Fig. 3.4, 880–920m; Fig. 3.18, Killoran 234). Where bog bean and rannock rush peats predominate, this indicates reasonable water flow in more or less open-water conditions in a northern direction.

Phase 1 of the gully is mapped in Fig. 3.16; Phase 2 in Fig. 3.17; Phase 3 in Fig. 3.18 and Phase 4 in Fig. 3.19. Fig. 3.20 provides a detail of it.

The increasing width of gully B, from about 15m to 60m (where a pool is shown in Fig. 3.18), indicates the supply of mineral-rich water from the Killoran upland mixed with bog water that was discharging from the forest peats to the west and the treeless raised bog peats to the east of the gully (as Fig. 3.20 shows). In this pool, with floating vegetation of bog bean and rannock rush, Killoran 75 (dated to *c.* 280 BC or *c.* 230 BC) was constructed. The fen peat deposits of the channel fill between *c.* 1600 and 600 BC have a thickness of 60–70cm. The bog bean and rannock rush deposits in the pool in question and in channel B are up to 40–50cm in thickness, dating from 600–100 BC. A water depth in this northern discharge channel of about one metre or even more is not unlikely.

The top of the bog bean/rannock rush peat in the gully (as depicted in Fig. 3.20) is dated to about 100 BC; this gully fill was overgrown by poorly and moderately humified *Sphagnum* peat, which most likely commenced growing about AD 100. It can be accepted that bog burst E caused the end of the discharging function of this stretch of the gully. After a gap of about two centuries, ombrotrophic peat growth could establish itself here. Regarding the impressive differences in spatial structures (as can be seen between Fig. 3.18 and Fig. 3.19), it can be assumed that about 100 BC–AD 100 important changes occurred in the peat-growing environment of Derryville Bog. One of these changes was the mass extension of poorly humified *Sphagnum* peat over the bog surface.

Southern discharge channel

Contrary to the former situation, the Cooleeny system concerns a raised bog complex with a centrally situated discharge gully.

The well-dated bog burst D (595 BC, see above) has affected the bog surface of the Cooleeny system to such a large degree that most, if not all, identifiable discharge features of the southern half of Derryville Bog pre-dating *c.* 600 BC have vanished. Nevertheless, it can be stated that both the bog bursts A (*c.* 2200 BC) and B (*c.* 1250 BC) have influenced the bog water table in the eastern stretches of the Cooleeny system, as can be concluded from the position of bog oak at Derryfadda 218 (3368 BC) and Derryfadda 218 itself (*c.* 2120 BC), both found at about the same level. This indicates bog burst A (Fig. 3.12). It also accounts for the levels and dates from Derryfadda 17 (*c.* 1160 BC), Derryfadda 311 (*c.* 1240 BC) and Derryfadda 203 (*c.* 290 BC). Therefore, the location and width of the different stages of the discharge channel as presented in Figs 3.16–19 are partly but soundly based on interpretations of later phenomena.

The bog surface of the Cooleeny system was dissected by a number of discharge channels and gullies since at least 595 BC. The main discharge channel of the Cooleeny system (Figs 3.16–19, channel A) can be seen in the low-lying area, filled with redeposited peat, as depicted in Fig. 3.18 (channel area A), representing the situation shortly after bog burst D. After *c.* 400 BC parts of this low-lying area became overgrown by very soft, poorly humified *Sphagnum* peat, as depicted in Fig. 3.19.

The first documented discharge system can be dated to about 5000 BC, when discharge occurred in a northerly direction (Fig. 3.13). The southern discharge system did not exist at that time. It commenced most likely about 4800–4500 BC. Up to about 3000 BC, when the discharge in a south-southwesterly direction increased greatly, there was a superficial flow system present (Fig. 3.14). The first well-dated bog burst, bog burst A (*c.* 2200 BC), must have affected this system, perhaps including the emptying at an early stage of the bog lake at the top of the fen peat in this area (Fig. 3.9, 60–218m). This lake is depicted in Fig. 3.15. It is supposed that this erosion caused a widening of the discharge channel to some degree (Fig. 3.15). Most of the southern raised bog complex, developing here since *c.* 1800 BC, discharged in a southwesterly direction (Fig. 3.16). Bog burst B (*c.* 1250 BC) seriously affected the raised bog surface of the Cooleeny system. In the eastern marginal areas the bog water table apparently dropped significantly, as trackways Derryfadda 17 (*c.* 1160 BC) and Derryfadda 311 (*c.* 1220 BC) indicated. This suggests an increase in the system of discharge gullies and the main channel (Fig. 3.17).

The large-scale erosion related to bog burst D (595 BC) caused a very extensive system of gullies to develop, which facilitated discharge over wide areas, especially downstream of the centre of the bog burst (Fig. 3.18), just where causeway Cooleeny 31 was broken and washed away (Fig. 3.9, 0–40m). A number of erosion gullies upstream of this location are also documented (e.g. gullies a–d in Fig. 3.18). Bog burst D caused

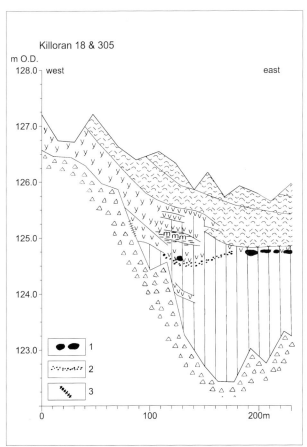

Fig. 3.8a Schematic west-east peat section through the western area of causeway Killoran 18/trackway Killoran 305. General key: see Fig. 3.7. Extra symbols: 1. stones of Killoran 18; 2. wooden section of the causeway; 3. location and level of the supposed landfall of Killoran 18.

a drop in the water table of about 1.5m in the surroundings of causeway Cooleeny 31, where large peat blocks collapsed. There was a drop of at least 1m in the northeastern area of the Cooleeny system at the location of trackway Derryfadda 13 (*c.* 590 BC) where the bog lake above trackway Derryfadda 17 (depicted in Fig. 3.17, symbol 15), which is dated from *c.* 700 BC to 595 BC, became drained. It can be calculated that over a distance of about 500m (between Derryfadda 13 and Cooleeny 31) the slope of the gully base was about 2.5m, a sloping degree of about 5 *promille*. This value is generally acceptable for the occurrence of superficial erosion in a south-southwesterly direction.

Bog burst D inverted the flow direction of the southern part of the northern discharge gully over a length of 250–300m in a southern direction (Fig. 3.18, stretch D of the northern discharge gully). It can be accepted that most of this discharge water travelled through gully a of Fig. 3.18.

The extensive discharge system as depicted in Fig. 3.18, and which is related to bog burst D, can be accurately dated to 595 BC. The bog burst took place in the window of time of eleven radiocarbon years between the construction of causeway Cooleeny 31 (778–424 BC), which caused the bog burst, and before the construction of trackway Derryfadda 13 (767–412 BC), built shortly after the bog burst. More exactly, very shortly before the beginning of the growth of bog oak E9 near trackway Derryfadda 13, which germinated in 593 or 592 BC after the bog burst, thus: 595 BC.

The southern discharge zone after bog burst D, as depicted in Fig. 3.18, covered a large low-lying area filled with very soft redeposited peat and dissected by many gullies, only partly filled with even softer peats. This zone had been inaccessible since 595 BC up to modern times, when all the peat was extracted in the framework of the construction of the TMF.

When the drainage of the southern raised bog complex had more or less re-established itself, large parts of the eroded area became overgrown by poorly humified to fresh *Sphagnum* peat, as mentioned above. This phenomenon narrowed the southern discharge channel to some degree, as depicted in Fig. 3.19. The very loose, and extremely water-permeable, redeposited fill of the erosion gullies retained their high water-discharging capacity and function.

Watersheds

The moraine ridge in the Derryville basin (Fig. 3.2) between the S gully and the N gully (Fig. 3.3) was, up to about 5000 BC, the watershed between the southern and the northern bog systems (Fig. 3.5, between 60–300m and 750–1330m, respectively, and Fig. 3.13). In the peat section of Fig. 3.5 this watershed is present between 300m and 480m or even to 750m. From *c.* 5000 BC to *c.* 3000 BC no clear watershed existed in the Derryville basin. Most probably the forest peat

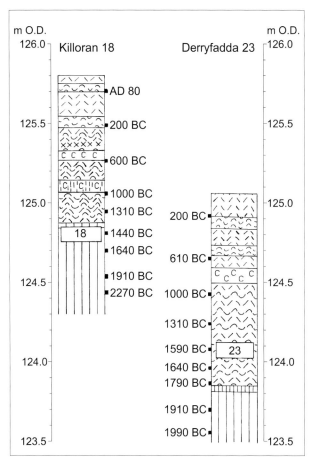

Fig. 3.8b Dated peat profiles through Killoran 18 and Derryfadda 23, respectively. General key: see Fig. 3.7. The mineral subsoil below causeway Killoran 18 is at c. 123.10m and below trackway Derryfadda 23 it is c. 121.90m.

deposits to the west and the east of the only remaining mineral ridge in the basin in Fig. 3.14 served as a watershed. Between c. 3000 and 2000 BC the basin again became divided in two hydrological systems, when discharge of the peats in the S gully in a south-south-western direction increased. The location of the related watershed is not very obvious.

The occurrence of forest peat deposits adjacent to the top of the mineral subsoil in the Derryville basin in Fig. 3.15 can be seen as an indication of the presence of the watershed here. From about 2000 BC on, the watershed existed as depicted in Fig. 3.16, as fen peat deposits. Up to about 1800 BC it divided the fen peat deposits in Derryville Bog into two hydrological systems (Fig. 3.16). In Fig. 3.5 this is represented by the reed fen peat zone between 630m and 710m. From 1800 BC to 1420 BC this fen peat watershed divided the two raised bog complexes (see Fig. 3.16 and Fig. 3.5 between 620m and 740m). As already discussed above, this watershed became flooded in a very short time by the acidic raised bog water flow from the southern raised bog complex (Fig. 3.16), caused by the construction of causeway Killoran 18.

In the meantime, the water supply from Killoran Bog significantly complicated the hydrological development in the western bog area, as has been discussed earlier. Here, the northern discharge channel developed. The watershed related to this is depicted in Fig. 3.17. The commencement of this extra water supply cannot be closely dated, but it was already fully in existence at about 1600 BC.

The flooding of the fen peat watershed by ombrotrophic water, at c. 1420 BC (see Fig. 3.16), did not result in the establishment of a single large hydrological system in Derryville Bog. Two systems can be distinguished bordering onto a curved zone, whose course fluctuated to some degree, relative to the strength of the successive bog bursts B (c. 1250 BC), C (c. 820 BC), D (595 BC), E (c. 100 BC) and a number of other, possible bog bursts between c. 1600 BC and AD 500 (see Fig. 3.11). In Figs 3.17–3.19 the location of the successive watersheds is indicated. A large number of characteristics of the established hummock and hollow systems have been recorded on which the spatial structures in question are based. In the central zone of Derryville Bog the presence of the watershed can be demonstrated in the peat face of Fig. 3.5 between 380m and 440m. The domed deposit of moderately humified *Sphagnum* peat was overgrown by poorly humified *Sphagnum* peat at about AD 100–500.

The significant shift of its western course between 1600 BC (Fig. 3.17) and 595 BC (Fig. 3.18) can be dated to 595 BC and was caused by bog burst D when the flow of the southern stretch of the northern discharge channel became inverted. The point of inversion, being the newly formed watershed, can be found near the trackway cluster Killoran 226/306/69 in the western bog margin (Fig. 3.18).

Between 100 BC and AD 200, when the growth of new poorly humified *Sphagnum* peats predominated, the character of the watershed became less clear and appears in a rather vague and broad zone (Fig. 3.19). This continued probably until about AD 500–700. From that time on, the organisation of the hydrological systems changed remarkably another time. The spatial aspects of this development could not be surveyed in detail.

The bog marginal forest peats

General aspects of the peat growth

From about 2500 BC both fen peat margins of the bog became forested (Figs 3.11 and 3.15). In general, these marginal forests had a rather strong wet–dry gradient but humid soils prevailed. At the wetter (bog) side, alder was the predominant tree, sometimes accompanied by birch. At the drier (upland) side, oak was often present. Here and there pine had established itself in these forests. With the exception of the drier locations tree growth was scanty. Poor-growing trees, with horizontal root systems, determined the appearance of most of the rather open forests. In the forests open spots occurred where fen peat growth continued and where tussocks with *Sphagnum*

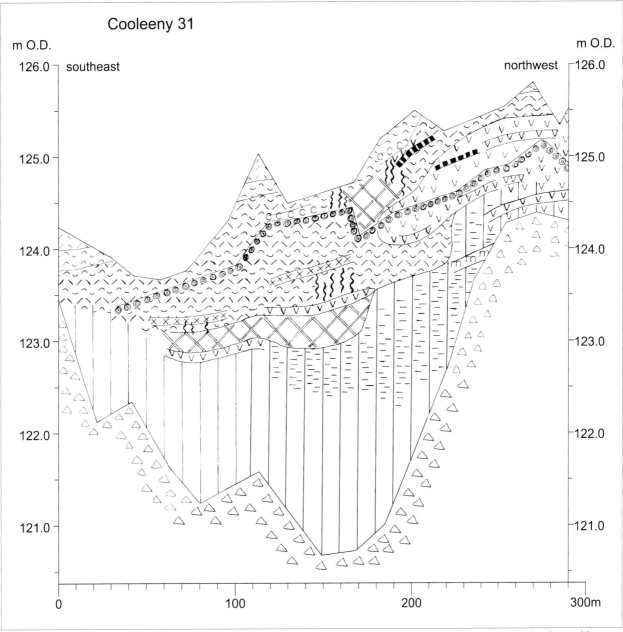

Fig. 3.9 Southeast–northwest peat section along the line of causeway Cooleeny 31. Vertical scale exaggeration 50x. General key: see Fig. 3.7.

peat or pools were present. During its upward growth, related to the rising bog water table (Fig. 3.12), it covered and buried the extant forests on the slopes of the mineral upland. The wood peat deposits also buried the numerous *fulachta fiadh* present on these slopes.

The most important feature of these deposits is the abundant presence of tree remains, including stems, trunks, branches, roots, bark and leaves. Highly humified herbacious remains form the next largest component. The peat is always very crumbly, caused by the abundant bark remains. The large fluctuations in the rise of the bog water table (Fig. 3.12), as seen in the occurrence of the bog bursts, have determined the development of these peats, especially the successive drier and wetter phases.

This brook peat, forest peat or carr was strongly influenced by the proximity of the rising mineral soil where it sloped towards the upland and by the prominent slope of ridges in the subsoil. Lowering of the water table after bog bursts or other desiccations caused collapsing of the peaty forest soil, inducing a good substratum for the growth of trees. Next to this, the water supply from Killoran Bog improved the conditions for forest peat growth largely in the western marginal area.

The width of the bog marginal forests was mainly about 40–60m. These forests were occasionally much wider, especially in the western area, as depicted in Figs 3.16–3.19. This can be attributed to the specific conditions of the Cooleeny area and the extra water

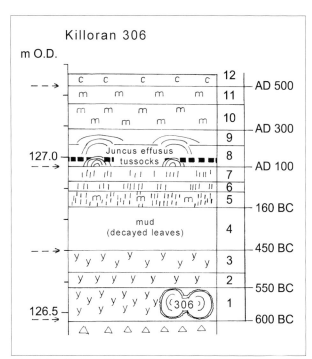

Fig. 3.10 Peat profile at Killoran 306. General key: see Fig. 3.7. These dates are partly derived from the ¹⁴C dates of Killoran 69/trackway Killoran 306 and are partly interpolated. The arrows with broken lines at the left side represent the hydrological effects of bog bursts.

supply from Killoran Bog. These features will be further discussed below.

The peat stratigraphy of the western marginal area is documented in Fig. 3.4. Complementary information can be derived from Figs 3.8–3.10. The peat stratigraphy of the eastern marginal area is depicted in Figs 3.6 and 3.7.

As previously mentioned, bog marginal forest peat started to grow about 2500 BC at a level of 124.20–124.40m OD; only in the northeastern part of the study area had forest peat growth already started at *c.* 3300 BC at a level of 123.80m OD. From 500 to 300 BC onwards, but more prominently after about AD 200, the bog marginal forest peat deposits became largely overgrown by ombrotrophic *Sphagnum* peat, indicating increasing acidity and oligotrophy of the peat-growing environment at levels of about 127.0m OD and higher.

The peat accumulation of the eastern bog marginal forest deposits could mainly be determined by the fluctuations of the bog water table and the life cycles of the trees. For the western marginal area it can be stated that the hydrological dynamics of the discharge systems played the dominant role in the recorded peat types, the peat accumulation as documented during the survey and archaeological excavations. For this reason the two deposits will be discussed separately below. Attention will also be given to a large number of trackways and platforms encountered in the bog marginal zones.

The eastern bog marginal forest peat deposits

The remarkable undulating eastern bog margin, caused by the direction, size and levels of the ridges of the eastern upland (see Fig. 3.2), is reflected in the location and course of the eastern bog marginal forest (as depicted in Figs 3.15–3.19). From *c.* 2500 BC onwards, at levels of roughly 124.20m OD, the eastern stretches of the bog were fringed by a narrow zone of forest, with mostly alder trees in the wetter places and oaks on the higher levels of the ridges. In Fig. 3.6, the forest peat deposits are present on the ridges of the mineral subsoil at 0–40m, 60–80m, 120–240m, 380–430m and 460–590m. In Fig. 3.7, the forest peat deposits in question are present between 170–200m, 250–340m and 450–510m. It is clear that the forest shifted in an easterly direction with rising water levels in the Derryville basin. The fluctuations of the bog water table, relative to the bog bursts, seems to have little or no influence on this eastward shift. Dependent on the relief and height of the ridges at *c.* AD 200, when the bog water level was nearly 127.0m OD, this shift varied from 50m to over 200m in an eastern direction, as Figs 3.15–3.19 indicate.

In the intervening period, at 400–300 BC, the plains of the ridges were covered with oak trees, a number of which were dated to between *c.* 650 BC and 300 BC. These became overgrown by peat. Below the level of 126.5m OD there was generally alder brook peat. Above this level it was nearly always raised bog peat from *c.* 300 BC onwards. From that time onwards, it can be accepted that the rise of the bog water table combined with the very acidic, very oligotrophic water quality was too rapid for the establishment of alder brook forest. The thickness of the bog marginal forest deposits varied from 0.4m to about 1.4m.

The rise of the bog water table was interrupted at least five times by the bog bursts. The effects of bog burst D (595 BC) are the most visible in the eastern margin. As mentioned above, an example of this can be illustrated by the occurrence at a level of about 125.4m OD of trackway Derryfadda 13 (Fig. 3.12, sample 15, dated to *c.* 590 BC) and the nearby bog oak E9 (Fig. 3.12, sample 10 at 124.75m OD), germinating in 593 or 592 BC. Here, causeway Killoran 18, dated to 1440 BC, indicates the bog water table at that time to be at about 124.90m OD. This indicates a drop in the bog water table in the eastern bog margin of over 1m, dated to 595 BC, from about 126.0m to 124.75m OD. Earlier, a serious drop in the water table caused by bog burst B can be explained in the same way by platform Derryfadda 213, dated to *c.* 1115 BC (Fig. 3.12, sample 6), and a bog oak under this trackway, dated to 2718 BC (Fig. 3.12, sample 5), both at the level of 123.90–124.20m OD.

Lamination of the forest peat is generally of limited extent, mostly related to horizontal root systems. Alder trees older than 40–50 years are rarely encountered. Long-lived oak trees, up to about 300 years old, reached such ages owing to improving growing conditions following drops in the bog water table, especially after bog burst D. Such large trees determined the local aspect of these forests to a high degree between about 500 and 300 BC. Remarkably, erosion phenomena are not present as gullies in the peat deposits in the eastern bog marginal forest. The sand content of the peat in question is minimal; there was only limited in-wash from the higher soils.

Access to the eastern bog marginal area

The eastern bog marginal forest zone provided access to the bog during its entire development, as is indicated by the presence of a number of trackways in this area. Between 2200 and 1800 BC (Fig. 3.15): Derryfadda 218 (*c.* 2120 BC), Derryfadda 207 (*c.* 2000 BC) and Derryfadda 204 (*c.* 1900 BC). Between 1600 and 1440 BC (Fig. 3.16): Derryfadda 23 (1590 BC), Cooleeny 22 (1517 BC) and Killoran 18 (1440 BC). Between 1250 and 800 BC (Fig. 3.17): Derryfadda 311 (*c.* 1240 BC), Derryfadda 216 (*c.* 1205 BC), Derryfadda 17 (*c.* 1170 BC), Derryfadda 213 (*c.* 1115 BC), Derryfadda 211 (*c.* 1100 BC), Derryfadda 216 (*c.* 1200 BC) and Derryfadda 209 (*c.* 880 BC). Between 600 and 400 BC (Fig. 3.18): Derryfadda 13 (*c.* 590 BC), Derryfadda 206 (*c.* 570 BC), Derryfadda 215 (*c.* 560 BC) and Derryfadda 210 (*c.* 430 BC). Between 300 and 100 BC: Derryfadda 9, Derryfadda 201 and Derryfadda 203 (*c.* 290 BC), Derryfadda 214 (*c.* 230 BC), Derryfadda 6 (*c.* 195 BC) and Derryfadda 208 (*c.* 180 BC). The locations of these trackways are indicated in Fig. 3.19, representing the bog surface between 100 BC and AD 300. In the latter window of time no trackways or platforms have been documented in this area. Clearly from about 200 BC or shortly after (more exactly 160 BC, as Killoran 226 Structure 2 in the western bog margin proves (see below)), it was impossible or at least very unattractive to access the bog surface. It seems that shortly after a bog burst (especially bursts B and D), but also shortly after 300 BC, increased access to the bog surface, at least into the marginal forest zone, occurred.

The western bog marginal zone and its forest peat deposits

The western bog marginal zone has been divided into three areas, the basin itself being the Derryville area, the south being the Cooleeny area and the upland Killoran Bog. Most of this margin runs roughly north–south (see Fig. 3.2). In this western zone, bog marginal forests of about the same type as in the eastern margin developed in Derryville Bog and in the Cooleeny area from about 2500 BC onwards (Fig. 3.15) at levels of about 124.20–124.50m OD.

Derryville Bog

The forest zone, producing wood peat, was mostly only 30–50m wide, depending on the relatively steep slope of

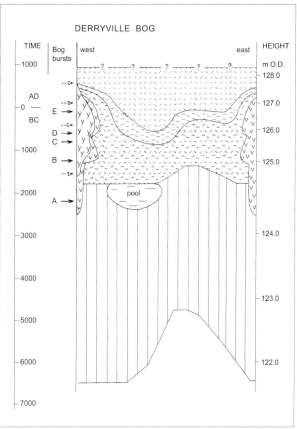

Fig. 3.11 Derryville Bog, highly simplified schematic overview of the bog development, plotted against time, focused on the area of the southern raised bog complex. The bog bursts A–E are indicated with arrows; the arrows with broken lines suggest additional bog bursts. General key: see Fig. 3.7. Further explanation in the text.

the western upland. This continued until about 2200–1800 BC (Fig. 3.15). Then the discharge system shifted to the western bog area before 1600 BC (Fig. 3.16). The northern half of the Derryville Bog marginal forest peat deposits became drowned. From about 1800 BC onwards, most of the western upland slopes became overgrown by the forest peat deposits of Killoran Bog, as depicted in Figs 3.16–3.19. In the southern stretches of the Derryville Bog marginal forest (from about the location of *fulacht fiadh* Killoran 304 and trackway Killoran 305; see Figs 3.15 and 3.16, respectively), the size of the forest increased significantly to 150–200m wide (Figs 3.16 and 3.17). This continued to about 800–600 BC. The deposit in question is not indicated in the north–south peat section of Fig. 3.4, for here only the bog marginal forest peats on higher levels are recorded, dated to 600 BC and later. In the west–east schematic peat section of Fig. 3.8a, the lowermost forest peat deposit is indicated between 110m and 140m (level *c.* 124.0m OD), where it can be dated to about 2500–2300 BC. The following extension of the Derryville Bog marginal forest peat is present in Fig. 3.8a, between 70m and 110m, at levels of about

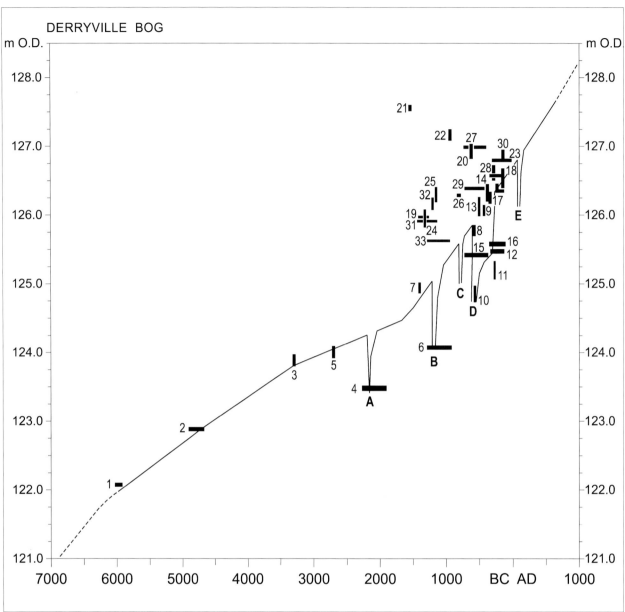

Fig. 3.12 Time-depth curve of the rising bog water table, based on ^{14}C and dendrochronological 'landfall' dates. Key: 1. Scots pine, GrN-21821; 2. fen peat, UB-4096; 3. bog oak over Derryfadda 218, Q9540; 4. Derryfadda 218, Beta-102759; 5. bog oak under Derryfadda 213, Q9401; 6. Derryfadda 213, Beta-102758; 7. Killoran 18, Q9349; 8. bog oak E8, start of growth, Q9126; 9. bog oak E8, end of growth, Q9126; 10. bog oak E9, start of growth, Q9127; 11. bog oak E9, end of growth, Q9127; 12. Derryfadda 9, Beta-102739; 13. bog oaks E1, E5, E6 and E7, start of growth, Q9119, Q9122, Q9124 and Q9125; 14. bog oaks E1, E5, E6 and E7, end of growth, Q9119, Q9122, Q9124 and Q9125; 15. Derryfadda 13, GrN-21934; 16. Derryfadda 203, Beta-102762; 17. bog oak E238, end of growth, Q9403; 18. bog oak E225, end of growth, Q9402; 19. Killoran 230, Beta-111368; 20. Killoran 234, Beta-111369; 21. Killoran 241, Q9542; 22. Killoran 243, Beta-9543; 23. Killoran 314, Beta-111376; 24. bog oak 306/69, start of growth, Q9489; 25. bog oak 306/69, end of growth, Q9489; 26. Killoran 69, UB-4184; 27. bog oak 248, Q9483; 28. Killoran 301, UB-4185; 29. Trackway 306, UB-4189; 30. Killoran 226, Q9475; 31. Killoran 265, Beta-111377; 32. Killoran 235, Q9541; 33. Killoran 253, Beta-111378. Further explanation in the text.

124.80–125.30m OD. It overgrew *fulacht fiadh* Killoran 304 and can be dated to about 1600–1400 BC. It is overlain by the forest peat deposits of Killoran Bog.

From *c.* 600 BC, the southern stretches of the bog marginal forest expanded further to the west, owing to the very shallow sloping subsoil of this area. This is recorded in the peat section of Fig. 3.4, between 230m and 450m. A bog lake developed in the forest, as is

recognisable between 380m and 530m by the lake-bottom sediment and dopplerite layer. The level, 125.70–126.20m OD, suggests an age of about 700–500 BC (probably to 595 BC). The mineral ridge between 580m and 870m became overgrown by bog marginal forest peat from about 600 BC onwards, as will be discussed below (Killoran Bog), in relation to the peat stratigraphy of trackway Killoran 306 (Fig. 3.10).

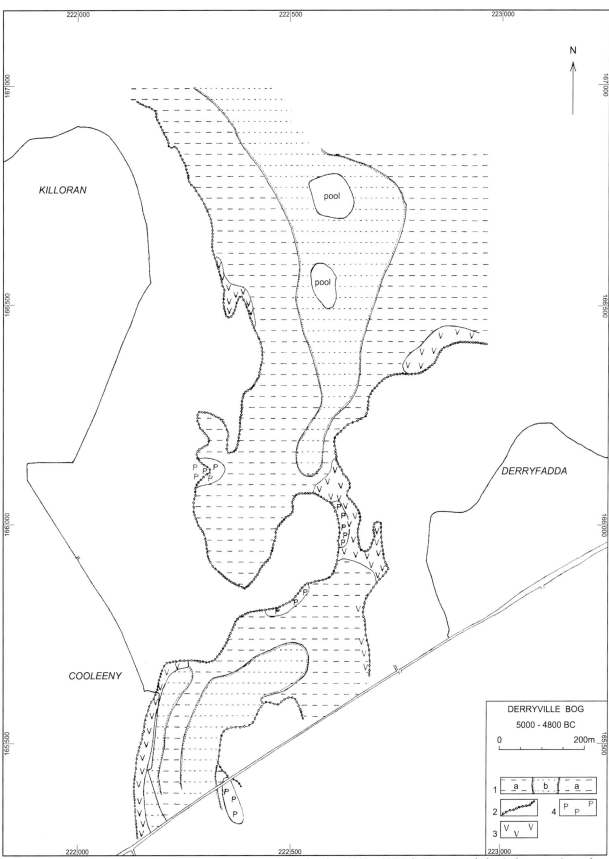

Fig. 3.13 Image of the bog surface, representing the situation dated to 5000–4800 BC. Key to symbols: 1. fen peat: 1a, rich in remains of Phragmites australis *(reed); 1b, poor in* Phragmites *remains; 2. bog margin 5000 BC (contour line mineral subsoil 122.75m OD); 3. forest peat; 4.* Pinus *(pine) forest peat and wood remains.*

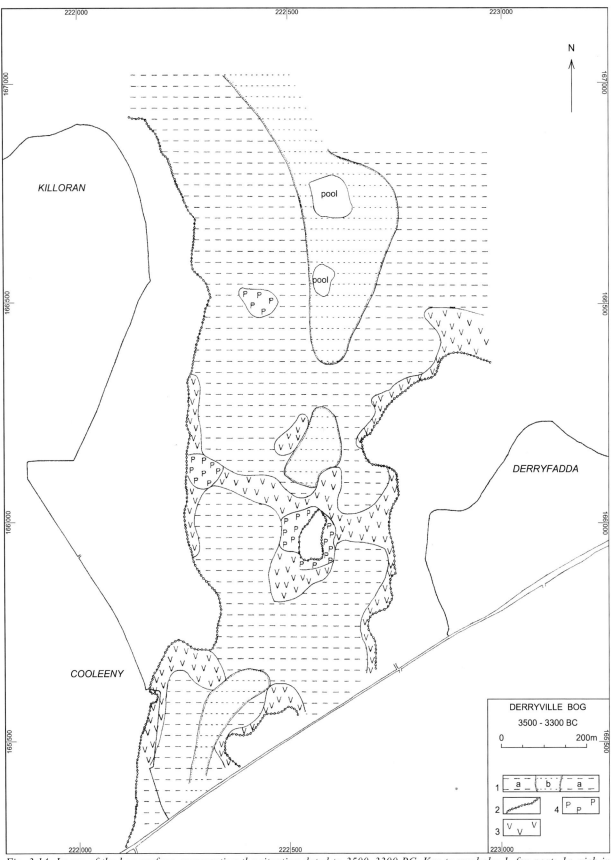

Fig. 3.14 Image of the bog surface, representing the situation dated to 3500–3300 BC. Key to symbols: 1. fen peat: 1a, rich in remains of Phragmites australis (reed); 1b, poor in Phragmites remains; 2. bog margin c. 3300 BC (123.75m OD); 3. forest peat; 4. Pinus (pine) forest peat and wood remains.

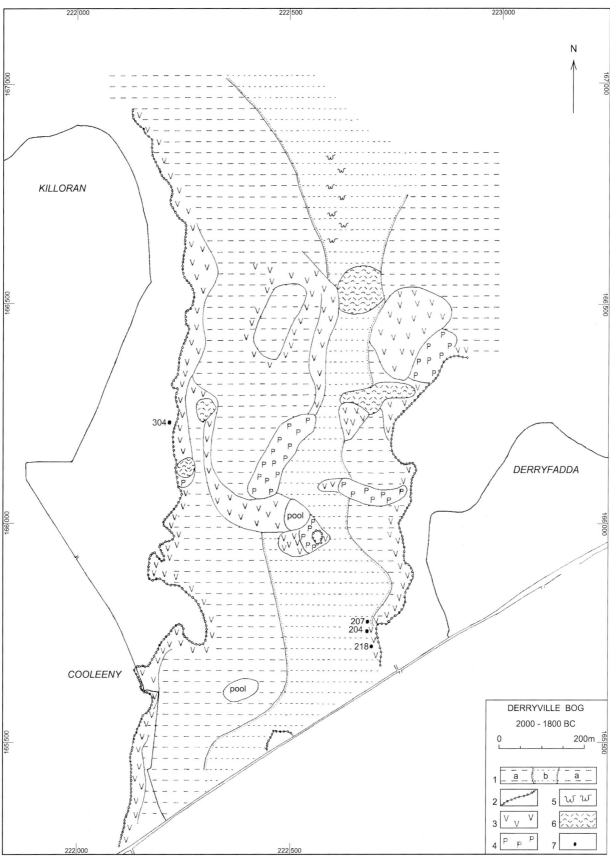

Fig. 3.15 The assumed bog surface, dated to 2000–1800 BC. Key to symbols: 1. fen peat: 1a, rich in remains of Phragmites australis *(reed); 1b, poor in* Phragmites *remains; 2. bog margin* c. *2000 BC (124.35m OD); 3. forest peat; 4.* Pinus *(pine) forest peat and wood remains; 5. water channels in the peat, probably the remains of pools; 6. raised bog peat accumulations in the fen peat; 7. location of trackways or other archaeological features with their numbers, dated to between 2200 and 1800 BC.*

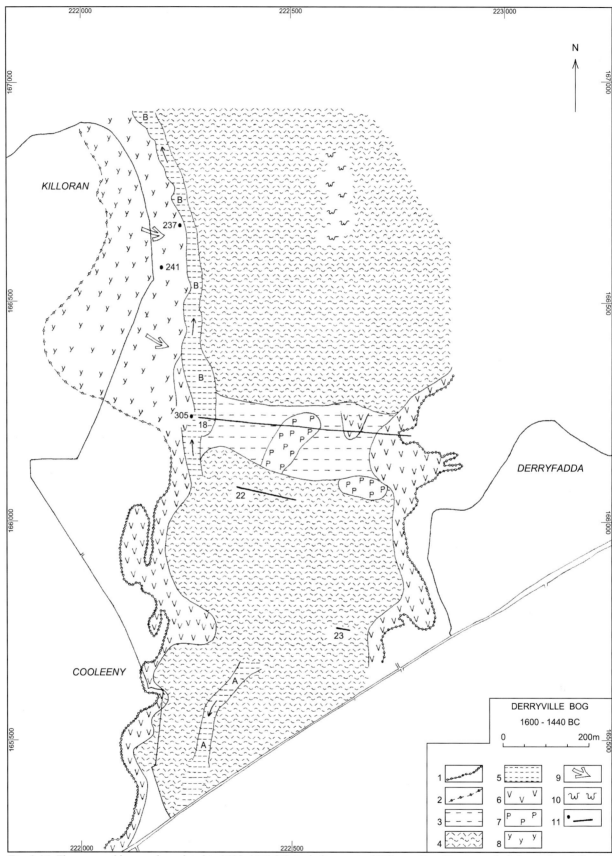

Fig. 3.16 The assumed bog surface, dated to 1600–1440 BC, including part of Killoran Bog. Key to symbols located on following page. Note: The discharge area B, at a level of about 125.0m OD, can be considered as the boundary between Killoran Bog to the west and the Derryville system to the east.

Fig. 3.16 Key to symbols: 1. bog margin c. 1500 BC (124.75m OD); 2. assumed western margin of the Killoran forest peat deposits (contour line 129.50m OD); 3. fen peat; 4. raised bog peat, spatial and peat-stratigraphical details are not indicated; 5. peat deposits of the discharge areas: A, southern discharge area (Cooleeny); B, northern discharge area (Derryville and Killoran), flow direction indicated by arrows; 6. forest peat, mostly bog marginal forest peat deposits; 7. Pinus (pine) forest peat and wood remains; 8. forest peat deposits of Killoran Bog between 129.50m and 124.75m OD; 9. location and direction of water flow systems of Killoran Bog; 10. water channels in the peat, probably the remains of pools; 11. location and course of trackways and other archaeological features with their numbers, dated to between 1600 and 1400 BC.

The thickness of the Derryville Bog marginal forest deposits varies from about 0.20m to 0.80m, with levels mostly below 126.0m OD. The deposits eventually became overgrown by raised bog peats between c. 400 BC and AD 500. From 400 BC onwards, the size of this forest was noticeably decreasing, as indicated in Figs 3.17–3.19.

Cooleeny area

Bog marginal forests have only been recorded on the northwestern margin, as is indicated in the cross-section through this area in Fig. 3.9, between 240m and 290m. The width is mostly restricted to about 40–60m, as is expressed in Figs 3.15–3.19. Only where the Cooleeny depression met the western upland had a very large area of brook forest developed (Figs 3.16 and 3.17) owing to the shallow sloping mineral soil, as is discussed above.

In the forest development three phases have been distinguished. In Fig. 3.9, Phase 1 is overlying the mineral ridge between 250m and 290m. Regarding its level, 124.40–124.60m OD, an age of about 2500–2000 BC is likely. Phase 2 concerns a substantial increase in its size, up to 110m in Fig. 3.9. Between 180m and 230m it covers raised bog peat deposits. It can be accepted that this extension is related to bog burst B, when the bog surface had collapsed and facilitated the germination of alder trees. Thus, Phase 2 can be dated to c. 1250 BC. Some time after the construction of causeway Cooleeny 31 (595 BC), of which the lengthways section is indicated in Fig. 3.9, the size of the bog marginal forest begins decreasing. Its southeast margin retreats from 180m to 215m. This is Phase 3. It clearly occurred under the influence of the increasing acidity and oligotrophy of the ombrotrophic *Sphagnum* peat growth. With regard to the flooding of the trackway and the southeastern edge of the alder forest between 160m and 200m, this decrease in forest size is related to the renewed rise of the bog water table after bog burst D. An age of 200–100 BC for Phase 3 is likely. The maximum thickness of the bog marginal forest peat is about 1.4m, from 124.20m to 125.60m OD in compacted condition. Originally, it was up to about 126.20m OD.

Regarding the very heavy, dam-like construction of causeway Cooleeny 31 (8m wide max.), it must be accepted that the builders were aware of the extra risks in bridging the discharge zone of the Cooleeny area (cf. Cross May *et al.*, pages 225, 341 and 348).

Killoran Bog forest peat

Although peat morphology, bog development and the size of this upland bog are only partially known, its influence on the peat growth in the study area was substantial, at least since about 1600 BC, with regard to the extent of the forest peats in the western marginal area of Derryville Bog (as depicted in Figs 3.16–3.19) and its effects on the bog hydrology in the Derryville basin. This has already been discussed above. The extension of the Killoran Bog deposits in these figures concerns only the easternmost deposits. The centre of the bog was situated further to the west of the forest peat deposits indicated in Figs 3.16–3.19. A large number of trackways, platforms and *fulachta fiadh*, present in or below these wood peat deposits, have been excavated and investigated.

The peat in question is recorded in the peat section of Fig. 3.4, between 750m and 800m, where it can be dated from about 600 BC to 100 BC, and between 850m and 950m (below the northern discharge gully deposits), dated from about 1600–1400 BC to 600 BC. In Fig. 3.4, the level of the Killoran Bog marginal forest peat (between 750m and 950m) is about 1m higher than the Derryville Bog marginal forest peat (between 230m and 550m). This is a good indication of the differences in water supply between both systems.

In the Killoran Bog marginal forest peat, three main water supply courses have been identified, indicated with open arrows in Figs 3.16–3.18. The southern flow must have influenced the Cooleeny system.

The higher levels of forest peat deposits of the Killoran Bog relative to the peat-growing levels in the Derryville basin are schematically depicted in Fig. 3.8a, between 0m and 120m. Here the deposits can be dated from about 1400 BC to about AD 500, as will be discussed below. The thickness of the deposits is more than 1m.

Detail around trackway Killoran 306

The water supply from Killoran Bog was very complicated, as can be demonstrated by two examples. Firstly trackway Killoran 306. Fig. 3.10 presents the detailed peat stratigraphy in twelve phases at the location of Killoran 306 (*c.* 580 BC), Killoran 226 Structure 1 (*c.* 450 BC) and Killoran 226 Structure 2 (161 BC). In Fig. 3.4 it is at 650m. This area of trackway clustering, Killoran 306/69, developed as a watershed since about 600 BC (Fig. 3.18). The peat sequence is characterised by a large number of peat types; the shifts from the one to the following type have been dated by means of the neighbouring trackways and a few peat levels. Between

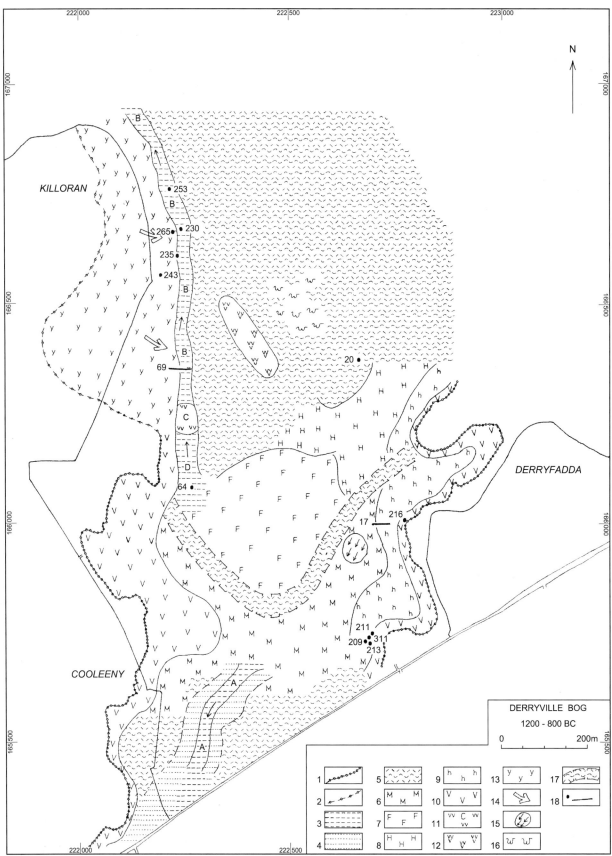

Fig. 3.17 The assumed bog surface, dated to 1200–800 BC, including part of Killoran Bog. Key to symbols located on following page. Note: Between c. 1600 and 600 BC the northern discharge channel shifted about 70–80m to the west. At c. 820 BC, when bog burst C affected this channel, most of Killoran 69 was situated in raised bog peat, indicating the further westward shift of the northern discharge channel.

Fig. 3.17 Key to symbols: 1. bog margin c. *1200 BC (level 125.10m OD); 2. assumed western margin of the Killoran forest peat deposits (129.50m OD); 3. peat deposits of the discharge channels: A, southern discharge channel with flow direction; B, northern discharge channel with flow direction; D, stretch of this channel blocked between 1440 and* c. *1200 BC; 4. area affected by bog burst B (*c. *1250 BC; eroded bog surface); 5. raised bog peat of which the spatial details are not indicated; 6–9. different types of raised bog complexes: 6. hummock and hollow system with moderately humified hummocks, between 1000 and 900 BC, overgrowing the highly humified* Sphagnum *peat deposits; 7. hummock and hollow system with traces of flooding and erosion in the hollows, grown since* c. *1000 BC; 8. hummock and hollow system with highly humified hummocks and moderately humified hollows, existing between* c. *900 and 200 BC; 9. highly and moderately humified peat, hummocks and hollows hardly or not visible, already in existence* c. *1100 BC; 10. bog marginal forest peat deposits; 11. wood peat accumulation in the blocked northern discharge channel (stretch C), 1440–*c. *1200 BC; 12. alder wood layer on top of highly humified* Sphagnum *peat, measuring 8cm in thickness and overlain by poorly humified S.* cuspidatum *peat, perhaps to be dated between* c. *1250 and 1000 BC (related to bog burst B); 13. Killoran forest peat deposits; 14. location and direction of water flow systems of Killoran Bog; 15. location of a system of erosion gullies with flow direction, cut into the underlying fen peat and filled with highly humified* Sphagnum *peat, very probably related to bog burst B (*c. *1250 BC); 16. water channels in the peat, most probably the remains of raised bog lakes; 17. watershed area between the northern (Derryville) raised bog complex and the southern (Cooleeny) raised bog complex, consisting of hummock and hollow systems; 18. location and course of trackways and other archaeological features, to be dated between 1200 and 800 BC.*

600 and 450 BC (Phases 1–3) water supply in the Killoran system was affected after bog burst D. From 450 to 160 BC a pool existed in which both phases of Killoran 226 were present (Phase 4). This flooding indicates a significant rise in the bog water table. Between 160 BC and AD 100 nutrient-rich water fed the peat-growing systems, surely originating in other sources than in the previous situation (Phases 5–7). In AD 100, the area dried out very severely and *Juncus effusus* (soft rush) could colonise the desiccated bare peat bog surface (Phases 8–9). Most probably, a serious drop in the bog water table as a result of a bog burst caused this desiccation. From AD 300 on, mineral-rich water, undoubtedly descending from the Killoran upland, flooded the area, giving way to the development of *Menyanthes* peat (Phases 10–11). About AD 500, the area became flooded again, now by very acidic bog water, as the deposit of poorly humified *Sphagnum cuspidatum* peat indicates (Phase 12). The minerotrophic peat growth shifted definitively in bog development under ombrotrophic conditions. Here, that was the final stage of the bog marginal forest peat deposition.

The peat profile, now 0.80m in thickness, had an original thickness of about 1.20m between 126.50m and 127.70m OD.

Detail around trackway Killoran 75

The other example is Killoran 75 (c. 280 BC), where a 14m-long hazel hurdle was used for accessing a deep pool with floating mats of *Menyanthes trifoliata*. Apparently, bridging the pool was a simpler way of approaching Derryville Bog than constructing a road in the surrounding bog marginal forest of Killoran Bog. In Fig. 3.20, the location is mapped, indicating the situation about 300–250 BC.

The pool, existing from c. 600 BC to 100 BC, was part of the northern discharge channel as illustrated in Fig. 3.18. Previously, this gully water flow was present from about 1600 BC, as the dated peat stratigraphy of this area proved. The presence of three *fulachta fiadh* (Killoran 253, c. 1125 BC; Killoran 265, c. 1270 BC, and Killoran 240, before 1547 BC) at the base of the peat

deposits suggests a relationship between the location of *fulachta fiadh* and water flow from the upland. The archaeological features in this region (partly trackways),Killoran 237 (c. 1525 BC), Killoran 230 (c. 1360 BC), Killoran 235 (1212 BC) and Killoran 229 (c. 1160 BC), illustrate the growth of Killoran Bog forest peat long before 600 BC. The nearby trackways/platforms Killoran 234 (c. 595 BC), Killoran 315 (c. 295 BC) and Killoran 312 (c. 280 BC) indicate that even during the full existence of the northern discharge channel this area was a preferable location for accessing the bog marginal forests just northeast of the Killoran upland. It cannot be proved that the tracks were intended to reach the treeless raised bog surface of Derryville Bog. Fig. 3.20 makes clear that the course of the northern discharge channel was a crucial event for the extension of the Killoran Bog marginal forests, of which the peat sequence measures up to c. 1m in thickness. The trackways mentioned above are also indicated on the related maps of Figs 3.16–3.19.

Access to the western bog marginal area

Considering the locations where trackways accessed the bog marginal forest peat areas, it is clear that accessing the peatlands occurred regularly in prehistoric times. Next to this, some locations seem to have been attractive. Here we see clusters of trackways/platforms, sometimes a (Late Bronze Age) *fulacht fiadh*. Examples of these are the trackways Killoran 75, discussed above, Killoran 240 and Killoran 306/69. Further can be mentioned between 2200–1800 BC (Fig. 3.15) the *fulachta fiadh* Killoran 304 (c. 2035 BC); between 1800–1400 BC (Fig. 3.16) Killoran 241 (1547 BC) and Killoran 64 (c. 1220 BC); between 1200–800 BC (Fig. 3.17) Killoran 243 (979 BC), Killoran 69 (c. 820 BC), and Cooleeny 62 (c. 650 BC); between 600–400 BC (Fig. 3.18) Killoran 306 (c. 580 BC), Killoran 246 (c. 560 BC), Killoran 226 Structure 1 (450 BC) and, bridging the Cooleeny area, Cooleeny 31 (c. 595BC) and Cooleeny 143 (c. 590 BC); between about 300–160 BC Killoran 301 (c. 324 BC), Killoran 315 (c. 295 BC),

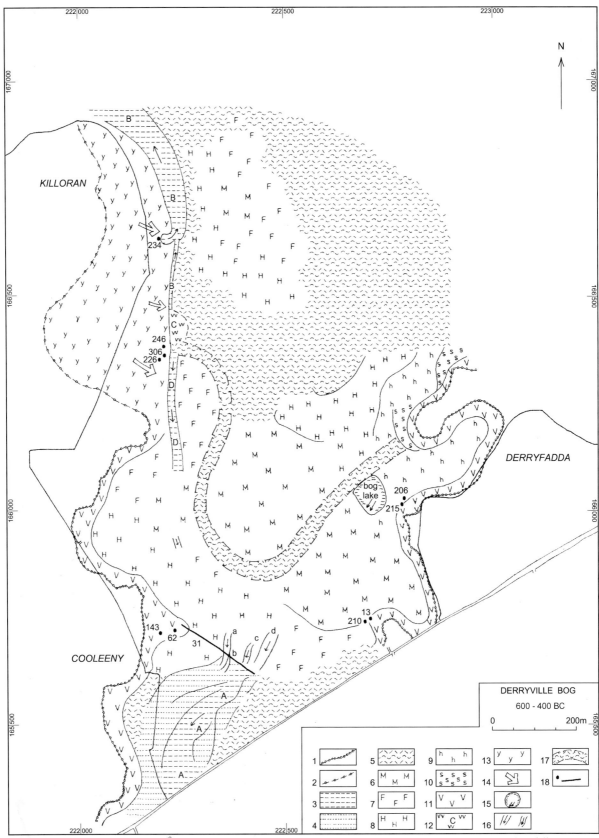

Fig. 3.18 The assumed bog surface, dated to 600–400 BC, including part of Killoran Bog. Key to symbols located on following page. Note: Major changes in the bog surface between c. 800 and 400 BC can be attributed to bog burst D (595 BC). The northern discharge channel became divided in three stretches: B, C and D. Owing to the lower drainage level in the Cooleeny complex the watershed shifted to some degree. With the increasing acidity and ombotrophy of the bog, the raised bog peat growth expanded greatly at the cost of the bog marginal forests. In Fig. 3.20 the bog surface surrounding the pool near Killoran 234 has been depicted in more detail.

Fig. 3.18 Key to symbols: 1. bog margin c. 600 BC (level 126.0m OD; for Cooleeny area: 125.50m OD); 2. assumed western margin of the Killoran forest peat deposits; 3. peat deposits of the discharge channels: A, southern discharge channel with flow direction; B, northern discharge channel with flow direction; D, stretch of the northern discharge channel with inverted flow direction (since 595 BC); 4. area affected by bog burst D (595 BC); eroded bog surface, washed-away peat deposits and erosion gullies; 5. raised bog peat of which the spatial details are not indicated; 6–10. Different types of raised bog complexes: 6. hummock and hollow system with moderately humified hummocks and poorly humified hollows; 7. hummock and hollow system with traces of flooding and erosion in the hollows, mostly poorly humified Sphagnum peat in the hollows, growing since c. 600 BC; 8. hummock and hollow system with mostly highly humified hummocks and poorly humified hollows; 9. highly and moderately humified peat, hummocks and hollows mostly hardly visible; 10. fields with abundant remains of Scheuchzeria palustris; 11. bog marginal forest peat deposits, since c. 600 BC decreasing in size; 12. wood peat deposits, established on the northern discharge channel deposits of stretch C, which lost its function; 13. Killoran forest peat deposits; 14. location and direction of water flow systems of Killoran Bog: the northernmost system supplied mineral-rich water, the southern system supplied bog water; 15. raised bog lake with floating Sphagnum cuspidatum vegetation mats from c. 700–595 BC, when it became drained by bog burst D; the flow direction is indicated by arrows; 16. erosion gullies with their flow direction; the gullies a, b, c and d belong to the bog burst D event, 595 BC; 17. watershed between the northern (Derryville) raised bog complex and the southern (Cooleeny) raised bog complex, consisting of hummock and hollow systems; 18. location and course of archaeological features, to be dated to between 600 and 400 BC.

Killoran 248 (*c.* 300 BC), Killoran 75 and Killoran 312 (*c.* 280 BC), Killoran 314 (*c.* 185BC), Killoran 226 Structure 2 (*c.* 161 BC); Cooleeny 169 (*c.* 290 BC), Cooleeny 178 (*c.* 285 BC), Cooleeny 141 (*c.* 280 BC) and Cooleeny 100 (*c.* 230 BC). The location of these features is depicted in Fig. 3.19, which in fact covers the period 100 BC–AD 300. Exactly paralleled by the situation in the eastern marginal area, in this period no trackways or platforms have been found here. As already mentioned, this can be attributed to the strong decrease in the accessibility of the bog surface from about 160 BC due to the increasing extension of the growth of poorly humified *Sphagnum* peat, probably after bog burst E (Fig. 3.12).

Noticeably, a large number of trackways were constructed just after 600 BC, indicating the large impact of bog burst D on the accessibility of the bog marginal area. Another interesting point is the presence of a number of trackways dating to shortly after 300 BC to 160 BC. Around 300 BC, the bog marginal forest deposits collapsed again after a serious drop of the bog water table. This possible bog burst is indicated in Fig. 3.11. After 160 BC poorly humified *Sphagnum* peat growth prevailed over bog marginal forest peat growth.

Summarising remarks concerning the western bog marginal area

The tree composition of the forest did not seriously differ from that of the eastern margin: mainly alder, some birch and, on the drier locations, oak. The greatest difference concerned the spatial occurrence of pools, channels and also gullies; the differences in height of the forest soil related to the Killoran peat levels and the greater hydrological dynamics, partly owing to the extra water supply of Killoran Bog and partly to the nearby presence of the discharge channels. The latter were in fact the impassable zone of Derryville Bog.

The Killoran upland area is much larger in size than the Derryfadda upland. The Killoran upland was occupied much more densely from the Middle Bronze Age on. The larger number of trackways and *fulachta fiadh* in the western zone can be ascribed to this activity.

In the southern half of the western bog marginal area, occasional silt layers have been encountered in the forest peat deposits, indicating in-wash from the higher soils. It can indicate ongoing deforestation, but it can also be explained as the effects of bog bursts. Two of these features are marked in Fig. 3.19. The southern one is well dated to between 180 and 80 BC. According to Caseldine *et al.* (this volume) the silt was, to a high degree, agraric soil.

The raised bog peats

The spatial organisation of raised bog development: the increasing size

The highly oceanic Irish climate with its large amounts of rainfall provides excellent circumstances for raised bog growth. All the fen peat deposits and most of the bog marginal peats eventually became overgrown by ombrotrophic peat. The beginning can be dated to 1800–1400 BC in the central zone of the bog (Figs 3.11 and 3.16). From then on, both margins shifted 250m, so the width of the bog increased from about 520m to over 1000m in the first centuries of the next millennium (Fig. 3.19). The bog water table rose from *c.* 124.50m OD to over 128.50m OD in a period of about 3000 years, from *c.* 1800 BC to *c.* AD 1200 (Figs 3.11 and 3.12). This represents a growth rate for the raised bog peats of 13.3cm per 100 years over the entire period.

Spatial structures

The ombrotrophic peat growth resulted in a large number of different growing conditions, recorded as identifiable spatial structures in the peat profiles studied. These reveal evidence for the hydrological systems and discharge channels already mentioned, for the development of different types of hummocks and hollows, the formation of pools and streamlets, the creation of erosion gullies, the growth of lawns, fields and floating mats. Most of these features could be surveyed, including mud-like lake-bottom sediments or pool fills, amorphous peat layers indicating local desiccations and dopplerite. The

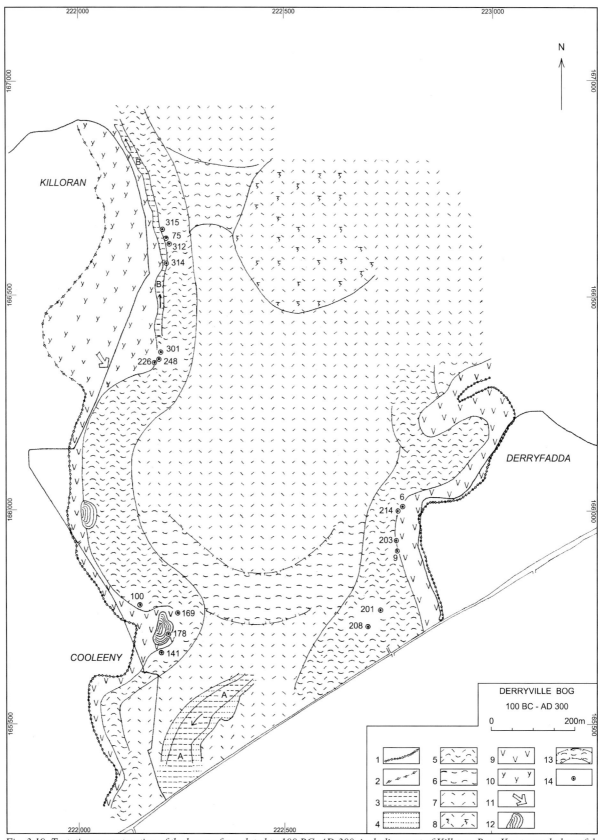

Fig. 3.19 Tentative reconstruction of the bog surface, dated to 100 BC–AD 300, including part of Killoran Bog. Key to symbols on following page. Note: The 60m-wide stretch of the northern discharge channel, active since c. 600 BC, became overgrown by highly humified Sphagnum peat c. 90 BC. The remaining narrow channel shifted between 90 BC and AD 400 by c. 60m to the west. From c. 400 BC on, the growth of poorly humified Sphagnum peat increased, smoothing many differences in spatial structures and topography of the bog surface. Clearly, regarding the absence of trackways and platforms, the poorly humified raised bog surface was not attractive to access.

Fig. 3.19 Key to symbols: 1. bog margin c. AD 100 (level 127.0m OD; for Cooleeny area: 126.50m OD); 2. assumed western margin of the Killoran forest peat deposits (level 129.50–127.0m OD); 3. peat deposits of the discharge channels: A, southern discharge channel with flow direction; B, northern discharge channel with flow direction, developing after 200–100 BC (after bog burst E); 4. area affected by bog burst D (eroded bog surface, washed-away peat and erosion gullies); 5–8. different types of raised bog growth: 5. hummock and hollow systems, with highly humified hummocks and poorly humified hollows, succeeding highly humified Sphagnum peat growth since about 600–500 BC; 6. systems with moderately to highly humified hummocks and poorly humified hollows, developed on a deposit of poorly humified, laminated Sphagnum peat; the laminated peat can most probably be dated to between c. 500 and 300 BC, the growth of hummocks and hollows of this type of peat can be averaged between 300–100 BC, then becoming overgrown by poorly humified Sphagnum peat from about 100 BC; superficial erosion in these deposits can be dated to roughly between 200 BC and AD 200; 7. poorly humified Sphagnum peat fields, to be dated from c. 300 BC on, covering a highly humified hummock and hollow system, to be dated between 600 and 300 BC. In many places, these highly humified deposits cover a layer of poorly humified Sphagnum peat, indicating wet conditions at the raised bog surface from c. 700–600 BC; 8. poorly humified Sphagnum cuspidatum peat, with many vertical cracks and replaced peat blocks, indicating erosion by bog burst E (100 BC–AD 100); 9. bog marginal forest peat; 10. Killoran forest peat; 11. location and direction of the flow system of Killoran Bog; 12. silt deposits in Cooleeny area, the southernmost dated to between 180–80 BC (see Chapter 6), the other deposit perhaps between 300 BC and AD 100; 13. watershed area between the Derryville and the Cooleeny raised bog complexes; 14. location of trackways etc. to be dated to between 400–100 BC, being the window of time just before the period as presented in this figure. Most remarkably, between 100 BC–AD 300 trackways and platforms are absent in these peat deposits.

latter is derived from humic colloids originating from higher levels in the peat that have percolated through a transition zone between two types of peat, mostly where a cavity existed. Causeway Killoran 18, dated to 1440 BC, is the best example of a structure embedded in dopplerite (see Chapter 10).

In Figs 3.16–3.19, the occurrence and, where possible, the distribution of the above-mentioned spatial structures has been presented rather than the degrees of humification of the *Sphagnum* peat, as will discussed below. In the figure captions, the characteristics of identified structures are described in brief.

Ombrotrophic peat formers—humification degrees

Derryville Bog does not differ from most other raised bogs in Europe in the way that the ombrotrophic deposits display the development from highly humified, via moderately humified, to poorly humified *Sphagnum* peat as the principal peat-formers. In highly humified peat, it mostly concerns the small-leaved *Sphagnum rubellum/fuscum*. The poorly humified peat is mostly formed from the large-leaved *Sphagnum imbricatum*. Remarkably, the typical hollow-dweller *Sphagnum papillosum* was not noted within the records (see Chapter 6). The absence of *Sphagnum magellanicum* in the records is not very surprising; this species only proliferated in the late medieval period. *Sphagnum cuspidatum* (water peat moss) occurred at many locations, indicating extreme wet conditions as pools and flooded hollows.

Moderately humified *Sphagnum* peat is mainly a deposit of more humified large-leaved *Sphagnum* species, such as *Sphagnum imbricatum*, or less humified small-leaved *Sphagnum rubellum/fuscum*. In this study, the moderately humified *Sphagnum* peat deposits can also be partially seen as the system where structures with both highly humified peat, such as hummocks, and poorly humified peat, such as hollow fill and *Sphagnum*

cuspidatum accumulations, occurred next to each other. Precisely these latter conditions presented the best peat-stratigraphical and spatial information in relation to bog development. The spread and expansion of such raised bog growing systems in the peat sequence and over the bog surface is depicted in peat sections Figs 3.4–3.10, as well as in the maps Figs 3.16–3.20. Generally, all raised bog complexes are treeless.

Palaeoenvironmental indicators

Important indicators of trophic and hydrological conditions that were recorded are *Eriophorum vaginatum* (hare's tail grass), *Calluna vulgaris* (heather or ling), *Scheuchzeria palustris* (rannock rush), *Menyanthes trifoliata* (bog bean), *Phragmites australis* (reed) and *Juncus effusus* (soft rush). Their presence and distribution in the peat deposits was recorded to gain a more detailed insight into the peat growth and hydrological development.

Trackways

Only a small number of trackways were intended to make the treeless raised bog surface accessible. The lightly constructed oak plank footpaths, Derryfadda 23 (1590 BC) and Cooleeny 22 (1517 BC), trackway Derryfadda 17 (*c*. 1170 BC), trackway Killoran 69 (*c*. 820 BC) and causeway Cooleeny 31 (595 BC) are examples of such. Some other trackways and platforms are situated in bog marginal peat but overgrown by raised bog peat. The location of these features is indicated in the related maps Figs 3.16–3.20.

Highly humified *Sphagnum* peat—some characteristics and dates

General remarks

In the central zone of the bog the thickness of the highly humified *Sphagnum* peat could reach 2m. In Fig. 3.5, at about 180m, this is about 1.50m thick in compacted form;

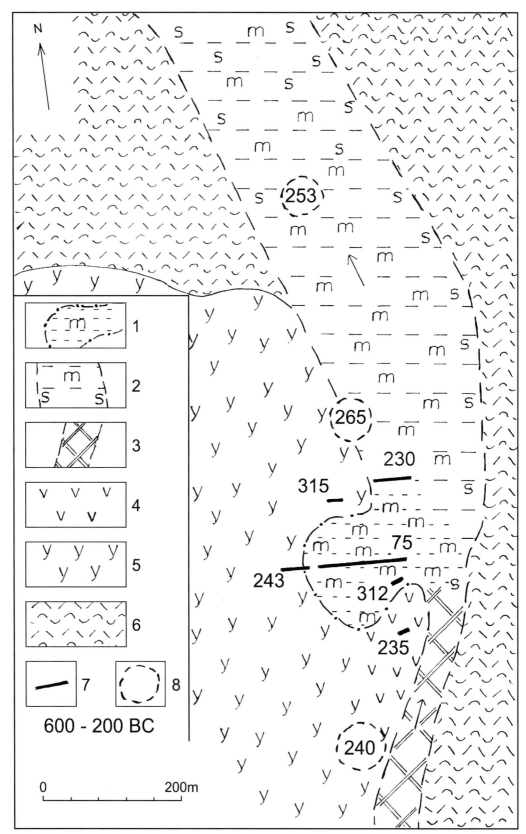

Fig. 3.20 Detail of the bog surface near the western bog margin, 600–200 BC, with the northern discharge channel near Killoran 75. Key to symbols: 1. pool with Menyanthes *peat; 2. northern discharge channel with* Menyanthes-Scheuchzeria *peat (m=*Menyanthes, *s=*Scheuchzeria*); 3. pool fill, indicating the northern discharge channel; 4. bog marginal forest peat; 5. Killoran Bog forest peat; 6. poorly to moderately humified* Sphagnum *peat bog surface (treeless); 7. location and course of trackways in the vicinity of Killoran 75; 8. location of (Late Bronze Age)* fulacht fiadh *at the base of the fen peat deposits.*

at 350m it is about 1.05m. That was in the southern raised bog complex, in the Cooleeny area. Further to the north, the thickness in compacted form measured up to not more than 0.80m. Here it was grown in extremely wet conditions, so the difference between highly and moderately humified peat could not always be exactly recorded. At both bog margins (Figs 3.4 and 3.6) the highly humified *Sphagnum* peat deposits were much thinner, sometimes less than 0.10m, or even absent.

The commencement of the highly humified *Sphagnum* peat growth in the southern area is dated to *c*. 1800 BC, at a level of about 124.50m OD. It is assumed that in the northern area this beginning can be dated to the same period. In Fig. 3.16, the spread of both complexes is depicted. In the intermediate fen peat watershed, raised bog peat growth started here at about 1420 BC, at a level of 124.90m OD, as previously discussed.

Highly humified peat growth continued in the central area to shortly after 950 BC near trackway Derryfadda 23, to *c*. 300 BC near causeway Killoran 18 (as demonstrated in the dated peat profiles of Fig. 3.8b) and in large stretches of the Cooleeny area. It became mostly overgrown by moderately humified peat. At the bog margins this transition occurred much later, up to the first centuries AD (see also Fig. 3.11).

The shift from fen peat to raised bog peat can be attributed to the increasing influence of the direct rainfall on the peat-growing environment. Most likely this was not initially due to climatic influences but was essentially caused by changes in hydrology, though not desiccation or, to the contrary, flooding. Here and there the ombrotrophic peat has overgrown filled lakes or ponds at the top of the fen peat. The peat section of Fig. 3.9 gives such an example between 60m and 180m. Underneath trackway Cooleeny 22 (Fig. 3.16) the transition from fen peat to raised bog peat also occurred via a pool.

Wood levels in this peat represent mostly localised drying episodes. Levels with heather remains in the peat faces reflect drier conditions on the bog surface, where the concentrations of hare's tail grass in the peat faces indicate relatively large fluctuations in the bog water table. The highly humified peat deposit in the peat section of Fig. 3.9 exemplifies these environmental indications. Layers of *Sphagnum cuspidatum* peat stand for the fill of pools and hollows with open water, as is recorded in Figs 3.4–3.7. In many places remains of rannock rush were surveyed and recorded. The presence of rannock rush is only depicted in the detailed profile in Fig. 3.8b (Derryfadda 23). The same applies to bog bean. Both plant species represent very wet conditions, primarily differentiating in the amount of nutrients in the supplying water.

Although this was not easy to determine in all the peat faces, the highly humified *Sphagnum* peat growth occurred mostly in hummock and hollow systems. This could be established by a number of trackway excavations, as will be discussed. In Fig. 3.16 there is no distinction presented in the different types of hummocks and hollows up to about 1400 BC. Only the presence of water channels in the peat is indicated (see below). From *c*. 1200 BC on, a number of different types of hummock and hollow systems could be established peat-stratigraphically, as already recorded in the vicinity of the trackways Cooleeny 22 and Derryfadda 23. This is depicted in Fig. 3.17. The hummocks consist of highly humified *Sphagnum* peat while in the hollows moderately humified peat layers have been identified.

Detail around trackways Derryfadda 23 (1590 BC) and Cooleeny 22 (1517 BC)

Both trackways are single-plank oak footpaths, present in the highly humified *Sphagnum* peat, rich in the remains of *Eriophorum vaginatum* (hare's tail grass) and of *Calluna* (heather). This represents an only moderately humid raised bog complex, with a relatively large fluctuating bog water table. Trackway Cooleeny 22 is situated in the central area of the southern raised bog complex. Trackway Derryfadda 23 lies near the eastern stretches of this complex (Fig. 3.16). The cross-section of Cooleeny 22 is cut at 510m in Fig. 3.5. At both locations the commencement of the raised bog growth can be dated to about 1800 BC (Fig. 3.8b). Based on peat-stratigraphical considerations, the transition to moderately humified *Sphagnum* peat most likely commenced in the central area of this complex before 1200 BC. In the surroundings of trackway Derryfadda 23 this transition is well dated to 900–850 BC. Here, in compacted form, the deposit has a thickness of about 0.65m (Fig. 3.8b, levels between 123.85 and 124.50m OD). The light form of the construction technique of both footpaths would be expected in an only moderately humid growing bog and the highly decayed condition of the oak planking also testifies to long exposure to the air. The excavations of both trackways made it clear that the bog surface was covered with a hummock and hollow pattern, with large flat hummocks and not very clear, shallow hollows, in which open water did not occur. The hummocks, rich in hare's tail grass, measure up to 4–5m.

Most of the water discharge from this complex was via the northern discharge channel but the southern stretches of this complex probably drained via the Cooleeny area in a southwesterly direction (Fig. 3.16). It could be proved that the drainage level of both channels was lower than the raised bog surface of the southern complex from at least 1590 BC to after 1200 BC (see below).

Water channels in the peat

For the northern highly humified *Sphagnum* peat complex, as depicted in Fig. 3.16, less information is available regarding the peat-growing environment and botanical composition. It is assumed that raised bog growth commenced here about 1800 BC, probably under wetter conditions. Large water accumulations in the peat are recorded in the northern stretches of Derryville Bog, such as the survey of the peat face of Fig. 3.5, 970–1260m. These most probably involved channels filled with water in the water-saturated amorphous peat. Though it is not known in exactly which form peat growth occurred here, and serious erosion can not be excluded, it in all probability indicates that highly humified peat growth occurred on floating mats in large pools or even bog lakes. The evidence for erosion mentioned above occurs in association with the presence of open water in the peat deposits. The possibilities for dating the existence of the open water accumulations are poor, especially with reference to redeposited peats derived from erosion. It is clear, however, that around 1800 BC such pools or bog lakes would have been present in the northern stretches of Derryville Bog. About 1200 BC or so it can be assumed, on the grounds of peat-stratigraphical considerations, that these phenomena must have ended.

From that time on, a large raised bog lake developed somewhat further to the south (Fig. 3.17). In this case the water channels are included in highly humified *Sphagnum cuspidatum* deposits, in which little or no erosion could be identified. To the southwest of these deposits a stretch of alder wood peat could be identified, about 10–15cm thick. Its location corresponds exactly with a higher level of the mineral subsoil (Fig. 3.2). So this lake very likely developed after a serious drop in the bog water table related to bog burst B (about 1250 BC), where the wood peat layer indicates the southwestern margin of this lake. The other shore areas could not be mapped. The recorded thickness of the alder wood peat and the *Sphagnum cuspidatum* deposits suggests that this lake existed for perhaps one or two centuries.

Detail around causeway Killoran 18 (1440 BC)

Highly humified *Sphagnum* peat growth started in the former watershed area between the two raised bog complexes to 1420 BC, caused by the construction of causeway Killoran 18 (Figs 3.8a and 3.8b), as discussed above (see also Fig. 3.16). From this time on, the two complexes became united in one large raised bog complex but they belonged to two different hydrological systems (Fig. 3.17). Highly humified *Sphagnum* peat growth continued here to about 200 BC (Fig. 3.8b), indicating the effect of the former watershed and hydrological activities in the northern discharge channel on peat growth conditions. Here in compacted form the deposit has a thickness of about 0.60m (Fig. 3.8b, levels between 124.87m and 125.47m OD). The peat stratigraphy will be discussed in more detail in the next section (the peat face: causeway Killoran 18). Near the eastern starting point of causeway Killoran 18, highly humified *Sphagnum* peat, rich in flooding and desiccation indicators, commenced about 950 BC with flooding, and continued onto about 200–150 BC. Its thickness in compacted form is 0.70–0.80m.

In contrast to the highly humified *Sphagnum* peat in the southern raised bog complex, *Eriophorum vaginatum*

(hare's tail grass) and *Calluna* (heather) are only barely present in the deposits in question but *Scheuchzeria palustris* and highly humified *Sphagnum cuspidatum* layers are clearly present between *c.* 1200 and 500 BC. This suggests fairly wet conditions in this period. From *c.* 500 BC highly humified *Sphagnum* peat growth continued under somewhat drier conditions, with increasing amounts of hare's tail grass and heather.

Detail around trackway Derryfadda 17 (c. 1150 BC)

Trackway Derryfadda 17 is one of the few trackways in Derryville Bog that are mainly constructed of stones, covering a highly humified *Sphagnum* bog surface with a clear hummock and hollow pattern (Fig. 3.17). A substructure is absent. This indicates an only moderately humid bog surface about 1150 BC. Bog burst B, dated to *c.* 1250 BC, influenced the bog surface to such a degree that this surface could be accessed with the help of a lightly constructed trackway, where, in fact, only the hollows needed a pavement with stones. Based on the excavation results, the size of the rather pronounced oval to round hummocks can be estimated to have been 3–5m (see Chapter 10). Hollows measured up to about 2–3m and were not very wet.

Summarising remarks

When highly humified *Sphagnum* peat started to grow *c.* 1800 BC, its extent increased dramatically at least from 1440 BC, as Figs 3.16–3.18 show. Expansion in a western direction was hampered or even blocked by the location of the northern discharge channel and its impact on the raised bog hydrology. About 100 BC most of the northern discharge channel became overgrown by moderately and poorly humified *Sphagnum* peat, as Fig. 3.19 indicates. The remarkable course of the watershed and its fluctuations during the period of highly humified peat growth can be attributed to the effects of three identifiable hydrological systems, acting since at least 1600 BC. The bog bursts B (to the south), C (to the north) and D (to the south) influenced the hydrology and thus the highly humified peat growth. This caused the development of a large number of different hummock and hollow systems, *Sphagnum* lawns and pools, as are indicated in Figs 3.16–3.18.

There are clear peat-stratigraphical indications that the peat-growing environment was not noticeably wet until about 1400 BC. The two trackways intended to cross the entire raised bog (Derryfadda 23 and Killoran 18) can be dated to before 1400 BC. Up to about 1200–1000 BC it can be expected that the development of the new hydrological situation with its contemporary watershed coincided with the increasing humidity of the peat-growing environment.

As already mentioned, from *c.* 950 BC the raised bog system in the central area became more humid or even wetter, and the large-leaved *Sphagnum imbricatum* became the predominant peat-former, primarily in the hollows. The first occurrence of moderately humified peat can be dated to about 1200 BC but, in general, between 900 and 850 BC is a more convenient date. In the marginal areas it lasted to about AD 500 before all the highly humified *Sphagnum* peat was overgrown by moderately or poorly humified *Sphagnum* peat.

Somewhat surprisingly, highly humified *Sphagnum* peat growth continued in the Cooleeny area to somewhat later than 600 BC (Fig. 3.9). This can very likely be attributed to the more sloping bog surface, resulting in more hydrological dynamics such as superficial desiccations with expansion of heather, hare's tail grass or even alder seedlings over the bog surface and floods giving rise to the presence of pools, as were surveyed in this area. Such events were obviously not favourable for the establishment of the environmental conditions giving way to the growth of poorly humified peat.

Most likely, the Killoran Bog water supply had no or only very minimal influence on the growth of highly humified *Sphagnum* peat in Derryville Bog.

Moderately humified *Sphagnum* peat—some characteristics and dates

General remarks

The term 'moderately humified *Sphagnum* peat' refers to ombrotrophic peat deposits forming the transition from highly humified to poorly humified peat in which many stratigraphic, microtopographic and spatial structures could be identified, as has already been mentioned. For this reason this type of peat can be regarded as favourable for peat-stratigraphical survey. Its main presence is in the central zone of Derryville Bog. Many features have been documented in this survey; see Figs 3.17–3.19. As the peat face of Fig. 3.5 indicates, in most places moderately humified peat growth succeeded the highly humified *Sphagnum* peat growth. Its thickness and levels vary conspicuously, from 0.4m to about 2m between about 123.00m and 127.50m OD. This can partly be ascribed to the compaction of the peat but also to the problems of identification. The main reason for the varying thicknesses is the significant differences in the peat-growing environment. In Fig. 3.5, the thickness is up to about 1.2m between 180m and 260m in the southern gully and can be ascribed to very wet conditions just outside the main discharge gullies of bog bursts B and D. The deposit between 370m and 430m, with its summit at 400m, at a level of 126.10m OD represents the watershed as depicted in Figs 3.17–3.18. The thick deposits in the northern stretch of Fig. 3.5, between 830m and 1010m, up to 1.40m thick, indicate the really wet peat-growing conditions in this northern gully from *c.* 1800 BC. Not all the details of

its occurrence can be understood and explained. Near both the western bog margin (Fig. 3.4) and the eastern bog margin (Fig. 3.6) the moderately humified peat deposits are present but only at a lesser thickness, to about 0.5m at the most. The extension of the identified spatial structures, as already mentioned for the moderately humified *Sphagnum* peat, is depicted in Fig. 3.18 and for the marginal areas in Fig. 3.19. The basis for these structures had already occurred earlier (Figs 3.16 and 3.17).

The occurrence of layers or deposits of moderately humified *Sphagnum* peat can largely be dated to between 900 and 200 BC but mostly from *c.* 600 BC on (Fig. 3.11). In the marginal zones the deposits can be dated from about 200 BC to AD 300. The microtopography was studied at many locations. Details will be presented for two of these sites, near trackway Derryfadda 23 and near causeway Killoran 18, respectively (Fig. 3.8b). At both locations the moderately humified deposits consisted of a laminated system of more humified peat and poorly humified peat layers, respectively, with intervening *Sphagnum cuspidatum* layers and with concentrations of *Scheuchzeria palustris*, *Eriophorum vaginatum* and *Calluna*. It can be concluded that the recorded sequence can largely be attributed to the draining effects of the bog bursts and the ensuing recovery of the bog water table.

The absence of trackways

Although many trackways and platforms can be dated to the same period, as is presented for the moderately humified *Sphagnum* peat deposits (between *c.* 900 and 200 BC, up to about AD 300 in the marginal zones), wooden structures are absent from these deposits. This is due to the impossibility of safe access to the discharge channels, lakes, pools and gullies, and also the hummock and hollow systems, as present in Derryville Bog in this period and depicted in Figs 3.17 and 3.18. Only a few trackways can be accepted or assumed to have been intended to access the treeless raised bog surface but there are only weak indications—trackways Killoran 69 (*c.* 820 BC; see Fig. 3.17), the eastern stretches of which were removed by the eroding water flows of bog burst C (see Chapter 10), and Killoran 75 (*c.* 280 BC; see Figs 3.19 and 3.20), present in the pool of the northern discharge channel (see Chapter 10). In all probability it can be doubted that the latter trackway reached the raised bog surface east of the discharge channel. It cannot be proved that moderately humified *Sphagnum* deposits were present in the line of the two trackways in question.

Peat face: trackway Derryfadda 23

A 20m-long stretch of one of the peat faces in which trackway Derryfadda 23 (1590 BC) was evident was documented in great detail. A schematic representation of this peat face is given in Fig. 3.8b, provided with the timescale as designed by Caseldine *et al.* (see Chapter 6). The highly humified *Sphagnum* peat growth (123.85–124.50m OD) has been dated to between 1800 and *c*. 950 BC. The deposits of the moderately humified *Sphagnum* peat (124.50–124.92m OD), dated to between *c*. 950 and 200 BC, commenced with the flooding of a hollow, resulting in the *Sphagnum cuspidatum* layer (124.50–124.60m OD) between about 900 and 700 BC. The poorly humified peat (124.60–124.67m OD), dated to 700–600 BC, represents the expansion of *Sphagnum imbricatum* hollow peat over the slope of the hummock flank. The moderately humified peat layer (124.67–124.72m OD), dated to shortly after 600 BC, indicates the desiccating effect of bog burst D on this area. The poorly humified peat (124.72–124.87m OD) represents the rehydration of the bog surface after the recovery of the bog water table, dated to between 500 and 300 BC. The moderately humified *Sphagnum* peat (124.84–124.90m OD), to be dated shortly before 200 BC, can perhaps be related to bog burst E. This marks the end of the moderately humified peat deposits at this location.

Peat face: causeway Killoran 18

A 14m-long stretch of one of the peat faces in which causeway Killoran 18 was evident was documented in great detail. The peat profile here is presented in Fig. 3.8b in the same way as for trackway Derryfadda 23 above. The highly humified *Sphagnum* peat (124.88–125.47m OD), dated from 1420 BC to *c*. 200 BC, represents two floods. The lower one (125.08–125.13m OD) resulted in the deposition of highly humified *Sphagnum cuspidatum* peat with *Scheuchzeria* and can be dated to about 900–700 BC, most likely the same feature as is recorded in the Derryfadda 23 peat face. The upper *Sphagnum cuspidatum* layer (about 125.30m OD) of *c*. 600 BC can be ascribed to the flooding which caused bog burst D. The *Calluna*-rich layer (125.33–125.38m OD) stands for the drier circumstances of the bog surface, up to about 420 BC, after bog burst D.

Moderately humified *Sphagnum* peat growth commenced here about 200 BC and continued at least up to AD 80. Fluctuations in the water content of the peat-growing vegetation determined the humification degree of the peat in question. The moderately humified layer (125.48–125.54m OD) can perhaps be related to bog burst E, about 200–100 BC; the upper moderately humified layer (125.70–125.76m OD), dated to about AD 80–150, represents perhaps one of the unlabelled bog bursts indicated in Fig. 3.11. The poorly humified layer between both moderately humified peat layers represents the overgrowth of *Sphagnum imbricatum* under increasingly moister conditions for large stretches of the moderately humified bog surface following the recovery of the bog water table after bog burst E.

It must be noted that the distance between both profiles is only 400m (Fig. 3.3).

Fluctuations in the bog water table

Taking an overview of the peat deposits of the type characterised as 'moderately humified *Sphagnum* peat' in the entire Derryville Bog area, it is apparent that the main feature of the peat sequence is the alternation of more humified and less humified peat layers, both horizontally and vertically. This concerns either the extension of hollow peat over hummock flanks, or vice versa, or the succession of poorly humified *Sphagnum cuspidatum* pool fill layers by highly humified peat layers, assuming much drier environmental conditions. Altogether, a broad scale of peat-growing structures based on environmental differences has been encountered in the field and depicted in Figs 3.16–3.19. This peat-stratigraphical development can be dated from about 1200 BC on, continuing until at least 200 BC in the central zone. Fig. 3.8b presents a specimen of this type of peat growth. The bog margins can be dated to about AD 500 or later (Fig. 3.10). Based on the palaeohydrological investigations of Ben Gearey it can be concluded that these alternations in humification and microtopography indicate considerable fluctuations of the bog water table caused by the bog bursts (see Chapter 6). This concerns not only the bog bursts indicated as B–E in Fig. 3.11 but also those indicated by arrows with broken lines.

In the northern discharge channel, where from 600 BC on the *Menyanthes–Scheuchzeria* deposit is present between the western Killoran Bog forest peat deposits and the raised bog peat to the east (Figs 3.18 and 3.20), this channel fill is intercalated with three thin layers of raised bog peat. These ombrotrophic intercalations mostly resemble moderately humified *Sphagnum* peat dated to between 600 and 50 BC. Thus, prominent fluctuations in the bog water table are also apparent in the discharge channel. One of these fluctuations can be dated more precisely. Between 250 and 200 BC ombrotrophic bog water dominated the bog surface in the western bog area. Perhaps this flooding gave rise to bog burst E, dated to about 100 BC.

The watershed

As already mentioned, the boundary between the two raised bog ecosystems, the Derryville system to the north and the Cooleeny system to the south, appears to be a complicated feature, not fixed as a permanent line but flexible in its location and character. Where dated to 1440 BC, the fen peat stretch of causeway Killoran 18 acted as a watershed for the two raised bog complexes (Fig. 3.16). Around 1200 BC, the time of the first commencement of the moderately humified *Sphagnum* peat growth, this watershed was mostly swept away (Fig. 3.17). Dated to *c*. 600 BC, and most likely related to bog burst D, another shift of its location occurred (Fig. 3.18). It involves the northward move of the westerly stretch of the watershed to the area of Killoran 226 and Killoran 69, as could be proved by the excavations of the trackways of this cluster. Paul Stevens (see Chapter 10) could prove the inversion of the water flow in this range of the northern discharge channel to about 600 BC. It is likely that the watershed that developed here was very weak in relation to the large fluctuations of the water table in the discharge channel. This can also be concluded from the later development of the bog surface, as depicted in Fig. 3.19.

Summarising remarks

Two aspects make the moderately humified *Sphagnum* peat deposits even more interesting. The first is the recognition of peat growth mechanisms allowed by the clearly documented differences in the microtopography of most of the identified peat types. Important for this point is the immigration of large-leaved *Sphagnum imbricatum*, dated to shortly before or around 1200 BC. This moss settled primarily in the hollows. In the long term it overgrew most of the hummock and hollow systems, giving way to the peat type to be discussed in the following section: the poorly humified *Sphagnum* peat. In the central zone this can be dated to about 200 BC, in the marginal areas to between AD 200 and 600. The hydrological effects of the bog bursts were of great influence on this development.

Despite all the catastrophic discharges of masses of water, the increasing wetness of Derryville Bog can doubtless be concluded from all the identified spatial structures. The main period of increasing wetness can be dated to about 600–400 BC, after bog burst D.

The other interesting feature of these deposits is the absence of trackways, as has been mentioned already, whereas a large number of trackways and platforms, dated to between about 900 and 160 BC, were encountered in the surrounding bog marginal forest peat deposits and the Killoran Bog forest peats. Two conspicuous clusters of dates for these bog marginal trackways are present. Between 600 and 550 BC a number of such constructions are built in both bog marginal areas, including the Cooleeny area. Between 300 and 250 BC a second cluster of dates can be seen, relating to such constructions near both bog margins. In both cases the serious drop of the bog water table will have facilitated the access to the bog but not to the treeless *Sphagnum* peat-dominated central area.

Poorly humified *Sphagnum* peat— some characteristics and dates

General remarks

This type of peat, recorded in nearly all the surveyed peat sections and most of the peat profiles (Figs 3.4–3.9 and 3.11), represents the last stage in the basic development of raised bog peat. In practice it can be regarded as the expansion of large-leaved *Sphagnum imbricatum* peat growth in the hollows over the more humified peat deposits of the

hummocks. The arrival of this moss species can be dated to about 1200 BC, as has been mentioned previously.

A first expansion has been dated to about 950–900 BC with increasing wetness of the bog surface; the bog water table was at about 125.40m OD (Fig. 3.12). The second and most extensive expansion occurred from *c.* 600 BC onwards, after bog burst D. The bog water table dropped from about 126.0m to 125.0m OD or somewhat lower. As will be discussed below, this expansion can largely be seen as an effect of the bog burst. Nearer to the bog margins a date of *c.* 300 BC for the following expansion has been established. From *c.* 200 BC poorly humified *Sphagnum* peat is the predominant peat type. The level of the bog surface was about 126.50m OD.

Owing to the extraction/milling activities of BnM the upper 1–2m of peat has been removed. Up to about AD 750 (Killoran 54) this was most likely poorly humified *Sphagnum* peat. What survived, when the peat stratigraphy was surveyed between 1995 and 1998, represents the peat growth to between *c.* 500 BC and AD 800. It can be estimated that the thickness of what survived of this peat type was from 0.5m up to 1.0m. The original bog surface dated to AD 800 was at about 127.80–128.00m OD.

The increasing wetness of the bog can be considered as having initiated the ombrotrophic development as discussed in this paper. Next to this, and even more important, was the increasing acidity of the peat-growing ecosystems of the water to oligotrophic conditions. Both phenomena were favoured considerably by the bog bursts. Bog burst D, in particular, had the largest impact. The sudden and extremely extensive discharge of masses of nutrients dissolved in the bog water and the recovery of the bog water table after the bog burst resulted in a bog surface suitable for the luxurious growth of large-leaved *Sphagnum imbricatum*. Regarding the blanket-like covering of the moderately humified structures by poorly humified to fresh *Sphagnum* peat, such an intensive discharge of nutrients must have taken place at some other time, perhaps to be dated to about 200–100 BC. The absence of trackways from about this time onwards (see Fig. 3.19) suggests a steep decrease in the accessibility of the bog surface by this development.

In the poorly humified *Sphagnum* peat, numerous thin layers of more humified peat occur. This can be ascribed to fluctuations in the water content of the peat as has been convincingly demonstrated by Ben Gearey (Chapter 6). Though many of these layers are quite small, it can be assumed that these fluctuations are based on environmental oscillations within much larger spatial events. The bog bursts and their related shifts in spatial structures can be seen as causes of major changes in the bog water table. To what extent possible climatic oscillations have influenced the type of peat grown in Derryville Bog cannot be established. The influence of the bog bursts on the fluctuations of the bog water table will have been much larger.

Spatial structures

Unusually, occasional poorly humified peat occurs at remarkably low levels in the peat sequence. In Fig. 3.5, between 120m and 160m, this peat is present from levels of about 123.50m OD. This is redeposited peat representing the fill of the erosion gullies and related to bog burst D. This bog burst left large areas with an eroded surface in Cooleeny, as indicated in Fig. 3.18. This bare bog surface was the base for the spread and extension of poorly humified *Sphagnum* peat growth, as Fig. 3.19 displays for this area.

In Fig. 3.5, between 1000m and 1300m, poorly humified peat can be seen just above the fen peat deposits. What this peat sequence indicates is uncertain, although peat growth definitely occurred under extremely wet conditions. The peat deposits in question may represent the remains of floating mats and erosion phenomena before about 1200 BC.

Erosion features have been surveyed in most, if not all, spatial structures of Derryville Bog, especially in hummock and hollow systems, as presented in Figs 3.16–3.18. In Fig. 3.19, representing the spatial structures of the bog surface between about 100 BC and AD 300, the distribution of a number of hummock and hollow systems, *Sphagnum* fields and the extension of areas with many pools has been recorded, together with the discharge systems and the bog marginal peats, as previously discussed.

In the first place, it can be stated that between *c.* 300 and 100 BC most of the clearly visible hummock and hollow systems of the 'moderately humified peat deposits' became replaced by systems in which the differences between hummocks and hollows were hardly visible owing to the lack of differences in the degree of humification of the component elements. The poorly humified *Sphagnum* peat deposit appears as a blanket. This explains the spatial and structural differences between the bog surfaces as depicted in Figs 3.18 and 3.19, respectively. The absence of trackways since about 160 BC can most likely be attributed to this development.

In large parts of the central zone of Derryville Bog (Fig. 3.19), south of the line of causeway Killoran 18 (depicted in Fig. 3.16), poorly humified peat growth occurred in large fields, very likely as floating mats. To the south, in the direction of the Cooleeny area, these pool-like situations were bordered by a hummock and hollow system with a complicated, laminated structure. A lower, poorly humified layer covered by a highly humified layer that was overgrown by poorly humified peat could be identified. Most likely this was the watershed between the Derryville system, with floating mats, and the Cooleeny system. Superficial erosion phenomena have been recorded in this poorly humified layer.

In stratigraphical terms, the watershed situation including the floating mats at its north side, as depicted in Fig. 3.19, can be dated from *c.* 300 BC to AD 100. The recorded superficial erosion occurred after AD 100.

In the northeastern stretches of the study area, an enormous flooding (a huge *Sphagnum cuspidatum* layer, 10cm thick, perhaps formed between 300 and 100 BC) followed by a drop in the bog water table has been recorded. This event was succeeded by an extensive inundation with subsequent erosion phenomena present as peat blocks and gullies. The area affected by this development is indicated in Fig. 3.19 by symbol 8.

There is a large difference in time between both events. The first drop in the water table can be related to bog burst E, about 100 BC–AD 100. The latter can perhaps be dated to the fourth or fifth century of this era. Both features had a wide extension, as is indicated in Fig. 3.19.

The bog surface in relation to stake row Killoran 54 (c. AD 750)

A medieval stake row, Killoran 54 (not indicated in Fig. 3.19), in the northeastern part of the study area, marks the bog surface over a length of about 400m, exactly following a strong change in superficial wetness. To the northeast of the stake row an extremely wet and dangerous area occurs, with large pools, lakes and collapsed peat blocks, which is dissected by numerous vertical cracks filled with very soft peat, dating from about AD 400 onwards. It is the same area as mentioned above: peat symbol 8 of Fig. 3.19. To the southwest of the stake row these features are absent. Reconstructing the original level of the bog surface between AD 400 and 750 suggests a difference in height of perhaps about 0.5m developing slowly since about 300 BC.

The line of the stake row forms the southwestern border of the wet area in question (Fig. 3.19, peat symbol 8). This southwestern border coincides very well with the cliff-like drop in the subsoil relief from 122.0–121.0m OD (Fig. 3.2; the southwestern flank of the N gully, as indicated in Fig. 3.3).

It is now believed that the subsoil relief was reflected in the bog surface. It can be assumed that, about AD 750, this line still indicated a strong difference in superficial wetness and that the stake row thus served as a fence. This suggests that shortly before the construction of this fence a drop in the bog water table must have facilitated the access to the higher stretches of the bog surface. It could not be established that either a bog burst dated to AD 700–750 or artificial drainage caused this feature.

Summarising remarks

A very striking feature in the growth of the poorly humified *Sphagnum* peat deposits is the dominating role of water in the peat-growing environment and on the development of spatial structures (Fig. 3.19), as was already taking place during the growth of moderately humified peat (Figs 3.17 and 3.18). The southern stretches seemed to have been more a lake with floating mats than a raised bog. In the central zone this process was already in development by *c.* 400 BC and was completed about 200–100 BC.

Large parts of the northern stretches became flooded after *c*. 300 BC. The discharge in a northern and north-western direction created many erosion gullies, probably after about 100 BC (bog burst E) and continuing up to at least AD 600–700, when the northern area was affected by inundation and the serious disturbance of the bog surface. Passage of the bog surface between 300 BC and AD 700 seems to have been impossible. About AD 750 these conditions must have changed fundamentally.

In the southern zone, the Cooleeny area, the bog bursts B (*c*. 1250 BC) and D (595 BC) determined the discharge structures of this area, largely filled with soft redeposited peat. This appeared to be an excellent sub-stratum for the luxurious growth of poorly humified *Sphagnum* peat. These deposits were so seriously affect-ed by the modern peat extraction works that it was impos-sible to reconstruct the spatial pattern of these *Sphagnum* peat complexes. It could only be proved that the area remained impassable from about 600 BC onwards.

Between *c*. 200 BC and AD 800, the period in which this peat deposit is placed, a noticeably small number of trackways and platforms were constructed in the margin-al area of Derryville Bog. These were, in fact, only con-structed before 160 BC and related to the previous period of 900–200 BC. It can be said that after 300–250 BC, when the bog marginal areas seemed to have been rela-tively accessible, the still rising bog water table seriously reduced the attractiveness of these zones. Most likely this situation changed in one way or another shortly before AD 750, as stake row Killoran 54 suggests.

Epilogue

The availability of sufficient amounts of water in a mod-erately sloping landscape with a nearly impermeable mineral soil can be seen as the main stimulus for the start of Derryville Bog, between 7000 and 6500 BC. The aim of this study was to achieve, in as much detail as possible, an understanding of the palaeoenvironmental development of this bog.

The key conclusions of the study proved that many influ-ences on the peat development, including those precipitated by human activity, could be documented and understood in relation to both the archaeological chronology and the nature of the human response to the peat bog environment.

The peat growth of Derryville Bog seems to have suf-fered mainly from the surplus of water supply, so large that the water accumulations could damage the peat-growing systems considerably. Only under exceptional situations, such as sudden drops in the bog water table, were the mar-ginal areas of the bog accessible for a relatively short peri-od. In this way, the numerous short trackways, platforms, stake rows and the failed longer trackways can be explained in a satisfactory way.

4. Chronology

Sarah Cross May, Cara Murray, John Ó Néill and Paul Stevens

Introduction

Sixty-six wetland structures and thirty-three dryland sites in Derryville Bog and the surrounding townlands were investigated by this project. This chapter discusses the results in a chronological framework in order to facilitate the integration of the material with the wider, existing archaeological record. Wetland structures have yet to be demonstrated to show morphological or functional variation within a chronological pattern, as is discussed in Chapter 9.

This chapter discusses continuity and change through the use and occupation of Derryville Bog over time. For each period all excavated areas are discussed in relation to clusters of dates, broad environmental patterns, spatial pattern within the bog and typological variation, where relevant. Detailed descriptions of each site may be found in the catalogues of the excavated sites in order of townland (see Chapters 10 and 11). The excavated sites are referred to here by the names under which they are described in the catalogues.

The sporadic nature of the evidence

The large number of sites in the bog leads to an initial impression of intense activity. When viewed over the full time-scale of peat growth, however, this picture is more diffused. Even sites that are chronologically quite close to each other may have one or two centuries separating their construction. Within this framework, clusters of material and associated activity may cross chronological boundaries.

Continuity, both in the types of use and the places of occupational importance within the bog, is one of the strongest patterns that emerge. With such long periods between the construction of structures and broad changes in peat development, it is hard to see such persistence of place relative to function as a coincidence. The reuse of places, and sometimes of wood, from many centuries earlier indicates an intimate knowledge of the bog. Visitors to Ireland in the eighteenth or nineteenth century noted that people went to the bogs early in the morning and looked for buried timber in the places where the dew evaporated quickly (Feehan and O'Donovan 1996, 444). This continuing sense that the bog constituted a resource, including the structures of past human activity, provides the background to the spo-

radic but persistent pattern of use that was seen at Derryville. It is also possible that structures were sited at the end of dryland routeways and that certain locations at the margins of the bog were exploited and reused time and time again.

Dating and chronology

Classifying structures and sites by period is a conventional approach for building a chronology but is problematic when dealing with wetland structures. The same construction techniques are used repeatedly by different groups at different times. The single-plank trackway Cooleeny 22, dating to 1517 BC, illustrates the persistence of a particular wetland structural type. An exact copy was excavated in the Lemanaghan Bogs in County Offaly and dated to AD 665±9 (Bermingham 1997, 93), demonstrating that single-plank trackways were built for at least 2,200 years. Thus, in the absence of a thematic or morphological distinction between structures of different periods, categorisation can only be achieved through reference to the dated samples.

The chronological framework for the project results was constructed from a number of sources. The primary data, in the form of radiocarbon and dendrochronological dates from excavated sites, were used alongside independent dating of peat and pollen sequences. The integration of these series provided controls for the dating of individual sites, particularly when a specific archaeological or environmental episode could be dated by dendrochronology and that date could be extrapolated across the whole bog. Radiocarbon dates were provided by laboratories in Miami, Belfast and Groningen, while the dendrochronology dates were obtained in Belfast. All the dates are listed in Chapters 17 and 18.

A total of 141 dates were obtained from sites of archaeological or environmental significance. The sequence of thirty-one environmental dates begins in 6080 BC, after the start of peat development, and includes seventh-century AD dates for the surviving surface levels (see Chapter 3). Dendrochronology dates from 3368 BC to 161 BC helped to date various expansions of oak wood into the wetland margins. Coupled with the series of dates from the pollen cores, the dating evidence has provided a sequential picture of the dryland flora and woodland cover from the end of the third millennium BC onwards (see Chapter 6).

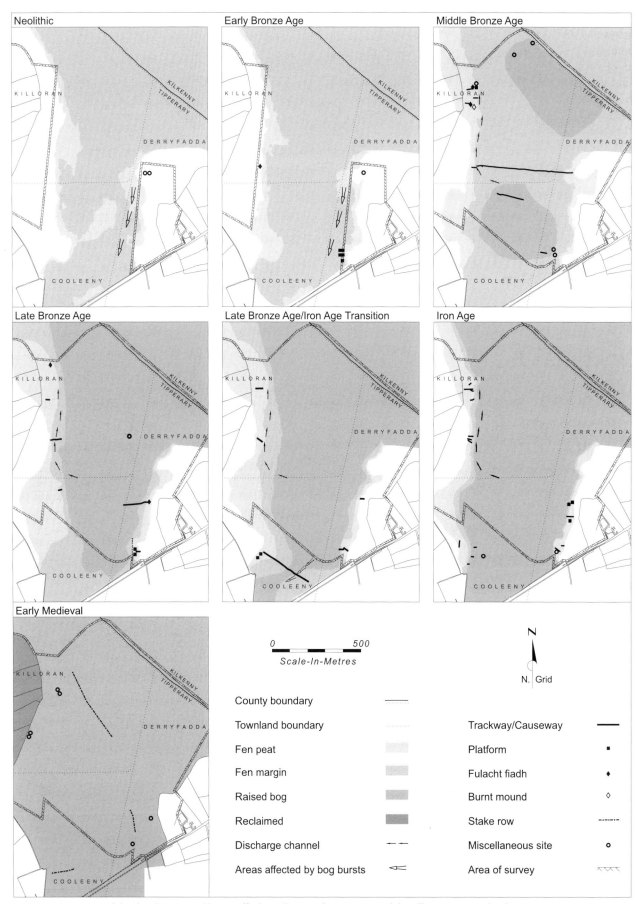

Fig. 4.1 Overview of the development of Derryville Bog. For a colour version of this illustration see back cover.

Overall, the dates obtained for the causeways, trackways and platforms in Derryville Bog do not fit neatly into the existing wider national pattern in that they fall more substantially into the first millennium BC. The general scarcity of Neolithic dates in wetland areas has been noted elsewhere (e.g. Raftery 1996), although in Derryville Bog it may be a product of the constraints on the research (see Chapter 1). The small number of dates did not allow for an examination of the supposed 2350 BC boundary between the Neolithic and the Early Bronze Age in Ireland (Brindley 1995, 5).

The broad date-ranges, and clustering, within the Bronze Age reflects the irregularity of the calibration curve at a number of points, in addition to discontinuities in the record (*ibid.*). The typical Bronze Age structures in Derryville Bog, *fulachta fiadh*, platforms and trackways, begin to appear in the Early Bronze Age. The small number of Early Bronze Age dates, like the small number of Neolithic dates, may reflect the constraints outlined above, but this too is a visible trend on a national level.

No structures occurring within the study area were dated from around 1900 BC until the beginning of the Middle Bronze Age (1700–1200 BC), following the limits of that period as defined by Cooney and Grogan (1994). After an initial phase in which two trackways and the causeway Killoran 18 were built, the pattern of activity took on a wider focus in the late fifteenth century BC, continuing until around 1000 BC. The occurrence of a bog burst in 1250 BC (bog burst B) aided the dating of thirteenth- and twelfth-century BC sites and, by extension, defined a separation between the Middle and Late Bronze Age around 1200 BC. The dates of the three largest Middle Bronze Age structures, Killoran 18, Cooleeny 22 and Derryfadda 23, fit within the mid-second millennium cluster of dates in the national pattern.

The picture from the *fulachta fiadh* closely matches that of the published distribution of dates (Brindley *et al.* 1989–90), with the earliest (Killoran 17) and the latest (Killoran 26) sitting comfortably within the parameters of the national dates that directly relate to *fulachta fiadh*.

The distribution of dates after 800 BC is unique in that there is a range of dates between 750 BC and 450 BC. Many structures were built during this period, including the 8m-wide causeway Cooleeny 31. The dating sequence from the period between 650 BC and 450 BC allowed the end of the Late Bronze Age (1200–650 BC) to be studied in some detail. This period formed a recognisable sequence within the woodworking assemblage. By cross-matching the profile of the axe marks on individual structures against the dating evidence (see Chapter 13), a significant distinction could be identified. Thus, a major change became evident in the mid-first millennium BC at the end of the seventh century BC. Comparison of measurable analogues representative of the properties of the axes showed continuity between the period following 600 BC and the succeeding

rather than the preceding period. On this basis, the axes in use from the seventh century BC onwards were considered to be made of iron.

These dates are from the 800–400 BC gap, which is bridged by the 2400 BP plateau in the radiocarbon calibration curve. The problem of dating sites to this period by radiocarbon analysis has been noted before (Bowman 1990). This technical difficulty is compounded by a general scarcity of dendrochronological dates from this period, although a number were obtained from structures in Derryville Bog.

The dating of bog burst D to around 600 BC and the series of dendrochronological dates from this period meant that greater chronological definition could be achieved for this project. With some certainty Killoran 306, Cooleeny 31, Derryfadda 206 and Derryfadda 13 can be placed just before 600 BC; Derryfadda 210 and Derryfadda 234 between 600 BC and 460 BC; and Derryfadda 215 and Killoran 226 to around 450 BC.

The gap in the record between AD 50 and AD 650 sees our early medieval period beginning in the seventh century AD, with the latest site dating to AD 1024–1162. Two later sites, Killoran 223 and Killoran 224, were assigned to the post-medieval period.

The sequence of dates obtained from Derryville Bog permitted the reconstruction of the peat development from the seventh millennium BC to the modern period. The archaeological sequence begins in earnest with the Bronze Age and ends around AD 900, reflecting both the current level of attrition of the midland raised bogs due to industrial milling and the safe depths to which sites in deep peat can be excavated.

General clusters of dendrochronological dates for archaeological sites in wetland contexts have been noted in the sixteenth, tenth and second centuries BC (Baillie and Brown 1996, 399). Examination of the clustering of all the dated wetland structures in Ireland shows that they can be grouped into a broader band between the seventeenth and tenth centuries BC than the dendrochronological dates indicate. In Derryville Bog the dendrochronological dates cluster mainly in the sixteenth and fifth centuries BC but contain outliers in other centuries. The pattern of the radiocarbon dates, as for the national pattern, is markedly different, with clusters between roughly 1200 to 1000 BC, 700 to 450 BC and 400 to 200 BC. As the clusters of dendrochronologically dated structures in Derryville Bog do not match the radiocarbon date clusters, the tendency of the sites producing dateable oak wood does appear to be an archaeologically derived, rather than an environmentally derived, trend.

The investigation of, and opportunity to excavate, continuous blocks of the prehistoric margins of Derryville Bog can explain this date spread. This created the opportunity to recover a series of dates from contexts that included lower order structures, such as short trackways. The sequences of dates show that local events,

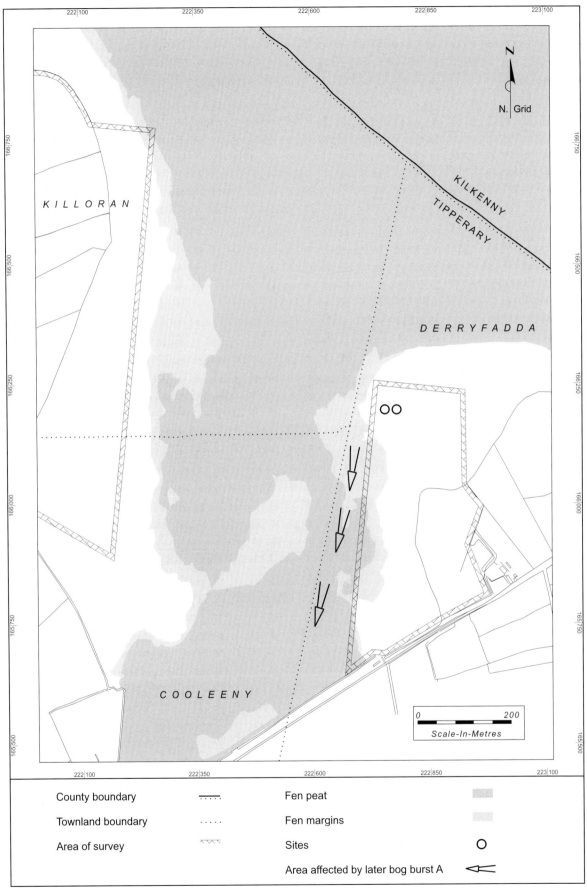

Fig. 4.2 Derryville Bog in the Neolithic.

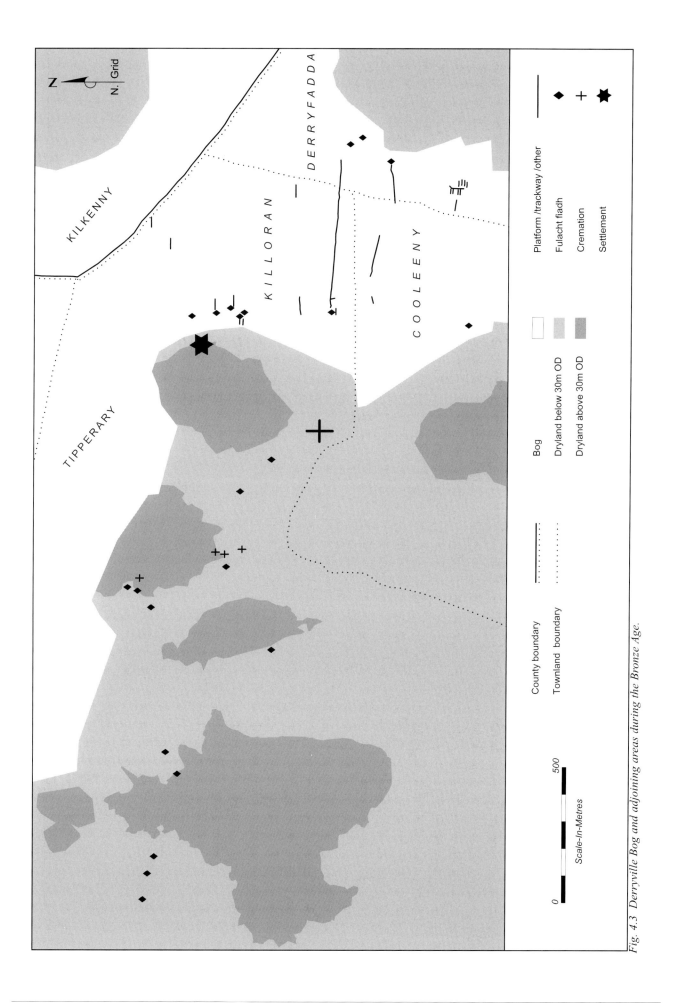

Fig. 4.3 Derryville Bog and adjoining areas during the Bronze Age.

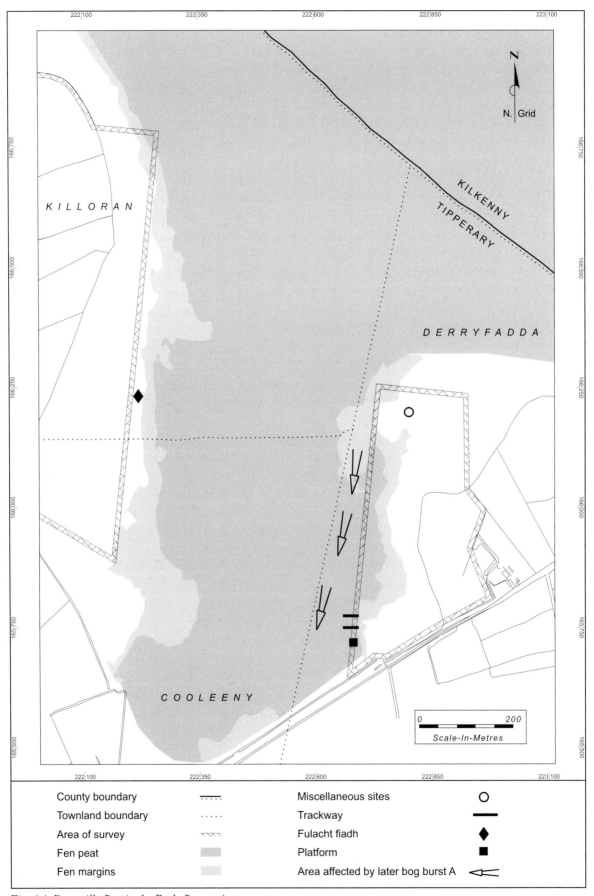

Fig. 4.4 Derryville Bog in the Early Bronze Age.

including bog bursts, were not necessarily driven by meteorological fluctuations but rather by the patterns of human activity in the area.

The majority of dates on record from outside Derryville Bog are derived from large, bog-crossing structures discovered in the series of drains cut in bogs being prepared for production. The absence of drains in the present bog margins or differential rates of attrition (owing to peat milling) can obscure activity in these areas. The work of the IAWU, in spite of these factors, has shown that the majority of wetland structures are located within the margins of bogs.

Comparison of the structures in Derryville Bog and those in Derryoghil Bog in County Longford shows that the availability of dated tree-ring samples does not necessarily reflect the pattern of oak use but is rather an element of that pattern. As such, the clustering of dendrochronological dates would seem to reflect a human influence or possible woodland management rather than a pattern dictated by the environment.

Patterns

General

The strongest chronological pattern seen in this study is continuity throughout the prehistoric material and a separation between this and the historic material. The pattern is seen in the site-types, the environmental contexts and the spatial groupings of sites.

Within the broad continuity of prehistoric material, there are some developments and anomalies. The intensity of use increases fairly steadily from the Neolithic to the Iron Age. The construction and use of *fulachta fiadh* ceases at the end of the Bronze Age, despite the increase in the intensity of usage of the bog. Structures bridging the bog, tying together two localities, are only constructed in the Middle Bronze Age and in the transition between the Late Bronze Age and the Iron Age. The possible social relevance of these events is addressed in Chapter 15, which discusses the concept of Derryville Bog as a vernacular landscape.

Neolithic (3500–2350 BC)

Late Neolithic dates were determined from archaeological contexts within the study area (Fig. 4.2). No evidence for archaeological activity between the commencement of peat development (*c*. 6200 BC) and the late Neolithic was revealed. This lack of evidence is likely to be a product of three factors: the nature of Neolithic evidence in Irish bogs, environmental disturbance and restrictions on the scope and depth of excavation in some locations where Neolithic levels occurred in dangerously deep peat. None of the Neolithic material recovered was structural. All of it came from split timbers, perhaps suggesting use rather than management of the bog as a landscape.

No material of Neolithic date was recovered from the western or southern sections of the study area. This is partially because investigation of the lowest levels, potentially yielding Neolithic material, was precluded by the depth at which it occurred and the safety conditions of the project. In the south and southeast of the study area, bog burst A (*c*. 2200 BC) disturbed and truncated the peat deposits in which Neolithic material might have been found. Settlement in the Irish Neolithic tends to be widely scattered and focussed on well-drained soils. The recovery of Neolithic material from Irish bogs is fairly rare and while this will also have been conditioned by research constraints, it fits patterns of Neolithic occupation.

The earliest date produced from the bog of 3339–2924 BC (UB-4082) related to a degraded, half-split ash timber (Killoran 18: C27) found within sedge fen located on the northernmost glacial ridge. The timber was located within 0.25m of dark purple amorphous sedge and reed fen peat, indicating that fen peat was developing in this area at this stage of the Neolithic. A second, tangentially split oak timber (Killoran 18: C37) was recovered during the excavation of Killoran 18, 30m east of the C27 timber. This timber was radiocarbon dated to 3020–2613 BC (UB-4097), crossing the Neolithic and Early Bronze Age date ranges. The timber was located in 0.20m of charcoal-flecked peaty, silty clay filling shallow hollows in the glacial till. Analysis of this clay revealed it to be naturally formed by fluvial/colluvial in-wash at the margins of the bog. These two dates demonstrate that there was human activity in the area during the late Neolithic but they do not inform an understanding of that activity. If people were using the developing fen as a resource, there is no evidence of how much effort went into that use or how it fitted with a wider settlement of the area. No Neolithic material of any sort was found in the dryland areas stripped during the site development work.

Bronze Age (2350–650 BC)

By far the largest concentration of excavated and surveyed sites recorded by the project dated to the Bronze Age. A total of forty-seven sites were radiocarbon dated, with a further twenty-five sites typologically or stratigraphically dated to this period. Twenty of these were undated *fulachta fiadh* or burnt mounds; the remainder were isolated cremation pits, flat cremation cemeteries and linear ditches (see below). In total, twenty-eight *fulachta fiadh* and burnt mounds were recorded, representing over half the sites recorded from this period. The majority of these were recovered from the dryland excavations and survey. The distribution pattern of Bronze Age sites was uniform throughout the study area (Fig. 4.3), with notable clusters of material relating to their topographic location. The degree of recognised terrain sensitivity was greatest during this period (see also Chapter 14).

The Bronze Age landscape of Derryville Bog and the adjacent dryland was one of fen and marginal forest or scrub surrounded by fertile pockets of land. The land to the west of the study area in Killoran consisted of low glacial ridges or knolls surrounded by shallow hollows or plateaus, overgrown with *Sphagnum* bog growth (see Chapter 3). Substantial settlement evidence from this period was revealed over this large tract of Killoran and the margins of Derryville Bog. Excavation and survey from the western Killoran dryland revealed settlement activity, including burials located on dry islands of raised ground overlooking and surrounded by the marshy blanket bog of the Killoran Bog system. Associated clusters of *fulachta fiadh*/burnt mounds were located on the slopes of glacial knolls or in peat surrounding the bog margins. Trackways were constructed in the bog margins near areas of settlement and/or the *fulachta fiadh*. A single causeway spanned the bog for a time during this period, with several other trackways constructed within the bog forming part of a routeway across it.

The Bronze Age activity represents evidence for quite expansive early settlement across the dryland area of Killoran associated with the marginal use of the bog and perhaps attempts to cross it. Further settlement was not in evidence here until the later Iron Age and early medieval period, when it was of a different nature to the relatively high level of activity in Derryville Bog in the Bronze Age. This is discussed in detail below.

Early Bronze Age (2350–1700 BC)

Six sites were dated to between 2350 and 1700 BC: two trackways, two *fulachta fiadh*, a platform and an area of burning (Fig. 4.4). The results from this period suggested a greater degree of recovery of archaeological material from the bog and gave rise to a greater confidence in the evidence for the discussion of human activity in the area. The Early Bronze Age was characterised by low-level, small-scale, marginal activity within the fen margins of Derryville Bog. Although sites were found on both sides of the bog there were no attempts to cross it in this period as it was mostly fen.

The palynology from this period indicates small-scale clearance of trees on the surrounding dryland, which at this time was still covered by primary woodland. The presence of the beetle *Prostomis mandibularis* in wood from the Middle Bronze Age track Cooleeny 23 demonstrates that primary woodland was still present in the area at a later date. This suggests that the low-level of activity in the bog may reflect a sparse human presence in the area as a whole.

The fen continued to serve as a marginal resource but no longer served as a boundary or obstacle, and it could now be accessed and crossed in places under favourable conditions. The presence of settlement on the dryland reinforces the palynological evidence and suggests that the Early Bronze Age marked the first long-term settlement of humans in the area. The nature of the bog material suggested episodic settlement throughout the period.

A cluster of activity in the southeast of the study area included a large irregular platform, Derryfadda 218, on the fen margin dated to 2290–1935 BC (Beta-102759). A number of sherds of coarse, undecorated pottery were found associated with the platform. Close-by were two short trackways, Derryfadda 207, dated to 2205–1875 BC and 1805–1795 BC (Beta-102757), and Derryfadda 204, dated to 2120–2080 BC and 2050–1755 BC (Beta-102764). These structures represented transient access into and use of the marginal woodland. They were constructed to fill pools within the marginal woodland growing on a low glacial ridge in the southeast of the study area.

Further north, on another ridge on the eastern side of the bog, there was evidence of fire. A deposit of charcoal underneath the eastern landfall of causeway Killoran 18 was dated to 2133–1548 BC (UB-4095). This burning could have represented scrub clearance by burning in that area or may have been an informal hearth.

One site, Killoran 304, was located in the west of the study area, at the edge of the fen. This small *fulacht fiadh* was dated to 2138–1935 BC (UB-4186). The small scattered nature of the firing debris suggested that the site was not in use for a long period of time. An earlier *fulacht fiadh*, Killoran 17, in the northern corner of Killoran townland, on dryland in an area of reclaimed bog, was dated to 2585–2195 BC (Beta-117547). This site consisted of three intercutting troughs located amongst a small ploughed-out arc of burnt mound material.

The range of sites utilising the bog margins in this period established a pattern that continued until the end of the Iron Age. The core of the use of Derryville Bog as it developed from fen to raised bog included the construction of narrow trackways for access into the marginal woodland, platforms within the margins and *fulachta fiadh* on the bog margins and dryland.

Middle Bronze Age (1700–1200 BC)

Nineteen sites dating to this period were excavated in the study area, including three sites on the dryland to the west of the bog (Fig. 4.5). This large body of material is in contrast to the relatively sparse archaeological record in Ireland as a whole from this period (Cooney and Grogan 1994, 123; Brindley 1995, 9). The sites within the study area revealed a continuing pattern of low-level casual access into the bog margins. Changes in emphasis and function also occurred during this period, with an intensification of activity. A wide spatial distribution across the study area, as well as the construction of the first causeways, shows an intensification of the human use of the bog. This activity can be broken down into episodes of activity throughout the period.

A concentration of archaeological structures, dated to the early part of the Middle Bronze Age, was constructed in the sixteenth and seventeenth centuries BC. This concentration included three narrow trackways, a large *fulacht fiadh* and at least three round houses. Derryfadda 23, a single-plank trackway, was located in

Plate 4.1 Excavated section of Killoran 18.

the southeast of the study area and was dated to 1606±9 BC or later (Q9369) and 1590±9 BC or later (Q9370). A similar single-plank trackway, Cooleeny 22, located in the south central part of the study area, was dated to 1521±9 BC (Q9544), 1517±9 BC (Q9547), 1526±9 BC (Q9548) and 1517±9 BC (Q9549). A large *fulacht fiadh*, Killoran 240, located on the northwestern margin within the study area, was dated to the later sixteenth century BC, pre-dating the construction of the short, narrow trackway Killoran 241, dated to before 1547±9 BC (Q9542).

Killoran 8 was dated to 1860–1845 BC or 1775–1430 BC (Beta-117553). This site to the west of the northwestern bog margin on the Killoran dryland headland, incorporated the remains of up to three houses. To the south, and possibly associated with these structures, was Killoran 2, an undated site consisting of narrow ditches and charcoal spreads that yielded a flint flake. Further west, the *fulacht fiadh* Killoran 5, dated to 1750–1410 BC (Beta-117545), was located on the western slope of a glacial ridge in the centre of the Killoran dryland.

The fifteenth century BC saw the construction of the long complex causeway Killoran 18 (Plate 4.1), which crossed the bog in the centre of the study area and was

dated to before 1440±9 BC. This structure was constructed in two phases and repaired after twenty to forty years. Several dates were determined. The construction of Phase 1 consisted of a double stake row and brushwood track dated to 1745–1405 BC (Beta-102752) and 1405–1000 BC (Beta-102750). The latter date may lie outside the date range of two standard deviations for the former date. The second construction phase consisted of a stone and wood causeway. It was dated to 1534±9 BC or later (Q9470) and 1605–1410 BC (Beta-102751). A repair to the causeway was dated to 1440±9 BC (Q9349) and 1542±9 BC or later (Q9349). The lack of sapwood in two of the samples dated by dendrochronology resulted in the date range of ±9 years or later. This could represent a range up to 100 years later (D. Brown, pers. comm.). Further features that appeared to be associated with the causeway were an area of burning and a narrow wooden trackway. The area of *in situ* burning was located close to the stone-paved eastern terminus of the causeway and was dated to 1620–1410 BC (UB-4083). The trackway Killoran 305, located close to the western terminus of the causeway, was dated to 1510–1310 BC (UB-4187) and 1436–1264 BC (UB-4188) and possessed a recognisable woodworking signature that corresponded to Killoran 18.

Plate 4.2 Killoran 253 during excavation, with the contemporary ground surface removed on three sides. The original structure lined a pit or trough dug into the surface of the fen.

A number of structures dated to the later stages of this period were revealed in the northwest of Derryville Bog. A complex shift in the bog hydrology in this area during the early fifteenth/late fourteenth century BC resulted in the deposition of a layer of alluvial silt within the bog. Soon afterwards, several structures clustered in the bog margin were built in association with the location of the main discharge channel. A small platform, Killoran 237, located in a small outlet channel directly over the alluvium was dated to 1685–1400 BC (Beta-111370). A *fulacht fiadh*, Killoran 265, located on the glacial till to the west of the channel had a large carved oak trough and was dated to 1425–1120 BC (Beta-111377). This latter site pre-dated trackway Killoran 230, which was close to, but unassociated with, the *fulacht fiadh*. It was located within the discharge channel, overlying the alluvium, and was dated to 1500–1195 BC (Beta-111368). To the north was a deposit of archaeological wood, Killoran 229, consisting of a scatter of worked roundwoods lying within the discharge channel and overlying the alluvium. It was dated to 1385–930 BC (Beta-102747). A later site in the northwest of the study area, Killoran 235, was a trackway located within marginal woodland dated to before 1212±9 BC (Q9541). Beyond the western margins of Derryville Bog, located on a dry plateau, a very large flat cremation cemetery dating to 1435–1215 BC (Beta-117546) was revealed. This site contained over twenty-eight simple, unmarked individual cremation pits.

A number of archaeological wood deposits (Derryfadda 15, Killoran 41, Killoran 56 and Killoran 57) and the burnt mound Killoran 316 were also dated to this period on the basis of their stratigraphic positions within the peat. The end of this period was marked by a bog burst that occurred in the eastern corner of Derryville Bog during the mid-thirteenth century BC (bog burst B). This occasioned a drop in the water table across the bog and also may have swept away archaeological material from portions of the eastern bog margin at the time.

The Middle Bronze Age sites were broadly distributed across the whole study area, with notable concentrations to the northwest and south but including structures in the centre of the bog. This spatial distribution reflects two main patterns in the environmental contexts.

The first trend seen in the material is a continuation from the Early Bronze Age of transient, casual activity accessing the bog margins and on the bog margins themselves, with the intensity of activity increasing especially in the earlier part of this period. The bog marginal woodland contained structures such as short narrow trackways and platforms, mostly located in the northwest of the study area. In addition to this, several loose, scattered deposits of archaeological wood were also recorded. These did not possess any obvious structure but formed part of destroyed structures or represented more casual wood clearing activity within the bog.

The second pattern is represented by causeways constructed across a variety of environmental contexts to cross the entire bog. Included among these are the narrow, but monumental, stone-built causeway and two earlier single-plank trackways, which, although not crossing the bog, almost certainly formed part of a route across the bog. Structures crossing the bog are extremely rare in Derryville. This is partly because of the long-term instability of the surface of the fen and partly due to the role that the bog played in the local landscape. These structures reflect a perception of the bog as an obstacle rather than a resource.

This phase indicates settlement in the hinterland of the bog and access to its resources. This occurred in association with the clearance and exploitation of the Killoran dryland during this period (see Chapter 6). The cemetery located close to the bog also provides evidence of the increased intensity of activity during this period. The creation of *fulachta fiadh* continued throughout this period, culminating in the construction of a particularly fine example, Killoran 240. This *fulacht fiadh* was constructed for long-term use and was likely to be associated with settlement on the dryland to the immediate west. Simple *fulachta fiadh* also continued to be built throughout this period.

In general, the Middle Bronze Age saw an expansion of human activity both in Derryville Bog and its hinterland to the west. The beginning of the period was marked in the pollen record by significant clearance, notably of elm and ash, which continued with increased severity up to about 1250 BC (see Chapter 6). This was largely borne out by the archaeological record. The activity on the bog was largely a continuum from the Early Bronze Age but with an increased size and scale. The construction of labour-intensive monumental structures, such as Killoran 240 (with obvious long-term design and function) and Killoran 18 (a clumsy but impressive causeway), as well as houses such as Killoran 8 and the cemetery Killoran 10, suggest a period of increasing settlement and investment in the locality.

The bog continued to serve as a marginal resource but no longer served as a boundary or obstacle; it could now be crossed with assistance and favourable conditions. The presence of settlement on the dryland reinforces the palynological evidence and suggests this period marked the first long-term settlement of humans in the area, while the nature of the bog material suggests episodic settlement throughout the period.

Late Bronze Age (1250–650 BC)

Thirteen sites dating to the Late Bronze Age were excavated in the study area (Fig. 4.6). The activity was characterised by an episode of concentrated activity that declined and later increased at the end of the period. The environmental context of the sites showed an increased sensitivity to water table fluctuations as the hydrology of the bog became unstable. The sites revealed a continuing pattern of low-level casual access into the bog margins, but no further structures built to cross the bog. The spatial distribution within the bog displayed continuity from the previous period, despite long gaps in the use of established zones of activity. Outside the study area, the unexcavated material surveyed within the Cooleeny Complex suggested a possible concentration of Late Bronze Age activity. The spatial distribution shifted further west to include the Killoran dryland headland.

On the southeastern and the western side of the study area, the stone trackway Derryfadda 311, dated to 1450–1030 BC (Beta-111375), and the brushwood and roundwood trackway Cooleeny 64, dated to 1420–1020 BC (Beta-111374), were both constructed from the dryland to the edge of the marginal woodland. On the eastern side of the study area, a second stone track, Derryfadda 17, dated to 1315–980 BC (Beta-111272), was also constructed from its dryland landfall through fen into the raised bog. To the north of this, a deposit of archaeological wood, Killoran 20, dated to 1305–940 BC (Beta-111373), was uncovered in an area of raised bog that was otherwise devoid of archaeological material, probably because it was previously too wet. These sites can be quite tightly dated on the basis of the 1250 BC bog burst (bog burst B), as none could have been used in the very wet environment that existed prior to the significant drop in the water table occasioned by the burst.

On the eastern side of the study area, a *fulacht fiadh*, Derryfadda 216, dated to 1400–990 BC (Beta-102305), was constructed on a glacial ridge west of the landfall of Derryfadda 17. On the northwestern side of Derryville Bog, the later *fulacht fiadh*, Killoran 253, dated to 1305–940 BC (Beta-111378), was constructed in marginal woodland and fen (Plate 4.2). Further west on the Killoran headland, two *fulacht fiadh*, Killoran 27, dated by dendrochronolgy to 932 BC (Q9698), and Killoran 26, dated to 1145–795 BC (Beta-117549), were constructed within the Killoran Bog system. The latter was the latest use of a *fulacht fiadh* in the study area.

In the southeast of Derryville Bog, the platform Derryfadda 211, dated to 1265–910 BC (Beta-102753), was constructed within a reed-filled pool in the marginal woodland, extending the western side of the ridge. A second platform, Derryfadda 213, dated to 1315–915 BC (Beta-102758), was constructed further downslope to the south in fen within the marginal woodland.

There was a subsequent lull in activity across the entire study area as it became wetter and water levels rose to the pre-bog burst levels. In the northwestern area, the trackway Killoran 243, dated to 979±9 BC (Q9543), was constructed in the marginal woodland, within the hollow formed by the earlier site Killoran 240. The evidence from all these sites indicates that the entire bog was becoming increasingly wet in the lead up to the bog burst of around 820 BC (bog burst C). A stake row, Derryfadda 209, dated to 990–770 BC (Beta-102749), was constructed along the edge of the steeply sloping ridge in the marginal woodland, and Killoran 69, dated to 838–799 BC (UB-4180), crossed the marginal woodland and fen from the west to the edge of the discharge channel (channel B, see Chapter 3), which remained active at this time.

Further west, beyond the Killoran Bog system, an isolated cremation pit, Killoran 6, was dated to 1145–900 BC (Beta-117548). This pit was located on the summit of a low glacial ridge overlooking the bog systems to the east and north. It was one of a number of such cremations on the ridge immediately east of the earlier *fulacht fiadh* Killoran 5. The coincidence of location demonstrates a pattern of reuse similar to that seen in the bog.

The two main clusters of dates from the Late Bronze Age sites display a pattern that is not reflected in the pollen record. A drop-off toward the end of the period follows a serious concentration of dates in the early part of the period. The pollen on the other hand suggests a lull in clearance from 1310 to 1000 BC, followed by a serious impact on the woodland in the period between 1000 and 600 BC. There are two possible reasons for this disjuncture. The first is that the wetland sites are not representative of the activity on the surrounding dryland. Perhaps the intensity of the fen use was conditioned by different parameters than those dictating patterns of general settlement, although there is a continuation of dryland activity, albeit further to the west of the bog. The second possible reason is that non-invasive woodland management techniques may have influenced the pollen record.

Part of the increase and diversity of activity on the bog and its margins at the beginning of the Late Bronze Age was due to the significant lowering of the water table caused by bog burst B in *c.* 1250 BC, which marked the beginning of the period. As a result, areas that had previously been too wet to access came into use, leading to a concentration of construction on the bog surface in these newly accessible areas. The methods of construction used during this phase of activity would have been ineffective in the pre-burst hummock and hollow system of the

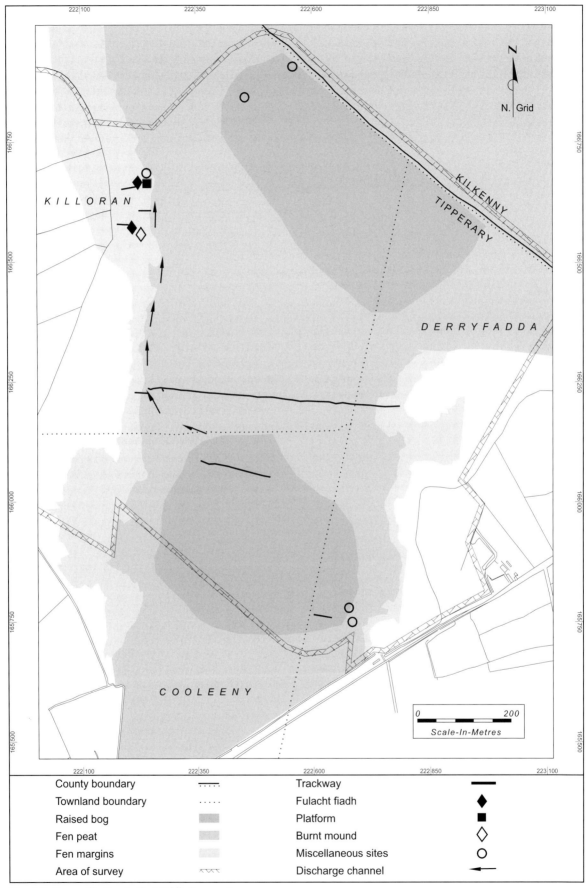

County boundary	--·--·--	Trackway	━━━
Townland boundary	·······	Fulacht fiadh	◆
Raised bog		Platform	■
Fen peat		Burnt mound	◇
Fen margins		Miscellaneous sites	○
Area of survey	⌄⌄⌄	Discharge channel	⭠

Fig. 4.5 Derryville Bog in the Middle Bronze Age.

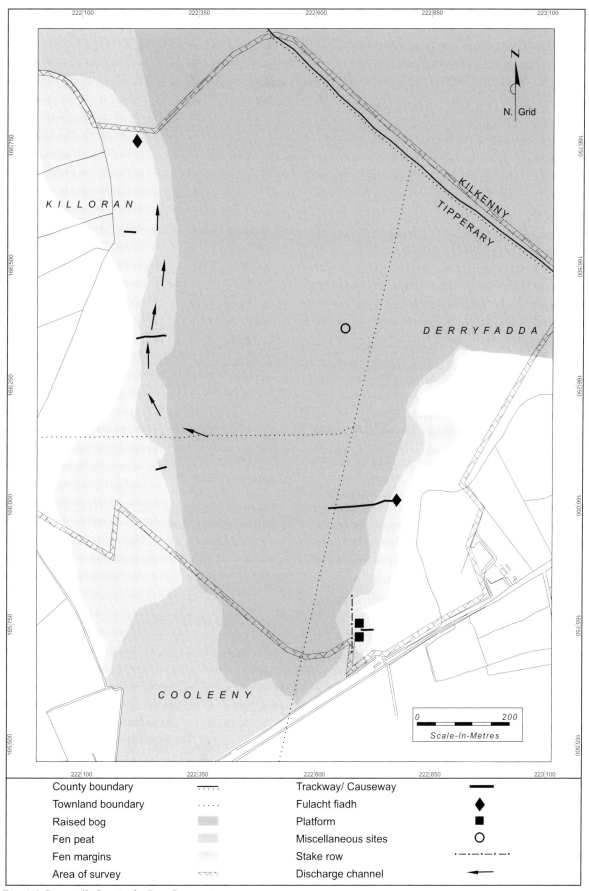

County boundary	——†—†—	Trackway/ Causeway	■■■
Townland boundary	· · · · ·	Fulacht fiadh	◆
Raised bog		Platform	■
Fen peat		Miscellaneous sites	○
Fen margins		Stake row	·–·–·–·
Area of survey	⌄⌄⌄⌄	Discharge channel	◄——

Fig. 4.6 Derryville Bog in the Late Bronze Age.

raised bog. All the sites securely dated to this episode could not have been constructed in their particular locations if the bog had been wetter.

Sites that were constructed in the final phase of the period were constructed in areas where there were subsequent concentrations of activity during the later transitional stage between this period and the Iron Age. After the bog burst it was possible to construct a substantial 69.5m-long trackway, Killoran 69, in the northwestern area, although its eastern end was washed away during the subsequent rise in water levels.

The stake row Derryfadda 209 was the earliest evidence of demarcation within the study area, and the site was probably constructed to mark off an area that could be safely accessed. The localised, small-scale nature of the constructions can perhaps be explained by the fact that all this activity occurred prior to bog burst C, *c.* 820 BC.

During this period, the extent of the study area that was occupied varied in relation to how dry the environment was. After the initial bog burst of 1250 BC, longer tracks were constructed across a variety of environments from landfall through marginal woodland and fen and out into the raised bog. Isolated structures were also discovered within raised bog peat, suggesting that, for a short period of time, parts of this environment were dry enough to use with only limited constructional deposition. As the water table re-established to pre-bog burst levels, activity was restricted once again to the marginal areas. At this point, *fulachta fiadh* as a site-type went out of use. The water levels continued to rise until it appears it was considered necessary to demarcate an area prior to bog burst C. Again, after the burst, more extensive and lengthy tracks could be constructed across a variety of temporarily drained environments. The end of this period was marked by a technological development: the introduction of socketed iron axes.

Late Bronze Age–Iron Age transition (650–450 BC)

Nine structures were ascribed to this period, dated by radiocarbon determinations, woodworking evidence and stratigraphic relationships to bog burst D of 600 BC (Fig. 4.7). Four of the structures were dated to the time of the bog burst: the lower levels of trackway Derryfadda 13, dated to 767–412 BC (GrN-21943) and 790–395 BC (Beta-102756), causeway Cooleeny 31, dated to 778–423 BC (GrN-21822) and 790–380 BC (Beta-111367), trackway Derryfadda 206, dated to 785–375 BC (Beta-102755), and trackway Killoran 306 (Plate 4.3), dated to 756–407 BC (UB-4189) and 756–412 BC (UB-4190). The trackway Derryfadda 210 was dated to 515–365 BC (Beta-102740), and the trackway Killoran 234 was dated to 795–395 BC (Beta-111369).

The transition from the Late Bronze Age to the Iron Age in Derryville Bog was marked by the introduction of iron axes on sites dating from the seventh century BC.

Plate 4.3 Killoran 306 after excavation.

The pollen record indicated low levels of human activity between 600 BC and 200 BC.

In the seventh century BC short trackways were built in the marginal woodlands on the western and eastern sides of the bog. This century also saw the construction of the short-lived causeway Cooleeny 31, which crossed the Cooleeny Bog system. It was built just before the 600 BC bog burst in that area, which destroyed a section of the causeway. The dated samples from structures in the unexcavated Cooleeny Complex, in marginal woodland and raised bog around the western end of Cooleeny 31, suggested that a number of these structures may also be ascribed to the seventh to fifth centuries BC. Some other structures, such as Derryfadda 210, showed similarities with structures dating before the fifth century BC.

The lower-order structures of this period, such as the trackways, appear in areas of marginal woodland, which show recurring use throughout the Bronze and Iron Ages. This includes both the western side of the bog and the Derryfadda headland on the eastern side. The same pattern may be apparent in the more sophisticated structures that cluster around the western end of Cooleeny 31. These structures produced dates ranging from the eighth to the second centuries BC and extend into the early medieval period.

Plate 4.4 Trackway Killoran 315, including a hurdle panel.

As no detailed knowledge of the lower levels of the peat in this area was available, evidence of the use of this area in the Late Bronze Age or earlier is only indicated by the presence of a *fulacht fiadh*, Cooleeny 119.

Many of the sites in this area may be Iron Age in date. Too much emphasis cannot be placed on the association between Cooleeny 31 and the other structures in this area. In fact, there is reason to believe that its social context may be a more regional one. A similar 7m-wide causeway has been recorded in Longfordpass to the south (Rynne 1965). Cooleeny 31 could have been constructed as part of a more regionally focused activity on the north–south chain of bogs that includes the Templetouhy and Littleton Bogs.

The large number of structures in such a short date range indicates that the intensity of activity in the bog was still rising. Construction took place both before and after bog burst D, indicating that the human use of the bog may need to be considered as part of the wider chronological pattern and not just as a particular response to an environmental event. Causeway Cooleeny 31 represents a departure from the normal pattern of construction used in Derryville Bog. This will be discussed in more detail in Chapter 14. The continuation of small-scale casual access to the fen margins on the other hand remains the dominant feature of the Late Bronze Age–Iron Age transition.

Iron Age (460 BC–AD 450)

Eighteen excavated sites in the study area showed evidence of Iron Age activity (Fig. 4.8). In addition, four sites recorded in the initial survey, but located outside the study area, were dated to this period, and over forty sites

in a dense cluster surrounding these sites are considered to form a complex in the same date range. There are clear breaks in the clustering of dates, with one group clustering in the fifth century BC, one in the fourth and third centuries BC and another in the second and first centuries BC. There were no sites recorded on the bog between AD 50 and AD 450, although one site on the dryland was dated to this period.

All of the areas showing evidence of early sites were also active at this time. Two areas in particular were clearly used persistently for most of the period, while two other areas had shorter bursts of activity. While, in general, the sites were very loosely constructed of round-woods, hurdles were used for the first time in this bog in the northwest of the study area. Given the predominance of raised bog in this period, it is striking that most of the sites occurred in marginal woodland and fen, making use of the same environments that were important throughout the human use of Derryville Bog.

The first cluster of dates was derived from tracks and platforms from marginal woodland and fen environments on both sides of the bog. The track Derryfadda 13 was constructed in the transitional period between the Late Bronze Age and the Iron Age but continued in use during the Iron Age; one portion of the track was dated to 380–100 BC (Beta-102736). The trackway Derryfadda 215 was dated to 760–635 BC and 560–370 BC (Beta-102754), with a date of 457±9 BC (Q9400) also being obtained. The two different calibration ranges for the radiocarbon dates point to the difficulties of dating sites from this period. Structure 1 at

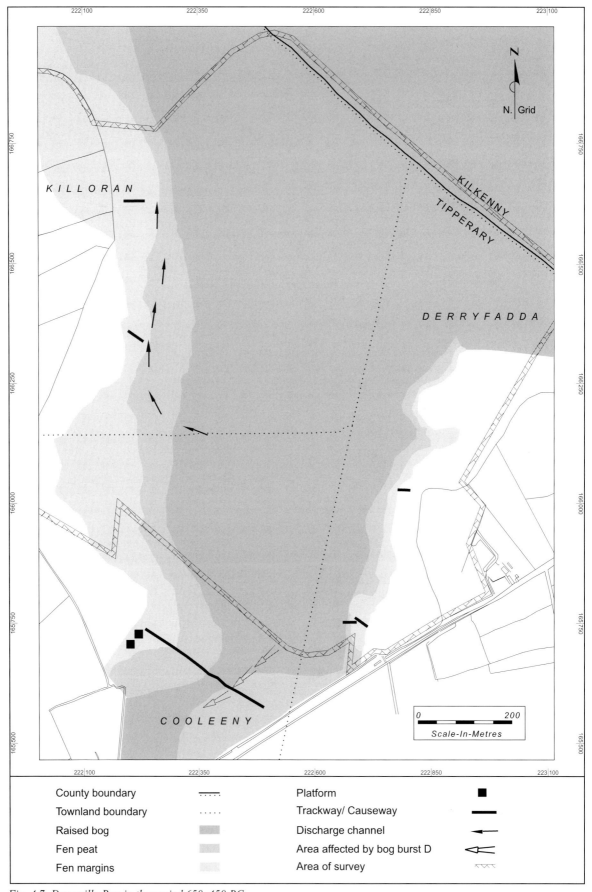

Fig. 4.7 *Derryville Bog in the period 650–450 BC.*

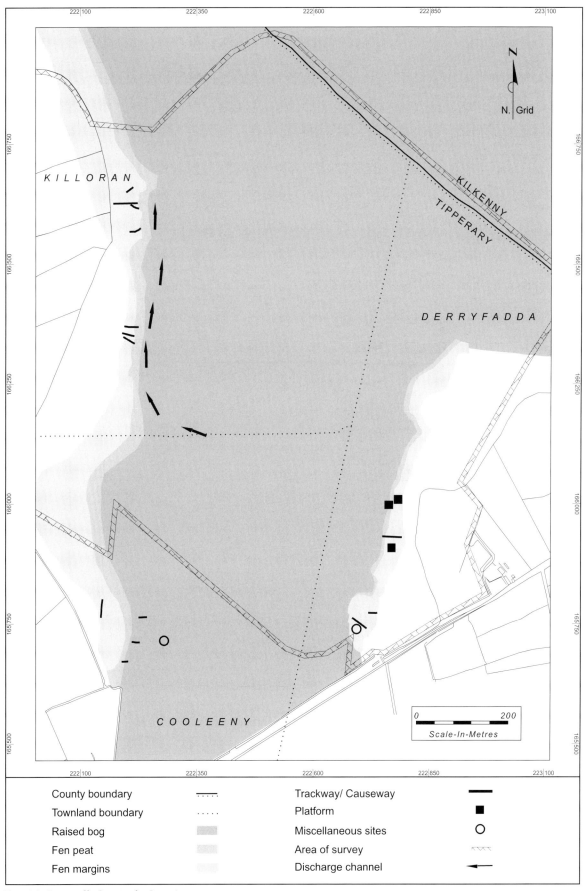

Fig. 4.8 Derryville Bog in the Iron Age.

Killoran 226 was a trackway dated to 542±9 BC or later (Q9477M), 541±9 BC or later (Q9479M), 475±9 BC (Q9480), 460±9 BC (Q9478M) and 450±9 BC (Q9476M). These dates, taken together, give a date in the fifth century BC.

Four substantial structures dating to this period were noted in the initial IAWU survey but were not threatened by development and so remained unexcavated. The site noted as 95DER0100 was dated to 351–120 BC (GrN-21820), 95DER0141 was dated to 368–190 BC (GrN-21949) and 95DER0178 was dated to 372–194 BC (GrN-21817). These three sites were recorded as possible platforms. A substantial trackway noted in the survey as 95DER0169 was dated to 388–207 BC (GrN-21950).

These latter sites formed part of a very dense concentration in Cooleeny townland. A combination of dates and stratigraphic associations has led to the suggestion that the majority of the forty sites in this cluster were likely to be Iron Age in date (IAWU 1996a, 11). The dates suggest that the complex could be fifth century BC in date. Sites excavated during the main field sections of the project reinforce this impression. The scatter of brushwood at Cooleeny 325 was 0.25m above Cooleeny 31, which was dated to 790–380 BC (Beta-111367). As the rate of raised bog growth (approximately 0.10m per 100 years) would have been delayed by bog burst D (see Chapter 3), Cooleeny 325 is likely to have been deposited in the last half of the first millennium.

The sites from the fourth and third centuries BC display very similar date ranges. The trackway Derryfadda 201, in the southeast marginal woodland, was dated to 390–190 BC (Beta-102738). The platform Derryfadda 9 returned a date of 395–180 BC (Beta-102739). On the other side of the bog, the trackway Killoran 248 was dated to 394–199 BC (UB-4183). A date of 641±9 BC or later (Q9483) from this site represents reused wood, possibly from another nearby structure, and underlines the persistent use of this area. The nearby trackway Killoran 301 was dated to 367–194 BC (UB-4184 and UB-4185). Further north, the trackway Killoran 315 (Plate 4.4) was dated to 405–180 BC (Beta-111371).

The last group of dates is distinguishable because of broader date ranges, many of which point to construction dates within the second century BC. Once again, the dates from all marginal areas of the bog are quite similar. On the east side, platform Derryfadda 6 was dated to 380–5 BC (Beta-102737); trackway Derryfadda 203 was dated to 385–50 BC (Beta-102761); three worked pieces of roundwood at Derryfadda 208 were dated to 365 BC–AD 5 (Beta-102746) and the platform Derryfadda 214 was dated to 309–75 BC (Beta-102741). In the northwest, trackway Killoran 75 was dated to 368–190 BC (GrN-21947), 380–4 BC (Beta-102766), 385–50 BC (Beta-102763) and 185 BC–AD 130 (Beta-1027651). These dates, taken together, suggest a date in the second century BC. The trackway Killoran 312 was

connected to Killoran 75 and so should be considered similar in date. The trackway Killoran 314 was dated to 370–5 BC (Beta-111376). Further south, Structure 2 at Killoran 226 was dated to 161±9 BC (Q9475M) and 165 BC–AD 8 (UB-4181). This structure represented a new episode of activity at the site of Killoran 226 and suggests a persistent use of the same zone.

A small area just north of causeway Killoran 18, on the western side of the bog, indicated a persistent use of similar routes through the marginal woodland and into the fen. Killoran 69 crossed this area during the Bronze Age and Killoran 306 provided access during the transition to the Iron Age. In the fifth century BC, Structure 1 at Killoran 226 provided a short stretch of substantial trackway across a pool in the bog marginal woodland. The trackway at Killoran 248 continued the pattern of use in the fourth or third centuries BC. It ran along a similar line to Killoran 226, about 5–6m to the north. Another similar track lay a further 15m to the north at Killoran 301. In the second century BC, movement shifted south again to Killoran 226, where Structure 2 followed the same line as Structure 1 but was more than twice as long. The area had flooded more acutely by this time, so a substantial trackway was required to reach the carr marginal woodland, which may have been a focus for grazing. All of these trackways were similar in structural detail and were degraded due to long exposure. While there are gaps between the dates of these sites, they show a continuity of similar activity along the same route in the same area.

A similar persistence was evident on the eastern side of the bog, though in this case the activity was represented by a mixture of trackways and platforms. Trackway Derryfadda 206 was constructed in the transition between the Late Bronze Age and the Iron Age. The construction of trackway Derryfadda 215 was followed by the platform Derryfadda 214 and finally by the platform Derryfadda 6. These sites all overlap each other, making use of the same place within the marginal woodland, which was becoming increasingly wet. The platform Derryfadda 9 and trackway Derryfadda 203 may represent a less intense version of the same phenomenon. They were constructed in a similar environment 30m to the south of the tight grouping that began with Derryfadda 215.

Most of the sites in this period were constructed in a fairly casual fashion. Most sites did not use pegs to secure the structures, and the distinction between trackways and platforms is almost completely based on the length of the structures. The sites in the western cluster are distinct because they used tightly packed roundwoods and were clearly in use for as much as twenty years. In the east and southeast, the structures were less well organised and appear to have been in use for less time.

In the northwest, all four structures in this loose grouping made use of hurdles. At Killoran 75, a remarkably long hurdle panel appears to have been constructed specifically for the track. At Killoran 312, 314 and 315,

Plate 4.5 Remains of Iron Age house at Killoran 16.

the hurdles were probably used expediently to deal with the particularly wet nature of the fen at this time. These more designed structures, making use of hurdle panels, may indicate an integration between the wetland and the dryland landscape. Woven panels are a widely used and efficient element in landscape management. Hurdles were woven in spring and autumn and could be used for fencing and other types of enclosure and could be moved for different uses. The hurdles found in the bog most likely formed part of this kind of pattern. Their use in bog tracks shows that, by the Iron Age, activity on the bog was an extension of a settled and managed landscape.

The dominant environment in the study area during the Iron Age was raised bog. The overall system was becoming wetter and the raised bog was encroaching on an established oak wood off the Derryfadda peninsula. In other midland bogs at this time, many causeways were built for crossing raised bog (Raftery 1996), but in the study area the focus remained on fen and marginal woodland. The trackways in the northwest were built to access the pools where the Killoran Bog system met the main Derryville Bog system. The western cluster crossed increasingly wet fen to utilise marginal woodland further into the bog. The eastern sites seem to be focused around the mixed marginal woodland there, allowing felling and general access to the marginal woodland as a resource. Only one excavated site from this period is found in raised bog, Cooleeny 325, which shows no real structure and may in fact be an accidental deposit. The pattern, therefore, shows continuity in the use of Derryville Bog as a resource and a focus in itself rather than as an obstacle to regional or local movement.

While the Iron Age sites in the bog group into three clusters of dates, the gaps in those dates do not seem to represent discontinuity in use, and the Iron Age sites continue a pattern of use that is already well established in the area. Nor do the gaps in these dates reflect the lulls in clearance activity that are seen in the pollen record for the area (see Chapter 6). The lull recorded in the fourth century BC may indicate increasing wetness, with the structures in the bog responding to that change by making increasing use of the wetland resources available. This suggests an increasing intensification of land use rather than a fall back in landscape management. The persistent reuse of specific places within the bog also emphasises this sense of continuity.

The six-hundred year gap between the Iron Age activity in Derryville Bog and the early medieval activity is paralleled by the indications of woodland regeneration and little human activity evident in pollen zone VIII (see Chapter 6). The only dated site spanning this period is the post-built round house Killoran 16 (Plate 4.5), which was dated to 180 BC–AD 425 (Beta-117551). This site is located on the dryland in the far west of the study area, on a glacial knoll. The site may in fact relate to undated wetland material discovered in a pilot survey by the IAWU (1996b, 93) in bog to the north in Killoran townland. The next group of dated structures appears in the seventh century AD, which sees the return of the utilisation of the raised bog, an activity that was not noted after the end of the Bronze Age, when a small number of structures were attempted on this type of terrain. This gap may reflect the previous preference for fen as a resource; the final expansion of raised bog peats, at the expense of the fen, would have removed the reason for the utilisation of this tract of wetland. When structures reappear on the surface of the raised bog, it is in a different pattern than in the fen (as discussed below).

Early medieval period (AD 650–1250)

Nine excavated and two surveyed sites were radiocarbon dated to this period. A further eight excavated sites were stratigraphically assigned to this era (Fig. 4.9). A large concentration of possible medieval sites identified in the IAWU survey lay outside the development area, in the townlands of Killoran and Baunmore (RMP ref. KK007-003), northwest and north of the bog study area.

Archaeological activity recommenced following a lull of roughly six centuries. Much of the later medieval material in the bog was removed, to varying depths, by BnM industrial milling. The excavated material fell into two clusters: sixth to ninth century AD and tenth to twelfth century AD. All of the wetland material occurred in raised bog and had a very different spatial focus to the prehistoric material. This suggests a major cultural or economic change in the use of the bog.

The earliest date in this period, AD 415–630 (Beta-117555), was derived from a linear ditch on the Killoran headland at Killoran 3. The site consisted of two large, parallel ditches and a cluster of possibly later pits, evidence for Early Christian drainage of the Killoran Bog system for agriculture.

Also on the Killoran headland, at the southern extent of the townland, was an early ecclesiastical settlement within a large circular enclosure (Killoran 31). This site was located at the southern end of a 150m diameter glacial ridge and enclosed on three sides by reclaimed bog and to the east by the Moyne Stream. The excavation of a small internal area revealed linear features and evidence of iron working dating to AD 450–690 (Beta-120521). The site is referred to in the townland name Killoran from the Irish *Cill Ódhráin*, the church of St Odran (O'Flanagan 1930). St Odran is referred to in the Life of St Columba as one of Columba's first companions at Iona. He died shortly after arriving in AD 563 and the cemetery there bears his name. By Irish tradition, St Odran was the abbot of Meath and founder of a monastery at Latteragh, County Tipperary (Farmer 1987, 323). The two traditions almost certainly refer to the same person and it is, therefore, likely that Killoran 31 was founded before AD 563.

The first phase of activity on the bog is broadly spread throughout the study area. The long stake row Killoran 54, dated to AD 668–884 (GrN-21944), consisted of two roughly parallel rows running northwest–southeast for about 406m. Further south, Derryfadda 19 was a second, less substantial, stake row dated to AD 640–890 (Beta-102760), measuring about 100m in length and running northwest–southeast. The stake row 95DER0048,

surveyed outside the study area, to the southwest, was dated to AD 672–853 (GrN-21823) and ran roughly east-west for a distance of about 105m. It is highly likely that these three sites were contemporary, in a phase of land division. The two stake rows represented the earliest extensive land division on the bog and possibly represented provincial or ecclesiastical boundaries associated with the monastery at Killoran 31.

Two settlement sites located in the western bog margin were dated to this period. A possible house site at Killoran 8, located on the dryland immediately west of the bog, was dated to AD 685–985 (Beta-117554). Killoran 66, a hut site on Derryville Bog, dated to AD 775–887 (GrN-21945), is evidence for possible settlement on the raised bog surface.

Killoran 23, an unrelated site in the far west of the Killoran dryland, consisted of a pit with fire-cracked stone and charcoal and was dated to AD 660–880 (Beta-117550). Close to this site was a large back-filled pit at Killoran 16, dated to AD 890–1040 (Beta-117552). The pit was unrelated to the earlier house site and both appeared to relate to isolated activity across the dryland.

A second phase of activity was determined both from the cluster of material in the northwest of the bog study area, close to the Killoran headland, and from some of the dryland material. It represented activity in a large area off the northeast headland of Killoran townland. A trackway, 95DER0098, identified in the preliminary survey was dated to AD 1024–1162 (GrN-21948), giving the youngest date recorded for the milled bog surface. Part of a varied group of twenty-six associated tracks clustered around the northern Killoran headland, this trackway incorporated six possibly contemporary stake rows that formed a circular boundary mirroring the shape of the Killoran headland. This early medieval activity may have formed part of a much more widespread group of elements across the bog. Within the dryland material, a burnt pit containing charcoal and silt at Killoran 3 was dated to AD 980–1270 (Beta-117556). Similar pits were excavated at Killoran 11 and Killoran 15 and at two sites at Killoran 16. All seem to indicate a pattern of sporadic casual use of the dryland in this period.

The survival of archaeological sites in the bog from this and later periods has been significantly reduced because of industrial peat milling by BnM over the past twenty years, and the removal of the upper peat layers made it impossible to determine the full extent of archaeological activity in this and later periods. The extent of peat removed across the bog was not uniform, however, so the survival of archaeological material was patchy, with the best preservation of early medieval material occurring outside the study area in the Killoran Complex, located around the top of Killoran headland. Of these sites, the latest date recorded, AD 1024–1162 (GrN-21948), was for 95DER0098.

As a result of milling, comparatively little material survived from this period within the study area itself. Traces of sites were found within the raised bog in the southeast at Derryfadda 212 and Derryfadda 42, although both of these sites were almost completely destroyed. Similarly, in the southwest, the trackway Cooleeny 44 was traced for 11m but most of the material was displaced. In the northwest of the study area, the hut site Killoran 66 may have been associated with the nearby Killoran 65.

While the truncation of the peat profile made it difficult to assess the spatial distribution of the later material, it is obvious that it differed significantly in character and pattern from the prehistoric material. The shift in distribution was associated with a shift in environmental context and site types. The return of human activity to Derryville Bog showed a significant change associated with the demise of the fen. The stake rows simply divided the raised bog into possible land holdings, possibly indicating connections with regional land division, perhaps connected with ecclesiastic centres to the south and west (Manning 1997); the other sites indicated a wide range of activities.

Discussion and conclusions

The combination of a long series of absolute dates and their relationships in the context of a developing environment facilitated the construction of a chronology for the excavated material. This chronology cross-cuts both construction techniques and environmental context. Its close correlation with tool use also facilitates the appraisal of the structures within a more conventional artefact-based typological framework.

Comparing the material across prehistory, the continuity from one period to the next is immediately striking. Throughout the prehistoric period, a similar range of site types is present, including a predominance of casual structures. Generally, these sites were constructed in marginal areas of the fen and raised bog. The clusters of dates within periods appear to relate to changing environments and changing uses of the bog.

The study area yielded a unique group of dates in the mid-first millennium BC. These have afforded an opportunity to sequence the material and sites recorded within a time period that is usually obscured by dating problems. On the basis of the similarity of the tools in use from the seventh century BC onwards, it appears that iron was in use from this date. A mid-seventh to mid-fifth century BC transition period was interpreted in the record at this point, based on two changes in the tool marks made by the axes used on the wood from the structures. These dates straddle the end of the Bronze Age and the beginning of the Iron Age. However, it should be borne in mind that defining Bronze Age or Iron Age at this date, in any context, is fraught with difficulties.

Conventionally, many of the typical activities of the Late Bronze Age, including metal industry and hoard deposition, are conventionally regarded as coming to a conclusion at the end of the seventh century BC (Cooney and Grogan 1994, 179; Champion 1989, 291–2). The recurrence of hoard deposition around 300 BC has led many commentators to suggest that there was some form of social collapse in the interim (Mallory and McNeill 1991, 140).

In reality, the evidence is not so clear cut. The implications from some pollen diagrams (Weir 1987) suggest that there is a certain amount of continuity throughout the first millennium BC. Where settlement evidence exists, such as at Navan Fort and Rathtinaun, the indications are of continuity rather than collapse. In Derryville Bog, the overall impression, *fulachta fiadh* excepted, is of continuity in the way the bog was used, despite the periodic abandonments around the time of the bog bursts. At Rathtinaun, objects normally dated to the Dowris phase of the Late Bronze Age were found with iron objects, including a shaft-hole axe (Raftery 1994, 34). When the practice of depositing hoards restarts, with objects dating from around 300 BC, the pattern seen in the Late Bronze Age continues (Cooney and Grogan 1994, 195–9), with the addition of iron objects, which were absent in the deposition pattern until that time.

The overall impression created is that the end of the Bronze Age may not represent a collapse into a dark age but a continuation into a period of slow transition. This fact is obscured by the problematic dating of sites around this time and the small quantity of associated relevant finds. It may be the case that historically, the association between the Iron Age and the Celticisation of Ireland (Waddell 1995, 166) has led commentators to expect to find dramatic changes in the archaeological record.

If the transition is considered to be gradual, against the background of the possible continuity identified in the little evidence that is known, other explanations can be put forth. For instance, the continuation of hoard deposition of Dowris material in this period would see the number of hoards deposited each year decrease to five or six, as seen in Table 4.1; with eight per year in the Bishopsland phase and nine per year after 300 BC. The absence of iron objects in these hoards may reflect the nature and meaning of the depositional activity. It is unlikely, given the complete reappraisal of resources that

Period	Duration (years)	No. of hoards	Deposition rate
Early Bronze Age	600 (to 1700 BC)	50	1 in 12 yrs
Middle Bronze Age	500 (to 1200 BC)	5	1 in 100 yrs
Bishopsland	200 (to 1000 BC)	25	1 in 8 yrs
Dowris	700 (to 300 BC)	130	1 in 5 yrs
Iron Age	200 (from 300 BC)	22	1 in 9 yrs

Table 4.1 Hoard deposition rates for different periods.

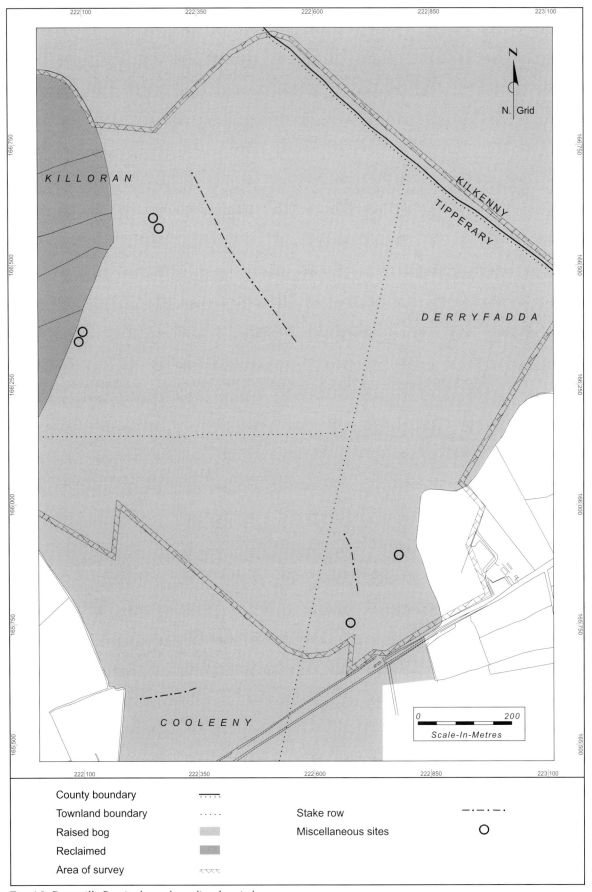

Fig. 4.9 *Derryville Bog in the early medieval period.*

would be brought about by the emergence of ferrous technologies, that iron objects would be deposited in hoards unless 'due process' demanded as such.

It could be that, after the introduction of iron, we see a change in the focus of hoard deposition. If the emergence of high-status objects made of iron is correlated with the re-emergence of hoard deposition, it may be that the appearance of iron in hoards comes with the manufacturing of high-status iron objects. This would obscure the development of iron objects in the interim period in which many individual finds are very difficult to date.

After the introduction of iron axes to the locality of Derryville Bog in the seventh century BC, we see their continued use until the end of the first millennium BC. For an area particularly lacking in high-status monuments in its immediate environment, the suggestion that iron was introduced as a basic domestic tool at this early date and continued in use in perpetuity reinforces the idea of gradual transition rather than dramatic social collapse.

The 600-year gap in our sequence at the end of the Iron Age seems to relate to a major change in the bog and its use. The entire basin had become engulfed by raised bog at this point and the use of this changed environment is substantially different from the patterns of use in prehistory. This is true for both the site types and the spatial groupings.

5. The palaeoenvironmental studies

Chris Caseldine

The nature of the project

The Lisheen Archaeological Project provided a unique opportunity to undertake palaeoenvironmental work in combination with extensive archaeological investigations on a scale not previously undertaken in Ireland. The comparable studies at Corlea, County Longford (Raftery 1996), although undertaken over a longer time period, did not involve the same scale of investment in terms of time and resources, but did provide a valuable model for the Derryville research. An important constraint on the work described below was time, in that, in effect, only two years were available from the final planning to the presentation of the final report. Hence, all the work had to be planned and carried out with this in mind, with completion relatively shortly after final excavation. This inevitably led to a bias of sampling in favour of sites excavated earlier in the project, with extremely interesting sites such as Cooleeny 31 receiving insufficient attention, although the design of the project overall allowed such late work to be placed within an environmental context. In addition to Wil Casparie's peat stratigraphy studies, presented in Chapter 3, a range of palaeoenvironmental analyses were carried out—pollen, macrofossils and testate amoebae by Caseldine *et al.* (Chapter 6); wood identification by Stuijts (Chapter 7) and Coleoptera by Reilly (Chapter 8).

The principal objectives of all the palaeoenvironmental analyses were to provide supporting environmental evidence for the specific archaeological sites excavated and to contribute towards an understanding of the evolution of the bog and the surrounding dryland landscape. Perhaps the most important overall contribution to the project was the reconstruction of the evolution of the bog, for Derryville Bog proved to have an extremely complex and variable history, which was of major significance for any understanding of the archaeological record, both in terms of individual sites and in a wider context. Unlike the archaeological record, which comprises a series of snapshots, albeit often in detail and potentially covering decades or even centuries, the palaeoenvironmental sequences have the benefit of being a continuous record and thus provide a backdrop for the cultural record. The interpretation of these sequences in a cultural context, i.e. to what extent they represent the degree and nature of the impact of prehistoric (and historic) communities on the landscapes of Derryville, may be open to debate, but they

are continuous. Caseldine *et al.* have outlined a number of key points that directed the approach to the project and that, in retrospect, appear to have been particularly significant in determining the overall success of the work, especially given the time and funding constraints when faced with an area of peat approximately 750m by 1km and over one hundred archaeological sites, sixty-six of them within the wetland area. These points may be summarised as follows:

Need to examine both archaeological and non-archaeological contexts

When faced with such a plethora of archaeological sites, it can be difficult to integrate non-archaeological contexts. However, as a major aim of the work was to discern the environmental context of the sites, examination of profiles for the prime purpose of answering environmental questions was seen as just as important as the archaeological contexts. This had the benefit of ensuring that a very full understanding of the history of the bog itself could be achieved, something without which many important findings would not have been made, particularly the history of bog bursts, which were a relatively common and extremely significant aspect of the environmental and archaeological record.

The importance of supporting evidence at different scales

It is inevitable that in such a wetland context it is possible to derive data applicable at a range of temporal and spatial scales. Temporally, it was of value to have detailed short-term evidence of water table variations obtained through the testate amoebae analysis to set against the broader temporal scale changes in peat hydrology derived from the extensive peat-stratigraphic analysis. Similarly, by concentrating on the production of a series of analyses on sites covering a good cross section of the bog area, it proved possible to determine significant spatial variations in bog and dryland landscapes over the timescale covered.

Independence of approach

Although the research design was constructed in a co-ordinated manner, with clearly defined overall aims, each specialist was given a degree of independence in terms of sampling strategy and prioritisation of post-excavation analysis. This had the benefit of creating independent data sets, which could then be drawn together to establish an environmental history against which the archaeological record could be evaluated, without pressure to reinforce findings already determined. An excellent example of this was the way the peat-stratigraphic work and the analysis of testate amoebae independently demonstrated the occurrence of periods of low water table associated with bog bursts. Not all profiles were studied by each specialist, but those examined were chosen as the most likely to contribute to the overall understanding of the site, in the view of the specialist concerned and after discussion with the archaeological team.

Varying suitability of environments

The surviving bog at Derryville, although mainly ombrotrophic raised bog peat, also comprised extensive areas of marginal fen or carr-woodland peat, as well as reed swamp. The peat-stratigraphic studies (Chapter 3) show how in earlier stages of bog development these latter peat forms dominated the environment. Different techniques varied in their applicability between peat types. Thus, testate amoebae proved only of value in the strictly ombrotrophic peat, as to a large extent did pollen (the fen peats being dominated by very local wetland taxa), whereas Coleopteran analyses proved far more informative in these marginal deposits, with relatively little information deriving from the raised peat. Thus, despite the presence of extensive and very interesting archaeological features within the marginal areas, very little pollen analysis was carried out in these environments given their taphonomic limitations. It should be noted that information on the dryland areas could be obtained from pollen profiles in peat away from the immediate dryland edge, whereas Coleopteran records of dryland taxa came primarily from the marginal areas.

Importance of chronology

Although the dating of the archaeological features provided a sound chronological basis, integration with the environmental record necessitated an independent chronology, as discussed below.

Developing a chronology

The successful determination of sequences of changes, both on and off the bog, required a firm chronological base. The approach to developing a chronology was based on combining three elements: radiocarbon dating of peats and wooden archaeological remains; dendrochronology, principally of timbers used in archaeological features associated with the palaeoecological profiles, but also from non-archaeological contexts; and correlation of pollen assemblages reflecting what were believed to be synchronous dryland vegetation changes. Because of the availability of some dates in dendrochronological years BC/AD, all ^{14}C dates were calibrated to yr cal. BC/AD (and are expressed merely as BC/AD in the text) using Stuiver and Pearson (1993) and Pearson and Stuiver (1986). Where dendrochronologically dated features occurred within the profiles, they were used as anchor dates to evaluate the radiocarbon calibrations above and below. This led to the rejection of some of the radiocarbon dates but provided a firmer basis for the chronology.

Similarly, boundaries between dryland pollen assemblage zones, usually reflecting either significant woodland removal or recovery and clearly synchronous for the area around the bog, once securely dated for a particular profile, were used as time markers for other profiles in which they occurred. For parts of the records between such dated levels, ages were estimated by extrapolation,

Lab. code	Site/Field No.	Depth (cm)	Altitude (m OD)	δ¹³C per mil.	Age BP	Calibrated intercept	Calibrated range 1σ
Beta-100937	DER18(E) Field 41	7–11	127.33–127.29	−24.5	1740 ± 60	AD 330	AD 240–395
Beta-100938	DER18(E) Field 41	27–31	127.13–127.09	−27	2180 ± 70	195 BC	365–115 BC
Beta-100939	DER18(E) Field 41	58–62	126.82–126.78	−26	2360 ± 60	400 BC	415–380 BC
Beta-100940	DER18(E) Field 41	82–86	126.58–126.54	−27.6	2960 ± 70	1145 BC	1275–1030 BC
Beta-100941	DER18(E) Field 41	104–108	126.36–126.32	−25.1	3050 ± 70	1295 BC	1400–1200 BC
Beta-100942	DER18(E) Field 41	142–146	125.98–125.94	−26.7	3820 ± 70	2270 BC	2465–2030 BC
Beta-100943	DER23 Field 41	9–13	124.91–124.87	−26.1	2150 ± 80	180 BC	355–290, 230–250 BC
Beta-100944	DER23 Field 41	34–38	124.66–124.62	−25.4	2490 ± 60	760, 670, 550 BC	780–485, 465–425 BC
Beta-100945	DER23 Field 41	56–60	124.44–124.40	−27	2850 ± 60	1900 BC	1065–915 BC
Beta-100946	DER23 Field 41	76–80	124.24–124.20	−27.1	3070 ± 60	1315 BC	1405–1260 BC
Beta-11084	DER75 Field 9	8–10	127.42–124.40	−27.5	1510 ± 60	AD 575	AD 530–630
Beta-11085	DER75 Field 9	20–22	127.30–127.28	−27.3	1740 ± 60	AD 330	AD 240–395
Beta-11086	DER75 Field 9	36–38	127.14–124.12	−27.1	1930 ± 60	AD 85	AD 25–135
Beta-11087	DERSILT Field 13	5–7	125.98–125.96	−17.5	2080 ± 70	60 BC	60 BC–AD 5
Beta-11088	DERSILT Field 13	24–26	125.79–125.77	−27	2150 ± 80	180 BC	355–290, 230–50 BC

Table 5.1 Radiocarbon dates from palaeoecological profiles at Derryville.

assuming uniform peat accumulation. Whilst this was not likely to always be the case, most firmly dated levels were within a few centuries of each other, and the additional error was considered relatively small. Certain parts of the sequence were more firmly dated than others, usually where both dendrochronology and radiocarbon dating (more than one date) reproduced similar ages. One major problem was the concentration of peat profiles from the first millennium BC covering the Hallstatt radiocarbon plateau. As there were no dendrochronological dates from the palaeoecological profiles for this period, most estimations between *c.* 800 and 200 BC were extrapolated dates, whilst radiocarbon results from this period provided long ranges (e.g. see Beta-100944 above). A summary of the radiocarbon dates used for the profiles is presented in Table 5.1. It is also possible to estimate ages from the altitude of any position in the peat stratigraphy once the radiocarbon framework has been established. Casparie (Chapter 3) used this approach in his reconstruction of the former lateral extent and surface conditions of the bog.

In all the diagrams, the general chronology is used rather than dates specific to that profile, except where they occur between zone boundaries or where dendrochronologically dated features are present. It should also be noted that because the zones defined on palaeohydrological grounds do not always correspond to the dryland pollen zone boundaries, the results are discussed within varying time periods.

Palaeohydrology

Understanding of the palaeohydrological development of Derryville Bog is based on two main lines of evidence: extensive field description of the peat stratigraphy and analysis of testate amoebae, pollen, plant macrofossils and peat humification from selected peat profiles. Underpinning this evidence is the reconstruction of the subsurface topography of the bog, based on a dense network of depth readings and accurate surface survey (see Chapter 3). The importance and value of this element of the work should not be underestimated, for it forms the basis for a lot of the detailed interpretation of how and why Derryville Bog grew as it did. The underlying topography proved complex and influenced heavily the way peat grew and how discharge varied over time, changes that affected the way people interacted with the bog environment. Thus, not only did the area include a major peat divide with drainage occurring in opposite directions, but the location and strength of this drainage pattern changed through time, with the western drainage system proving an important element of the bog surface.

The peat-stratigraphic studies showed sequences of hydroseral change across the bog and were able to indicate the relatively late onset of ombrotrophic raised bog peat growth, compared with studies elsewhere in Ireland,

as at Corlea for instance (Raftery 1996), as well as facilitating the construction of a series of maps of the bog for different time periods, delineating the highly variable surface communities and conditions (see Chapter 3). This work provided key information for understanding both the remaining palaeoenvironmental data and the archaeological evidence, and is a study unparalleled in the Irish archaeological and palaeoecological record.

The peat-stratigraphical data was supplemented by a series of palaeohydrological profiles based on the analysis of testate amoebae, plant macrofossils and pollen and spores, which showed local changes in surface conditions at various locations across the bog surface within a well-controlled chronological framework, some profiles incorporating archaeological features, others not. Testate amoebae are a group of microscopic protozoa characteristic of moist and freshwater environments. Their shells, or tests, are preserved in anaerobic environments. Following recovery and identification (Hendon and Charman 1997), sequences of assemblages can be translated into estimates of water table depths using a transfer function based on current relationships. Although such estimates have error terms, they are extremely valuable in tracing changes in water tables, and, in the Derryville record, it was the identification of a series of rapid lowerings in the water table identified in this way that supported peat-stratigraphic evidence for bog bursts. Furthermore, the network of profiles allowed an estimation of the relative impact of these bursts across the bog surface, highlighting those areas particularly affected. The gradual recovery in the water table following the bursts was also seen in the testate record, allowing estimation of the recovery time of the bog. Raised bog peats have been used widely to identify major changes in late Holocene climate, usually in the form of 'wetness shifts,' periods when climatic change was sufficient to lead to persistent higher water tables (Aaby and Tauber 1975; Barber 1981; Hendon *et al.* 2001; van Geel *et al.* 1996), but the Derryville analyses have shown the importance of understanding local autogenic changes in what can be complex systems.

Pollen analysis

The analysis of pollen and spores from peat profiles at varying locations across the bog allowed the development of two separate sequences. Pollen and spores from species growing on the bog surface supplemented the peat-stratigraphic research and clarified sequences of autogenic change on the bog surface itself. Of more value from the pollen record was the establishment of patterns of vegetation change on the dryland areas surrounding the bog, reflecting the varying impacts of human communities through the record. Although bogs the size of Derryville reflect a very general view of the surrounding landscape, i.e. pollen trapped on the bog surface may be derived from quite a wide source area, the patterns of change seen in the record, especially oscillations between woodland taxa and taxa indicative of open land, proved to be relatively uniformly represented. This allowed a sequence to be developed that, with confidence, could be interpreted in terms of what the vegetation was like beyond the immediate margins of the bog.

The interpretation of such pollen records involves a range of assumptions about the likely source area of the pollen and what changes in pollen curves mean for human activity. By comparing the pollen results with the Coleoptera analyses, it was clear that the former reflected an extremely general picture of the surrounding environment, unable to discern differences between the slopes to the west and east, differences that were significant in the Coleoptera evidence, especially in the early survival of undisturbed 'primary' woodland. Points in the record when pollen from tree taxa recovered were assumed to reflect periods of lessening human impact on the dryland. However, contrasts in the timing of such changes between the pollen and the archaeology may indicate changes in the way the landscape was used rather than widespread abandonment and woodland regeneration. There were periods when all records were in agreement, indicating a new wave of clearance and construction or a period of genuine abandonment. The advantage of conducting a range of types of analyses is that points of contention can be isolated. At Derryville, the low level of resolution for landscape reconstruction based on the pollen evidence alone stood out, but the ability of all the records across the bog to show a reproducible sequence of change provided an extremely valuable series of time markers for the site as a whole.

Wood identification

The huge amount of wood produced by the excavations afforded a very detailed and complete analysis of the character of woodland from both bog marginal environments and the dryland. Whilst there must always have been an element of selection in the use of wood for construction purposes, the changing nature of the species used for different time periods also reflected the availability and nature of the trees growing locally. The difference between earlier periods, when there were still remnants of the original woodland cover, and later periods, when secondary and even scrub woodland dominated, was evident in the record. Examination of the character and preservation of the wood also added further to the reconstruction of the landscape and contributed to the archaeological understanding of wood as a construction element.

Coleoptera

The analysis of Coleoptera from selected profiles proved of particular importance in two areas: firstly, in reinforcing inferences concerning the surface characteristics of the peat at various periods, especially the acidity and nature of surface water in the marginal peats; and, secondly, in revealing in greater detail than the pollen, local landscape variability, especially differences between the western and eastern slopes around Derryville. The identification of species originating from the nearby dryland in marginal samples also enabled a sharper definition of what these slopes were like at various times, again in greater detail than was apparent from the pollen record. The persistence of primary undisturbed woodland in Bronze Age levels was also inferred from the Coleoptera data, a feature implied, but not easily proved, from the pollen record.

Conclusion

Although presented as separate chapters, the palaeoecological results provide an integrated backdrop for the archaeological findings. They are important in their own right and are discussed as such, but it is in the integration of the findings with the archaeological evidence that their full value is seen.

6. Pollen and palaeohydrological analyses

Chris Caseldine, Jackie Hatton and Ben Gearey

Aims

The pollen and palaeohydrological investigations undertaken at Derryville were designed to achieve two principal aims:

(1) To determine conditions on the bog surface and how they changed through time, complementary to the extensive stratigraphic studies undertaken by Wil Casparie, and to inform the archaeology concerning interpretation of the relationship between trackway (or feature) construction and the bog environment.

(2) To develop a sequence of vegetation and land-use changes on the dryland areas surrounding the bog broadly covering the period from the earliest archaeological features to the end of the existing peat record.

These two aims required differing but related analyses, and the ensuing discussion concentrates first on the bog surface conditions and changing palaeohydrology, before moving on to the wider environmental changes inferred for the surrounding dryland areas.

Sampling programme

The overall sampling programme was designed primarily to achieve the broad aims defined above, covering both the eastern and western areas of the bog and also attempting as wide and detailed a chronological coverage as possible. In all, six major site profiles were analysed at varying sampling intervals, and covering as full a range of analyses as necessary to provide the required results. These are summarised in Table 6.1 and will be discussed in detail later.

Two long profiles through the eastern and western areas crossed by Killoran 18 were used for the main pollen and palaeohydrological diagrams, with a further detailed profile through Derryfadda 23 in the southeastern area of the bog. Near the western edge, a profile was taken through Killoran 75 and, to the south of this, a short profile incorporating a well-defined silt layer originating from the higher dryland to the west. Where Killoran 18 terminated at the eastern end, a short profile through the basal layers of the overlying peat and the underlying material at the terminus was also sampled.

Following experience at Corlea, County Longford (Caseldine, Hatton and Caseldine 1996; Caseldine *et al.* 1996), emphasis was placed on utilising ombrotrophic raised bog peat as the main sediment sampled. This had the advantage of providing a well-defined and restricted local flora that could be separated from the dryland elements of the pollen record and that adequately recorded pollen from the drier margins; this is unlike fen and carr peat, which often comprise trees and shrubs and effectively filter out the longer travelled pollen from the dryland. The ombrotrophic peats also provided a good testate amoebae record, with tests becoming virtually absent from the fen and carr. Thus, the marginal areas, especially to the east, which were predominantly fen or carr, often with extensive local bog woodland, were avoided despite their high archaeological potential. Pollen preservation can also be poor in such marginal conditions, with much higher nutrient status and pH due to inflowing water from the surrounding limestone tills.

Reconstructing palaeohydrological changes at Derryville

Techniques

Various proxy measures were utilised to assess changes in palaeomoisture conditions at Derryville. These includ-

Site	Pollen	Testates	Humification	Macros	Dates	TOC/Nitrogen
DER18 (East) Field 41	2cm (1.5m)	2cm (0.85m)	2cm (1.5m)	2cm (0.85m)	6	Yes
DER18 (West) Field 18	2cm/4cm (1.85m)	2cm (1.4m)	-	2cm (1.4m)	-	Yes
DER23 (above) Field 41	2cm (0.94m)	2cm (0.94m)	1cm (0.94m)	2cm (0.94m)	4	Yes
DER23 (below) Field 41	2cm (0.5m)	2cm (0.2m)	1cm (0.5m)	2cm (0.2m)	-	Yes
DER75 Field 9	2cm (0.76m)	2cm (0.44m)	2cm (0.44m)	2cm (0.44m)	3	Yes
DERSILT Field 10	2cm (0.4m)	-	-	-	2	Yes
Landfall DER18 (East)	2cm (0.22m)	-	-	-	-	-

Table 6.1 Summary of profiles examined for palaeoecological analyses.

ed analysis of testate amoebae, macrofossils and humification and are described below. The analysis of testate amoebae in the Derryville sequences facilitated the reconstruction of fluctuations in the water table of the bog. Alongside the stratigraphic record and the analysis of pollen, peat humification and *Sphagna* macrofossils, this led to a fuller picture of the hydrological changes in the bog ecosystem. This was of interest not only for elucidating local changes in bog hydrology in relation to the archaeological sequences, but also for addressing wider questions of climatic change, such as the proposed climatic deterioration in Western Europe around 2650 BP that may be recorded in palaeoenvironmental sequences in bogs from the Netherlands and elsewhere and that has been implicated in the abandonment of settlement in certain areas (van Geel *et al.* 1996).

Testate amoebae

Testate amoebae are a group of protozoa of the superclass Rhizopoda that are commonly found in moist and freshwater environments, including standing water, soils, mosses and peats. The shells of these rhizopods are preserved in anaerobic environments, such as peats, long after their occupants have died and are generally taxonomically diagnostic to the species level. These shells may be recovered and concentrated for identification and counting (Hendon and Charman 1997; Charman *et al.* 2000). The relationship between testate faunas and microenvironment, particularly moisture conditions, has long been known (e.g. Heal 1962; Meiserfeld 1977), and studies have in the past utilised fossil assemblages to reconstruct moisture changes in peatland ecosystems (Tolonen 1986). However, it is only recently that ecological studies utilising multivariate analysis have allowed the relationship between modern species distribution and soil water conditions to be quantified more precisely and applied to fossil data (Tolonen *et al.* 1992; 1994; Charman and Warner 1992; Warner and Charman 1994; Buttler *et al.* 1996). These studies have shown that depth to local water table is the main environmental variable that governs the composition of testate amoebae faunas. Recent work in the British Isles has produced a set of transfer functions that permit fossil testate amoebae faunas to be utilised to reconstruct past depth to water table (Woodland 1996).

Humification

Humification is a measure of the degree of decay of peat and is thus utilised as a proxy measure of changes in mire hydrology (Blackford 1993). The wetter the conditions that peat accumulates under (i.e. the more anaerobic), the less the decay of the peat-forming constituents; if the peat is relatively dry, then more decomposition will take place. A number of measures have been devised to quantify the extent of decay, ranging from a simple estimate of the structure of the peat-forming constituents (von Post 1924) to the colour of the water squeezed from peat samples (Troels-Smith 1955). More recently, techniques have sought to extract humic acids, compounds produced by the decomposition of organic materials that give humus its brown colour, and to utilise the concentration of these as an estimate of the degree of decay (Blackford and Chambers 1991). This procedure involves the extraction of the insoluble residues from a known weight of dried peat by boiling the sample in NaOH (sodium hydroxide), the most widely used extractant (e.g. Aaby 1976; Chambers 1984; Blackford and Chambers 1991). The resultant solutions are measured by the percentage of light transmitted through the extract at 540nm. The higher the percentage of light transmitted through the solution, the lower the concentration of humic and other acids in solution and thus the lower the extent of decay; the lower the percentage transmission, the higher the level of decay. Wet/dry shifts can then be inferred from the changes in the recorded transmission curve. It should, however, be noted that recent work has questioned some of the underlying assumptions behind this technique (Caseldine *et al.* 2000).

Macrofossil analysis

Macrofossil assemblages from ombrotrophic mires, especially those including *Sphagnum* species whose habitat preferences are strongly related to water table levels (e.g. Ivanov 1981), can be used to reconstruct palaeomoisture changes (e.g. Barber 1982; Barber *et al.* 1994). Shifts in species composition can be interpreted in terms of the movement of the water table due to factors ranging from localised hydrological variations to the effects of climatic change. The reconstruction can be either fairly simple and approached through the assessment of the relative moisture preferences of different taxa present (e.g. *Calluna*—dry bog surface; *S. cuspidatum*—very wet/pool) or more sophisticated using a 'calibration' or 'training set' approach that uses detailed knowledge of the modern bog ecology to produce a statistical transfer function to elucidate the record. Examples of this approach include that by Dupont (1985; 1986), who assigned a 'weight' to each *Sphagnum* species ("... a certain arbitrary value that is used to separate ecologically different species groups" (van der Molen and Hoekstra 1988, 222)) that could then be converted to a humidity index to afford a picture of wet-dry shifts for the site in question. More recently, studies by Stoneman (1993) and Barber *et al.* (1994) have extended this approach, with the specific aim of identifying the climatic signal contained in the macrofossil record.

Methods

Testate amoebae

Testate amoebae preparations followed a modified procedure of that recommended by Hendon and Charman (1997). The monoliths were sub-sampled at 2cm intervals. One millilitre was measured by fluid displacement. The sample was then placed in a beaker with approximately 25ml of water and inoculated with *Lycopodium clavatum* tablets (Stockmarr 1971) to permit the calculation of testate concentrations. The sample was then heated on a hot plate and boiled until the peat was disaggregated. After it had been allowed to cool, the contents of the beaker were washed with water through 125μm and 355μm sieves and backsieved onto a 10μm mesh. The 125μm and 10μm portions were retained. The 10μm fraction was then concentrated by centrifuge and mounted in glycerine for counting. Hendon and Charman (1997) recommend staining with saffranine-O, but this was not found to be necessary. The use of the extra sieve (125μm) removed much extraneous organic material that both obscured tests and reduced the effective concentration of tests on the slides, which in cases where concentrations were low, would have considerably increased counting time per sample. Testate amoebae were identified using a draft copy of Charman *et al.* (2000) and Corbett (1973).

Some testate amoebae are larger than 125μm (e.g. *Bullinularia indica* and some species of *Arcella*), and the 125μm fraction for every sample was scanned to ensure that such species were not being excluded from the count. In every case, it was found that even though some larger species were present (generally *B. indica*), the numbers were always so small relative to concentrations in the counted sample that even had the extra sieve been omitted, the larger taxa would not have been expected to comprise above 1% of the count. This would have made little difference to the water table reconstruction.

The samples were counted on a Dialux 20 EB microscope at a magnification of x500. The 125μm fraction was scanned under a WILD M5C binocular microscope, and any testate amoebae present were picked out with tweezers and mounted in glycerine for identification on the light microscope.

Humification

The technique used for humification determinations was that of Blackford and Chambers (1994): 2cm contiguous samples of peat were cut from the monoliths. Any matted *Eriophorum* was cut up with scissors. The peat was dried, then ground up in a mortar and pestle, and 0.2g of peat was weighed and placed in a 150ml beaker with 100ml of 8% (w/v) NaOH. The beaker was heated on a hot plate until the solution boiled. The temperature was then reduced and the samples were simmered gently for *c.* one hour. The contents were allowed to cool, transferred to a 200ml beaker and topped up to the mark. The samples were then filtered through Whatman Qualitative 1 paper. Fifty millilitres of the solution was diluted 1:1 with water in a 100ml flask and the mixture was shaken. The transmission was measured in a colorimeter at 540nm. Each set of measurements was taken at an equal time interval after initial mixing.

Macrofossil analyses

The sediment samples from which the testate amoebae and pollen were extracted were retained and examined under a binocular microscope to assess the macrofossil composition of the sample. The macrofossil content of the sampled levels was assessed on two scales: the relative percentage of the different groups of macrofossils and the percentage of the leaves of different *Sphagnum* species that may be present in the sediment. A sample may thus have two different measures for *Sphagna*: an estimate of the proportion of *Sphagna* in the sample overall compared with the vegetative remains of other plants and, if present, a percentage of the different *Spahgnum* species or group making up the leaves. Four main groups of macrofossils were recognised: undifferentiated monocotyledons—grass, sedge rootlets and leaves; *Sphagna*—branch and stem leaves; ericaceous rootlets—rootlets of *Calluna* and other members of this family; and *Eriophorum*, usually in the form of densely tufted masses with sclerenchytamous spindles from the leaf bases. No attempt was made to assess the percentage of unidentified organic matter (UOM) in the sample, sometimes used as a proxy measure of peat humification (Barber 1981).

Sediment composition was assessed as an approximate percentage on a five-point scale (1–20%, 20–40%, 40–60%, 60–80% and 80–100%), a modification of the system used by Barber (1981). If *Sphagna* leaves were present, then a sample of these leaves (usually at least fifty, although in some samples very low numbers of leaves meant a smaller count was necessary) was picked out of the sediment and mounted in glycerol on slides. When viewed under the light microscope, characteristics of the branch and stem leaves were used to identify either the species or the broad family group that the leaves belonged to. Identifications were made using the key by Daniels and Eddy (1990). The leaf counts were converted to a percentage for each sample.

Results

The results of the testate amoebae were calculated as percentage total testate amoebae and are presented as diagrams produced using the computer programmes TILIA* and TILIA*GRAPH (Grimm 1987). Water table reconstructions derived from the testate assemblages utilised the transfer functions produced by Woodland (1996). The reconstructions are presented as diagrams produced using the program ORIGIN. Macrofossil analyses are included on the same diagram as the testates. The results of the humification determinations are presented as graphs of raw data produced using TILIA*GRAPH and also as graphs with the data smoothed using a five-point moving average.

Killoran 18 (East), Field 41 (Figs 6.1 and 6.2)

Testate amoebae and macrofossils

Tests do not tend to be present in fen peat, and if they are it is in very low concentrations. Therefore, the test record at Killoran 18 did not extend below 85cm, although one sample at 115cm did contain tests in sufficient quantities to be counted. This may reflect incipient ombrotrophic peat growth (W. Casparie, pers. comm.). The depth of sediment covering the causeway level did not contain testate amoebae, probably due to the disturbance of the bog microenvironment by the track construction, which apparently led to the initiation of raised bog growth in this area. The lowest section of the Killoran 18 (East) diagram, the period immediately following track construction, is one of generally low water table, with values of -10cm to -15cm below the surface being recorded from 85cm to 71cm, although the large standard errors on the samples means that this may be an over- or underestimation. The testate amoebae fauna were dominated by *Hyalosphenia subflava* (55–60% at 85–75cm), with *Trignopyxis arcula* also present (15% at 85cm, its highest value for the diagram).

The macrofossil analysis demonstrated that the sediment from 150cm to 81cm consisted of monocotyledons with some ericaceous rootlets, although *Eriophorum* dominated the sample at 149cm. The level of the causeway at *c.* 100cm depth was marked by the presence of mineral matter and also by the presence of some fragments of wood, presumably originating from the structure of the causeway. The macrofossil record indicated that *Sphagnum* first became a component of the local peat-forming vegetation at 81cm, although the leaves in the samples were few in number and badly preserved, making further identification impossible. Monocotyledons still formed around 95% of the sediment matrix, whilst some ericaceous rootlets were also identified at 76–81cm.

There was a change in the testacean communities after 71cm. *Amphitrema flavum* increased to 55%, *Amphitrema wrightianum* was recorded for the first time and reached a peak of 20% at 67cm, whilst *Arcella discoides* suddenly peaked at over 30% at 63cm. Depth to water table derived from transfer functions predicted generally high water tables (-2 cm to -6cm below the surface), with brief periods of standing water possible between 67cm and 63cm. *Sphagnum* mainly attributable to the Acutifolia became a significant component of the vegetation after 71cm, forming around 50% of the sample at 71–67cm. *Sphagnum* mainly attributable to the Cuspidata and *S. cuspidatum* were recorded at 71cm and 65cm, although the latter record was in a sample consisting of 95% monocotyledons.

There was a change in testate faunas at 61cm, with *Difflugia pulex*-type demonstrating a marked increase to 15–35% from 55cm, after which it remained a consistent component of the testate fauna. *Hyalosphenia subflava* reached 50–70% at 61–41cm. *Amphitrema flavum* and *A. wrightianum* were restricted to low and sporadic percentages. *Assulina muscorum* increased steadily from below 10% at 55cm to 20% by 40cm, at which point *Arcella discoides* became significant, attaining nearly 30% by 35cm. The species assemblages translate to a phase of low water

table, with mean values generally between -10 cm to -15cm below the mire surface, although there were large standard errors on the reconstruction. Following the increased percentages of *A. discoides* after 39cm, the trend was for the water table to move slightly closer to the surface.

Sphagnum apparently disappeared from the local peat-forming vegetation between 61cm and 57cm, whilst monocotyledons remained dominant. *S. imbricatum* was first recorded at 49cm alongside Acutifolia *Sphagnum*, and, from this point until 30cm, percentages of *Sphagnum* fluctuated between 5% and 90% of the total sediment composition. Ericaceous rootlets were also present at 61–44cm.

Sustained increases in *Amphitrema flavum* from 35% to 50% were recorded between 24cm and 18cm, whilst *A. wrightianum* was consistently present at 15–20% between 30cm and 16cm. *Arcella discoides* also remained present in high percentages, whilst the disappearance of *Hyalosphenia subflava* from the palaeorecord was recorded after 28cm. *Assulina muscorum* demonstrated a steady decline in percentages after 35cm. The reconstructed water table showed that this was a period with high water levels, probably less than 1cm below the active growing *Sphagnum*, with standing water possible at the sampling site from 35cm to 12cm. *Sphagnum imbricatum* was the dominant species from 30cm to 12cm, with *Sphagnum* forming 95% of the sample composition. Acutifolia *Sphagnum* was also present between 30cm and 16cm, comprising at a maximum, just less than 20% of the identified leaf remains.

The dominance of *Amphitrema* sp. was reduced after 12cm, with values falling gradually to below 5% by the top of the diagram. This taxon was replaced by *Hyalosphenia subflava*, which increased to over 50%, despite some fluctuation to lower percentages. *Difflugia pulex* reached its highest representation for the diagram of 50% at 6–8cm, with *Arcella discoides* remaining a component of the fauna despite reductions in its representation after 14cm. These changes translate to a fluctuating water table with depths of 3–15cm below the surface for the remainder of the diagram. The dominance of *Sphagnum* was reduced at 10cm, with monocotyledons once again forming up to 95% of the vegetative remains. This situation was reversed by the close of the diagram, with *Sphagnum imbricatum* recorded at 90% in the final sample.

Humification

Percentage transmission values remained below 20% for the basal section of the sequence, from 150cm to 108cm, at which point a high value of 36% was recorded. Percentages then fell and showed some oscillations, but remained generally over 20% until 58cm. There was a fall in values after this point to the lowest recorded value for the diagram of 6% at 52cm. Percentage values increased steadily following this to reach a peak of 38% at 36cm. There was a reduction to values below 20%, before a

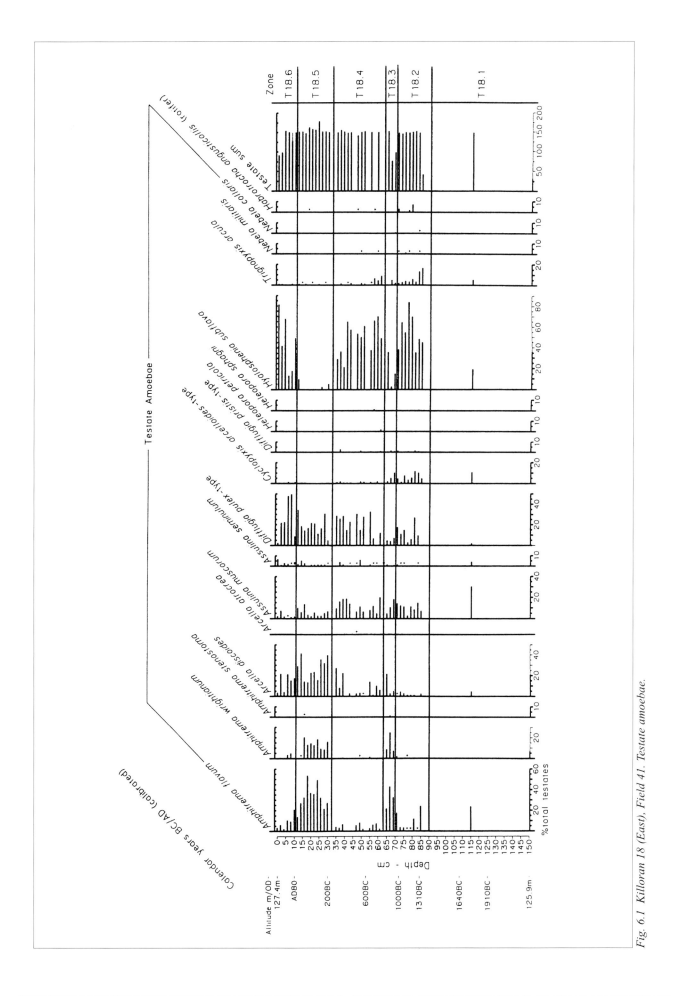

Fig. 6.1 Killoran 18 (East), Field 41. Testate amoebae.

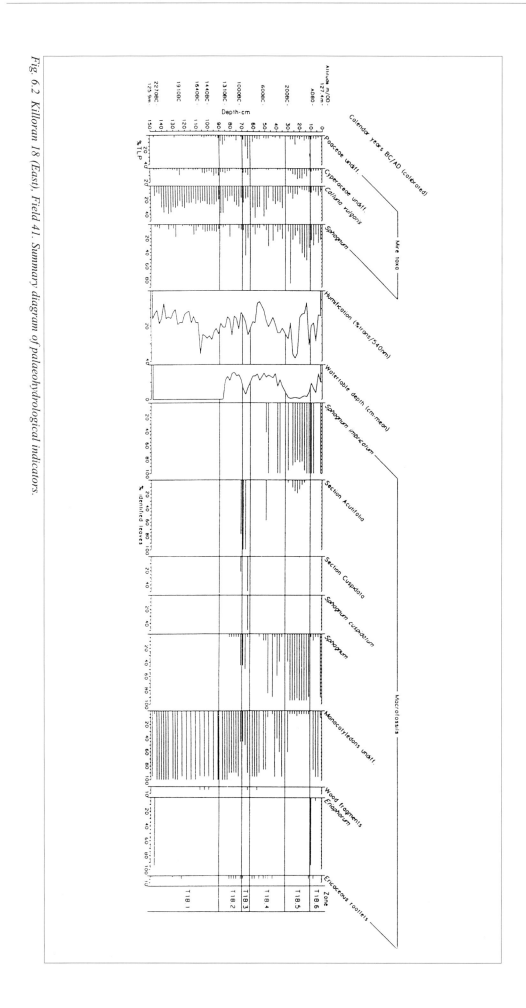

Fig. 6.2 Killoran 18 (East), Field 41. Summary diagram of palaeohydrological indicators.

recovery to over 35% between 32cm and 38cm. Another fall was recorded after this, to 6% at 14cm. Percentages fluctuated for the remainder of the samples, between below 20% to above 30%, with percentage transmission of less than 15% recorded for the top of the sequence.

Killoran 18 (West), Field 17 (Figs 6.3 and 6.4)

Testate amoebae and macrofossils

The testate amoebae record began at 138cm in the Killoran 18 (West) sequence, just above the fen peat, which was bereft of testates. *Amphitrema flavum* was present at nearly 70% in the basal sample, but by 132cm had declined to 1%, before recovering slightly to 20% by 125cm. *Hyalosphenia subflava* increased to over 40% prior to declining to 1% over the same samples and was only recorded in occasional samples at trace values between 114cm and 100cm. *Arcella catinus*-type appeared at over 20% at 136cm, following which it was only recorded once for the remainder of the diagram. *Assulina muscorum* was present from the base of the diagram and peaked at over 40% at 124cm; this species was consistently present for the whole of the Killoran 18 (West) sequence but showed little variation and rarely increased to much above 20%. *Difflugia pulex* increased suddenly to 40% at 32cm, whilst *Difflugia pristis*-type showed a steady rise to 20% at 134–124cm.

The macrofossil record indicated that the samples consisted of monocotyledonous remains from the base until 118cm, after which point there was a change to *Sphagnum*-dominated sediment consisting of *S. cuspidatum* and Acutifolia *Sphagnum*, with some Cuspidata *Sphagnum* also present. The transfer functions indicated a high water table of -1cm to -3cm from 138cm to 134cm, with a sudden and marked fall in mean depth to -11cm at 132cm, before an upward movement to -6 cm by 124cm.

Amphitrema flavum was dominant again between 118cm and 100cm, with *A. wrightianum* a significant presence at 20–30% during this phase. *Amphitrema stenostoma* increased from 120cm to peak at just over 5% at 112cm, before declining, and was only sporadically recorded after 108cm. *Difflugia pulex* fell to low values at 110cm, before showing an unsteady rise to 40% at 98cm. At this depth, *Hyalosphenia subflava* reappeared and replaced *Amphitrema flavum* as the dominant taxon after 92cm, peaking at 70% at 84cm, with the former species fluctuating between 5% and 25% until 76cm. After this point, until 56cm, this species was consistently present, although it fluctuated between 5% and 60%. *Difflugia pristis*-type and *Cyclopyxis arcelloides*-type were also present across the same samples, but at no point did either reach much more than 5%. *Arcella discoides*-type demonstrated a distinct rise from 76cm to a peak of 50% at 64cm, after which it declined and had disappeared from the record by 50cm.

Sphagnum alternated with monocotyledonous remains as the most significant component of the macrofossil samples between 104cm and 74cm. *Sphagnum cuspidatum* disappeared, to be replaced by Acutifolia *Sphagnum* after 100cm, which was the only identifiable *Sphagna* until the appearance of *S. imbricatum* at 80cm, at which point, until the close of the diagram, *Sphagnum* became the dominant component of the sediment matrix. These changes all translate to a phase of raised water table of -1cm to -4cm depth from 118cm to 96cm, at which depth there was a sharp fall to -10cm. There was a brief recovery to -2cm depth by 90cm prior to a fall to the deepest mean water table recorded in the diagram: -14cm at 84cm. The water table remained comparatively deep, with values of -7cm to -11cm between 84cm and 76cm.

Amphitrema flavum recovered at 50cm to peak around 60% from 50cm to 40cm, from where, until the top of the diagram, it remained a significant component of the species assemblage at 40–50%, despite some brief declines to 5–10% around 24cm and 8cm. *Amphitrema wrightianum* was only sporadically present, and only at low values from 90cm until 64cm, but increased after this point to fluctuate between 25% and 40% until 42cm, before declining and disappearing from the record by 32cm. *Difflugia pristis*-type was also significant at 40–60% from 40cm until the close of the diagram. This part of the diagram also demonstrates a reappearance of *Arcella discoides*-type and an uneven rise to 25% by 10cm before a sudden fall. *Hyalosphenia subflava* was not identified from 56cm until 20cm, at which depth it displayed an isolated peak of 15%. These changes translate to a phase of generally high water table, generally -1cm to -3cm deep, although brief dips to -6cm are evident at 66cm, 62cm and 18–20cm. The macrofossil record was characterised by fluctuating percentages of *S. imbricatum* and Acutifolia *Sphagnum* from 44cm until the close of the diagram. *S. cuspidatum* and Cuspidata *Sphagnum* were generally absent or present in very low quantities, apart from a small pocket between 50cm and 42cm.

Derryfadda 23, Field 41 (Figs 6.5 and 6.6)

Testate amoebae and macrofossils

There is some fluctuation in the dominant testate taxa in the bottom 10cm of the diagram, with alternating peaks of both *Amphitrema flavum* and *Hyalosphenia subflava*. The former reached nearly 70% at 92cm and 84cm, and the latter species peaked at 95% at 88cm. *A. wrightianum* and *Trignopyxis arcula* were also recorded in smaller percentages. *Assulina muscorum* was consistently present, although it did not attain more than 15%, whilst *A. seminulum* was sporadically represented at less than 5%. This variation was reflected in the range of water table depths derived from the transfer functions for 94–84cm, with a mean water table depth of below -

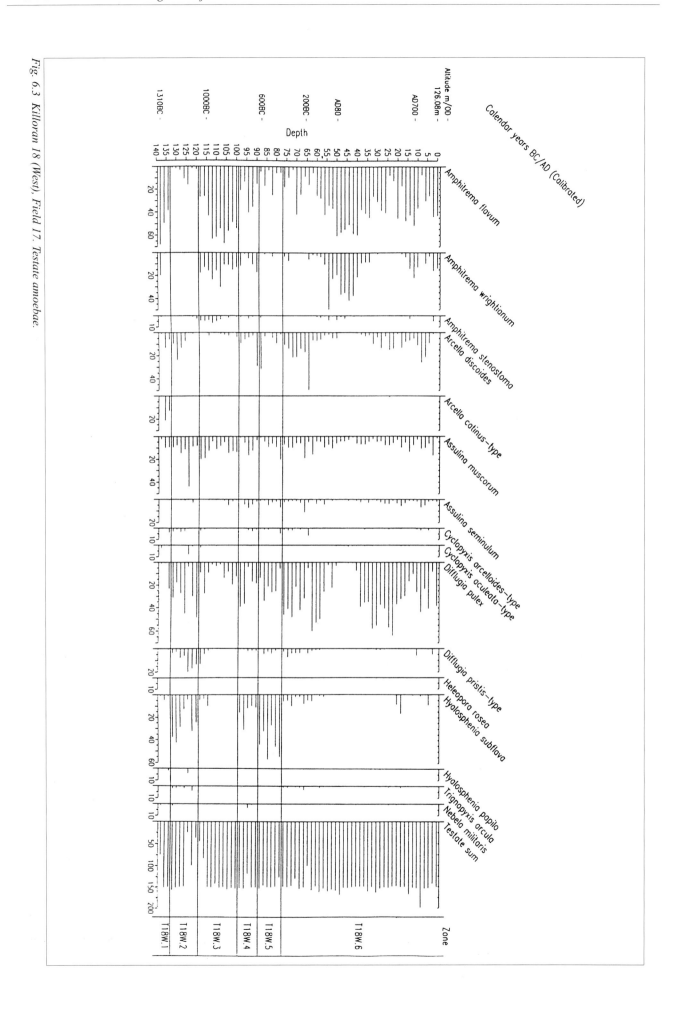

Fig. 6.3 Killoran 18 (West), Field 17. Testate amoebae.

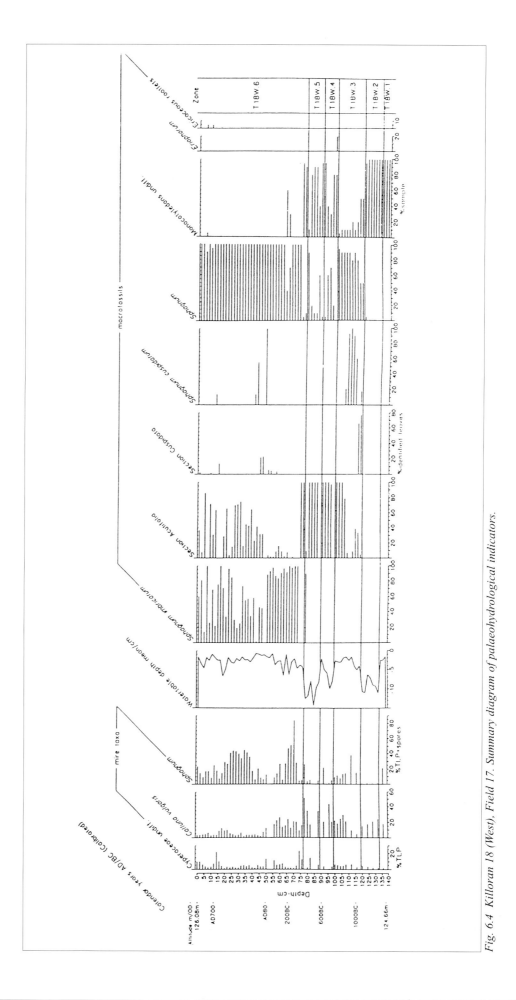

Fig. 6.4 Killoran 18 (West), Field 17. Summary diagram of palaeohydrological indicators.

15cm and with periods of raised water table of -3cm to -5 cm indicated at 92cm and 84cm.

From 80cm to 56cm, the testate fauna was dominated by *Hyalosphenia subflava*, at percentages of over 70%. The only other species present as more than the occasional appearance was *Assulina muscorum*, which recorded at around 10%. The water table remained below -15cm from the surface until 58cm, although there was a comparatively large standard error in the reconstruction due to the high representation of *Hyalosphenia subflava*. The macrofossil record was dominated by monocotyledons from 94cm to 54cm, although some badly preserved *Sphagnum* leaves were recorded in a few samples.

Sudden increases in *Amphitrema flavum* to 50% and *A. wrightianum* to 40% were recorded at 54cm, accompanied by a reduction in *Hyalosphenia subflava* to 20%. This marked a raising of the water table, with mean depths between -0.3cm and -5cm from the mire surface, although there was some fluctuation to a deeper water table at 45cm, where there was a peak in *H. subflava* of 55% and reductions in *A. flavum* and *A. wrightianum*. Identifiable *Sphagnum* leaves were present from 54cm. Some samples consisted of up to 80% *Sphagnum*, but generally monocotyledons dominated the samples, with *Sphagna* forming around 10% of the peat matrix until 24cm. The leaves were mainly identified as Acutifolia *Sphagnum*, with Cuspidata *Sphagnum* significant between 48cm and 54cm.

The most marked feature of the diagram from the 40–20cm zone is the low concentration of testate amoebae—some of the counts were very low. No testate amoebae were present in the samples at 30cm, whilst test concentrations were very low in some samples (see Table 6.1). The water table data must therefore be utilised cautiously, with more reliance on the other proxy measures. *Amphitrema flavum* increased from 36cm to reach a peak of over 65% at 32cm, at which point *A. wrightianum* was present at 20%. This translates to a mean water table of -2cm depth. The high percentages of *Hyalosphenia subflava* at 34–36cm and 28cm reflect a deep water table of -13cm and -16cm, respectively. *Amphitrema flavum* and *A. wrightianum* increased at 24cm, whilst *Hyalosphenia subflava* disappeared from the record until 16cm. The water table moved to within -1cm to -3cm of the surface between 24cm and 18cm. *Sphagnum* was the dominant component of the peat-forming vegetation from 24cm, with Acutifolia *Sphagnum* forming 60% of the macrofossil samples. *S. imbricatum* became significant after 14cm, after which *Sphagnum* formed at least 40% of the remains, increasing to 80–100% from 12cm.

The uppermost 15cm of the diagram is characterised by increases in *Amphitrema flavum*, after a brief reduction in its values at 14cm. *Assulina muscorum* displayed a peak of 55% at this point, after which it steadily declined to the close of the sequence. This pattern was followed by *A. seminulum*, although it was present at a maximum of just over 5%. *Hyalosphenia subflava* reappeared at 16cm and showed an increase to 40% by 10cm, before falling again and then peaking abruptly at 75% at 4cm. These assemblages translate to a fall in water table to -11cm below the mire surface at 14cm, followed by an erratic increase to a water table -5cm deep by 6cm and another dip to below -13cm deep at 4cm. The uppermost sample indicated that the water table had risen to a depth of -6cm by the close of the diagram.

Humification

Transmission values were generally over 20% in the bottom 20cm section, with some fluctuation to lower values. After 74cm, there was a steady fall to 10% by 68cm. Values then stabilised at 16% until 62cm, after which a brief fall to 11% was followed by a sustained increase to 25% by 54cm. A steady and sustained decline in values was recorded from 50cm, with percentage transmission reaching a low point of 5% at 30cm. Following this, there was a recovery in values to a peak of 26% by 24cm. An erratic fall in values was evident until 16cm, after which values stabilised at over 20% before reaching the highest value recorded for this diagram of 34% at 8cm. Percentage values dropped below 15% at 4cm, before increasing to 25% by the close of the sequence.

Killoran 75, Field 10 (Figs 6.7 and 6.8)

Testate amoebae and macrofossils

Testate amoebae were not found in the basal section of the Killoran 75 sequence. The lower sediment consisted of *Menyanthes* and *Scheuchzeria* remains, indicating very wet/pool conditions. As with fen peat, testates are not recovered from very wet microenvironments. *Hyalosphenia subflava* is the dominant species at the base of the testate diagram, the transition to ombrotrophic peat, present at nearly 90% at 42cm but falling steadily to just over 20% by 36cm. *Amphitrema flavum* was present at very low values from 44cm but rose consistently from 40cm to attain a peak of 75% by 32cm. *Arcella discoides* also increased across the same samples but reached a maximum of 20%. *Assulina muscorum* and *Difflugia pulex*-type were recorded from the opening of the sequence and remained steady at around 20%, showing little variation in values. *Cyclopyxis arcelloides*-type followed a similar pattern at values of below 10%. *Trignopyxis arcula* increased from the base of the diagram and reached a high point of over 10% at 38cm. *Nebela militaris* also appeared at this level but was only present in low and fluctuating percentages. The water table fell from -12cm at the base of the sequence to -16cm at 42cm, before rising steadily to -3cm below the surface by 30cm.

Amphitrema flavum declined from its peak after 32cm and had disappeared from the record by 26cm. The water table fell to a depth of -15cm by 24cm, with *Hyalosphenia subflava* becoming the dominant taxon.

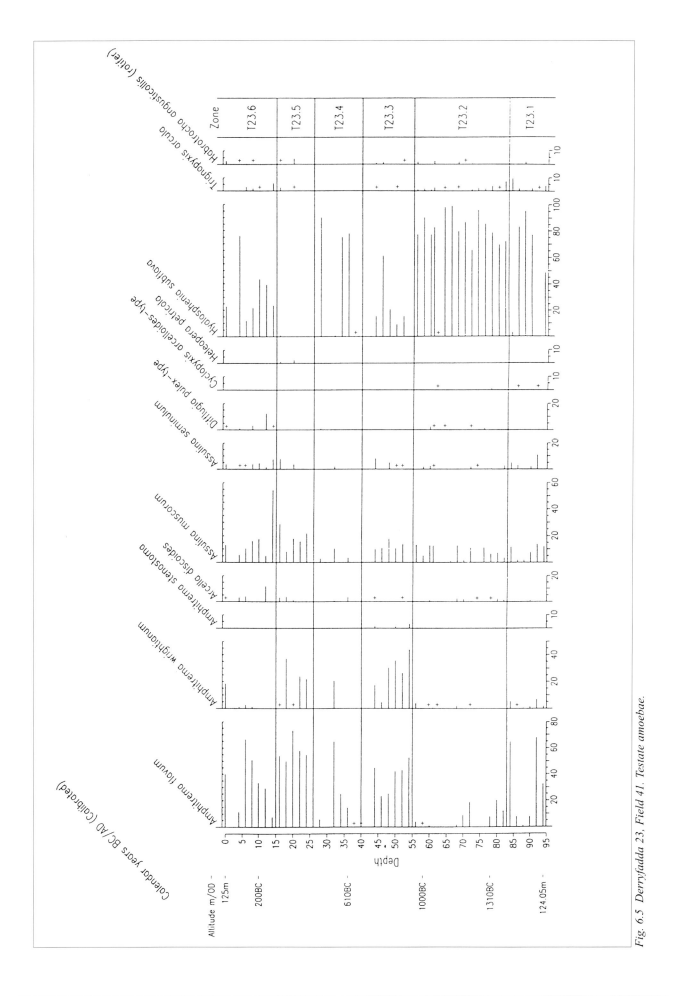

Fig. 6.5 Derryfadda 23. Field 41. Testate amoebae.

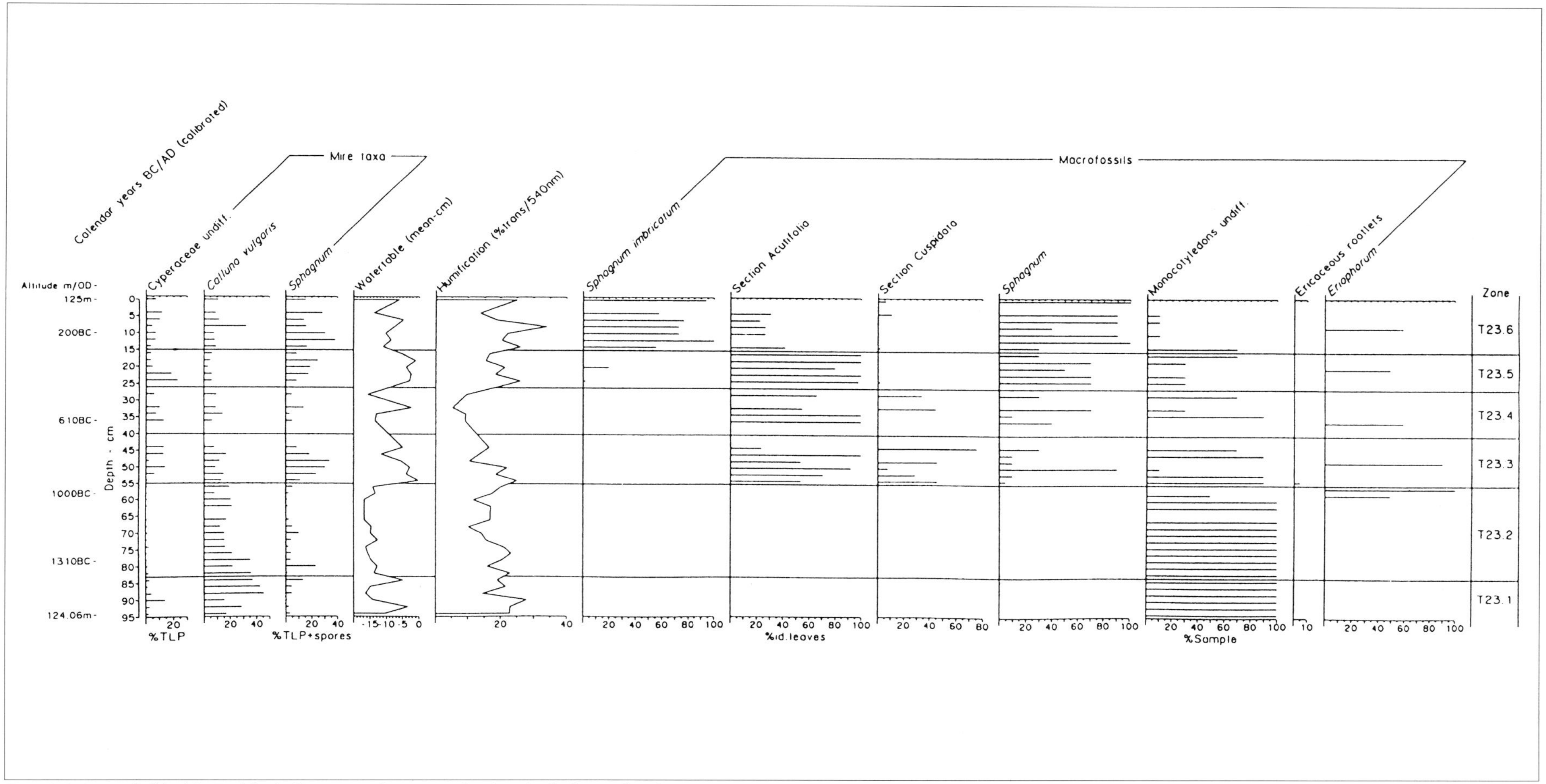

Fig. 6.6 *Derryfadda 23, Field 41. Summary diagram of palaeohydrological indicators.*

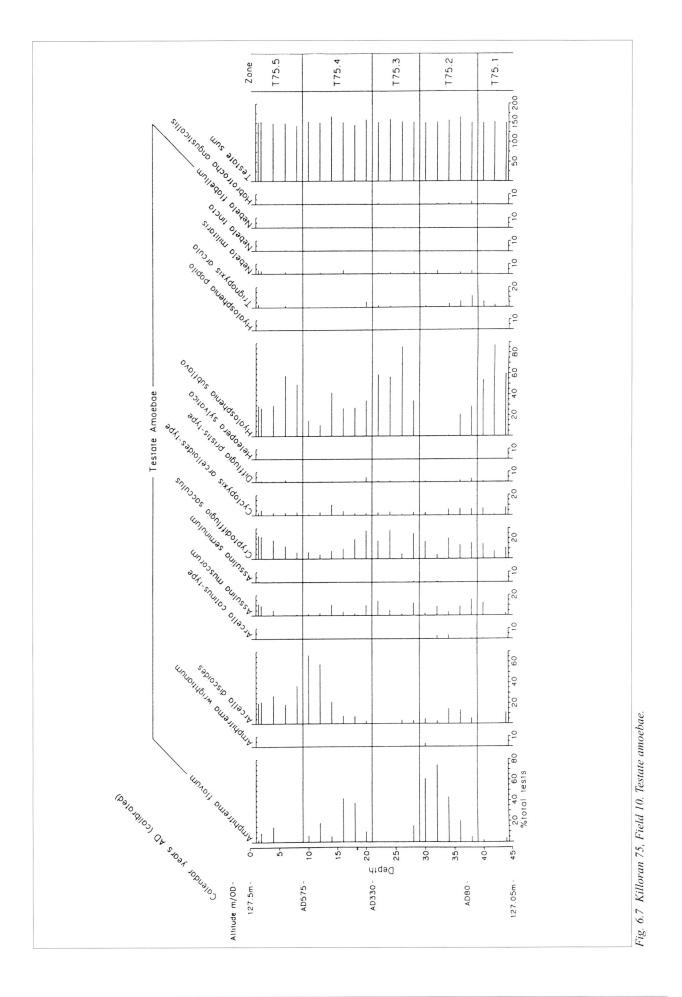

Fig. 6.7 Killoran 75, Field 10. Testate amoebae.

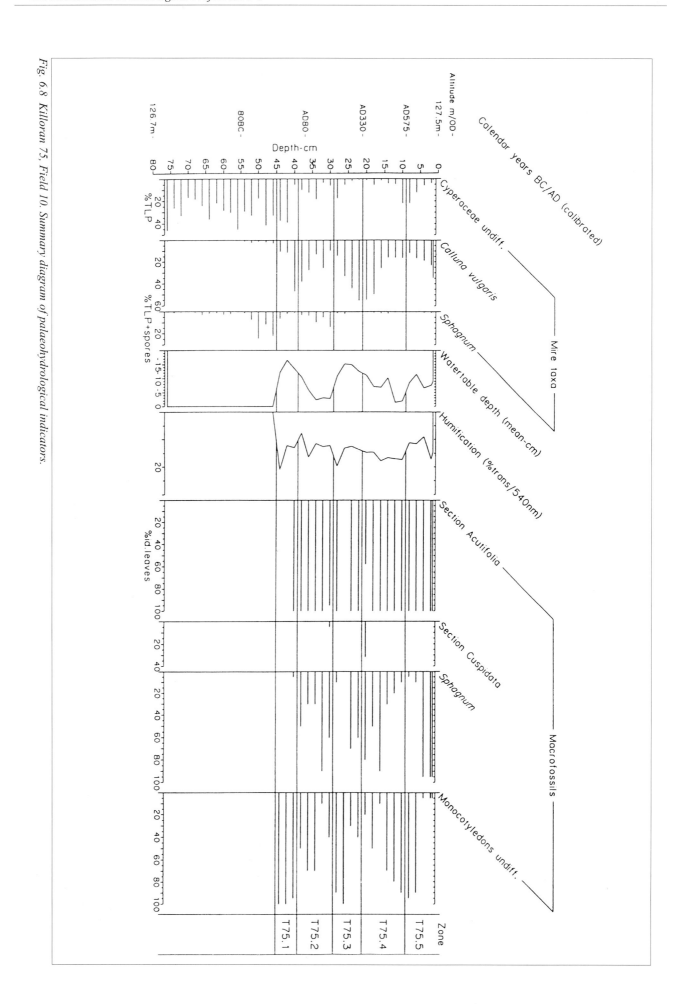

Fig. 6.8 Killoran 75, Field 10. Summary diagram of palaeohydrological indicators.

Trignopyxis arcula also displayed a low peak of 5% at 20cm. After this level, percentages of *H. subflava* were reduced, whilst *A. flavum* recovered to 40% by 16cm. *Arcella discoides* was present from 20cm, rising to a peak of 65% by 10cm. Over the same depth, the *Difflugia pulex*–type fell from over 20% to 5%, and both *Cyclopyxis arcelloides* and *Assulina muscorum* showed a low peak of 5% at 14cm. There was a rise in water table from -7cm depth at 22cm to -2cm depth by 10cm.

Percentages of *Arcella discoides* fell from 8cm to 20% at the close of the diagram, whilst *Difflugia pulex* rose from less than 5% to 20% across the same samples. *Hyalosphenia subflava* increased to over 55% by 6cm, before stabilising at 30% by the top of the sequence. The water table fell to -11cm by 6cm before stabilising at -6cm to -8cm for the final three samples.

The macrofossil samples for the Killoran 75 sequence consisted of monocotyledons and Acutifolia *Sphagnum*. Cuspidata *Sphagnum* was present in two samples, 5% at 30cm and 30% at 20cm. *Sphagnum* was the dominant peat former at 36–30cm, 24–16cm and 4–0cm, with monocotyledons significant at 44–40cm, 28–26cm and 14–6cm. *Sphagnum imbricatum* was not found in the macrofossil samples, probably due to the specific water chemistry in this part of the mire, where run-off from the uplands would have resulted in a less acid microenvironment.

Humification

There was no clear trend in the bottom 10cm of the sequence, with values fluctuating between 8% and 20%. The highest value for the diagram of 27% was recorded at 31cm, after which values fell before stabilising around 11–15%, with a gradual rise to 21% by 11cm. A decline in values to 9% at 4cm was then recorded. Values rose again to 17%, falling slightly to 15% by the close of the diagram.

Discussion

Integrating the palaeohydrological proxies: testate amoebae, humification and macrofossil records

This section will discuss the integration of the proxy records of mire hydrology: the water table data derived from the transfer functions (Woodland *et al.* 1998; Woodland 1996), the humification data and the macrofossil record. Additional information is also provided by the pollen record, especially that of taxa typical of the local bog environment, principally *Sphagnum*, *Calluna* and Cyperaceae. These data are included on a composite diagram for each sequence (see Figs 6.2, 6.4, 6.6 and 6.8), with a chronology derived from the correlation of radiocarbon dates and pollen stratigraphy used for the pollen diagrams. The diagrams are zoned on the basis of the movement of the water table and are thus not equivalent to the pollen diagram zones. The zones are also linked to the peat-stratigraphical data.

Killoran 18 (East), Field 41

150–87cm; 2270–1300 BC

The palaeoenvironmental record at Killoran 18 (East) begins just before 2270 BC, at which point the sediment is fen peat. The dominance of stem, root and leaf remains of monocotyledons is demonstrated by the macrofossil record, with *Eriophorum* and ericaceous remains rare. This environment would have been very wet, with a water content of perhaps 90% (Casparie, this volume), a contention that is supported by the analysis of Coleoptera from the fen peat (Reilly, this volume). Humification values for the bottom 40cm of the sequence tend to range from 10–20%, but it is difficult to establish whether the fluctuations in the curve are related to actual hydrological changes in the absence of testate amoebae or more detailed macrofossil data, and humification is not a particularly reliable indicator in fen peat. The microenvironment appears to become wetter after about 1550 BC, as humification values increase to 20–30%, and there is a rise in *Sphagnum* spore and Cyperaceae pollen percentages. *Calluna* pollen percentages are consistently recorded at 30–40%, suggesting there were drier areas quite near to the sampling site where heather could flourish.

T18-E1; 87–70cm; 1300–1000 BC

The transition from fen peat to raised bog growth began at this point for Killoran 18 (East), at *c.* 1400 BC. Casparie (this volume) has shown that the local environment was very wet around the time of causeway construction, with the monocotyledonous remains probably representing *Phragmites*, with some *Scheuchzeria* also recorded. The water table reconstruction begins *c.* 1300 BC, at which point the water table was some 10cm deep. The accumulation of highly humified peat seems to have continued under drier conditions, with the water table remaining at -13cm to -14cm.

There is little correspondence between the water table curve and the humification in this zone, with a peak in transmission of 25% at 79cm possibly suggesting a wet shift at a time when the testate record indicates that the water table was falling. However, the microenvironment seems to have been particularly dry between 1250 BC and 1100 BC: the water table falls to below 15cm and the disappearance of *Sphagnum* spores and reductions in Cyperaceae pollen are recorded, whilst *Calluna* pollen is present at up to 30% and ericaceous remains are found in the macrofossil samples. The evidence for alder seedling in the period following causeway construction further supports the contention of local dryness (Casparie, this volume). The absence of any significant increase in *Alnus* pollen can probably be explained by the fact that tree growth only lasted for a few years under marginal growth conditions, with saplings producing little pollen.

T18-E2; 70–65cm; 1000–800 BC

Shortly after 1000 BC, identifiable *Sphagnum* remains increase in the samples, including Section Acutifolia and *S. cuspidatum*. The *Sphagnum* macrofossils correspond with the layers of *S. cuspidatum* identified by Casparie (this volume) in this peat face. By just before 800 BC, the water table rose to a level of less than -1cm. The increase in the humification transmission curve to a peak of 25% also suggests increasingly anaerobic sedimentary conditions after *c.* 900 BC. Higher percentages of *Sphagnum* spores are recorded, up to a peak of 45% TLP +spores (TLP=Total Land Spores), with a less marked increase in Cyperaceae evident. The record of *Myriophyllum alterniflorum* pollen indicates the presence of areas of standing water nearby. The predicted high water table in this zone is therefore supported by the other proxy measures, and a period of increased surface wetness may confidently be inferred.

T18-E3; 65–36cm; 800–320 BC

There is strong evidence for a drier mire surface after 800 BC, with the water table apparently reaching its lowest point of below -15cm just before 600 BC. After this, there is an erratic upward movement of the water table and evidence from the other proxies for an increasingly damper local microenvironment.

Percentages of *Sphagnum* spores and Cyperaceae pollen show sustained falls at the opening of the zone, supporting the idea of reduced local peat moisture and the deep water table of -12cm to -13cm suggested by the testate amoebae. The growth of highly humified peat continued under wetter conditions after this, with *Sphagnum*, *S. imbricatum* in particular, significant in the macrofossil record after *c.* 600 BC, although the samples remained an accumulation of mixed *Sphagnum* and cyperaceous/undifferentiated monocotyledonous remains. Increases in *Sphagnum* spores at 49cm correspond closely to the appearance of *S. imbricatum*.

There is a fall in transmission values to 5–7% at the opening of the zone, paralleling the fall in the water table, and then an increase in transmission values to over 15% after 49cm, corresponding to the appearance of *S. imbricatum* and the upward movement of the water table from below -15cm to -11cm. The record of ericaceous rootlets in the macrofossil record in the lower half of the zone is probably referable to *Calluna vulgaris*, as percentages of *Calluna* pollen also peak at this point. Heather species were apparently growing locally, reinforcing the idea of a relatively dry bog surface between *c.* 800 BC and 600 BC. Ericaceous remains disappear from the macrofossil record shortly after the increase in *Sphagnum* in the pollen and macrofossil record. Casparie (this volume) identified hummock development in this period, with evidence of damp hollows until 600 BC.

T18-E4; 36–10cm; 320 BC–AD 100

By *c.* 320 BC reduced percentages of *Calluna* pollen and increases in *Sphagnum* spores and Cyperaceae pollen parallel the upward movement of the water table inferred from the testate record. Humification data for this period generally reflect a reduced concentration of humic and fulvic acids, consistent with a wetter bog surface, with values showing some fluctuation but increasing to 30%, some of the highest percentages recorded for the diagram. However, transmission values fall between *c.* 330 BC and 200 BC, a period when *Sphagnum* declines in importance and monocotyledons dominate the macrofossil record for a period of perhaps 100 calendar years. This may be interpreted as reflecting a brief dry shift that is not apparent in the water table reconstruction.

Both the water table and humification data indicate that the wettest period is between *c.* 200 BC and 100 BC, when *Sphagnum imbricatum* was the dominant peat-forming constituent locally, with some *Sphagna* attributable to Section Acutifolia also present. Another fall in transmission to below 10%, which may be interpreted as indicating a dry shift between *c.* 100 BC and AD 100, is not reflected in either the water table reconstruction or macrofossil data. Casparie (this volume) has identified layers of *Sphagnum cuspidatum* in the period *c.* 200–100 BC. These layers are shown to be discontinuous in the section drawings, perhaps representing ephemeral pools, which probably explains why this taxon is not well represented in the macrofossil record in the samples.

T18-E5; 10–0cm; AD 100–375

A fall in the water table and humification data at the opening of the zone suggests a drier episode, followed by a trend to wetter conditions and fluctuation to drier conditions by the close of the diagram. *Sphagna* remains are sparse, although some *Sphagnum imbricatum* leaves were identified in the samples. Cyperaceous and other monocotyledonous species seem to have been the major peat-forming constituents, with *Eriophorum* also present at the opening of the zone, when the water table curve falls to below -9cm. *Eriophorum vaginatum* tends to prefer drier conditions and usually grows on the top of hummocks. Hummock development in Field 41 at this time is supported by the stratigraphic data (Casparie, this volume).

Killoran 18 (West), Field 17

T18-W1; 138–134cm; 1310–1260 BC

The Killoran 18 (West) diagram opens around 1310 BC. The dominance of monocotyledonous remains in the macrofossil samples with the high water table of -1cm to -2cm suggest a wet, transitional environment with vegetation typical of the underlying fen peat. The pollen sampling intervals are quite coarse at this stage, but the peak for *Sphagnum* also supports the suggestion of a wet microenvironment, although *Calluna* is also present at nearly 20%, suggesting heather (ling) was present near to the sampling site.

T18-W2; 134–120cm; 1260–1070 BC

A sudden fall in the water table to -11cm marks the opening of this zone. There is some fluctuation to a higher water table of -6cm in the middle of the zone prior to a further fall at the close of the zone. The low levels of *Sphagnum* pollen and increase in *Calluna* support the idea of generally drier conditions.

T18-W3; 120–99cm; 1070–715 BC

The upward movement of the water table to -1cm to -2cm corresponds to the first appearance of *Sphagnum*, with *S. cuspidatum* and Cuspidata *Sphagnum* particularly well represented in the middle of the zone. There is some evidence for slightly drier conditions in the upper half of the zone. The mean water table depth falls slightly to between -2cm and -3cm, and Acutifolia *Sphagnum* replaces *S. cuspidatum* as the dominant bryophyte. This interpretation is also supported by the pollen record: *Calluna* pollen is reduced with the higher water table at the opening of the zone, Cyperaceae pollen is recorded in its first significant quantities, whilst *Sphagnum* peaks at 40% TLP+spores at 112cm. With the increase in Acutifolia, this situation is reversed, with *Calluna* increasing to 25–30% and *Sphagnum* decreasing to 10%.

T18-W4; 99–89cm; 715–600 BC

The opening of this zone is defined by a fall in the mean water table to -10cm. Again, this is reflected in the pollen record, with a reduction in *Sphagnum* to trace levels at the same point and a peak in *Calluna* pollen of 40% at 96cm. This is also indicated in the macrofossil record, with an increasing dominance of monocotyledons at the expense of *Sphagna*. The record of *Eriophorum* in the macrofossil samples at 98cm also supports the contention of a lower water table. The upward movement of the water table to -2cm in the upper section of the zone is indicated by the increased representation of *Sphagnum* in both the pollen and macrofossil records.

T18-W5; 89–77cm; 600–380 BC

The recovery in the water table at the close of the last zone is followed by a further reduction in mean water table depth at the opening of this zone to the lowest value recorded for the sequence: -14cm at 84cm. *Sphagnum* remains are recorded in reduced quantities at this point, and whilst there is a gap in the pollen record in the middle of the zone, the increase in *Calluna* and low *Sphagnum* percentages further demonstrate these changes. The peak in *Sphagnum* spores at the close of the zone corresponds to the appearance of *Sphagnum imbricatum* in the macrofossil record.

T18-W6; 77–0cm; 380 BC–AD 720

The water table moves upwards at 77cm and remains generally at -1cm to -3cm for the remainder of the diagram. The sediment matrix is almost pure *Sphagnum* for this zone, with the dominance of *S. imbricatum* from 77cm to 65cm marked by a peak in *Sphagnum* spores of 80% at 70cm. Cyperaceae pollen reaches its highest value for the diagram of nearly 25% at 77cm. This pollen taxon remains at generally fairly low percentages for the remainder of the diagram, which alongside the absence of monocotyledons in the macrofossil record, reflects the lack of sedges at the sampling site.

There is evidence in all the proxies for some fluctuation toward slightly drier conditions between 77cm and 55cm, with brief falls in mean water table corresponding to increases in Acutifolia in the macrofossil record and a fall in *Sphagnum* spore percentages and pockets of monocotyledon-dominated sediment between 62cm and 64cm. *Sphagnum cuspidatum* and Section Cuspidata support the contention of periods of very wet microenvironment.

The water table remains high at -1cm to -2cm from 50cm to 24cm. *Calluna* is reduced to less than 5% and *Sphagnum* spore percentages increase to reach over 45% between 38cm and 24cm. A short but distinct phase of drier conditions is evident around 20–16cm: there is a dip in the mean water table to -6cm, which is paralleled by an increase in *Calluna* pollen to 15% and a reduction in *Sphagnum* to just over 5%. The end of this drier episode is indicated by an upward movement of the water table to between -1cm and -2cm, a recovery of *Sphagnum*, a peak in Cyperaceae pollen and a decline in *Calluna*. *S. cuspidatum* and Section Cuspidata are also recorded.

There is some evidence for slightly drier conditions towards the close of the diagram, with dips in the mean water table from -1cm to -4cm occurring alongside increased representation of Section Acutifolia. Ericaceous remains are also present in the macrofossil samples from this part of the sequence, although there is no clear response from the *Calluna* pollen curve to indicate that there was anything more than sparse heather growth in the vicinity.

Derryfadda 23 (Above) Field 41

T23-1; 94–82cm; 1570–1400 BC

Highly humified peat growth began around 1800 BC at Derryfadda 23, reflecting the transition to ombrotrophic conditions in this part of the mire. The Derryfadda 23 diagram opens just after 1500 BC, the sediment consisting of monocotyledons with sparse *Sphagnum* remains, which is consistent with low *Sphagnum* spore percentages. The early stage of raised bog vegetation was apparently very similar botanically to the underlying fen peat (cf. Casparie, this volume).

There is evidence from the proxies for fluctuations in the bog water table at this time. Percentage transmission values range between 14% and 23% and show some agreement with the movement of the water table derived from the testate assemblages. Casparie (this volume) interprets the presence of extensive remains of *Eriophorum* in the peat sequence near Derryfadda 23 (Fields 40–41) as indicative of relatively large fluctuations in the bog water table between 2200 BC and 1000 BC.

T23-2; 82–55 cm; 1400–950 BC

The low percentages of Cyperaceae pollen, *Sphagnum* spores and identifiable *Sphagnum* remains support the contention of a deep local water table between 13cm and 15cm below the surface. The dominant peat-forming vegetation remained monocotyledonous taxa. There is a steady fall in transmission values across this phase to a low point of 15% just after *c*. 1200 BC, supporting the suggestion of a relatively dry local environment and the accumulation of highly humified peat. Increases in recorded transmission values after this are not supported by the other proxies. The water table apparently reached its deepest point of below -16cm just after 1250 BC, at which point *Sphagnum* spore percentages fall to very low values. The response from the humification determinations are not as marked. There is some increase in local wetness after 1100 BC, with a slight upward movement of the water table and humification curve, accompanied by a recovery in *Sphagnum* spore percentages.

T23-3; 55–40cm; 950–700 BC

Conditions become significantly wetter after 950 BC. The increases in transmission values parallel the upward movement of the water table, with a peak in transmission of 25% at *c*. 930 BC, corresponding to a high point for the water table of less than 0.5cm below the surface. *Sphagnum* becomes a more significant component of both the macrofossil and microfossil record from the opening of this zone, demonstrating the increase in peat moisture, as does the increase in Cyperaceae pollen. Cuspidata and Acutifolia *Sphagnum* are both present in the macrofossil record.

These changes correspond to a transition from highly humified *Sphagnum* to poorly humified *Sphagnum*, identified as *S. cuspidatum* layers by Casparie (this volume), who suggests that these deposits reflect the development of shallow pools in wet hollows. The water table data certainly support the possibility of standing water, although the presence of Acutifolia indicates some tendency to drier conditions. The decrease in *Eriophorum vaginatum* in the Derryfadda 23 peat face after 950 BC also indicates the higher and less fluctuating water table portrayed by the palaeohydrological proxies for this zone.

T23-4; 40–26cm; 700–500 BC

Testate concentrations are very low in this zone, so the water table reconstruction should be approached with some caution. *Sphagnum* disappears from the macrofossil record at the opening of the zone, and there is a fall in both *Sphagnum* spores and Cyperaceae pollen. *Sphagnum* macrofossils reappear at 38cm, where the majority of the identified leaves are of Section Acutifolia, although pockets of Section Cuspidata are present at 28cm and 32cm. Transmission values fall steadily across the zone to reach their lowest recorded value at 30cm.

Taken collectively, the data suggest a period of a low water table at the opening of the zone around 700–600 BC and again towards the top of the zone around 500 BC, although the continuing record of *Sphagnum* macrofossils indicates that there cannot have been too severe a desiccation of the bog surface. The low levels of testate amoebae can probably be attributed to the dry microenvironment prevailing locally.

The pockets of Cuspidata *Sphagnum* and testate amoebae data indicate that the bog surface was wetter for a brief period in the middle of the zone, around 550 BC. The evidence for periods of increased surface wetness demonstrated by the macrofossil and water table data in this and the previous zone may be correlated with the lenses of highly humified *S. cuspidatum* peat identified by Casparie (this volume) in various profiles and dated by him to sometime between 1200 BC and 500 BC.

T23-5; 26–15cm; 500–300 BC

Testate counts in the lower half of the zone are low, so once again the results should be interpreted cautiously. The *Sphagna* macrofossil record remains dominated by Section Acutifolia, although *Sphagnum* spore percentages show a sustained increase from the opening of the zone. The fall in Cyperaceae pollen percentages corresponds to the increasing dominance of *Sphagnum* as a local peat-forming constituent relative to sedges, with *S. imbricatum* macrofossils recorded for the first time in the diagram at *c*. 380 BC. There is an upturn in the percentage transmission curve to 25% at the opening of the zone, reflecting the wetter conditions that accompanied the establishment of *Sphagna*. The transfer functions predict that the water table moves closer to the surface, with mean values of -1cm to -3cm. The data thus all support an increasingly wetter local environment in this period.

T23-6; 15–0cm; 300–70 BC

The water table fluctuates quite widely in this zone between -4cm and -13cm. This is paralleled to a certain extent by the behaviour of the *Sphagnum* spore curve, which oscillates between 15% and 40%. There is some agreement with the humification curve, which increases to over 30% around 200 BC, just prior to an upward movement of the water table. The presence of both *Sphagnum* and monocotyledonous remains in the samples and the dominance of *S. imbricatum* in the *Sphagna* macrofossils, but with Section Acutifolia and occasional Section Cuspidata present, supports the contention of an unstable water table during this period. Stratigraphical evidence for short-term fluctuations in the humidity of the bog surface at this point in the Derryfadda 23 section in the form of thin layers of moderately humified peat is also highlighted by Casparie (this volume).

Killoran 75, Field 10

T75; 88–45cm; 500–45 BC

Killoran 75 is situated on the western edge of the bog in an area identified as a discharge channel that developed from *c.* 500 BC (Casparie, this volume). The peat stratigraphy in this sequence is characterised by the presence of a thick lens of *Menyanthes* and *Scheuchzeria* peat, indicating an area of open water. Tests are not present in this sediment or in the underlying fen peat, but very wet/pool conditions are evident from the stratigraphy and also the pollen data. The water table reconstruction covers the top 44cm of the sequence only.

T75-1; 44–35cm; 45 BC–AD 120

The palaeohydrological data indicate that the very wet/open water conditions present on the site at the time of trackway construction are followed by a phase of drier surface conditions. There is a marked rise in the *Calluna* curve to values of over 40% TLP: Evans and Moore (1985) have studied modern surface pollen spectra and suggested that values for *Calluna* of over 20% total land pollen indicate the local presence of heather. The contention of a relatively dry bog surface is supported by the testate record, which begins at 44cm (*c.* 25 BC). At this point, the transfer functions show that the water table was around -12cm. This is clearly a fairly brief dry interval, as the water table rises steadily across this zone to -6cm. Local sedimentary conditions were evidently becoming more anaerobic, with *Sphagnum* increasingly significant in both the macrofossil and microfossil records. The percentages of Cyperaceae pollen suggest that sedges were also becoming better established locally. Percentage transmission figures show a degree of fluctuation between 10% and 20% and do not show the clear trend to an increasingly wet local environment suggested by the other proxy measures.

T75-3; 35–29cm; AD 120–215

The upward movement of the water table culminates in a phase of very wet local environment, with mean water table depths of only -2cm to -3cm below the surface. The dominant peat-forming vegetation is *Sphagnum*, mainly of Section Acutifolia, although a small percentage of Section Cuspidata at 30cm reflects the very wet local conditions. This depth sees *Calluna* pollen fall to a low percentage of 10%—heather was clearly restricted in its local distribution. These changes are not closely paralleled by the humification curve, with transmission values remaining similar to those recorded for the previous zone.

T75-4; 29–19cm; AD 215–380

The phase of high water table ends shortly after AD 215, after which the average water table depth falls to between -10cm and -15cm. Once again, this is reflected in the recovery in percentages of *Calluna* pollen.

Monocotyledons become more significant as a local peat-forming constituent, although Acutifolia *Sphagnum* remains at around 50% of identifiable vegetative remains. The only apparent inconsistency is the record of Cuspidata *Sphagnum* at 20cm, where *Calluna* pollen has fallen slightly from a high point but is still present in high percentages and the water table reconstruction indicates a water table depth of at least -11cm. This indicates that there may have been some brief fluctuation in the water table towards wetter conditions that is not reflected in the other proxy measures. A slight fall in humification values is recorded at the opening of the zone, but the magnitude is not as significant as what might be expected given the reduction in water table depth.

T75-5; 19–9cm; AD 380–570

The water table begins to rise after 22cm, again reflected in falling *Calluna* pollen percentages and a slight increase in *Sphagnum* spores, although *Sphagna* are apparently becoming a less significant constituent of the peat-forming vegetation, with cyperaceous elements dominating both the microfossils and macrofossils by the period of highest water table between AD 520 and AD 555. The water table depth during this latter phase was less than 2cm below the surface. Percentage transmission reaches a peak of 20% at this point, indicating increasing concentrations of humic and fulvic acids, consistent with a wetter bog surface.

T75-6; 9–0cm; AD 570–730

Another fall in water table level to a low point of -16cm is followed by steady falls in Cyperaceae and a recovery in *Calluna* pollen. The water table reconstruction indicates a slight rise in the water table, concurrent with a renewed growth of Acutifolia *Sphagnum* toward the top of the diagram. The humification curve shows some agreement with the other proxy measures, falling from 17% to 11% at the opening of the zone, prior to increasing to 15% by the close of the diagram, and the slightly higher water table evident at the close of the record.

Integrating the proxy measures: some general comments

Interpretational and methodological problems

All the proxy measures have particular methodological and/or interpretational problems connected with them. The degree of change in the main controlling variable necessary (i.e. the water table) to produce a signal in each of the palaeorecords, or what may be termed the threshold of response, may not be the same for each proxy. In the case of humification determinations, the extent to which the signal is controlled by variables other than moisture has not been convincingly explained or quantified. It also remains unclear as to how much this signal varies with, for example, the species composition of the peat. A change in local temperature can also have the same effect on humification as a rise in groundwater, potentially complicating the interpretation of the record. The fact that humification is measured on a nonlinear scale means that interpretation of the record often has an essentially subjective element.

The water table reconstruction is produced using a transfer function derived from the analysis of modern testate populations and their relation to hydrological conditions and other environmental variables. There is thus a strong uniformitarian assumption behind this approach. However, although there is no reason to assume that testate ecology has significantly changed over the Holocene, the reconstructions are dependent on the presence of modern analogues for the species assemblages identified in fossil samples. Some species are found less frequently in contemporary situations than others. *Hyalosphenia subflava*, a common taxon at Derryville and typical of dry conditions, is comparatively rare in modern samples, mainly due to the fact that monitoring has not been carried out on the drier peatlands where this species is most typically found (D. Charman, pers. comm.). This means that in statistical terms, the samples that are dominated by this species have a large standard error on the water table reconstruction, which should be borne in mind in interpretation. Some fossil species assemblages may therefore not have exact contemporary analogues.

The macrofossil samples are those from which the testates were extracted and are thus comparatively small (1cm³). This is effectively a 'spot' sample, and it should be remembered that the assessment of the composition of the peat is based on a smaller sample than the 10cm³ employed in other macrofossil studies (Barber 1981). There are other problems connected with the use of macrofossils as palaeohydrological indicators. Differential decay of species may lead to a false impression of moisture changes. It has been established, for example, that hummock and hollow *Sphagnum* species can decay at different rates (Johnson and Damman 1991; Belyea 1996). Changes in the habitat preferences of species can also affect the interpretation of the palaeoecological record: *Sphagnum imbricatum*, for example, is presently very restricted in its distribution in the mires of Europe. Palaeoecological studies have demonstrated that this species was previously much more common, suggesting that it is currently occupying only part of its former ecological niche and has been out-competed in certain habitats, probably by *Sphagnum magellanicum* (Stoneman *et al.* 1993). The application of the present ecology of this species to the interpretation of the palaeorecord is thus significantly curtailed. The sensitivity of the mire system itself to the effects of climate change may also be significant.

It is by combining the different strands of evidence that a coherent picture may emerge. The greater the degree of correspondence between the different proxy measures, the greater the probability that the inferred palaeohydrological change in question was an actual event at the sampling site. Likewise, changes that are recorded at more than one site at the same time may be accepted with a greater degree of confidence as mire-wide phenomena, as opposed to localised water table fluctuations. It is clear that reliance on such a 'multi-variable' approach is the best way to proceed in studies of records of palaeomoisture derived from the peat bog archive (cf. Charman 1997, 480; Chambers 1993, 254).

Using the three data sets, it is therefore possible to identify synchroneity or diachroneity in the moisture record. Based on the above assumptions, the following periods of the record from Derryville are suggested as being synchronous across the bog (Fig. 6.9).

1400–1000 BC

Raised bog growth in the vicinity of Derryfadda 23 began *c.* 1800 BC, some 400 years prior to 400m further north at Killoran 18. The period of the later second millennium BC to the start of the first millennium BC appears to have been one of a relatively low water table at Derryville, although there is some fluctuation towards brief periods of higher water table at Derryfadda 23. Water table reconstructions at both Killoran 18 East and Derryfadda 23 indicate that the mean water table was deeper than -10cm, although the dominance of *Hyalosphenia subflava* in the samples means that this may be an over- or underestimate. The macrofossil record is dominated by sedges and other monocotyledonous remains. *Sphagnum* is very sparse in the samples, possibly due in part to the differential decay of *Sphagna* due to the relatively dry conditions prevailing locally, although low percentages of *Sphagnum* spores in the pollen diagrams suggest that *Sphagna* were not major peat-forming taxa at this time.

The sequence at Killoran 18 West begins just before 1300 BC, at which point the western side of the bog was evidently much wetter than the eastern side, with a water table -1cm to -3cm deep, compared with below -10cm at Killoran 18 East and Derryfadda 23. There is a sudden, marked fall in water table in 1250 BC from -1cm to over -10cm. This clearly represents the effect of bog burst B, identified in the Cooleeny area by Casparie (this volume). The effects of this catastrophic event are also evident at Killoran 18 East, where the water table falls to -15cm by 1100 BC and at Derryfadda 23, where it reaches -16cm at the same date.

1000–800 BC

The water table reconstructions demonstrate that it took the mire over 300 years to recover from the effect of the 1250 BC bog burst. Water table reconstructions show rising water tables at all three sites from *c.* 1000 BC, with mean depths of 0cm to -2cm attained by 900 BC. The macrofossil data show that this period corresponds to the

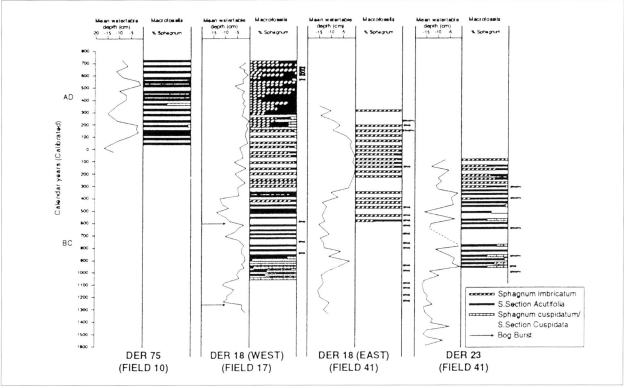

Fig. 6.9 Moisture record for Derryville Bog.

first significant growth of *Sphagnum* at the sampling sites. The groups present include both Cuspidata and Acutifolia *Sphagnum* and *S. cuspidatum* at Killoran 18 West. Cuspidata and *S. cuspidatum* are typical of very wet conditions, with a mean height of moss surface above the water table of 0–1cm for the latter species (van der Molen and Hoekstra 1988; Ivanov 1981). Hammond *et al.* (1990) have reported a water table range of 0–8cm for this species from raised bog in County Kildare, with 90% of occurrences around 5cm depth. Section Acutifolia probably represents *Sphagnum fuscum*, which has a mean level of 37–25cm (Ivanov 1981) or *Sphagnum rubellum*, which has a mean level of 25cm above the water table (van der Molen and Hoekstra 1988). The record of both types of bryophytes suggests some fluctuation in water levels during the intervals represented by the sample depths. This phase corresponds to the poorly humified *S. cuspidatum* layers identified in Fields 40 and 41, probably representing flooding episodes (Casparie, this volume).

800–400 BC

The wet phase ends shortly before 800 BC at both Killoran 18 East and Derryfadda 23 (signalling bog burst C), and the subsequent period is one of generally lowered water table at both these sites, with mean values of below -10cm at Killoran 18 East and a gap in the water table record at Derryfadda 23 from 850 BC to just before 600 BC. These changes correspond to the disappearance of *Sphagnum* from the macrofossil record in both cases, with cyperaceous and undifferentiated

monocotyledons typical peat formers. There is evidence that there was some fluctuation toward a higher water table at Derryfadda 23 in particular, with lenses of Cuspidata peat corresponding to brief periods of raised water table, although this is not reflected in the humification data. The situation at Killoran 18 West is somewhat different. There is a slight fall in the water table after 850 BC, but the major drop in mean water table from -2.5cm to -10cm is dated to 700 BC and is clearly not synchronous on comparison of the dryland pollen record between sites. This is followed by a recovery prior to another marked fall in the water table from -2cm *c.* 600 BC (bog burst D) to -14cm by 500 BC, the deepest mean water table recorded in the Killoran 18 West sequence.

Although there is a slight fall in mean water table around 820 BC at Killoran 18 West, the most significant reduction does not occur until 700 BC. This is difficult to reconcile with the fact that the bog burst occurred on the western side of the bog and so would be expected to be clearly represented in the water table reconstruction. The macrofossil record indicates the disappearance of *Sphagnum cuspidatum* and its replacement with Acutifolia *Sphagnum* around 800 BC. It is possible that the water table reconstruction is inaccurate at this stage, whilst the replacement of hydrophilous bryophyte species with a macrofossil assemblage typical of drier conditions is demonstrating the effects of the bog burst of *c.* 800 BC. The accuracy of the water table reconstruction elsewhere in comparison with the stratigraphical data may be seen to mitigate against this interpretation (see below). If this is accepted, then either the variations in local hydrological conditions at Killoran 18 West were such that the effect of the bog burst was not registered until some 100 years after the burst, or there is a problem with the chronology at this point. The latter is considered highly unlikely on palynological grounds.

Casparie (this volume) identifies another bog burst (bog burst D) in the southern discharge system around 600 BC, the event clearly reflected in the water table at Killoran 18 West, where the mean water table falls after this date to reach its lowest recorded point for the diagram by *c*. 500 BC. Although conditions were already drier at Killoran 18 East prior to the bog burst and although there are large errors in the water table reconstruction due to the high percentages of *Hyalosphenia subflava*, there is a fall in the mean water table from around -11cm at *c*. 650 BC to below -13cm by 600 BC. There is a gap in the testate record at Derryfadda 23 from 850 BC to just before 600 BC, and test counts tend to be low, so the events in this part of the sequence are hard to establish clearly. The presence of Section Cuspidata in the macrofossil record and the raised mean water table around 550 BC suggest that the effects of the burst may not have been as marked at this location, a contention also advanced by Casparie (this volume) on the basis of peat stratigraphical data.

400 BC–0 BC/AD

The records diverge somewhat after this point, but generally show increasingly wet conditions. The dry phase at Derryfadda 23 ends at around 500 BC, whilst the predominating peat moisture conditions do not become substantially wetter until nearer 300 BC at Killoran 18 East, although there is an erratic upward movement of the water table after 600 BC. The former shift corresponds to the appearance of the large-leafed *Sphagnum imbricatum* at *c*. 400 BC and the transition from highly to moderately humified peat (Casparie, this volume). *S. imbricatum* also appears at 400 BC at Killoran 18 West and, as at Killoran 18 East, marks an upward movement of the water table. At Killoran 18 East, the appearance of *S. imbricatum* occurs earlier (*c*. 600 BC) and marks the more erratic raising of the water table mentioned above. At Killoran 18 West, the local environment becomes wetter after 400 BC, with conditions remaining generally wetter, with water table depths of -2cm to -3cm recorded. The wetter conditions, once established, persist until the end of the millennium at Killoran 18 East, whilst at Derryfadda 23, there is a fall in the water table just before 300 BC, with some fluctuation of water table depth until the close of the diagram just after 100 BC. A similar pattern is recorded at Killoran 18 West in the period 300–100 BC. In summary, the period from 500 BC until the end of the millennium should be seen as one of increasingly wet conditions at Derryville, although the

stability apparent at Killoran 18 East and the fluctuating water tables at Killoran 18 West and Derryfadda 23 suggest some intrasite variation.

Later changes: the first millennium AD at Derryville

The palaeoenvironmental record at Derryfadda 23 ends in the first century BC. The period into the first millennium AD is therefore covered by Killoran 18 East, Killoran 18 West and Killoran 75. The water table, humification and macrofossil data for Killoran 75 cover the period from the first century AD, although the pollen record extends further back than this.

The wet phase represented by the facies of *Menyanthes* at Killoran 75 extends back as far as at least *c*. 500 BC, supporting the other records suggesting increasingly wet conditions at Derryville from *c*. 400 BC, albeit with some fluctuation toward drier conditions, as demonstrated by the drop in the water table curve at the opening of the Killoran 75 diagram. This may also be reflected in the water table data at both the Killoran 18 sites, where there is a fall in mean depths at the turn of the millennium. This is more pronounced at Killoran 18 West, which may indicate that the western side of the mire was more affected by whatever factors caused the drier conditions. The phase of flooding that led to the deposition of the silt layer on the western edge of the bog, now well dated to between *c*. 200 BC and 120 BC, also coincides with this wetter period.

The proxy measures demonstrate that conditions became wetter at all three sites by the start of the first millennium AD. The water table curves show that the period of the later first century AD was one of very wet conditions at Killoran 75, Killoran 18 East and Killoran 18 West, with mean water table depths of -1cm to -2cm. There is some divergence after AD 200, with Killoran 18 East indicating some variation from the two sites in the western half of the mire.

There is a fall at Killoran 18 East from less than –2cm in AD 100 to a depth of nearly -10cm by AD 160. There is another fall in mean water table after AD 240 from -4cm to -15cm by the close of the diagram at *c*. AD 360. There is also a fall in the water table at Killoran 75, with a lower water table recorded between AD 240 and AD 320, after which the water table begins to move closer toward the surface again. This corresponds to a less marked phase of lowered mean water table at Killoran 18 West, from a depth of above -1cm to a depth of -3cm. These drier shifts in all three sequences around AD 200–250 may represent the hydrological effects of the bog burst identified by Casparie (this volume) in the northern zone of the northern discharge system, dated to AD 100–300.

The drier phase at Killoran 75 is followed by a high mean water table of -1cm to -2cm between *c*. AD 520 and AD 550 and a final reduction in water table depth to between -6cm and -11cm for the remainder of the diagram. A similar pattern is again evident in Killoran 18 West, where there is little change in local palaeohydrology until a fall in mean water table from -2cm to nearly -7cm in AD 500, followed by a generally high water table of -1cm to -2cm from AD 550 until the close of the Killoran 18 West diagram just after AD 700. The impression, therefore, in that part of the first millennium AD covered by Killoran 18 West and Killoran 75 is that the water tables at the two sites often follow a similar direction but not magnitude of change, with the Killoran 18 West site showing less severe fluctuations. This may be due to the fact that Killoran 75 is close to the western edge of the mire and therefore more susceptible both to changes in run-off from the surrounding dryland and the effects of reductions in the direct atmospheric supply of water.

Summary of palaeohydrological changes: autogenic changes and the identification of climatic change

Barber (1981), Barber *et al.* (1994), Blackford (1993), Charman *et al.* (2001), Dupont (1986) and Hendon *et al.* (2001), among others, have all suggested that the palaeoecological record from raised bogs and ombrotrophic peats, particularly that derived from macrofossils, may contain a '... strong climatic signal' (Stoneman *et al.* 1993, 20). Records for the late Holocene now show a number of periods of suggested 'wet shifts,' such as around 2000 to 1900 BC, a more pronounced decline at 1500 to 1400 BC prior to a sudden deterioration to cooler and/or wetter conditions sometime around 900 to 600 BC and another shift around 100 BC. Recently, a significant climatic deterioration in AD 550 has also been proposed on the basis of humification data (Blackford and Chambers 1991). These sequences do, however, vary from site to site and are not that well dated in some profiles due to the problems of radiocarbon dating.

The palaeohydrological changes at Derryville are summarised in Fig. 6.9. Significant dates in later Holocene history can be identified as 1250 BC, 1000 BC, 800 BC, 600 BC, 400 BC, AD 250 and, perhaps, AD 550. The first, second, fourth and sixth dates refer to dry phases caused by bog bursts, whilst the second and fifth dates mark wet phases following the recovery of the bog ecosystem from the effects of the bursts. The 800 BC date refers to the dry phase recorded in the eastern half of the mire. AD 550 may mark a mire-wide wet phase.

Casparie (this volume) points out that the water supply was never the limiting factor to bog development at Derryville. The bog bursts demonstrate that there was a surplus of water from the earliest phases of ombrotrophic peat development, whilst the wet phases from 1000 to

800 BC and 400 BC into the first millennium AD can be essentially attributed to the recovery of the bog ecosystem from the earlier catastrophic loss of water. Other factors that may explain localised fluctuations include changes at sampling sites with relation to bog surface topography, such as the development of hummock and hollow systems, pools and discharge channels and erosion gullies (*ibid.*). There may be a climatic influence in bog hydrology, but this is hard to disentangle from the effects of autogenic changes and adjustments. Derryville Bog is clearly not a 'sensitive' enough mire system for palaeoclimatic reconstruction in the terms of Barber (1994).

There is evidence for a period of drier conditions from perhaps as early as AD 200–500, followed by a shift to wetter conditions around AD 550. This is most apparent at Killoran 75 as a fall in water table and rise in *Calluna* pollen in AD 200, followed by an upward movement of the water table and a peak in Cyperaceae pollen at the expense of *Calluna* pollen around AD 500. At Killoran 18 West, there is a slight reduction in water table from AD 250, with a more marked drop just before AD 500, accompanied by a low peak in *Calluna* and a fall in *Sphagnum*. Wetter conditions are evident after AD 550, with a rise in the water table synchronous with a peak in Cyperaceae and the reappearance of *Sphagnum cuspidatum* in the macrofossil record.

Climatic data are rather thin for this period, although Lamb (1977; 1982) has highlighted contemporary sources that can be interpreted as indicating that a period of warm and dry conditions in Britain ended around AD 400–500 and was followed by marked climatic deterioration in the mid-first millennium. Closer to the study area, Baillie and Munro (1988) point out that the period of narrowest tree rings in the range 13 BC to AD 894 occurs around AD 541. Blackford and Chambers integrate these lines of evidence with peat humification analyses of sequences from blanket mires in the British Isles to hypothesise a climatic deterioration around AD 550, possibly related to volcanic activity, resulting in '... short-lived climatic cooling or more prolonged climatic wetness ...' (1991, 66). The problems with using the Derryville sequences for palaeoclimatic study have been discussed above; it is outside the scope of the present study to draw any conclusion beyond the fact that there was apparently a period of wetter conditions at Derryville in the mid-first millennium AD.

Proxy measures of palaeohydrology: some comparisons

This is the first study carried out in Ireland to utilise multiprofile, close-sampled testate amoebae analysis using recently developed transfer function estimates of water table change, alongside detailed stratigraphic, macrofossil, pollen and humification analyses. The close agreement between the palaeohydrological data and the stratigraphical interpretation of the Derryville deposits has demonstrated the potential for this approach and affirmed the potential of testate amoebae analysis as a powerful palaeohydrological proxy.

Testate assemblages have been criticised as not always reflecting changes in palaeohydrology established by macrofossil analyses, with humification preferred as a measure of moisture changes. However, this has not been found to be a problem in this study. If anything, humification determinations seem to be the more insensitive proxy. An example of this is in the Derryfadda 23 sequence, where a period of flooding represented by a lens of *Sphagnum cuspidatum* peat at 54cm is clearly reflected in the water table reconstruction but not in the humification curve. The catastrophic nature of the bog bursts are clearly reflected in the water table data, but whilst there is some fluctuation in the humification curve at the corresponding levels, this could easily have been rejected as 'noise' in the absence of other proxies. Connected with this is the fact that the lack of any sort of calibration of humification measurements means the interpretation is largely subjective and restricted to gross changes in the curve, with possible small-scale variations indiscernible from 'noise' in the system and intersite comparisons effectively limited to 'threshold' changes only. The use of transfer functions derived from macrofossil analyses remain hampered by the difficulty in establishing the palaeoecological niche of certain *Sphagnum* species (see below).

Possible problems with the lack of modern analogues for 'drier' testate amoebae assemblages (see above) were not encountered. Even in phases where the local peat environment was relatively dry, the sensitivity of the transfer functions to further reductions in moisture such as those resulting from bog bursts suggests that the technique is not restricted by problems of lack of contemporary analogues or cases where assemblages are dominated by low species diversity. Interpretation is strengthened where changes in the water table are paralleled by the behaviour of taxa such as *Sphagnum* and *Calluna* in the pollen record. The lack of testates in fen peat means it is not possible to employ such analyses in these situations, and their absence from very dry situations means restrictions at both ends of the moisture spectrum.

Where moisture changes can be inferred from macrofossil analyses, they generally agree closely with the testate amoebae, although there is often the problem that where *Sphagnum* is not present in the sediment, the interpretation is severely hampered. The use of larger samples and more detailed identification of monocotyledonous remains could partly remedy this situation.

Fluctuations in the curves of Cyperaceae pollen and *Sphagnum* spores generally parallel the changes recorded by the macrofossil and water table data. It has been suggested that *Sphagnum* percentages are closely related to the *Sphagnum* content of the peat (Tallis 1964), with spore production related to the wetness of the bog surface (Conway 1954). The correlation of percentages of *Sphagnum* spores with estimated *Sphagnum* content of the macrofossils suggests that this assumption may hold for the Derryville sequences. (Sample correlation coefficients for *Sphagnum* spore percentages–*Sphagnum* macrofossil sample percentages: Derryfadda 23, r = 0.65; Killoran 18 East, r = 0.62; Killoran 18 West, r = 0.6.) The exception to this is the Killoran 75 diagram, where there is no linear correlation between these two measures. *Sphagnum* percentages are very low in this sequence, rarely reaching much more than 2% TLP+spores, even when the *Sphagnum* content of the macrofossil samples reaches 80%. This may be related to depressed spore productivity of the dominant *Sphagnum* section, Acutifolia, as different *Sphagnum* species have different levels of spore production (Ivanov 1981; Boatman 1983). Low spore percentages are not generally observed in the other diagrams though when Section Acutifolia attains a high percentage of identified *Sphagna* remains, making this hypothesis difficult to sustain. Establishing the relative importance of a species-dependent signal compared with the increase in spore production that would be expected to result from a raised water table is problematic. Statistical analyses of the pollen, water table and macrofossil data may clarify this further. Generally, the Derryville data suggest that *Sphagnum* and Cyperaceae percentages may be cautiously used as rough estimates of mire surface wetness in the absence of other palaeohydrological proxies, as has been employed elsewhere (e.g. Tipping 1995).

Sphagnum imbricatum: palaeoecological inferences

Sphagnum imbricatum is the only large-leaved *Sphagna* found in the Derryville sequences and, as was observed above, there are inferential problems connected with the use of this species as a palaeohydrological indicator. Some palaeoecological studies (summarised in Stoneman *et al.* 1993) have suggested that, in the past, *S. imbricatum* occupied a wider range with respect to the water table than its present ecology indicates, with evidence for the species' presence in both wet lawn and pool and drier hummock environments.

In the Derryville sequences, where this species occurs on its own, mean water table depths range generally between less than -1cm to -6cm, suggesting a fairly wet habitat. In these phases, humification values also tend to be at the higher points of the transmission curve, suggesting comparatively damp microenvironment conditions during peat accumulation. There is some difference between sites where this species occurs alongside other *Sphagna*. At Killoran 18 West, it tends to be found most frequently with Acutifolia *Sphagnum*, although it is also recorded with *S. cuspidatum* and Cuspidata *Sphagnum*. Where it is found with the former, mean water table depths are generally fairly shallow, with depths of -1cm to -4cm. Associations with *S. cuspidatum* are always in very wet situations, with mean depths of -1cm to -2cm. At Killoran 18 East and Derryfadda 23, where its leaves are found in samples with higher levels of monocotyledonous remains, mean water table depths tend to be as deep as -10cm to -14cm. Taken collectively, these data suggest a wide habitat range for *S. imbricatum* with respect to the water table at Derryville. Where it is found in samples that represent pure stands, a moderately high mean water table is suggested, indicating a lawn or hummock base habitat. Its appearance with the hydrophilous *S. cuspidatum* also suggests its presence in wet hollow or pool habitats. Elsewhere in Ireland, a similar conclusion was reached by van der Molen and Hoekstra (1988), who found this species to be present in both hummock and hollow situations, but with a particular association with hummocks.

Concluding remarks on palaeohydrological reconstruction at Derryville

It was observed earlier that the analysis of multiple, proxy measures is increasingly becoming regarded as necessary in the characterisation of palaeohydrological changes in mire ecosystems. Integral to this, recent research has also tended to be towards the detailed identification and explanation of local changes in individual mire systems (e.g. Makila 1997). This study has demonstrated the importance of such a multi-proxy and multi-sequence approach. If the relative influence and importance of localised variations and changes in factors such as groundwater tables and flow conditions upon mire stratigraphy and the palaeorecords contained therein can be addressed, then, with the synthesis of intra- and inter-site records, the over-arching effect that may (or may not) be attributed to climate should emerge all the more clearly. The Derryville project design has permitted the close integration of detailed studies of peat stratigraphy and micro- and macrofossil data. Moreover, the model of mire development derived from the study of peat stratigraphy has been conducted independently of the palaeohydrological inferences drawn from the study of testate amoebae and macrofossils (although the sampling was driven at least in part by the original interpretation of the stratigraphy). The interpretation of each line of evidence was thus unencumbered by assumptions and inferences drawn from the other. The conclusions drawn from the integration of the data sets may be seen as all the stronger for this.

Dryland land use and vegetation change at Derryville: the pollen record

The composition of pollen assemblages from raised bog deposits

Pollen assemblages from raised bog deposits comprise pollen grains from a number of source areas that may be defined as local, extralocal and regional. The local component will include plants growing on and around the sampling site—grasses, sedges and aquatic plants, and perhaps also trees such as alder and birch. The extralocal component will include plants growing in the immediate vicinity of the bog, on the dryland and margins—trees, shrubs and herbs. The regional signal includes the pollen that has travelled from further afield and is 'rained out' and becomes incorporated in the peat.

Separating the three source areas is rarely straightforward, but some taxa can be assigned to a local context with a good degree of confidence: *Sphagnum*, Cyperaceae (sedges) and the pollen of aquatic plants will have derived almost exclusively from plants growing close to and on the sampling site. Other taxa identified in the record will definitely not have been growing on the bog itself: herbs such as *Plantago lanceolata* (ribwort plantain) are rarely found in wetland habitats, and, likewise, *Pteridium* (bracken) tends to be restricted to relatively better-drained soils. However, dispersal distances for the pollen of low-growing herbs are small, so where these species are encountered, it is likely that they were growing on the nearby dryland areas and probably within a few hundred metres of the sampling site. Trees such as *Ulmus* (elm), *Quercus* (oak) and *Fraxinus* (ash) will only be present on dryland areas, as will *Corylus* (hazel), which tends to be very common in many pollen records. The majority of arboreal pollen will be derived from an extralocal source area, but pollen from regional woodland will also be included, although this will form only a very small proportion of the total, as will other 'long distance' grains. However, as the openness of the bog increases through time with expanding peat coverage and declining woodland the ability to 'capture' a regional signal increases.

Other pollen types, grasses in particular, may be derived locally and/or extralocally. In such cases, the behaviour of other taxa with more secure source areas may be used to determine whether the pollen curve is likely to reflect changes on the dryland or on the bog. Factors of pollen production and dispersal are also relevant: the pollen of some taxa, especially those that are insect pollinated, will not travel far from source, so only small quantities of their pollen may illustrate their local or extralocal presence, such as *Salix* (willow). Taxa such as *Corylus* will produce large amounts of wind-distributed pollen, especially in an open situation rather than as a component of the forest understorey.

'Selected' or 'dryland taxa' diagrams

Pollen diagrams are usually constructed on the basis of a total dryland taxa calculation involving all tree, shrub and herb pollen grains counted, but excluding spores such as *Sphagnum* and pteridophytes such as ferns. Each taxon is then calculated as a percentage of the TLP sum. On the basis of such a calculation, the Derryville sequences are dominated by *Corylus* pollen, with *Calluna* (heather) also forming a large proportion of counted grains. *Corylus* was obviously a common shrub on the fringes of the bog, with heather present on the bog itself contributing a large percentage of the counted grains. The 'swamping' effect of these two taxa means that changes in the proportion of other species is obscured. In order to 'tease out' changes in the vegetation on the dryland and thus better investigate possible anthropogenic activity, dryland taxa diagrams were constructed. This involved removing taxa such as *Alnus* (alder), *Corylus*, *Calluna* and most of the herbs from the pollen sum and calculating a new sum that includes only *Pinus* (pine), *Ulmus*, *Fraxinus*, Poaceae (grasses) undiff., Poaceae >37µm[1], *Plantago lanceolata* (ribwort plantain), *P. major/media*, *Rumex* spp. (docks), *Hedera* (ivy), *Pteridium* (bracken) and occasional taxa of obviously dryland origin. The inclusion of these taxa, which were almost certainly present on the dryland, means that phases of human activity and disturbance to this vegetation may be better identified on the basis of falls in tree pollen and increases in grasses, bracken and so-called clearance indicators such as *Plantago lanceolata and Rumex*[2]. These clearance phases reflect the removal of tree cover and the spread of grassland, pasture and ruderal habitats. The identification of Poaceae >37µm[3]. may also indicate the presence of arable activity and cereal cultivation.

It is important to note that the location of activity relative to the sampling site cannot be determined from a single profile, hence the need for a network of sites with good dating control to help to locate possible areas of activity on the basis of the nature and extent of fluctuations of pollen curves in synchronous phases in different diagrams.

Methods

The preparation of samples for pollen analysis followed standard techniques for peats, as outlined in Moore *et al.* (1991), with the addition of HF for all samples to equalise the preparation of non-siliceous and siliceous samples when pollen grains were being measured. Samples were mounted in silicon oil. Pollen sums usually comprised a minimum of 500 dryland pollen taxa (all non-aquatic taxa), and diagrams were prepared using the computer programmes TILIA* and TILIA*GRAPH (Grimm 1987). Pollen diagrams for each profile were prepared using first a diagram with all pollen and spore taxa represented and then a second selected taxa diagram used to determine changes on the surrounding dryland areas (see above for explanation).

Fourteen diagrams were prepared, as follows:

Fig. 6.10 (a and b) Killoran 18 East, Field 41: Total pollen and spores.

Fig. 6.11 Killoran 18 East, Field 41: Selected dryland taxa.

Fig. 6.12 (a and b) Killoran 18 West, Field 17: Total pollen and spores.

Fig. 6.13 Killoran 18 West, Field 17: Selected dryland taxa.

Fig. 6.14 Derryfadda 23, Field 41: Below-track total pollen and spores.

Fig. 6.15 Derryfadda 23, Field 41: Below-track selected dryland taxa.

Fig. 6.16 (a and b) Derryfadda 23, Field 41: Above-track total pollen and spores.

Fig. 6.17 Derryfadda 23, Field 41: Above-track selected dryland taxa.

Fig. 6.18 (a and b) Killoran 75, Field 10: Total pollen and spores.

Fig. 6.19 Killoran 75, Field 10: Selected dryland taxa.

Fig. 6.20 Killoran 18 East: Landfall total pollen and spores.

Fig. 6.21 Killoran 18 East: Landfall selected dryland taxa.

Fig. 6.22 (a and b) DERSILT, Field 10: Total pollen and spores.

Fig. 6.23 DERSILT, Field 10: Selected dryland taxa.

Based on the dryland taxa diagrams, a series of pollen assemblage zones that summarise the sequence of vegetation changes found on the dryland areas, labelled DV I to DV IX, were defined, and these are identified on each diagram with an estimation of the age of the zone boundary. Rather than discuss each diagram separately, they are summarised in a single overview of the sequence. The zones and their ages are presented in Table 6.2.

Results: sequence of dryland vegetation changes and land use implications

DV I: c. 2400–1910 BC

The earliest deposits analysed for pollen come from the lowest section of Killoran 18 East and below the trackway at Derryfadda 23. A radiocarbon date of 2270 BC from the Killoran 18 section implies that the age of the basal pollen sample is probably *c.* 2400 BC. The longer record for this period comes from Killoran 18, but there is slightly better resolution from Derryfadda 23. Given the reduced extent of the bog at this time and the proximity, especially to Derryfadda 23, of marginal communities dominated by carr woodland and reed swamp, there

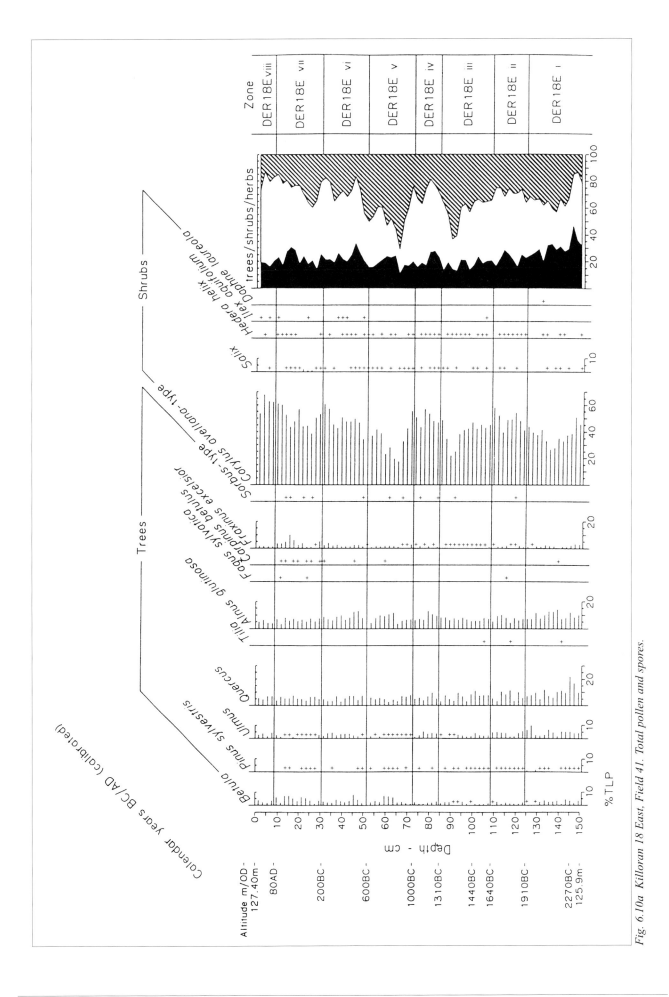

Fig. 6.10a Killoran 18 East, Field 41. Total pollen and spores.

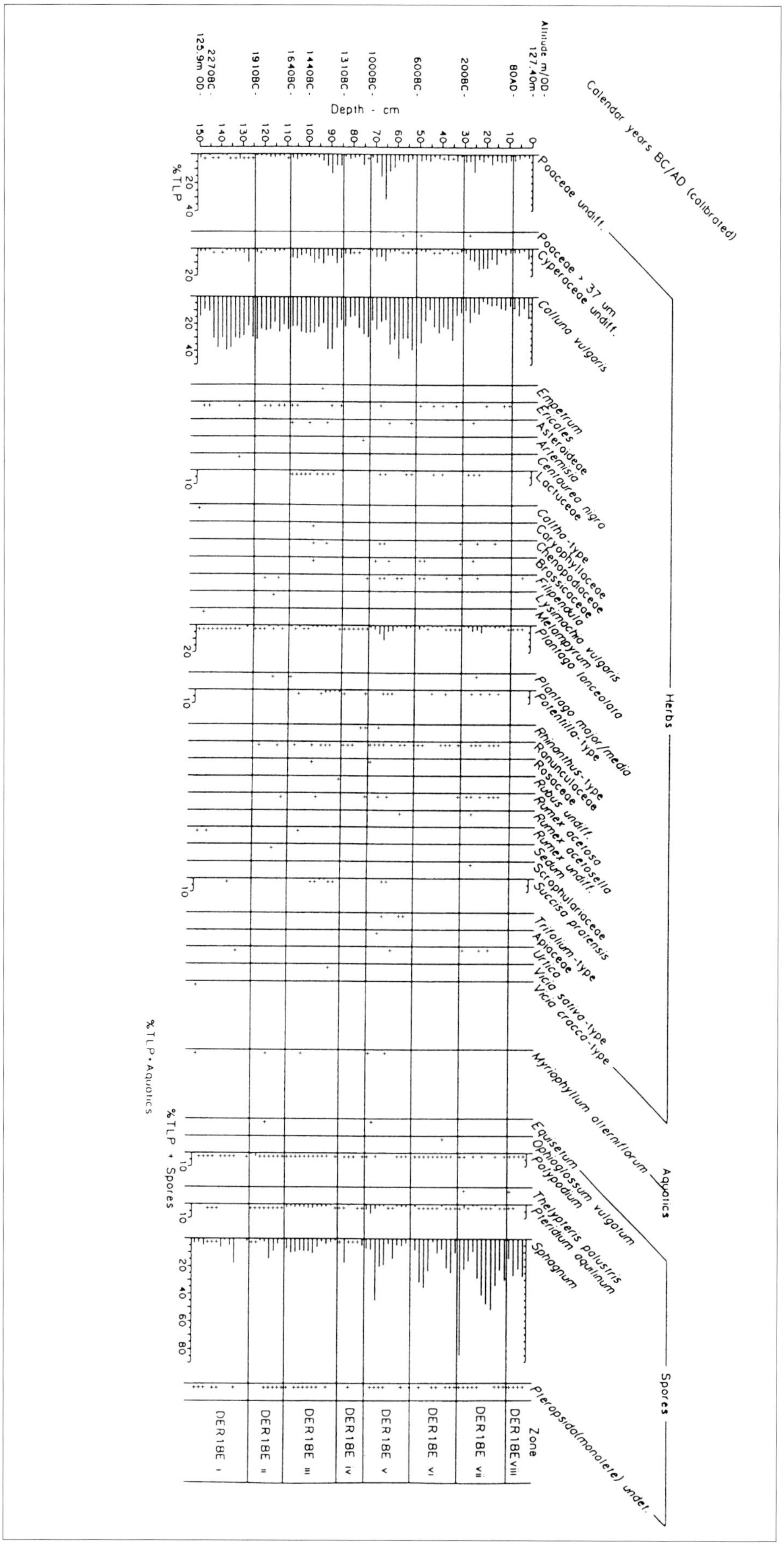

Fig. 6.10b Killoran 18 East, Field 41. Total pollen and spores.

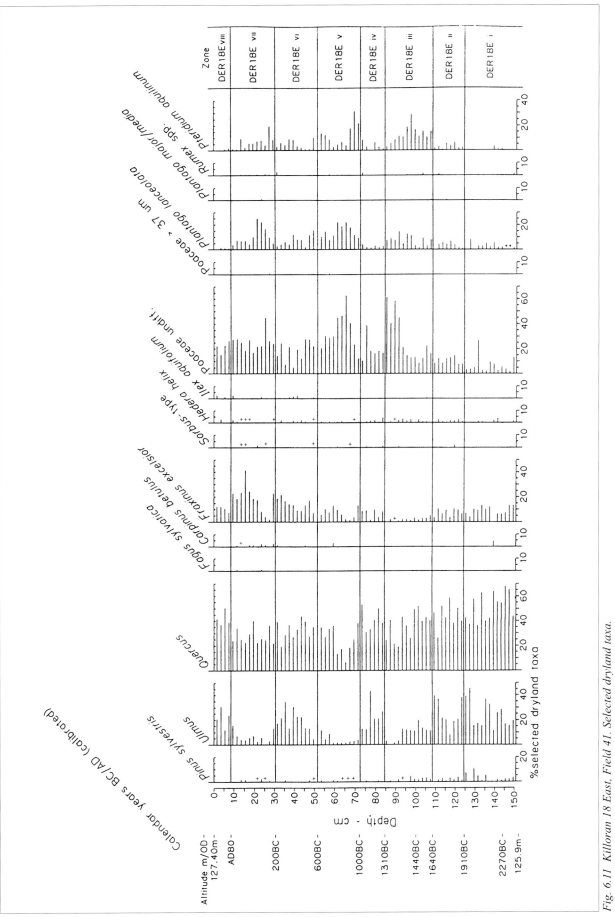

Fig. 6.11 Killoran 18 East, Field 41. Selected dryland taxa.

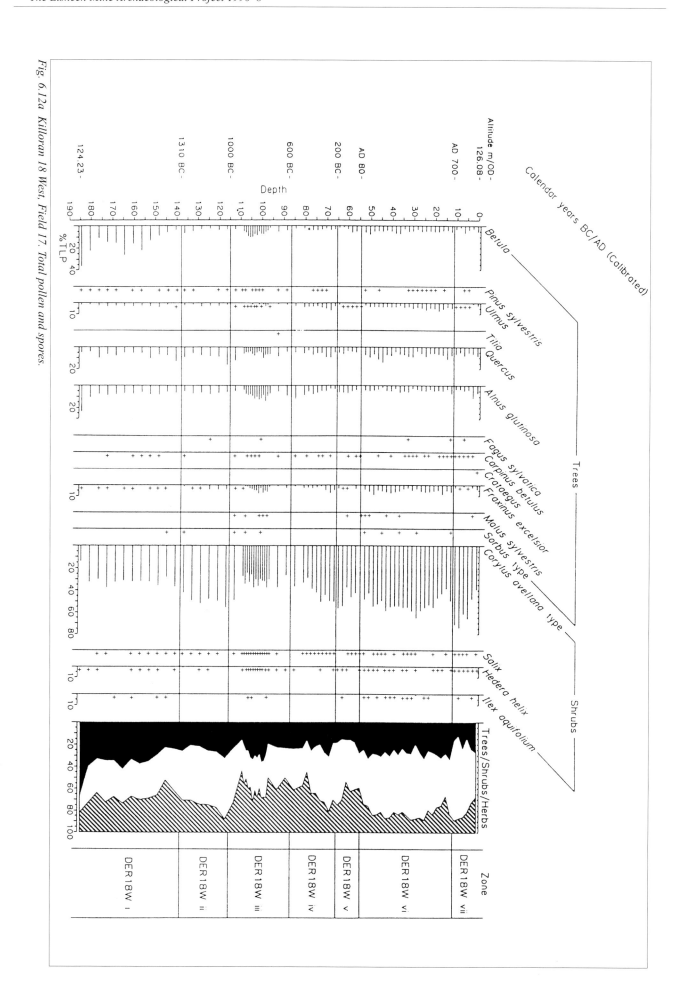

Fig. 6.12a Killoran 18 West, Field 17. Total pollen and spores.

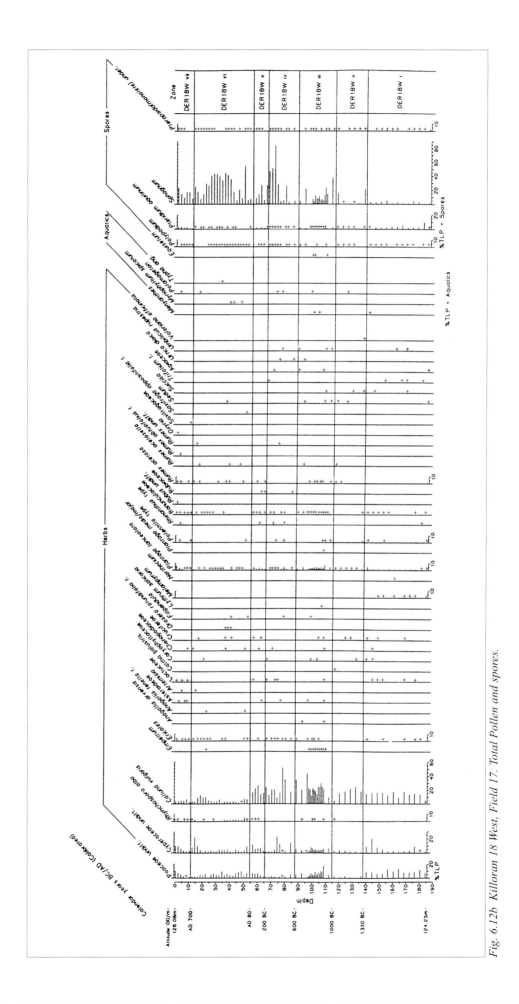

Fig. 6.12b Killoran 18 West, Field 17. Total Pollen and spores.

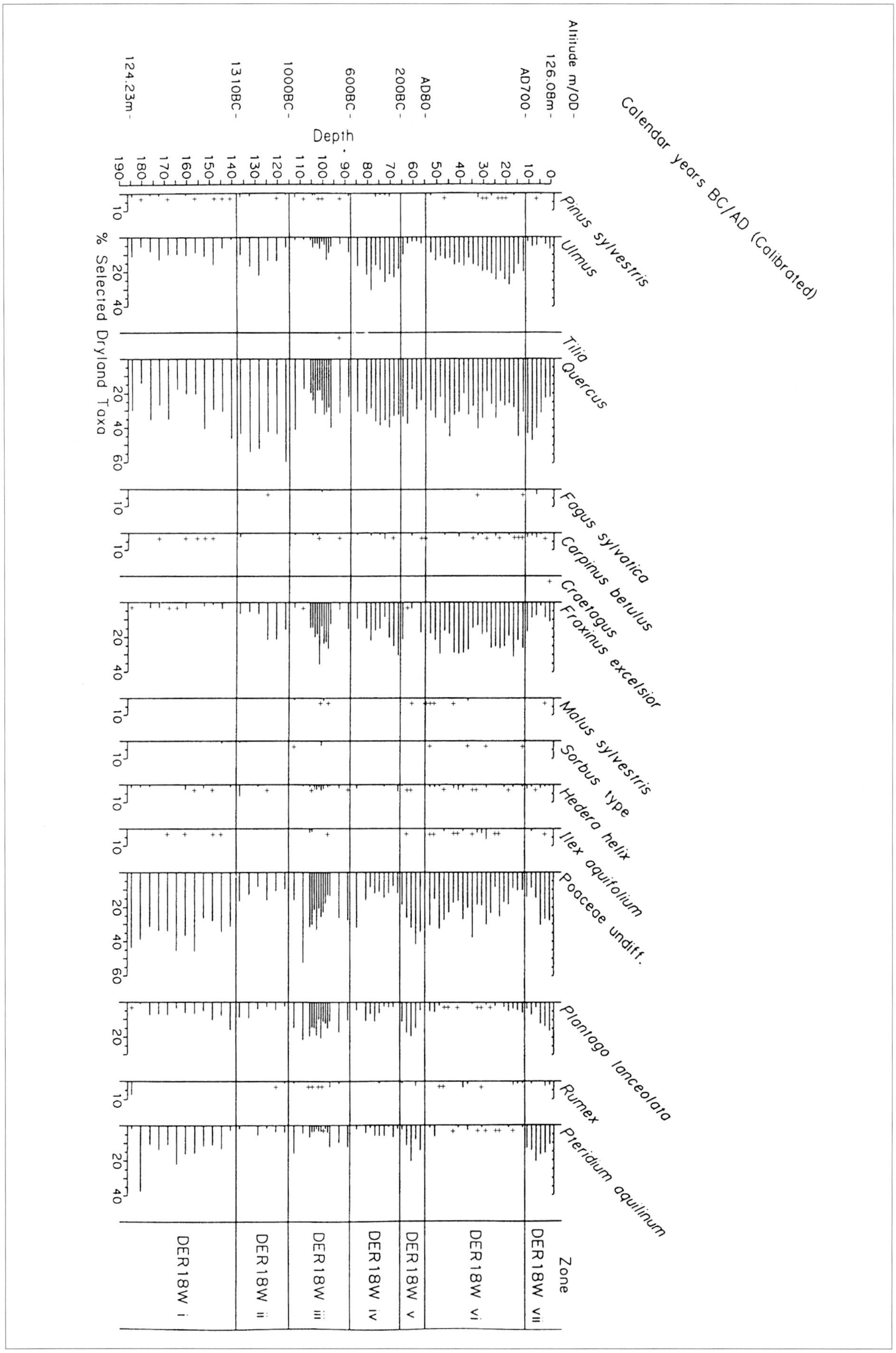

Fig. 6.13 *Killoran 18, Field 17. Selected dryland taxa.*

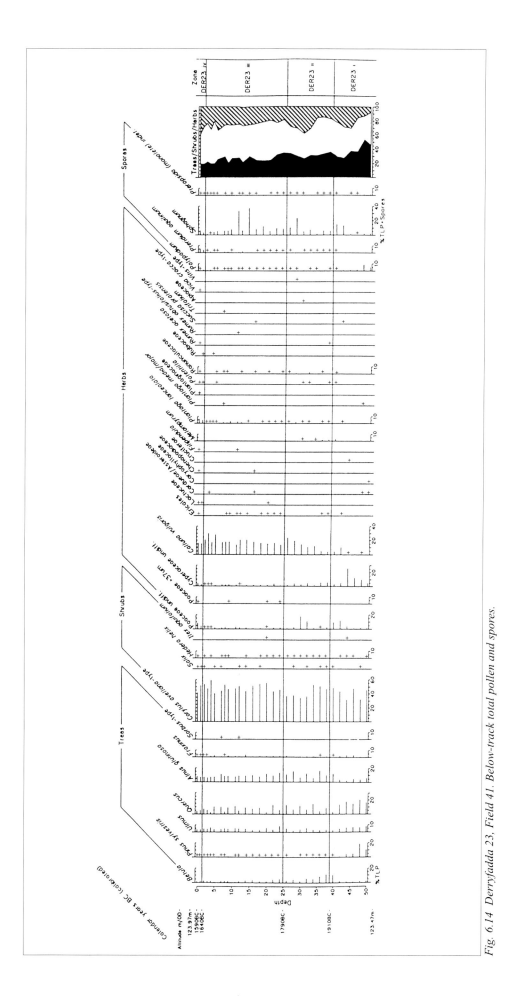

Fig. 6.14 Derryfadda 23, Field 41. Below-track total pollen and spores.

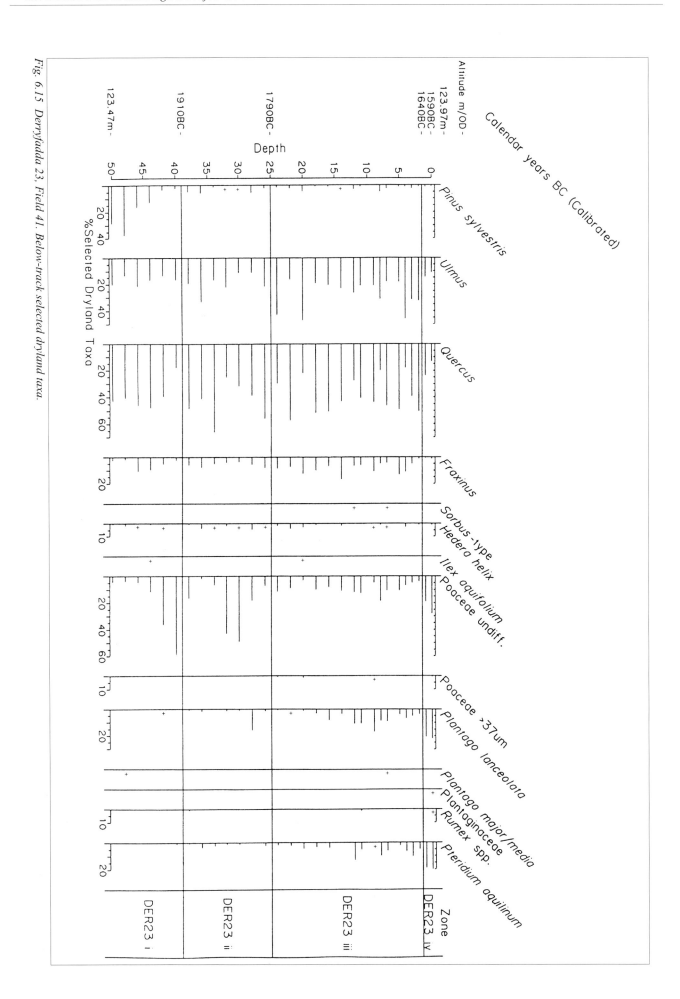

Fig. 6.15 Derryfadda 23. Field 41. Below-track selected dryland taxa.

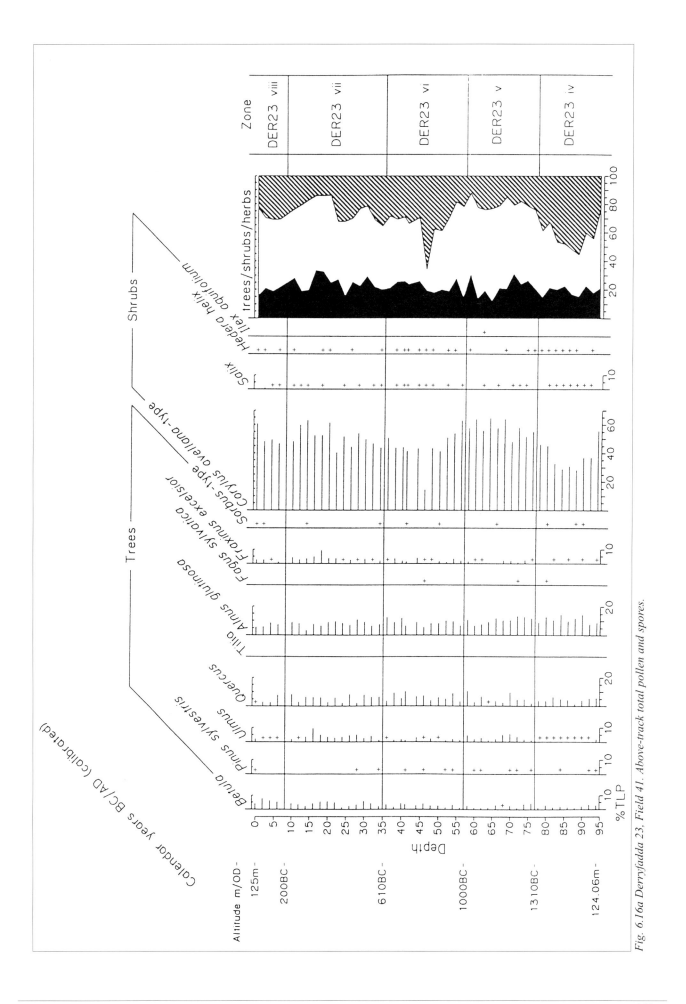

Fig. 6.16a Derryfadda 23, Field 41. Above-track total pollen and spores.

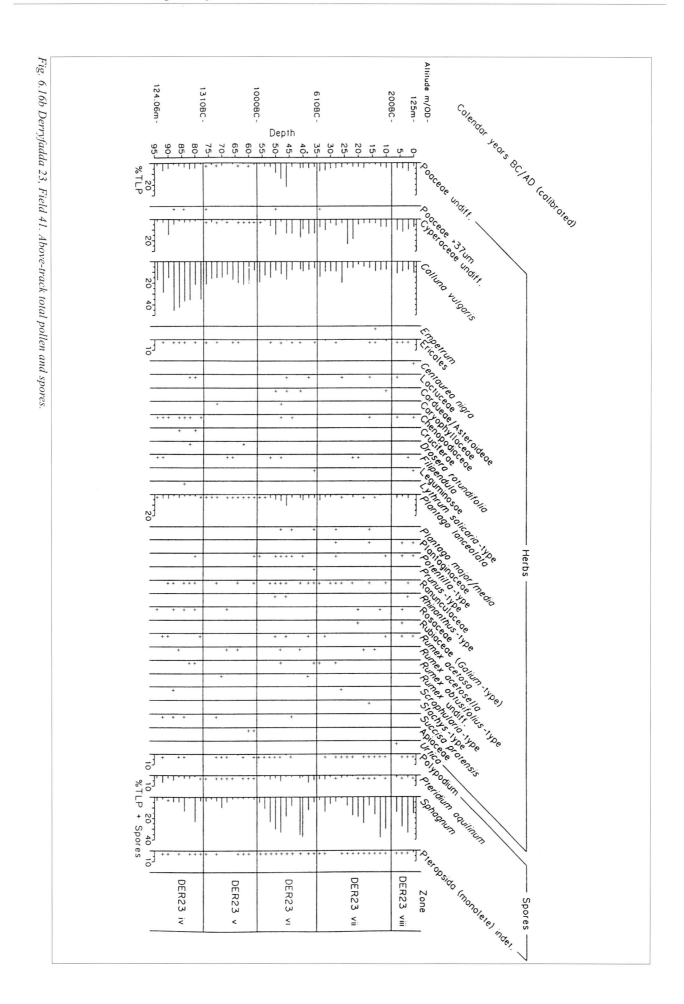

Fig. 6.16b Derryfadda 23, Field 41. Above-track total pollen and spores.

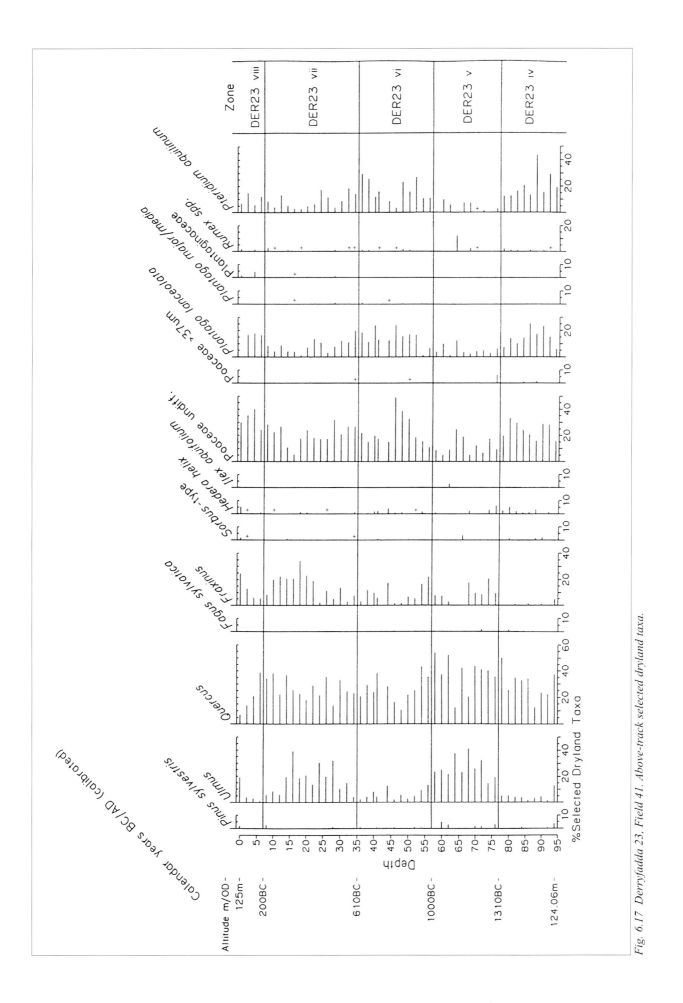

Fig. 6.17 Derryfadda 23, Field 41. Above-track selected dryland taxa.

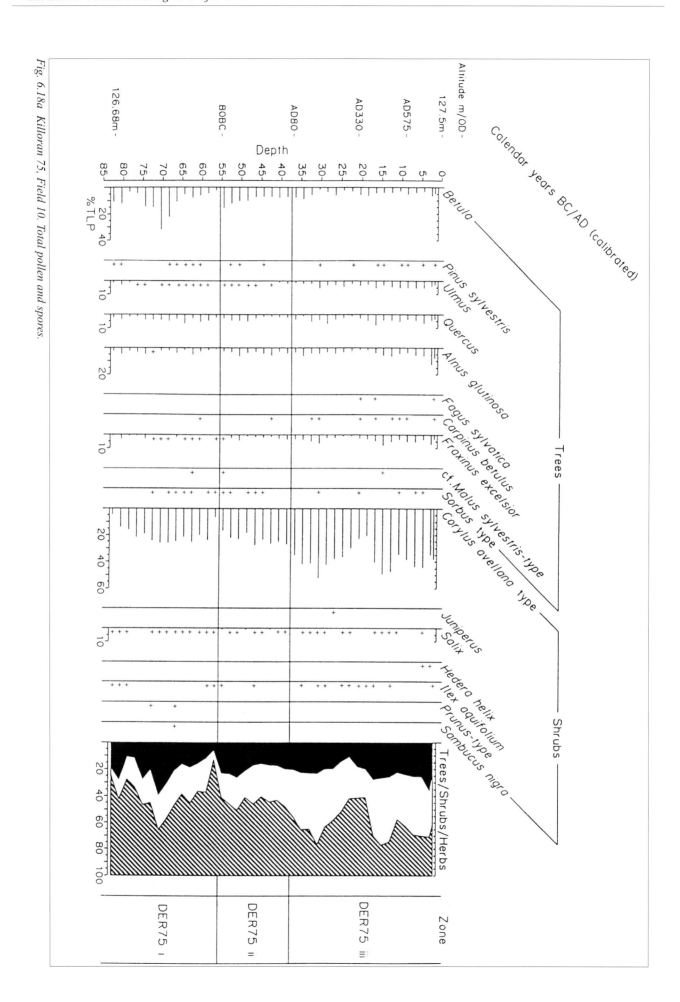

Fig. 6.18a Killoran 75, Field 10. Total pollen and spores.

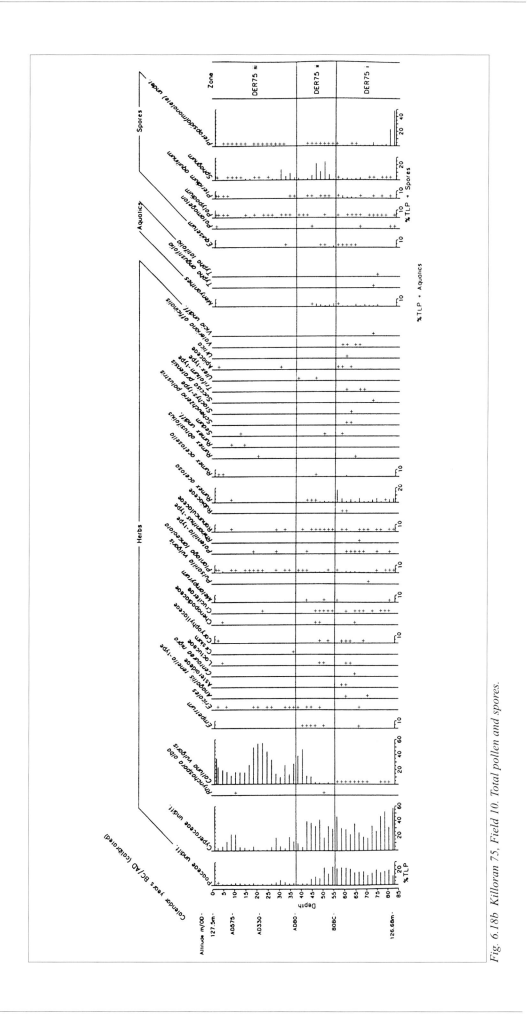

Fig. 6.18b Killoran 75, Field 10. Total pollen and spores.

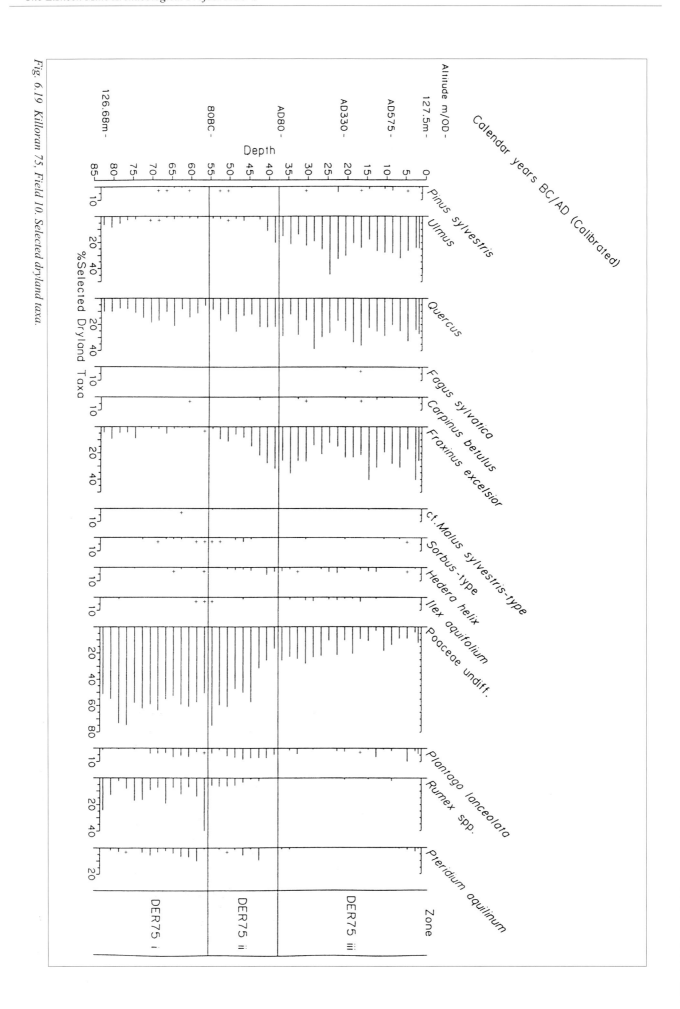

Fig. 6.19 Killoran 75, Field 10. Selected dryland taxa.

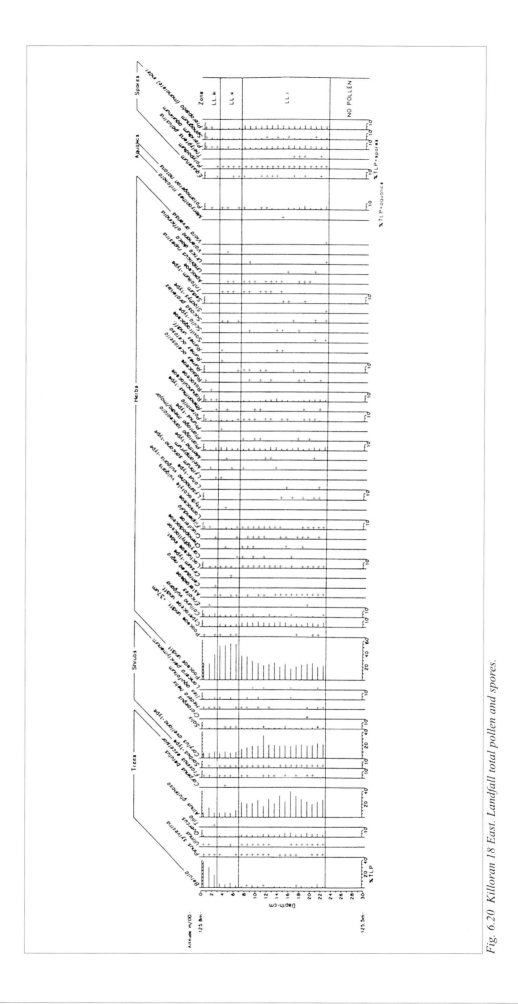

Fig. 6.20 Killoran 18 East. Landfall total pollen and spores.

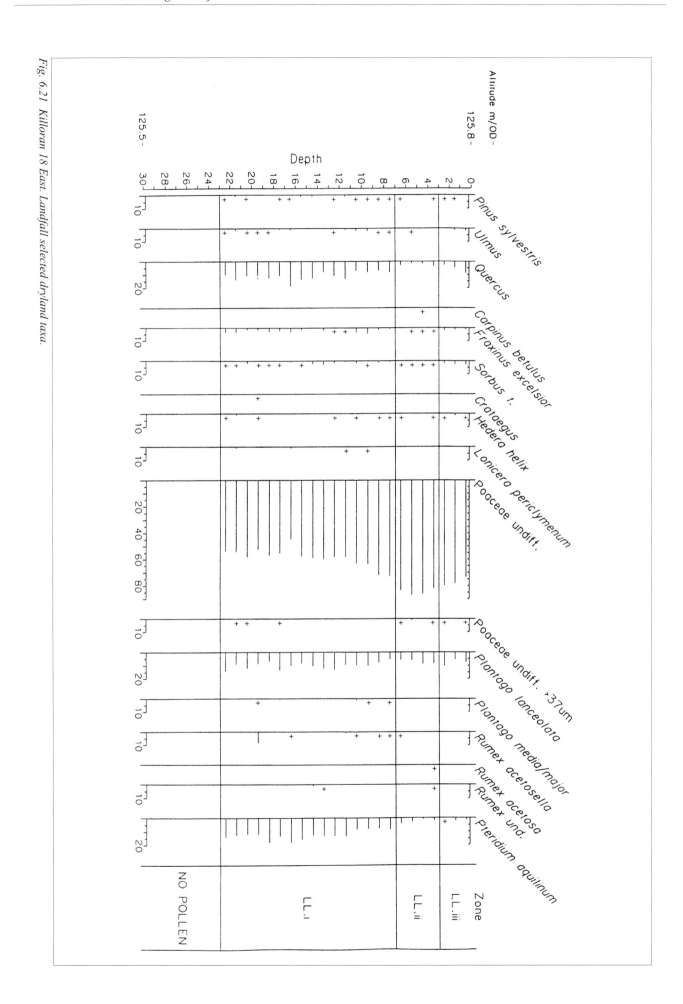

Fig. 6.21 Killoran 18 East. Landfall selected dryland taxa.

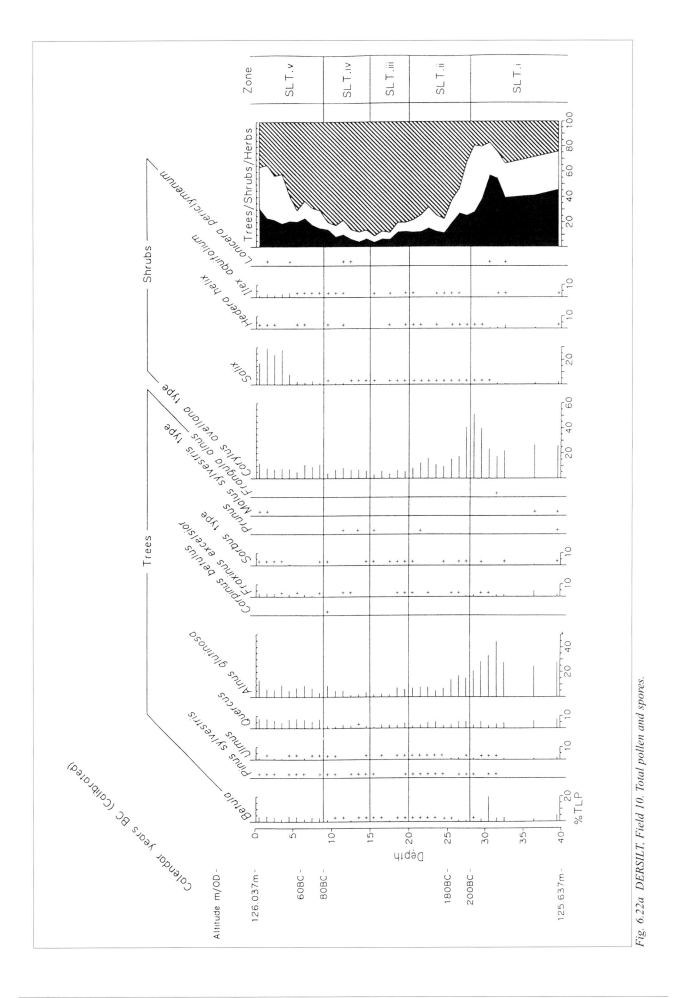

Fig. 6.22a *DERSILT, Field 10. Total pollen and spores.*

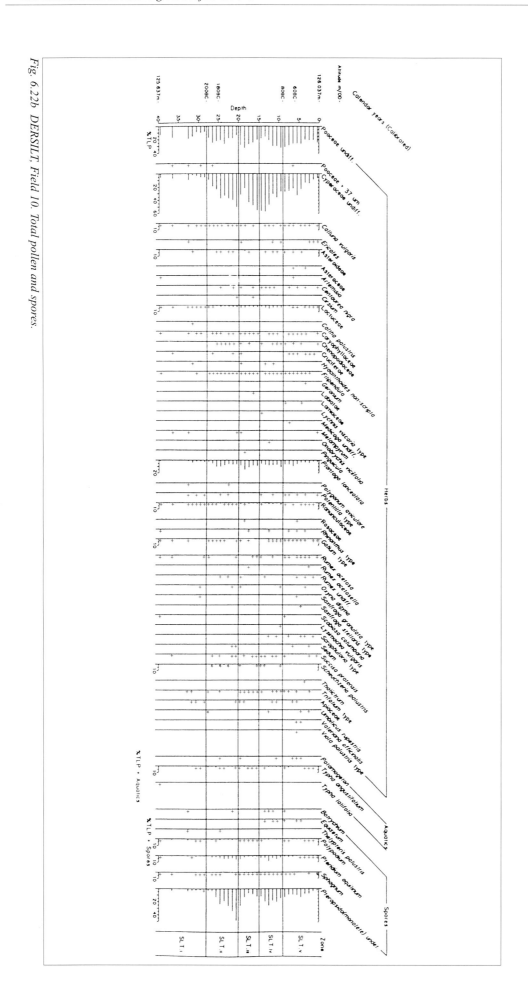

Fig 6.22b DERSILT Field 10. Total pollen and spores.

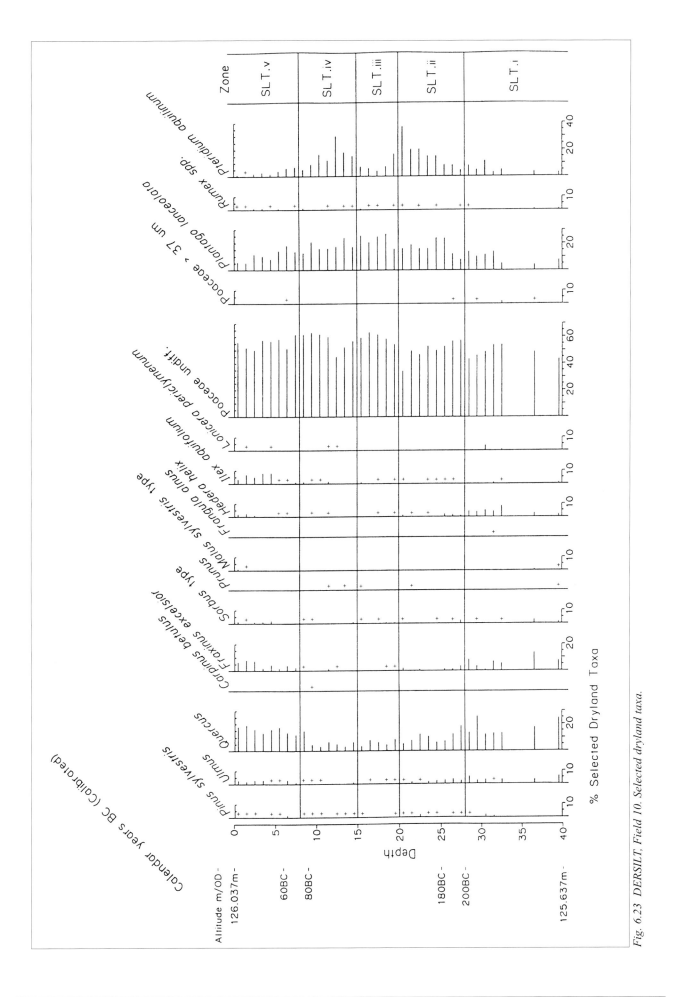

Fig. 6.23 DERSILT, Field 10. Selected dryland taxa.

Zone	Profiles including the zone	Age of upper boundary
DV I	DER23 (below); DER18 (East)	1910 BC
DV IIi	DER23 (below); DER18 (East)	1790 BC
DV IIii	DER23 (below); DER18 (East)	1640 BC
DV III	DER23 (below/above); DER18 (east); DER18 (West); DER18 Landfall	1310 BC
DV IV	DER23 (above); DER18 (East); DER18 (West)	1000 BC
DV V	DER23 (above; DER18 (East); DER18 (West)	600 BC
DV VI	DER23 (above); DER18 (east); DER18 (West); DER75; DERSILT	200 BC
DV VIIi	DER23 (above); DER18 (East); DER18 (West); DER75; DERSILT	80 BC
DV VIIii	DER23 (above); DER18 (East); DER18 (West); DER75	AD 80
DV VIII	DER (East); DER18 (West); DER75	AD 700
DV IX	DER18 (West)	-

Table 6.2 Dryland pollen zones.

is a considerable local pollen input, which tends to mask the dryland signal. The expansion of *Quercus* after the drying of the marginal surfaces with bog burst A of 2200 BC is registered in some of the highest values for *Quercus* pollen in any of the pollen diagrams. Similarly, the high counts for Poaceae may well have derived largely from *Phragmites* rather than dryland open communities, as the levels for other open-ground taxa from the drier areas are extremely low and nowhere match the high Poaceae levels, as might be expected were they from the same source.

Despite these caveats, it is clear that primary woodland dominated by *Quercus*, *Ulmus* and *Fraxinus* formed the majority of the woodland around Derryville Bog, particularly the first two trees, as pollen of trees and shrubs comprise 90% TLP in the lowest levels. Other species represented in the woodland would have been *Betula* (birch), *Corylus* and *Alnus*, but assessing the contribution of these trees is difficult. *Corylus* must have formed the fringing woodland on the dryland margins of the encroaching peat. *Corylus* pollen swamps most of the counts throughout the entire records, reflecting the importance of this highly efficient pollen-dispersing taxon. Both *Betula* and *Alnus* would have been components of carr woodland as it developed variously around the bog margins; hence, individual peaks for these taxa are most likely to reflect the changing nature of the bog marginal areas. As for understorey shrubs, these are notoriously poorly represented in pollen diagrams from large peat bogs in Ireland (Watts 1985). The most commonly occurring pollen taxa, *Hedera* (ivy) and *Ilex* (holly), occur occasionally, especially the former, and *Sorbus* (mountain *Fraxinus*)-type pollen is also represented. *Sorbus*-type includes a number of species, some of which can be separated on reference material (see discussion for later periods). There are no records for *Taxus* (yew), despite wood remains, but this is not unusual, as it is an extremely poor disperser of pollen and the presence of isolated trees is unlikely to be recorded. The position of *Pinus* is interesting in that this part of the pollen record shows some of the highest frequencies in any of the dia-

grams, with a regular decline towards 1900 BC; this is especially noticeable at Derryfadda 23 towards the south-eastern margin of the bog. The causes of the decline of *Pinus* in Ireland, and indeed northwest Britain, has been the subject of much debate, particularly as in most pollen diagrams it appears as a relatively sudden phenomenon (Blackford *et al.* 1992; Hall *et al.* 1994). At Derryville, scattered pines occasionally grew on the bog, but the period around 2100 BC appears to have been one in which it was able to expand in preferred locations before declining and effectively dropping out of the local tree flora. Hall *et al.* (*ibid.*, 82) have commented that by 2310 BC, "pine was rare or absent from the lowlands of the north of Ireland," so the decline in pine does appear to have been countrywide at this time.

Despite the dominance of primary woodland, there is evidence that human communities had already started to modify their local environment. Pollen of *Plantago lanceolata* and spores of *Pteridium* occur before 1900 BC at both Killoran 18 and Derryfadda 23. The sequence of changes in the pollen record in the lowest zone is not a simple pattern of clearance and abandonment, as tended to occur later. This is mainly due to the influence of local bog and fringing taxa, which dominate the pollen diagrams whichever way they are expressed. Values for clearance indicators are extremely low, but their presence in the local flora does reflect human activity. Values for *Fraxinus* are relatively low in comparison with later periods. It has been suggested that the expansion of *Fraxinus* was encouraged by human woodland manipulation, which created openings for the tree to colonise, as it is a relatively light-demanding species and one that could almost be classified as a weed (Caseldine and Hatton 1996). Here the rise is around a millennium after the initial expansion of *Fraxinus* in Ireland, dated at Corlea to 3000 BC (*ibid.*), a date not dissimilar to others across a range of sites in Ireland. It is probable that in the third millennium BC, the area of Derryville saw some local presence of human communities that had relatively little lasting impact on the woodland. The nature of the modification would have been either the creation of small openings or a form of forest farming *sensu* Göransson (1986). Thus, instead of clearing areas of woodland, the effect would have been a gradual modification of the woodland community, with reductions in canopy allowing the substitution of primary trees such as *Quercus* and *Ulmus*, and possibly *Pinus*, by *Fraxinus*, with sufficient open areas to allow the limited expansion of the clearance indicators seen in the pollen diagrams. It should, however, be pointed out that the earliest woodland clearances are likely to be underestimated in pollen diagrams because of the presence of remaining areas of primary, and indeed secondary, woodland, which would be producing the majority of pollen getting to the bog. Hence, it may well be that the relatively slight record in the earliest levels at Derryville is representative of more than occasional human impacts. The continuity in the presence of clearance taxa, even though only as occasional grains, is suggestive of continued human presence. In the absence of any evidence for cultivation, it is necessary to infer only pastoral agriculture of some form, but even had there been any cultivation in isolated clearings within a dominantly forested environment, it is highly unlikely that it would be seen in the pollen record.

DV II: 1910–1640 BC

The opening to DV II is defined principally where there are continuous and rising curves for the clearance indicators *Pteridium* and *Plantago*, especially the former. It is also coincident with low values for %C (percentage carbon) at Killoran 18 East—only 35% compared with average values of >50% for most of the peats (Fig. 6.24a) associated with Killoran 18, in levels separated from the low %C results associated with the paved trackway above. There is no comparable drop in %C at Derryfadda 23. There is also a gradual decline in *Ulmus*, most obvious at Killoran 18 East, which seems to be less affected by local/marginal pollen and has a more smoothed-out record due to the lower rate of accumulation compared with Derryfadda 23. At Derryfadda 23, the very high Poaceae values are almost certainly due to the nearby presence of *Phragmites*. The period between 1910 BC and 1640 BC can be separated into two distinct sub-zones at Derryfadda 23, where there is a relatively rapid accumulation rate (9 yr cm^{-1}) compared with the peat at Killoran 18 (15 yr cm^{-1}). This record indicates a probable brief increase in the impact of human activity on the woodland around 1850–1800 BC, before a more continuous and increasing effect from around 1750 BC, with both *Plantago and Pteridium* expanding as seen in the selected taxa diagram. Because of the effect of local bog marginal *Quercus*, the pattern of change is not very regular, but, at Killoran 18, there is a noticeable minimum for both *Ulmus* and *Fraxinus* that appears to coincide with this phase of increased representation of clearance taxa. The 'large' (>37μm) Poaceae grains at Derryfadda 23 are not referable to cultivated cereals but are considered to be wild grass species on the basis of their

pollen morphology. Hence, as with the earlier zone, it is not possible to identify any cultivation from the plant record. All herb taxa are clearance indicators rather than indicators of cultivated ground.

The presence of low %C values at the opening of DV II at Killoran 18 reflects in-wash into the fen sediments of silts, presumably as a result of erosion around the bog margin. Casparie (this volume) noted a visible layer of in-

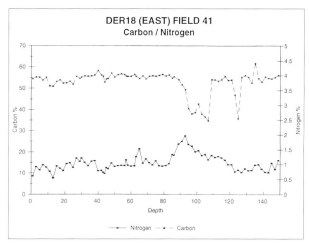

Fig. 6.24a DER 18 (East) Field 41. Carbon/Nitrogen.

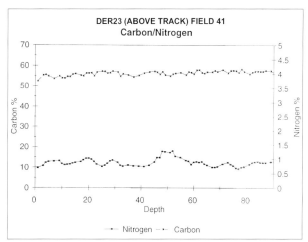

Fig. 6.24b DER 23 (above track) Field 41. Carbon/Nitrogen.

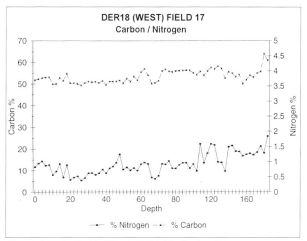

Fig. 6.24c DER 18 (West) Field 17. Carbon/Nitrogen.

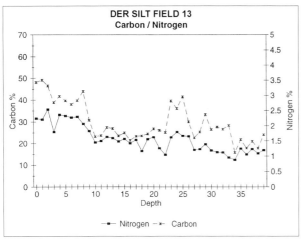

Fig. 6.24d DERSILT Field 13. Carbon/Nitrogen.

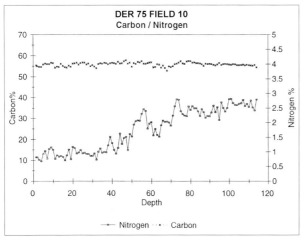

Fig. 6.24e DER 75 Field 10. Carbon/Nitrogen.

wash at other profiles at around 2000 BC, but, at Killoran 18, its presence only became apparent with the chemical analysis. The silt layers are predominantly found in the western marginal peats, deriving from sources along the western edge of the bog, in the Killoran and Cooleeny areas. The absence of any record at Derryfadda 23 in the ombrotrophic peat suggests that the silts may be restricted to fen areas with some degree of lateral in-washing, and the occurrence of such a subdued signal over in the eastern area of the bog emphasises a western origin. Should there have been a concentration of effects in the western area leading to erosion, then the relatively slight pollen signal could further implicate a western and more distant source for the preponderance of human activity. Such an interpretation has to remain speculative in the absence of a pollen record of similar age from the western bog. Nevertheless, by the end of DV II, at around 1640 BC, there was a brief phase of woodland regeneration, with *Ulmus*, *Fraxinus* and *Quercus* recovering. This period must have been short, as it is recognised in both pollen diagrams, but at Derryfadda 23 it covers at most 4–5cm, less than 50 years. It does, however, indicate the likelihood of almost complete abandonment of the dryland pollen source areas

contributing to the record of earlier clearance and is dated quite accurately through the stratigraphic position of the overlying trackway at Derryfadda 23 to around 1680 to 1640 BC on the basis of the peat accumulation rate.

DV III: 1640–1310 BC

This period marks the building of two major features, Killoran 18 and Derryfadda 23, and the pollen evidence from three profiles incorporating these provides an important basis for understanding dryland changes at this time. At Killoran 18 East, the position of the paved trackway is marked by the peak for mineral particles in the peat (Fig. 6.2), and the low %C figures (Fig. 6.24a). At Killoran 18 West, the largely wooden trackway comprises almost all the lowest zone from 124.63m OD to 124.27m OD, although the exact position at the top is variable and probably slightly lower, and there is no signal in the %C results. At Derryfadda 23, the trackway marks the top of the lower diagram and the base of the above trackway diagram. Dates for the trackways are 1606±9/1590±9 BC or later for Derryfadda 23 and 1542±9 BC or later, repaired in 1440±9 BC, for Killoran 18. The pollen results from Derryfadda 23, immediately below a plank of the trackway, show that there had been a significant attack on the local stands of *Ulmus*, *Fraxinus* and, to some extent, *Quercus* (if only briefly initially, as the severest effect on *Quercus* can only be seen in the higher resolution of the diagram from Derryfadda 23) prior to trackway construction, with a consequent rise in *Plantago* and *Pteridium*. Even allowing for some peat compaction immediately under the trackway, it must have been constructed only a matter of decades after this initial clearance. Assuming no loss of peat below the trackway, the construction would have taken place when there was only limited evidence for local primary woodland remaining around the immediate bog margins. *Corylus* was barely affected and may well have benefited from woodland clearance, and *Alnus* shows no signs of reduction. The effect on *Fraxinus* and *Ulmus* was long lasting in that, even by the end of this period, at around 1300 BC, neither had regenerated locally. Indeed, the period between 1350 BC and 1300 BC seems to be characterised by the virtual absence of both from all pollen diagrams.

There appear to be two distinct phases in this period, with the earlier period of trackway construction showing high *Pteridium* values and the latter part higher Poaceae, and possibly slightly higher *Plantago*. Again, none of the bog records provide any evidence for cultivation. The fall of the pollen of trees and shrubs to <50% by 1350–1300 BC implies considerable openness around the Derryville area, and, had there been cultivation, more indicators of this might have been expected to enter the bog system. Neither trackway Derryfadda 23 nor the wooden section of Killoran 18 East show the presence of pollen of taxa that could have come in with the wood, as occurred occasionally at Corlea, County Longford (Caseldine, Hatton and Caseldine 1996; Caseldine *et al.* 1996), for all recorded taxa are found above or below the trackway levels. The presence of *Carpinus betulus* (hornbeam) at a number of levels at Killoran 18 West probably reflects its presence to the west of the bog. Although there are records for *Hedera* and *Ilex*, these are no more than in non-trackway levels, and there is no increase in, for instance, epiphytes such as *Polypodium*, which could have come in in association with the wood. There is also very little variability across the bog in the percentages of the main clearance indicators. Assuming a predominantly westerly prevailing wind transporting pollen onto the bog, this may be taken as representative of pollen sources from both sides of the bog, with no significant variation in the amount of human activity between the western and eastern margins.

Samples taken through the sediments underlying the peat at the landfall of Killoran 18 on the eastern margin, which must date to the building of the paved trackway around 1500 BC, confirm the virtual absence of *Ulmus* and *Fraxinus*. Below 6cm to 7cm at the landfall site, the sediment is described as largely detrital, owing its origins to fluvial/colluvial activity at the margins of the expanding bog (Ellis 1999). The pollen within this sediment is largely uniform, suggesting relatively rapid accumulation of the sediment, and there is a wider range of taxa represented than on the bog surface. This would be expected from a relatively dry 'soil' site, with the potential for a greater number of plant species growing in the immediate vicinity. Of trees and shrubs, only *Quercus*, presumably deriving from the marginal woodland to the south, and *Corylus* are well represented. Occurrences of *Hedera* and *Ilex* are no greater than from the bog record, but there are occasional records for other understorey/shrub components, including *Lonicera* (honeysuckle) and *Crataegus* (hawthorn). It is in the range of herb taxa that a wider selection of species are represented, many only referable to family or genus (e.g. Caryophyllaceae; Chenopodiaceae; Compositae, including members of Lactuceae, probably mainly *Taraxacum* and Asteroidae; *Trifolium* and Cruciferae). Although specific weeds of cultivated ground are largely absent, there are occurrences of large Poaceae pollen referable to cereals in the lower levels. On size and morphological criteria, the example from 20cm to 21cm is probably *Triticum*-type (Table 6.3). The overall record from the landfall diagram in DV LL1, below 7cm, shows a wet open area fringed by *Alnus* and *Corylus*, with predominantly grassland, local pools (as seen in the presence of *Potamogeton*) and, in the vicinity, possibly some small amounts of cultivation. It is unlikely that the area of the ridge that formed the landfall was cultivated, but cereals may have been grown on the slightly higher land to the east. The lower levels include charcoal, as also noted in the soil descriptions (*ibid.*) and at the peat transition. At Killoran 18 West, there is also a single level

for charcoal at 180cm (Table 6.4), predating the construction of Killoran 18 but lying within the overall zone.

As peat overwhelmed the terminus of Killoran 18, eventually a wood peat dominated by *Betula*, several of the herb taxa are lost or severely reduced. However, the period of transition, DV LL2, shows considerable reduction of even the local shrub or bog woodland and an expansion of grassland, with more indication of cereals occurring locally, all still probably *Triticum* (Table 6.3). Given the age of Killoran 18 and the stratigraphical position of these levels, it is likely that this phase equates with the expansion of grasses seen at the end of DV III in Killoran 18 East and the minima for 'primary' woodland trees, dated at all profiles to around 1350 to 1300 BC. Thus, the latter part of DV III was marked by a period of perhaps a century or so, following the burial of Killoran 18 in the peat, of increased pressure on the local landscape and some cultivation, certainly on the eastern side, and most likely also on the west as well, although there is no direct evidence for this in the pollen record. This came after a period of about 200 to 300 years of

Profile & depth (cm)	Diameter of grain (μm)	Diameter of pore +annulus (μm)
DER18 Landfall		
0-1	46	10
2-3	48	12
3-4	46	10
5-6	46	12
6-7	47	13
DERSILT		
6-7	55	14
6-7	47	14
DER18 (East)		
27-28	49	12
49-50	48	12
57-58	48	12
DER23		
34	47	12
50	54	17
52	49	12

Table 6.3 Cereal-type pollen identifications.

Profile	Depths at which charcoal recognised
Landfall	5/6–14/15cm
DER18 (East)	None
DER18 (West)	52/53–62/63cm; 180/181cm
DER23	12cm
DER75	None
DERSILT	32/33cm; 36/37cm; 39/40cm

Table 6.4 Levels with countable frequencies of microscopic charcoal.

continuous local occupation, including the building of both Derryfadda 23 and Killoran 18, with human activity taking place on both the eastern and western areas bordering Derryville Bog. Comparison of the sequence at Derryville with that further to the north at Corlea shows a remarkably similar period of increased human activity between 1550 BC and 1330 BC, with a clustering of trackway tree-ring dates from the earlier Bronze Age, 1600–1350 BC (Raftery 1996), and examination of the dates for wooden trackways at Corlea (*ibid.*, Plate 51) demonstrates a significant increase in construction starting around 1600 BC. This comparability between the Derryville and Corlea sequences continues into the later prehistoric period (see below).

DV IV: 1310–1000 BC

DV IV represents a period of woodland regeneration and an apparent severe reduction in the amount of human activity around the bog. Values for tree and shrub pollen recover to over 80% TLP within a few centimetres at each of the three profiles, probably within a century or two of the previous main period of impact, and the level of recovery of both *Fraxinus* and *Ulmus*, particularly the latter represent a considerable colonisation of previously open areas by the main tree species of the original woodland. *Quercus* increases slightly, but so does *Corylus*, which must also have taken advantage of the opened areas that were abandoned. The record at Derryfadda 23 shows a phase in mid-zone where *Fraxinus* disappears and *Ulmus* and *Quercus* are reduced, whilst there is a peak for *Plantago* and *Rumex acetosella*. This is not apparent from the two Killoran 18 profiles, but that may be a combination of sampling interval and location. The estimated date for this is around 1100 BC. Correlation with the Hekla 3 eruption, which may have affected the landscape of Ireland for a short time, is possible, but in the absence of any corroborative evidence, this remains purely speculative. Furthermore, this period was a complex one in the history of the development of the bog, with a bog burst in the western area of the bog shortly before 1250 BC (bog burst B), a feature reflected in the water table evidence for Killoran 18, but not for Derryfadda 23.

It is difficult to see how these bog conditions could have had any effect on the dryland pollen record, unless the bog burst somehow influenced the abandonment of the area and allowed woodland regeneration to take place. There is variability in the average rate of growth of the three profiles: around 15 yr cm⁻¹ at Derryfadda 23 and Killoran 18 West and 26 yr cm⁻¹ at Killoran 18 East, and this affects the available level of time resolution. A number of archaeological features date to this period, but usually within a calibrated 2σ range. The possibility remains that some of these may have been from the end of the previous zone, i.e. nearer 1300 BC, or that they do reflect some extremely low local presence of human communities, barely noticeable in the pollen record.

From all the pollen data, it is, however, reasonable to assume that there was up to 300 years of, at best, extremely light local human interference with the vegetation, especially in comparison with the impact in the previous 300 years. An alternative explanation would require a radical change in the way human occupation related to woodland, with much more local woodland cover surviving than previously. Comparison with Corlea again shows remarkable similarities, with 1350 to 1050 BC seeing woodland regeneration very like the Derryville pattern. Apart from the possible poor climate around the Hekla 3 eruption, marked by a run of very narrow tree rings (Baillie and Munro 1988), there is no obvious climatic limitation on human activity during the period as a whole, the rising water tables towards 1000 BC probably being as much a function of a return to pre-bog burst conditions as to any really noticeable deterioration in climate.

DV V: 1000–600 BC

DV V opens with a further attack on the regenerated woodland, evident at all sites. Dating the end of this zone is difficult because of the radiocarbon plateau problems. However, the estimate of *c.* 600 BC is supported by the correlation of events with well-dated features elsewhere (see below). The initial attack affects all trees, particularly *Fraxinus*, *Ulmus* and *Quercus*, and values for *Plantago* are notably high (e.g. at Killoran 18 West). There are two occurrences of Poaceae referable to cereals at Derryfadda 23 and Killoran 18 East, both of similar size and morphology and probable *Triticum*-type (Table 6.3). By around 850 to 800 BC, the removal of the regenerated woodland had reached significant proportions, with pollen of trees and shrubs between only 30–40% at all sites, the value at Derryfadda 23 being the lowest recorded in the diagram by almost 20%. At Derryfadda 23 and Killoran 18 East, this point also marks a very brief but severe drop in the pollen of *Corylus*. Although close sampling was undertaken at Killoran 18 West, no comparable reading was found (it should be noted that continuation of the close sampling down the profile may well have located it nearer 110cm). At the time of the minimum for *Corylus*, Poaceae reaches very high levels, with a following drop that could almost mark a hiatus in the Derryfadda 23 diagram. There is some variation in the impact on trees and shrubs across the bog, with more severe reductions in the eastern sites. This could reflect a predominance of activity to the east, but the differences in the pollen frequencies are within margins that could have been affected by local conditions.

A comparable reduction in *Corylus* has recently been identified by van Geel *et al.* (1996) in the Netherlands and dated very precisely to around 800 BC by wiggle matching of radiocarbon dates. They argue for a climatic cause, citing the sensitivity of *Corylus* to rainy late winters and spring frosts and pointing to the close correlation with a ¹⁴C rise that is in part responsible for the dating problems of the time period. The clarity of this similar feature at Derryville, which can also be found in pollen diagrams of the same time from Corlea (Caseldine, unpublished data), also in a period of severe woodland reduction, and the position in the stratigraphy strongly suggest that the same event is registered in Ireland. The climatic shift to colder conditions was abrupt and has been cited as causing a crossing of significant thresholds for certain *Sphagnum* species, although, as argued earlier, this did not happen so synchronously or widely at Derryville as elsewhere. In Tipperary, the climatic signal was superimposed on an environment in which human interference was again becoming important and appeared to coincide with a relatively high intensity for such activity. The original reduction in *Quercus* may, however, have also been exacerbated by rising water tables overwhelming local bog marginal forests. The record at Derryville is further complicated by the western bog burst shortly before 820 BC (bog burst C), which led to a reduction in water tables after initially high levels in the earlier part of DV V. Towards the end of DV V, approaching 600 BC, regeneration is still only slight, and despite the climatic influences, the first half of the first millennium BC saw a resurgence of occupation around Derryville. The drier surface conditions after the bog burst may well have helped the regeneration of the *Quercus* in the bog and contributed to the rising *Quercus* curve towards 600 BC.

DV VI: 600–200 BC

The period to 200 BC was one of further local woodland regeneration. Frequencies for trees and shrubs reach over 80% TLP, comparable to values for 1300 to 1000 BC. Such values predominantly occurred after 400 BC and especially around 350 to 300 BC, when the lowest percentages of *Plantago* and *Pteridium* are found. This was a period of generally high water tables, although this need not have been a reason for largely abandoning the area. Open areas survived and were still an important part of the landscape for the first half of DV VI, but by the end they must have been restricted, and regenerated woodland and scrub would have been common. Killoran 18 West has a virtually continuous record for *Carpinus betulus*, which must have been growing on the dryland to the west, as it is barely recorded in the eastern sites. The lowest levels from Killoran 75 derive from later DV VI, but the record is affected by considerable local *Alnus* growth. Nevertheless, there is a strong local presence for *Hedera*, which must also have been growing in relative abundance close to the western edge of the bog. The lack of a good *Hedera* record towards the eastern end of Killoran 18 is perhaps not surprising, as the plant is insect pollinated and the pollen is poorly dispersed. A better representation at Derryfadda 23 probably represents a source to the immediate southeast at the bog margin.

The pattern of regeneration during the middle and later first millennium BC is reproduced at Corlea (Caseldine *et al.* 1996), and regeneration can be found elsewhere in Ireland at this time, e.g. at Carrownaglogh, County Mayo (O'Connell 1986; 1990) and Redbog, County Louth (Weir 1995), although the dating is not always as well defined as at Derryville. Although some archaeological features at Derryville within DV VI have good dendrochronological dates, they date mainly to the earlier centuries, when the pollen evidence still shows human presence. Although it is not impossible for archaeological material to derive from this later part of DV VI, several of the radiocarbon dates for features apparently of late DV VI age have wide calibrated ranges. Given the uncertainty over the precision of the overall chronology used, it is likely that they mainly relate to the later period, DV VII, and hence date to the first two centuries BC. Despite these uncertainties the pollen record does show a period of considerably reduced activity in the middle to later first millennium BC, preceding a very different period when human activity again become a dominant feature of the local landscape.

Towards the end of this zone at Derryfadda 23, in one sample, there is evidence for microscopic charcoal in the peat (Table 6.4). It also occurs in the lowermost sediments, probably washed in, at DERSILT, immediately before the main silt layer, from which it is absent. There may therefore have been some use of fire towards the end of DV VI, perhaps more as a feature of the transition to the next zone. There is very little charcoal in any of the profiles overall (no countable levels in Killoran 18 East or Killoran 75), implying that whatever burning took place was extremely light and not capable of registering a visible component, even at a microscopic level, in any of the bog profiles, except in a very few instances. This was therefore likely to exclusively reflect burning for domestic use, as seen from the settlement and *fulachta fiadh* evidence.

DV VII: 200 BC–AD 80

Although the late first millennium BC is recorded at both Killoran 18 sites and partially at Derryfadda 23, the profiles for Killoran 75 and DERSILT from the western area of the bog have much more detailed records and provide the main evidence. At Killoran 18 West, peat accumulation is very slow, at almost 28 yr cm⁻¹, whilst to the east it is only half this figure. At both DERSILT and Killoran 75, the annual accumulation rate is in single figures; hence, there is much more detail. The opening of DV VII appears sudden at the slowly accumulating sites, but, although more gradual at the western sites, the actual detail is to some extent confused by the local marginal influences, especially the influence of *Alnus* and *Corylus*. In view of the much more rapid rate of peat accumulation, the change from the high levels of regeneration seen at the end of DV VI to severe clearance of *Fraxinus* and

Ulmus probably took place within a century. At the profile through the well-developed silt layer DERSILT, the lowermost sediments, below 28cm, have %C values indicating the presence of in-washed material, prior to the increasing dominance of peat up to 24cm. It is during this period of peat development, which may be seen as the transition to largely ombrotrophic conditions, that *Fraxinus* and *Quercus* are reduced, the former leading to its local disappearance, with *Ulmus* also reaching low levels. The in-wash of the main silt layer took place here once *Fraxinus* had virtually disappeared, and the main silt depositional unit, bracketed between 24cm and 28cm in the profile, corresponds to the absence or very low values for *Fraxinus* and *Ulmus* in particular.

The dating for the onset of this major in-wash of silt from a well-defined area of the western dryland margin of the bog is well constrained by a number of radiocarbon dates. The initial woodland removal, which began shortly before the erosional episode, is dated to 200 BC, with erosion probably starting by 180 BC (although appearing precise, the actual dates should still be seen as mean values with associated error terms, but the reproducibility of these figures adds confidence to the figures quoted). The initial erosion also saw reductions in *Quercus* and the expansion of *Plantago* and *Pteridium* to very high levels. The Killoran 18 profiles do not show overall frequencies of trees and shrubs falling to levels experienced earlier, but both Killoran 75 and DERSILT identify occasional samples with <30% TLP. The evidence as a whole points to woodland removal in the specific catchment area of the western margins, with perhaps less effect over a wider area, especially to the east. Despite reflecting erosion of the surrounding soils, pollen preservation in the silts is very good. This suggests that erosion was largely a superficial overland flow of material without any removal of lower soil profiles. Had this occurred, there would have been evidence of poorly preserved pollen. The cause of the erosion was therefore a phase of vegetation removal allowing greater wash over the surface, perhaps for almost a century, given the thickness and assumed rate of peat accumulation at DERSILT. There is almost a doubling in the number of pollen taxa represented in the silt layer over the counts from the peat profiles, even greater than counts from the earlier landfall site, but there are no large Poaceae referable to cereals, and most of the clearance indicators are those of damp pasture. This probably means that soil cultivation was not a contributing factor to erosion at all.

The termination of silt in-washing is well defined by the %C values and occurs at a point where *Quercus* increases and *Fraxinus* and *Ulmus* form continuous curves, with the former expanding first, presumably due to the colonisation of local formerly cleared ground. This affected *Pteridium* more than *Plantago*, and there are two grains referable to *Triticum* type at this point from DERSILT, after erosion had ceased. Dating of the transition puts the point at which *Fraxinus* begins to

recover and *Pteridium* is reduced to around 80 BC, with the eventual end of DV VII occurring where both *Plantago* and *Pteridium* are virtually absent and *Fraxinus* and *Ulmus* both recover, at AD 80. The silt layer and the century or so of major local clearance coincides with the construction of Killoran 75 within the pocket of very wet *Scheuchzeria-Menyanthes* peat. The radiocarbon dates for this feature cover a calibrated range from 385 BC to AD 130. From the palynological evidence and the stratigraphic position of the trackway, a date after the opening of DV VII is likely, probably near the middle of the second century BC.

Microscopic charcoal occurs at Killoran 18 West throughout this zone, despite its absence at Killoran 18 East, suggesting a westerly source for the fires. Its absence from the in-washed silt at DERSILT further implies a different or wider source for the charcoal, but the continuity of the record at Killoran 18 over 10cm must reflect relatively continuous use of fire throughout the zone. The values are never very high, but, compared with the general absence of microscopic charcoal from the Derryville sites, they are significant.

Correlation with Corlea, is again striking. There was a brief period of woodland removal that saw the building of the main 'road,' Corlea 1, in 148 BC. At Corlea, the construction took place immediately after a renewed attack on the woodland, following relatively low human activity in previous centuries. The duration of the clearance episode was brief, less than 200 years. At other sites, because of the brevity of the impact, it is not always possible to define this particular phase, but Iron Age clearance is considered to be a widespread phenomenon, with a number of possible correlates, e.g. Whiterath Bog (Weir 1993a) and Loughnashade (Weir 1993b).

DV VIII: AD 80–700

From about AD 80, both Killoran 75 and Killoran 18 West show a remarkable degree of woodland regeneration. The complete period is recorded at the latter profile, and the percentage of trees and shrubs once again reaches around 80%. It is not so high at Killoran 75. This is probably a statistical feature due to the greater local pollen production of *Calluna*. Within this first half of the first millennium AD, clearance indicators almost completely disappear; they are far lower than in the previous two millennia, and regenerated woodland must have covered most of the area around Derryville, with very little human interference taking place. This secondary woodland included all the species originally found in the primary forest, but there is pollen evidence for a wider range of trees and shrubs, probably reflecting both a more scrubby nature for parts of this woodland and the availability of sites closer to the western woodland edge. Thus, both *Ilex* and *Hedera*, especially the former, are better represented than before, and *Sorbus*-type plus firm pollen identifications of *Malus sylvestris* come

DERRYVILLE				CORLEA	
Zone	**Vegetation/Land Use**	**Age**	**Age**	**Vegetation/Land Use**	**Zone**
IX	Significant clearance	AD 700	AD 740	Increased clearance	G_2
VIII	Regeneration— little human activity	AD 80	AD 380	Clearance	G_2
			AD 30	Regeneration— little human activity	F
VIIii	Continued human activity— some regeneration			Clearance, Corlea 1 built	E_2
VIIi	Clearance and silt deposit	80 BC			
VI	Regeneration but low level of human activity	200 BC 600 BC	200 BC	Regeneration but low level of human activity	E_1
V	Clearance and continued if variable activity	1050 BC	600 BC	Clearance and continuous activity	D_3
IV	Regeneration, reduced activity	1310BC	1000 BC	Regeneration, low level of human activity	D_2
III	Significant clearance greater than before	1640BC	1330 BC 1550 BC	Significant regeneration	D_1
IIii	Low human presence	1790BC		Regeneration	C_5
IIi	Some light clearance	1910BC	1850 BC	Some light clearance	C_4
I	Mainly primary woodland but some small areas of clearance	2300BC	1980 BC 2200 BC	Very low level of human activity, mainly primary	C_3

Table 6.5 Schematic correlation of the sequences from Derryville and Corlea.

from this zone. Areas of open ground survived, but there is no real evidence for any consistent local human occupation, and the overall impression is one of almost complete woodland recovery. This took the form of a secondary woodland comprising more *Fraxinus* than previously and a greater influence for understorey shrubs, probably a less dense cover but with shrubs rather than ground flora benefiting.

The abandonment of the Derryville area corresponds to the widely established Iron Age 'lull' (Mitchell 1965) seen in many of the pollen records throughout Ireland. At Corlea, this dates to 30 BC–AD 300 (see Table 6.5 for a comparison of the complete sequences for Derryville and Corlea), and the duration of the episode at Derryville is

one of the longest recorded. Precise dating of this period at other sites is often lacking. Weir (1993a) argues for 200 BC–AD 200 at three sites in County Louth, but the radiocarbon control for this is poor, although the sampling interval is very good. In a review of previously published sites, Weir points to sites showing regeneration across Ireland from Tyrone (Edwards 1985), Down and Antrim (Goddard 1971; Smith and Goddard 1991) in the north to Mayo and Sligo (Dodson and Bradshaw 1987; O'Connell 1986) in the west. Other sites in western Ireland in Galway and Clare do not, however, show this feature (O'Connell *et al.* 1988; Jelicic and O'Connell 1992; Molloy and O'Connell 1993). Although not radiocarbon dated, the pollen evidence from Littleton Bog, close to Derryville, which formed the basis for Mitchell's initial work, is very similar to that shown at the sites analysed here. Weir (1993a) points out that the impact on the landscape was more than the 'lull' often used to describe it; at Derryville, it was almost half a millennium of virtual abandonment. The length of time represented at Derryville is in excess of that normally found, and although the dating of the end of this zone is by extrapolation from lower dates it is unlikely to be less. Water tables were high and hence peat accumulation relatively rapid, but by comparison with accumulation rates lower in the profiles, it is unlikely that the date of the end of the zone was much earlier. The radiocarbon date that calibrates to AD 575 from Killoran 75 predates the end of DV VIII, which may be identified in the uppermost sample from this profile, and hence at the earliest, AD 650 is a possibility, but this would require an acceleration of peat growth after the radiocarbon date, a time when the water table was dropping.

The first half of the first millennium AD was also the time when there was extensive flooding of the bog area north of Derryville, and it may be that the increasing wetness of the area as a whole had some impact on the perception of communities considering recolonising the area in the Early Christian period.

DV IX: Post-AD 700

Clearance following the 'lull' is only recorded at Killoran 18 West, where the upper 10cm shows a major reduction in *Ulmus*, *Fraxinus* and, eventually, *Quercus*, and even, after an initial expansion, *Corylus*. Clearance indicators also expand, with Compositae occurring, but there are no records for cereals throughout the whole of the Killoran 18 West profile. This clearance probably marks the onset of extensive woodland removal and the intensification of cultivation seen during and following the Early Christian period in Ireland, with the introduction of a wider range of cereals, but the record for this is missing due to the milling of the peat.

Footnotes:

[1] The term should not be confused with the term 'Poaceae undiff.,' which refers to grass pollen not identified to species level.

[2] *P. lanceolata* is one of the most commonly used 'anthropogenic indicators' in pollen diagrams due to the species high light requirements and exclusion from closed woodland environments. Increases in its pollen reflect the spread of open habitats. As well as pasture and meadow, *P. lanceolata* occurs on wasteland, waysides, spoil heaps and rocky outcrops. *Rumex* species are also good indicators of open, disturbed habitats and are found in acid pastures as well as meadow and grassland and occasionally very damp environments.

[3] Cereal-type grains identified on the basis of size of grain and other morphological characteristics. This group can also include wild grasses such as *Glyceria* (sweet grass).

7. Wood and charcoal identification

Ingelise Stuijts

Introduction

The excavations in Derryville Bog created a unique opportunity to look in detail at the continuous and fundamental interaction between prehistoric populations in Ireland and their natural surroundings over a long period of time. Wood and charcoal formed the majority of finds in the peat sediments of the bog, and the large quantities of wood found at most sites presented an excellent opportunity to analyse each structure in detail, as well as compare the wood usage at the different sites. It was expected that a thorough analysis of the wood and charcoal would give information on the marginal forests and the way these forest resources were used in prehistory.

Due to the time constraints of the project, it was not possible to analyse all the material collected. However, in 1997 and 1998, almost 8,300 individual pieces of wood from fifty-six sites were analysed. A selection of eighteen charcoal samples from the bog and bog margins were also investigated. The eleven wooden artefacts found in Derryville Bog formed a small assemblage and the identification of these artefacts is discussed in Chapter 12.

Following sections on the scope of wood identification analysis and the sampling strategy and methodology, the main discussions in this chapter are divided into eight sections.

The first section describes the characteristics of the wood species (in alphabetical order) found in Derryville Bog. The second section presents the results of wood identifications on material collected from non-archaeological contexts. During the initial stages of the project, archaeologists collected wood from trees growing *in situ* without an archaeological context. These wood pieces were often sampled for dating purposes. They also provided direct information on the wood growth on or immediately bordering the bog. The third section presents the total results from the wood and charcoal analysis, regardless of age or situation. The fourth section compares the species identifications on the western and eastern sides of the bog. The excavated area of Derryville Bog covered an area of approximately 72ha. The surrounding landscape must have shown some differentiation due to small topographical variations in the soil, slope, moisture situation, drainage or discharge pattern etc. Local variations might have promoted or limited the growth of certain tree species. The compari-

son of eastern and western data was made to explore this possibility. The fifth section outlines the evidence for woodland management, with reference to the results from individual structures. This section includes discussion of the hurdles that were excavated at a number of locations. The sixth section looks at wood usage over a long period of time, during which the local bog changed dramatically (see Chapter 3). The seventh section compares the results of the wood research and the pollen research, arranging the wood identifications in the pollen sequence of Caseldine *et al.* (see Chapter 6). The final section compares the results of the wood research with reference to the archaeological chronology as discussed in Chapter 4.

A detailed description of the wood results of the individual sites can be found on the additional files available to download from www.mglarc.com. Aspects of woodworking form a separate chapter (Chapter 13).

Scope of wood and charcoal analysis

The purpose of the wood research in Derryville Bog was to obtain as much information as possible within the time limits of the project, with the aim of reconstructing the prehistoric landscape and human interference to as detailed a level as possible.

The identification of wood is of great importance for archaeological research because wood was one of the most important raw materials in prehistoric and early historic times. From a biological point of view, the anatomical study of wood gives information on vegetation history, especially woodland history.

Within a prehistoric community, wood was used for a multitude of purposes, indoors as well as outdoors. Large timbers that were used for the building of houses and farms were generally obtained as close to a settlement as possible to avoid transport problems. The choice of a specific wood species depended on its quality, strength, speed, regularity of growth and durability. In most cases, oak was used for large constructions.

Inside a settlement, many objects were needed for normal housekeeping, from tools to kitchen utensils and furniture. These objects were all used for specific tasks and therefore required particular qualities from the material from which they were made. The requirements included durability in dry and wet environments (ovens, hearths, fishing equipment, buckets and scoops), flexibility (axe handles and bows), smoothness (bowls and spoons) and beauty (the use of burr wood for bowls). It is clear that this group includes a greater variety of wood species than structural wood, and this variety reflects the selection of certain wood species. Usually, objects with the same function were made of the same wood species.

Firewood was collected as close as possible to or within a settlement. In most cases, firewood was gathered at random, but waste material (chips from the local felling of trees) was also used. It is for this reason that the wood species found within charcoal hearths provides information on the local vegetation directly surrounding settlements or activity areas. Larger pieces of wood for fuel were obtained by felling trees, but sometimes discarded building material was used.

In Derryville Bog, the main archaeological remains were not those of settlements but of small and large paths or features associated with the use of bog marginal areas. This might be one reason why the excavations produced very few artefacts. No remnants of buildings were expected in the bog (although a possible hut site, Killoran 66, was identified). Only in the latter phases of the project were some hut sites found on the dryland close to the bog, but these sites contained no wood remains. The charcoal content from contemporaneous cremations in a cemetery found close to the huts has not yet been investigated.

The wooden constructions in the bog margins, the remains of local fires on the bog surface and the *fulachta fiadh*, with their associated hearths and charcoal fills, all situated outdoors in the wetland area, formed the main source of information for the wood study. It was possible to obtain information on which wood species were used for the constructions in the bog margins, the selection of species, the application of the different wood species within the constructions and the origin of the wood. It was expected that the wood results would give information on the species composition of the marginal woodland locally and perhaps some indication of the landscape in the wider surroundings of Derryville Bog. Because of the scarcity of finds, it was not possible to investigate the selection of wood species for objects, although the few artefacts found were made of wood from local trees.

Sampling strategy and methodology

Wood

During the first year of the campaign, every individual piece of wood was sampled. By 1997, this method had already produced large amounts of samples. Due to the constraints of the project and the large amounts of wood to be expected in the following year of excavation, it was decided to restrict further sampling to the amount of wood that could be analysed within the duration of the project.

The sampling strategy was adjusted according to the local situation. The sampling procedure was often discussed with the excavators, depending on the archaeological findings. Within a site, generally one-third of an exposed surface was sampled. Individual features, such as wooden structures associated with *fulachta fiadh*, were fully sampled. Unfortunately, the relined trough of Killoran 240 could not be analysed. Hurdles consisting of panels of woven rods were sampled such that all sails and rods were represented.

In contrast to other palaeoenvironmental disciplines, the emphasis for the wood research was on sites excavated in the later stages of the project. Only a few hundred pieces from the 1996 campaign were analysed.

It was essential that the samples were packed properly and thus could be handled quickly. For the wood identifications generally, a piece the length of a hand was sufficient, as the archaeologists had already noted general features such as trimming, amount of knots and the total dimensions. Separate samples were taken for woodworking and dating purposes.

Although the identification work included more than 8,000 pieces, this number only represented a small portion of the total amount of archaeological wood in the bog. Probably at least 25,000 pieces of wood were present within the exposed areas, and the total in Derryville Bog must have been much greater and can be multiplied by at least ten.

The first observations on the wood were made under a stereomicroscope with low magnification, varying from 30x to 75x. A sliver was cut from each piece of wood with a razor blade and identified under a transcident light microscope (magnifications 40x to 400x). The age and growth pattern were established through microscopic cross-sections with magnifications of 40x. For identification, the manuals of Tjaden (1919), Greguss (1945) and Schweingruber (1978) were used.

Generally, there were no particular difficulties with the identifications, as they were made mainly to genus level, such as *Alnus* (alder), *Betula* (birch) or *Fraxinus* (ash). In some cases, the species was identified, such as *Pinus sylvestris* (Scots pine); in others, one species was considered the most likely represented, such as *Fraxinus excelsior* and *Alnus glutinosa*. Problems arose in the identification of the Pomoideae (Maloideae) group, which includes the genera *Malus* (apple), *Pyrus* (pear), *Sorbus* (rowan or whitebeam) and *Crataegus* (hawthorn). For this publication, wood identified as belonging to this group is referred to as Pomoideae type.

Charcoal

A microscope with incident-light optics (40–400x) was used to identify charcoal, after initial selection, under a low-powered binocular microscope. To analyse the charcoal, it was necessary to break the individual lumps to expose a fresh cross section.

The charcoal samples came mainly from *fulachta fiadh* and were used for the heating of stones. Some charcoal patches on the bog associated with archaeological structures were also examined. The charcoal was identified and some specific features and characteristics were noted, including whether the material was freshly burnt or stored wood or waste material such as wood chips.

Charcoal as a material does impose some limitations. It represents only a fraction of the material that was burnt, with the majority generally burning down to unidentifiable ashes. Also, wood species differ in their resistance to burning. Softwoods such as birch, alder, hazel and willow burn more easily to ashes than hardwoods such as oak. Therefore, oak tends to be over-represented in the lists of identified species. Smaller wood pieces such as branches and chips also burn more easily than thick logs, and this too affects the identification results.

In the identification process itself, a complicating factor is the fact that during carbonisation part of the characteristics of wood are lost. Some wood species are very difficult to identify in a charred condition or cannot be distinguished from other very similar wood species, as is the case with alder, birch, willow and poplar. Other species, such as oak, pine and ash, can often be identified even in severely burnt conditions. Despite these limitations, it was possible to study the considerable wood consumption associated with the Bronze Age *fulachta fiadh* tradition.

Presumably, most of the wood for the prehistoric fires came from the landscape surrounding the bog. The species identification would therefore reflect to some degree the tree vegetation of that area. The charcoal in the samples comprised chunks, small pieces and dust. For each sample, the volume of each identified wood species and the volume of the remaining unidentified material (residue) were measured. The total sample volumes and the identified wood species volumes varied greatly. For this reason, the identified amounts were not only given in volume but also in number of pieces analysed (the score).

Description of wood species

Acer campestre L. (field maple)

According to Scannell and Synnott (1987), all maple in Ireland is introduced. However, in Britain, field maple is a common tree in woodlands, showing a preference for calcareous or clay soils (Stace 1997), and is often associated with ash and hazel woods on neutral or alkaline soils (Milner 1992). Maple can also be found in hedgerows. It is well suited for coppicing or pollarding, and because it is often repeatedly cut, mostly only a shrub results. Normally, maple is seldom higher than 15m; very occasionally, a tree of 25m is found. Few trees, therefore, become big enough to produce timber. Another species of maple, *Acer pseudoplatanus*, the sycamore, grows into much larger trees, but this species is not indigenous to the British Isles. Seedlings

and the full-grown trees of field maple are resistant to considerable shade and are therefore often found as understorey trees. Maple would have formed part of the dryland vegetation and would not have been represented in the marginal forests close to wetland areas.

The wood of maple is pale brown, hard, tough and strong but not durable in the open. It can be used for turnery, especially bowls, furniture and musical instruments (violins and harps). The sweet sap produced in spring can be used for making wine or maple syrup. By boiling, sugar can be extracted from the wood. The wood is moderately good when used as fuel.

In Derryville Bog, field maple was only identified at one site. It is also absent from most other prehistoric sites in Ireland. In historic times, turned bowls made of maple are sometimes found within settlements, although these bowls could have been imported. Maple must have been virtually absent from the Holocene woodland vegetation in Ireland.

Alnus glutinosa L. Gärtner (alder or black alder)

Generally, alder does not like an acid soil where there is movement or seepage of water. It will tolerate sporadic but not prolonged waterlogging and is often found on the banks of streams and rivers and also in damp woodland (Orme and Coles 1985; Rackham 1995). On swampy ground, alder trees grow closely together, forming an alder carr. Alder often forms stilt roots under waterlogged conditions. It grows regularly in and on fen peat. A consistent and abundant supply of moisture is essential for germination and early growth. After germination, the seedlings will not tolerate much shade. Once the tree is established, the root system makes the tree less dependant on high water levels. The roots have nodules bearing the bacterium *Schinzia alni*, which enables alder to fix nitrogen. For this reason, alder is often used to enrich the soil by its fallen leaves and thus increase fertility (van der Meiden 1961).

Alder trees can grow up to 25m high and reach a maximum girth of up to 1m. Generally, growth is fast, and the tree is very suitable for coppicing. Growth diminishes after twenty years, and the maximum height is reached by sixty years. Alder trees generally reach ages of between eighty and one hundred years (Stortelder *et al.* 1998).

It is poor firewood, but the charcoal is of excellent quality; it formerly formed an important constituent of gunpowder. The wood quickly turns reddish after cutting. Once dried, it is water-resistant and does not split easily. The wood is very durable when permanently waterlogged; therefore, it has often been used for scaffolding poles, water pipes and piles under bridges and houses. Easy to carve, yet tough and water-resistant, it makes the best soles for clogs. Alder turns well and is especially suitable for making bowls. Small objects such as broomsticks and handles have often been made of alder (Rackham 1980). The bark can be used for tanning leather. The catkins and bark give an inferior black colouring known in medieval times as 'poor man's dye.'

Alder was one of the most important tree species occurring in Derryville Bog. It was mostly the dominant or one of the most common species. It occurred at forty-three sites, being the dominant species at thirteen. One artefact was made of alder wood and many of the charcoal samples included alder.

Betula pendula Roth (silver birch) *and B. pubescens* Ehrh. (hairy birch)

It is not possible to distinguish the wood from these two species. Silver birch is a slender tree up to 35m high. It is a quintessential pioneer, requiring light and showing a preference for dry, sandy soil. It avoids thin chalky soil and is seldom found in very wet soil. Hairy birch is a somewhat smaller tree, growing up to 25m in height. In contrast to silver birch, hairy birch avoids dry environments and mostly occurs in wood fen, on poor and wet soils (Orme and Coles 1985). It colonises raised bogs but also likes the wet marginal areas besides lakes.

Birch often colonises clearings and is often the first tree to establish itself on drying raised bogs. Because the tree has an open canopy, herbs and shrubs can expand underneath. The species is short-lived. Trees of 70–80 years (even 150 years) are found, but generally birch trees do not reach this age. The birch polypore fungus, *Polyporus betulinus*, which projects conspicuous bracket-shaped sporophores from the trunk of the infected tree, often rots the inner timber causing downfall after only sixty years growth (Stortelder *et al.* 1998). At Killoran 226, isolated polypore fungi were found, indicating that birch trees died an early death at this spot. Hairy birch, but not silver birch, can be affected by the ascomycete *Taphrina betulina* (Jahn 1979). The infected wood forms so-called witches' brooms or birds' nests. These galls are balls of many buds crowded together. This infection is found especially on free-standing trees.

Betula in general has excellent qualities as firewood but burns only a short time. The charcoal has been used to melt iron in factories. The wood is strong and tough but not durable out of doors, rotting easily. The wood is best suited for fine objects such as furniture, turned bowls, tool handles and broomsticks. The bark can be used for shoes or roofs. It was even used sparingly mixed with flour in bad times. In the past, the sap, tapped in spring, was a major source of sugar and was fermented to make beer and wine. Pitch to attach implements such as spears to shafts may be made from birch.

Betula was identified at thirty-three sites but was the dominant species at only two sites (Derryfadda 210 and Killoran 235). At some sites, such as Killoran 18 and Killoran 226, birch grew locally when the structure was built. Considering the bog environment, it is tempting to suggest that the wood represents *B. pubescens*, the hairy birch. However, *B. pubescens* and *B. pendula* can grow together.

Corylus avellana L. (hazel)

Hazel is mostly a woody, multi-stemmed bush that grows up to 6m in height. Nowadays, it is mostly found as understorey trees up to 8–10m or in hedges on relatively rich soils. It is often associated with oak (in the Boreal also with Scots pine). It was one of the first trees to grow widely after the last glacial period. Rackham (1980) mentions its natural association with ash and maple in England. Hazel tolerates a wide range of conditions (but not waterlogged situations) and generally demands light. It can reach seventy years of age, but being a major coppice tree can become much older. It can be cultivated in beds. The first fruiting occurs after some ten years. After forest fires, hazel can expand from root shoots.

It produces good firewood. The wood is soft and easy to split but remains tough and flexible at pole size. After one or two years exposure, it snaps easily, but if kept dry and stable, it retains its strength for many seasons (Orme and Coles 1985). Because hazel can be twisted and bent without snapping, it is very suitable for thatching spars and wattling, hurdles, hedges, rough baskets and barrel hoops. Hazel can also provide pea sticks, bean rods and faggots for kindling. Hazelnuts provide many proteins and fats.

In Derryville Bog, hazel occurred at thirty-five sites, but less often than alder, ash or willow; nevertheless, it was dominant at twelve sites thanks to its extensive and almost exclusive use in the manufacture of hurdles. Surprisingly, hazel was the dominant species at Cooleeny 31.

Frangula alnus Mill. (alder buckthorn)

Alder buckthorn is a small tree (up to 5m), more often a shrub growing in hedges, in woodlands and on bogs. It grows in marshy or wet places. Guelder rose also prefers this environment, but this tree will grow on more acid soils than alder buckthorn. Alder buckthorn is often found in the same situation as alder, hence the name (Orme and Coles 1985). It is a light-demanding species. It may grow as an understorey tree in woods given light, but also near lakes and bogs, where it may be associated with willow.

Alder buckthorn yields excellent charcoal and has been identified at several prehistoric sites in Ireland. The wood is of no specific use. Beekeepers use the branches to build the skeletons of their beehives and the bark is used for medicinal purposes.

The wood was only identified once in Derryville Bog. It probably grew in the wetter parts of the marginal woodlands surrounding the bog.

Fraxinus excelsior L. (ash)

Ash is a tall tree (up to 45m) of slender girth with an open crown, usually forming part of the canopy layer in woodlands. The open crown often encourages a rich ground flora. It flourishes on moist soils, preferring nutrient-rich environments. It avoids permanently waterlogged situations, in contrast to alder. Although ash is sometimes found in marginal forests and on stream banks, it is normally associated with mineral soils. According to Rackham (1995), ash often occurs in mixed ash, maple and hazel woods in England. It is a light-demanding species with a wide-spreading root system. This can be a nuisance when planted in hedges or near gardens. The seedlings, however, need shade and grow rather slowly at first. Ash can be coppiced. Its age seldom exceeds 150 years.

Fresh or dry, ash is considered the best firewood, and its charcoal is held in high esteem. The timber is valuable, although it may rot rapidly when exposed. It is relatively easy to fell and split using an axe and a wedge; tangential splits are especially easy to get. It can produce long posts. The wood is hard and elastic and therefore is often used for sports equipment (e.g. hurley sticks), handles, agricultural implements, furniture, shafts and spears. It is easy to use when green. Good straight-grained timber bends well when steamed. Young shoots can be used as winter fodder for cattle. Lathe-turned objects such as bowls and small ornaments are sometimes made of ash. The most valuable timber is that from quickly growing trees.

According to the pollen analysis (Chapter 6), ash may have been widespread in the Derryville area, albeit in low numbers, until prehistoric man opened the forest canopy. This may have favoured its expansion, because of the increased light. Decrease in pollen values could represent intensive use of ash. Ash occurred at almost all sites (forty-five) and was generally in good condition. It was normally found only as a minor component, but at seven sites, it was the dominant species and at two sites, it was as important as alder. Three artefacts were made of ash wood.

Ilex aquifolium L. (holly)

Holly is an evergreen, erect tree that grows from to 15 to 25m in heigh with individuals reaching great ages. It mostly occurs as an understorey tree or shrub because it can endure shade and is frequently a component of oak woodland. It is also found in hedges surrounding fields. Holly dislikes continental conditions and harsh winters; in Atlantic conditions, such as in Ireland, it abounds but avoids very wet soils. The seedlings are hardy, but grow rather slowly. The tree favours abandoned agricultural clearings.

Holly produces good firewood. The white, very fine-grained timber is hard and heavy, and does not appear to have had any specific uses other than for carving and turning (Orme and Coles 1985). It tends to distort when drying, so it is generally used in small pieces. It is unsuitable for outdoor use. Coppice shoots are frequently used for walking sticks. Birdlime, made of boiled and fermented bark, was formerly used to trap birds. The leaves, despite being prickly, are grazed by deer. The upper foliage of mature trees and the leaves of saplings provide nutritious and apparently highly palatable fodder for domestic animals such as sheep. Historically, it was coppiced and pollarded for this purpose.

In Derryville Bog, holly was identified at eighteen sites, generally in low quantities.

Pinus sylvestris L. (Scots pine)

Pinus is a light-demanding tree, often found with a dense undergrowth of grasses and bushes. Its maximum height is about 35m, its maximum age 150 years. Seedlings can be damaged by frost. By the time it reaches maturity, the side branches have fallen off (because of the increasing shade) and left the trunk clear. It is very wind firm and able to thrive in exposed and infertile environments, on bare mineral soils or sand. In the Boreal period, pine trees established themselves on peat. It was especially common before 6000 BC in Ireland, before the bogs started to expand and drier climatic conditions prevailed. Pine is a characteristic species of the transition from fen wood to raised bog (Godwin 1975). The buried trunks of large pine forests can be found under the peat bogs of Ireland. It is generally assumed that pine became extinct and was reintroduced some centuries ago.

Pine is not ideal firewood but is very often found in Mesolithic hearth pits. The wood is fairly hard and strong. The best quality is slow-grown pine. It is used for various construction and joinery purposes. Pine lasts very well in wet conditions and was formerly used for water-wheels and for building piles. However, only mature trees are of value. The resin of pine yields turpentine. Other products include rope from the inner bark, tar from the roots and a reddish-yellow dye from the cones.

In Derryville Bog, pine was only identified at one site. Pine trunks, however, were present in the peat, and one radiocarbon sample of such a pine described the base of the fen peat formation (see Chapter 3). These pines represented relics of the Boreal pine forests mentioned above. Casparie also found badly grown bog pines at the top of the fen peat, indicating the occurrence of small pine stands under marginal conditions. This would mean that pine was not totally extinct in Ireland at the time (see Chapter 6).

Prunus avium/padus L. (wild cherry/bird cherry)

It is impossible to distinguish these two species. *P. avium* is a medium to tall tree that reaches 30m or more. It is found in woodlands and hedges on light, somewhat fertile, well-drained soils. It prefers light, although it can tolerate some shade, and grows fast. According to Rackham (1980), it can occur in somewhat acidic ash-hazel woodlands. Witches' brooms are not uncommon in wild cherries and are sometimes seen in considerable numbers. The tree produces abundant suckers, and these can develop into small groups of trees. It produces inferior firewood. The timber is reddish-brown, rather heavy, tough and hard. Large trunks are very valuable because of their decorative timber. However, it is unsuitable for outdoor work, as it is subject to decay. Mostly, it is used for small pieces of furniture, turnery and similar work. The reddish fruits are quite edible, although not very sweet.

P. padus is a much smaller and slender tree or bush with a rounded crown and an erect trunk, which can grow to 15m in height. It is less common than *P. avium*, occurring especially in marginal forests, and is nearly always solitary. It can live for eighty years. The smell of the wood is unpleasant, and the colour is paler than that of wild cherry. It has no economic value, although sometimes it was used for barrels. The fruit is black, shiny, harsh and bitter.

The *P. avium/padus*-type of wood occurred at just four sites in Derryville, especially on the eastern side of the bog. The finds of roots suggested a local origin. It was also found in the charcoal samples, suggesting that bird cherry rather than wild cherry was involved.

Prunus spinosa L. (blackthorn or sloe)

Blackthorn is a spiny suckering shrub or tree of up to 5m in height, often found in woodlands where the canopy has been opened or in forest margins. It is common in scrub vegetation and along streams, where it grows only 2m high, sometimes with alder (Orme and Coles 1985). It is a very thorny shrub. Generally, blackthorns do not live beyond forty years, but they readily produce new shoots from their roots. Therefore, they were used to shield young trees to prevent grazing animals from eating the tender shoots. Eventually, these big trees would grow above the shrub, and because blackthorn is shade-intolerant, eventually the shrub was killed by the increasing shade produced by the expanding tree.

Blackthorn wood is ideal for walking sticks. It has been used to make the teeth of hay rakes. The bitter fruits are used to colour and flavour gin.

The wood from blackthorn was identified at two sites in Derryville. It was also present as charcoal, implying that it probably grew near the bog. The presence of thorns would have made blackthorn less attractive for track construction.

Pyrus pyraster Burgsd. (wild pear)/*Malus sylvestris* Miller (wild crab apple)

These two genera cannot be distinguished, and this type of wood (Pomoideae) also includes *Sorbus* (rowan and whitebeam) and *Crataegus* (hawthorn). In exceptional circumstances, these latter two genera can be distinguished. In the Lisheen Archaeological Project, wood pieces that were semi-ringporous were called cf. (conform) *Sorbus*. All these genera belong to the apple family, which is sometimes lumped into the Rosaceae family.

P. pyraster (wild pear) is mostly found as a thorny, erect, isolated tree up to 15m in height. It is very rare nowadays. The tree has an open, pyramidal crown with relative few main branches and mostly one stem. It reproduces generally through suckers. The wood is good as fuel. It is hard and fine grained, pale, pinkish-brown in colour and used for high-quality turnery. It is not durable out of doors. It is sometimes used instead of box (*Buxus sempervirens* L.) for engraving purposes. A wide range of musical instruments, such as clarinets, flutes and harpsichord, are made from pear wood.

M. sylvestris (wild crab apple) is a small tree or bush that seldom grows above 5m to 10m in height. This small tree rarely has a straight trunk and shows a wide, dense head of interlacing branches. Sometimes spines are present on small branches. Crab apple is common in open oak woods, but always in small numbers. It is a light-demanding tree, but in nature belongs to the woodland. The fruits of crab apple can be dried and stored for the winter season. Mixed with rowan berries, they make good jellies. By systematic pruning, the natural formation of dense masses of branches can be prevented and fruit production stimulated. The wood often reacts to pruning by producing curl wood (see additional files at www.mglarc.com). When dry, crab apple is one of the best woods to use as firewood. The wood is brown, hard, durable and of good quality, with a close and even structure. However, it is little used because of the crooked trunks and the small size of the logs. When it is used, it is for turned articles, such as bodkins, tool handles and screws, woodcarving and similar small work. Mistletoe is found more frequently on apple trees than on any other tree.

In Derryville Bog, Pomoideae was a common wood type, although never in great quantities. It was identified at thirty-four sites and one artefact was also made from this type of wood. The presence of much curl wood could suggest some kind of management, as a result of pruning. At Cooleeny 31, one fine round ball of curl wood, certainly derived from a heavily pruned apple tree, was found. It is important to note that it was not possible to exclude other members of the Pomoideae family, such as rowan (*Sorbus aucuparia*), whitebeam (*Sorbus aria*) and whitethorn (*Crataegus*) in the identification of some pieces.

Quercus robur L. (pedunculate oak) *and Q. petraea* Liebl. (sessile oak)

These species cannot be distinguished from each other. Since prehistoric times, oak has been the predominant timber tree throughout Europe. It is a pioneer tree, doing well on infertile and acidic soils. The trees can reach a very great age, sometimes over 500 years. The seedlings can grow in shade as well as in open land, although Rackham (1995) has concluded that the combination of damaging influences such as mildew, caterpillar defoliation and grazing pressure by deer has diminished the natural regeneration of oak in England. In more natural woodland, they would thrive in clearings caused by fallen trees.

Pedunculate oak grows up to 25m high, rarely 40m, attaining a girth of up to 9m. Its canopy is open enough to give undergrowth a chance because some light can reach the ground. Fertile acorns are produced when trees are ten years old, and these are more popular with animals than the acorns produced by the sessile oak. The nuts are poor in fats but rich in starch.

Sessile oak can reach heights of 40m and girths of up to 13m. Leaves often stay on the tree during winter. Acorns are first produced when the tree is about forty years old, and crops are seldom as regular and heavy as those of the pedunculate oak. The sessile oak favours lighter, more porous soils than the pedunculate oak, preferring sandy and stony, somewhat acid soils where there is abundant rainfall combined with good drainage. Pedunculate oaks would perhaps have grown on the heavier and wetter soils. Both species may well have been present around Derryville Bog.

Oak makes good firewood when well dried. The wood has generally been favoured as timber for buildings and for planking, and is easy to split using wedges when straight grown. Split roundwood has traditionally been used for hurdle work (sails) and coppiced oak is not unknown (Orme and Coles 1985). After felling, the outer thin sapwood zone of an oak stem rots in a few years, in contrast to the heartwood. The tannin in the heartwood prevents attack by fungi and is thus extremely durable. Oak is always chosen for the staves of beer barrels and sherry casks, the spokes of wooden wheels, the rungs of strong ladders and the timbers of buildings.

Fast-grown oak holds proportionally stronger summerwood than slow-grown oak and is therefore stronger. Since it also bends well and is impermeable, it is used for the building of boats. Bark strips from felled logs were formerly widely used for tanning hides to turn them into leather. The acorn is a welcome food for pheasants and was used formerly as a valued food for fattening pigs. Good crops of acorns only occur at intervals of a few years—in 'mast' years.

Oak was not an important species in Derryville, but many oak stumps were noted on the slopes surrounding the bog and several oaks were found in the peat, indicating drier conditions in the past. Some of these bog oaks were dendrochronologically dated, providing an important framework for the interpretation of the dynamics of the bog. Oak wood was present at thirty sites, and one artefact was made of oak. Oak was used almost exclusively in the single-plank trackways Cooleeny 22 and Derryfadda 23. The wood from Derryfadda 23 was identified by the IAWU and is not described here. Apart from Cooleeny 22, oak was the dominant species at four sites. Several of the sites also included roundwood, usually of relatively small diameter. In general, the wood was in reasonably good condition. At Killoran 69, however, the wood was of extremely bad quality and rotten to its very core.

Salix spp. (willows, such as *S. alba* L. (white willow) and *S. fragilis* L. (crack willow), both of which may be introduced to Ireland, and sallows such as *S. caprea* L. (goat willow) and *S. cinerea* L. (gray willow))

There are many species of willow and the species are themselves very variable. Prehistoric willow wood cannot be distinguished further into species. Generally, willow is not a natural woodland species but a solitary tree growing in open conditions. It mostly favours wet conditions and may be a pioneer species on wet soils, but not generally on peaty soils. The use of willow depends on the species concerned because some grow as shrubs and others as trees. Most willows are characteristic trees of lowland parts, lining the banks of rivers. To germinate, the seeds need fresh damp earth or mud directly after maturing. Generally, *Salix* species are divided into tree willows, osiers and sallows.

Willows are often pollarded and the vigorous shoots on pollards have been used for a variety of purposes. White willow is mostly grown in a rotation of 15–20 years. Sallows, on the other hand, can form secondary woodland and have strong pioneering qualities. They are readily coppiced. The osier willows, *S. viminalis* and *S. purpurea,* were possibly introduced into Ireland by the Normans. These small shrubs (3–6m high) produce very supple stems. On the Continent, willows were being cultivated in osier beds by the Iron Age (e.g. the hurdle trackway near Emmerschans, dated to 170±50 BC (Casparie 1986)).

Used as firewood, willow burns fairly well but sparks dangerously. The vigorous shoots of pollarded willows are particularly suitable for hurdles, wattle work and basketry. Poles from the crack willow and white willow may also be used to make stakes, hoops for barrels and tool handles. The roots of crack willow yield a purple-red dye. The wood is light, soft and very variable. It is used in the construction of fast sailing boats and coracles (by covering willow rods with hides), cricket bats, artificial limbs and toys. The flexibility of the sally shoots of coppiced sallows makes them ideal material for baskets, frames, ties, hurdles etc. Sallows are favoured by bee-keepers because they are among the first nectar producers in spring. The supple stems of osier willows are exceptionally suitable for basket making. More generally, the feathery covering of willow seeds has been used for stuffing mattresses. The foliage is good winter fodder. Bark mixed with oatmeal was used in times of famine in Scandinavia. The bark can also be used for tanning. The stem is bitter, due to the presence of tannin and salicin in the bark.

The following species are probably the most common in Ireland (all of them sallows except *S. pentandra* L.):

S. pentandra L. (bay willow), a 5–7m high tree, grows along nutrient-rich peats and lakes. According to Scannell and Synnott (1987), it is only native in the northern counties.

S. cinerea (rusty willow or gray willow), a 2–10m high shrub, is the most abundant species by far, growing in wet places, but not on peaty soils. Its wood is used for basketry.

S. aurita (eared willow), a 1–3m high shrub with many spreading branches, is especially found in wet places on acid soils (e.g. along the edges of bogs). Its wood is of no specific use.

S. caprea (goat willow or common sallow), a 6–7m high bush with smooth bark and a trunk rarely exceeding 0.5m, grows on the edges of ponds or patches of waste-land and along the edges of bogs. It is a source of 'palm blossom' and of nectar for bees. The wood has been used for clothes pegs but the timber is of no specific use.

S. repens (creeping willow), a 1m high dwarf form, grows on high-level blanket bog. It was one of the first shrubs to return to Ireland after the last glacial period. It is not likely to grow in Derryville Bog.

In Derryville Bog, willow occurred at thirty-seven sites but only at two sites as the dominant species (the non-archaeological site DER246 and trackway Killoran 248). Clearly, willow was not used for hurdle making in Derryville. Leaves of goat willow were identi-fied at the *fulacht fiadh* Killoran 240.

Sorbus aucuparia L. (rowan or mountain ash) and *S. aria* L. Crantz (whitebeam)

With current techniques, it is impossible to separate the wood of rowan and whitebeam when identifying prehis-toric remains. Often, these species are lumped into the Pomoideae type of wood, which also includes *Crataegus* (hawthorn). They all form part of the Pomoideae family (sometimes included in the Rosaceae family).

Rowan, also called mountain ash or quicken tree, is a short-lived small tree that grows up to 15m in height. It has a smooth, silvery-grey bark and an erect trunk. It is very common in open woodland and scrub, by mountain streams and in valleys. Rowan prefers moist areas and light soils, and avoids clays and limestone. It is a light-demanding species. Rackham (1995) mentions that it often occurs in oak woods in England, but also with horn-beam (*Carpinus betulus* L.) and hazel. The seedlings are hardy and grow fairly vigorously, giving the tree some pioneering qualities. It can germinate on dried raised bog surfaces. The seedlings are shade-tolerant.

The timber is hard and smooth and suitable for turn-ery. It can be used for cabinetwork, handles and other household utensils (such as spoons); cartwheels, planks and beams. The wood should be suitable for firewood. Coppice shoots are sometimes used for hoops and crates. The leaves are very palatable to grazing animals. The bark can be been used as winter fodder for cattle. The clusters of orange/red fruits have a high vitamin C con-tent and can be used for jellies (with apples) and wine. In years with poor rowan fruiting, the apples suffer because a caterpillar that lives on rowan berries then turns to apples instead.

Whitebeam is generally a rather small tree with a wide, dense growth; a smooth, dark-grey bark and oval leaves with a silvery underside. The fruits are bright red and very popular with birds. It grows up to 20m in height and prefers calcareous and limestone soils. It is very common along hedgerows and in small woods. It requires light and is intolerant of shade. The flowers are attractive for bees. The timber is not of economic importance because large trees are scarce. The wood is strong, fine-grained, very hard and white, and can be worked to a smooth finish. It can be used for turned bowls, plates and small decorative woodwork.

It is not certain whether whitebeam is native to Ireland but *S. hibernica* certainly is. This latter tree can easily be confused with whitebeam but is generally found as a 6m high shrub or small tree in woodlands, hedges, rocky places and scrubs on limestone soils. The berries can be eaten after the frost has turned some of the starch into sugar.

It is very likely, considering the preference of white-beam for chalk and limestone soils, that the *Sorbus*-type wood identified in Derryville Bog is rowan rather than whitebeam. *Sorbus*-type wood was found at eleven sites, but never in great quantities. As mentioned above, it also cannot be excluded that the cf. *Sorbus*-type is in fact anoth-er member of the Pomoideae group. The *Sorbus*-type was identified on the basis of its semi-ringporous cross section.

Taxus baccata L. (yew)

Yew is usually found as a freestanding, rounded, densely branched tree with a massive trunk. Normally, it is not very high, but sometimes trees up to 28m in height can be found. Yew will grow on any soil and also under the shade of other trees, but it is mostly solitary. When old, yews are always hollow. They grow to by far the greatest age of any native tree. Very old yew trees tend to create new trees where the main branches reach the ground and start to root. This property of layering can eventually result in the growth of a ring of trees growing after the original trunk has died. Yew also grows very slowly. During interglacial periods very large yew trees grew in many places; in the Holocene bog, yews preserved in the peat have been dug up in places in Ireland. All parts of the yew are poisonous to humans, but birds eat the berries. The seedlings grow slowly for the first two years, thenceforth making vigor-ous progress if conditions are suitable. Godwin (1975) describes the occurrence of yew in the buried forests of the fens of eastern England, its growth in the Neolithic in the Somerset Levels in fen carr dominated by alder and birch and its association with ash and alder in fen woods. The ancient fen wood habitat of the yew, therefore, seems well documented (Orme and Coles 1985). Several large yew trunks were noticed in Derryville Bog.

The timber is scarce in suitable sizes, but of consider-able strength; it is very tough and elastic, with a fine tex-ture and resists decay when exposed. It is first-rate fire-wood. The Vikings used yew wood for nails when build-ing their ships. Bows are traditionally made of yew. Other uses include turned domestic objects, such as the vessels common in eighteenth-century Ireland, bowls, dagger handles, furniture parts and fence posts. The bark can be used for braiding and weaving. Nowadays, many artists use bog yew for carving purposes. In Poland, yew is often found in churchyards. Originally, it was planted for roof-chip production (W. Casparie, pers. comm.). One tree was traditionally planted beside the path to the main doorway from the funeral entrance, the other by the path leading to the lesser doorway.

Yew was identified at fifteen sites in Derryville Bog, mostly as single occurrences and often representing non-archaeological wood. Two artefacts, however, were also made of yew. The wood survived well in the peat. Yew also occurred regularly in the charcoal.

Ulmus glabra Hudson (wych elm)

Elms are often very large trees, 30–35m high, with a dome-like crown. The trunk is seldom straight. They occur in woodlands, hedges and beside streams, but favour fertile soils rich in nitrogen. They are often found in oak and ash woods on limestone. Elms were coppiced and pollarded in the past. The wood is susceptible to rot and disease, notably Dutch elm disease, which is caused by a pathogenic fungus, the spores of which are transported trough various elm bark beetles (*Scolytus* species). This, however, is probably not the cause of the decline of elm noted in pollen diagrams some 5,000 years ago. It is very unlikely that during such a long period, no elm varieties with resistance to this disease developed.

Elm wood is tough and heavy when green but soon dries to a much lighter state. Straight-grown elm is easy to split, but twisted and knotted parts can be surprisingly difficult to cope with. Historically, it has been used for coffins, but also for water pipes and bridge piles, as long as it is used underground and waterlogged. It has also been used for wheels, chairs, mallets and longbows, where its high tensile strength at right-angles to the grain is of particular value (Rackham 1980; 1995).

Surprisingly, elm was not often found in Derryville Bog, occurring at just five sites, mostly as single occurrences. At Cooleeny 31, however, three pieces were found together, seemingly from the same tree. In the pollen diagrams, in contrast, elm was always an important constituent of the dryland vegetation. The minimal use of elm, despite its high pollen values and the amount of trees available for use, is difficult to explain. The low occurrence of elm might indicate that the dryland vegetation was hardly used at all and that most of the ash and oak used in the trackways was of local origin.

Viburnum opulus L. (guelder rose)

This small tree or shrub (up to 5m in height) is generally found as an understorey tree in alder forests or in hedges. It mostly occurs on damp, nutrient-rich soils, such as fen woods and overgrown stream banks. It may grow with alder buckthorn, but the latter tolerates more acid soils. Guelder rose easily forms suckers. It is poisonous and the wood has no specific value.

Guelder rose was identified at one site only. One artefact was also made of guelder rose.

The natural woodland situation around Derryville Bog

The growth of trees in and around bogs

Only a few trees were found surrounding Derryville Bog when excavations started. Most of the area was under cultivation, and the trees were mostly confined to hedges surrounding field plots. The bog itself had no tree vegetation. The excavations conducted during the Lisheen Archaeological Project yielded large amounts of wood, indicating that this situation must have been different in the past.

The non-archaeological wood in Derryville Bog was derived from fen peat or its marginal surroundings. Some of the wood, however, was found in raised bog deposits.

Generally, fen peat contains few wood remains. During relatively dry periods or after bog bursts, and wherever well-drained areas develop during the last stages of fen peat development (when organic material has almost filled up the depression), trees can become established (see Chapter 3). Most trees grow in the marginal areas surrounding the fen ('brook' forest), in pools or alongside streamlets. The main constituents of this marginal forest are alder in wetter parts, oak in drier areas and birch (generally *Betula pubescens*). Fen peat is not poor in nutrients but because of the rising water table, the conditions for tree growth are marginal, resulting in a rather open forest with stunted trees. In clearings, enough light can reach the ground to allow some ground vegetation.

There are generally no trees on raised bogs. The only exception is birch (*B. pubescens*), which colonises the peat as soon as the surface has dried out a little (oxidised). Raised bogs have acid water. The rather steep slopes on the sides of a dome-shaped raised bog create a dynamic environment where acid water from the bog can mix with nutrient-rich water from the slopes surrounding the bogs.

The local environment changes dramatically after a bog burst, which results in a considerable lowering of the water level (examples in Derryville Bog are 1m for Derryfadda 213 and Derryfadda 13 and 1.5–2.0m for Cooleeny 31; see Chapter 3). Marginal wooded areas expand considerably after such events, creating more space and nutrient-rich situations for forest growth.

Trees need nutrients to grow, therefore, whenever trees are found on or in the peat, this means that they must have been in contact with nutrient-rich water or underlying mineral-rich subsoil. Wherever the remains of yew or oak were found in Derryville Bog, the trees must have had contact with the underlying mineral soil. Most trees thus started growth before the peat developed and then survived less favourable conditions.

Trees have no difficulty establishing themselves on the fen peat during the last stages of fen peat development, when drier areas come into existence. Trees are not generally found on raised bog, but there are exceptions. Bogs are complex ecosystems with much local variation. On raised bogs, relatively dry hummocks alternate with wet hollows. During dry years, the dry patches will be accessible and, wherever there is access to nutrients, trees can gain a foothold, until situations become less favourable again and they die off. Considering the historical time span represented in the Derryville excavations, at most times (after the initial phases of fen peat development), at least some trees, dead or alive, would have been present on the bog itself. Most of the species identified, but especially alder, birch, Scots pine, oak, willow and yew, could have grown on the bog or its direct surroundings.

Identification of non-archaeological wood at Derryville

Fig. 7.1a summarises the results of all identifications of wood from non-archaeological sites in Derryville Bog, regardless of date or location. The wood pieces (n=315) were derived from the bog or the marginal forest directly surrounding the archaeological sites.

About 40% of the pieces were the remains of roots, clearly from trees growing *in situ*. Most of the roots (75%) were of *Alnus* (alder); the remainder (15%) were *Betula* (birch), *Salix* (willow) and *Fraxinus* (ash). Four alder branches were found in the Derryfadda area. The remainder of the non-archaeological samples represented parts of trunks or the upper parts of trees.

Almost half of the analysed wood samples were identified as *Alnus*, but *Fraxinus*, *Quercus* (oak) and Pomoideae must have been common too. Noteworthy low values for *Corylus* (hazel) were found; clearly, hazel did not like the marginal environment. *Taxus* (yew) also occurred in low percentages. It is possible that some of the yew was actually derived from trees that started their growth prior to fen peat development. *Pinus* (pine) was not identified. Casparie, however, noted the presence of pine stumps at the base of the fen peat dating to 6000 BC and also in the peat at a level corresponding to a date between *c.* 2500 BC and 2000 BC (see Chapter 3).

At Killoran 241 and Killoran 226, mature *Fraxinus* and *Quercus* trees had collapsed on top of the constructions. The presence of mature trees points to a dry period of considerable length in which ash and oak could germinate and grow. Their death indicates that gradually their growth circumstances became less favourable. In the bog environment, this is most likely explained by an increasing water level.

The eastern area, Derryfadda (Fig. 7.1b), yielded most pieces for the identification of trees growing *in situ*, and therefore the results resemble Fig. 7.1a. In the Derryfadda area, the relatively high values for Pomoideae (crab apple-type) are particularly interesting. This tree must have been common in the area, although it might have been suffocated by the peat that encroached on the marginal forest there. Values for *Salix* and *Betula* were low in the Derryfadda area, but *Fraxinus* was common.

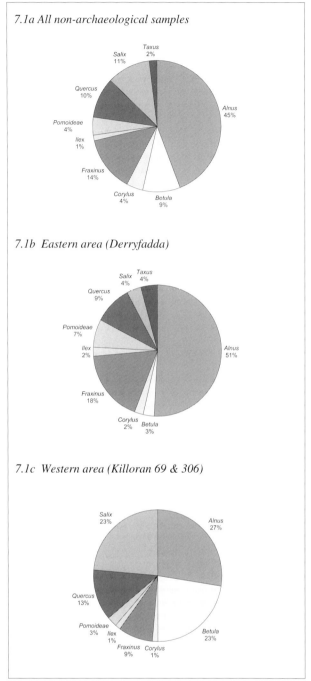

7.1a All non-archaeological samples

7.1b Eastern area (Derryfadda)

7.1c Western area (Killoran 69 & 306)

Fig. 7.1 Non-archaeological wood identification.

Although the number of identified pieces from the western area was much lower, two sites, Killoran 69 and Killoran 306, yielded somewhat more identifications. The results from this area (Fig. 7.1c) show a different picture than the east. Values for *Alnus* are much lower, and those for *Salix* and *Betula* are higher. *Salix* was largely found around the non-archaeological site DER246. The high values for *Salix* indicate a local flow of water (brooklets or streamlets from the discharge channel). The high values for *Betula* show that this tree grew locally on the peat, and hence the surface of the peat must have been somewhat oxidised at the time of growth.

Age of wood collected at Derryville

Besides species composition, age also gives information on the local woodlands. Therefore, some attention was given to the age of the wood from non-archaeological sites (Fig. 7.2).

Normally the height and girth of a tree increases with age until 'old age' begins and growth slows down. The circumference of a solitary tree 1.50m in height is often taken as a guide to its age. However, the amount of wood added each year depends on many factors, including the amount of foliage. A tree growing in dense forest adds less wood than a solitary tree. Trunks were seldom found in Derryville, and most of the samples collected were small pieces of roundwood or brushwood, in which annual rings are generally narrower than in trunks.

Most of the non-archaeological wood was derived from trees growing on the bog. A minor portion represented trees growing on the bog margins on mineral subsoil, in a marginal area that in the past extended further into the basin. The expansion of the peat over time (see Chapter 3) gradually pushed the margins and its wood cover outward. Trees from the bog surface probably grew in rather open situations, where light promoted growth to some extent.

Alder

The age of alder trees varied widely, but most were between ten and forty years old. The oldest sample was a piece of trunk approximately eighty-three years old. Generally, the diameter of the brushwood varied between 2cm and 5cm; only a few roundwoods with a diameter greater than 5.5cm were found.

Willow

Growth seemed to be quite variable, but generally not very fast. Willow normally grows quite quickly, but this was not apparent in the Derryville samples. The majority of the willow was between ten and forty years old (Fig. 7.2a, the line was drawn to facilitate interpretation).

Birch

Most of the birch trees might have grown on the bog itself. Their age varied between twenty and forty years. The oldest sample was seventy-five years old. The diameter of the samples varied widely but a good proportion was more than 5cm in diameter. Growth was variable but not particularly slow.

Pomoideae

The crab apple-type of wood (Pomoideae) almost never exceeded forty-five years in age; the majority were between twenty-five and thirty-five years old. Growth was generally slow: 4cm in thirty-five years. This wood type probably did not grow on the bog but in marginal situations in a crowded environment.

Ash and oak

Some of the ash and oak were remnants of mature trees older than one hundred years. Part of these remains point

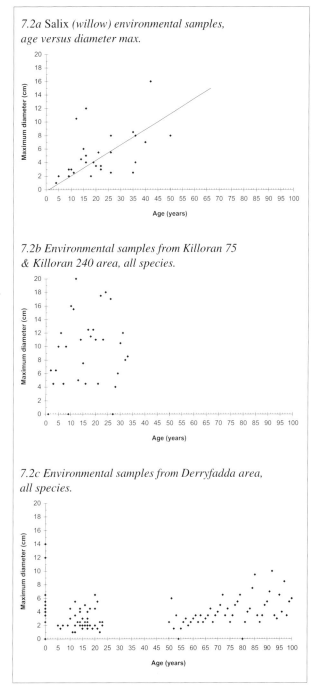

7.2a Salix *(willow) environmental samples, age versus diameter max.*

7.2b Environmental samples from Killoran 75 & Killoran 240 area, all species.

7.2c Environmental samples from Derryfadda area, all species.

Fig. 7.2 Age of non-archaeological wood samples.

to the presence of former dry conditions in the bog after bog bursts. The general growth of ash and oak was slow. Ash trees added only 5cm in forty years. The growth of oak varied but seemed to be faster than ash.

East versus west

Fig. 7.2b shows the age of all natural wood identifications from the western area of the bog, Killoran 75 and Killoran 240, plotted against their maximum diameter. The majority of the trees here were rather young—less than thirty years old—and showed a considerable variation in diameter. The growth rate varied considerably but was generally fast, indicating favourable conditions with lots of nutrients and light. This could point to a regenerating situation or a rather open vegetation. Whether the regeneration was the result of felling or changes in water level was not clear.

Comparison of the western data with those from the southeastern, Derryfadda, area (Fig. 7.2c), shows that conditions in Derryfadda were different. Although the majority of trees here were also between ten and thirty years of age, a considerable number were older than thirty years. Their diameters were generally smaller than those of the trees in the western area, indicating that growth conditions in the southeastern area were less favourable. This could mean fewer nutrients and/or less light. The latter would indicate that the trees were standing close to one another.

The slower tree growth in Derryfadda would support the idea that the southeastern area was a fairly undisturbed natural environment with mature trees and possibly less human interference than the western area.

Results of wood and charcoal identifications

Wood

The results of wood and charcoal identifications in Derryville Bog are presented in bar diagrams in which the total score for each species is given. Fig. 7.3a summarises the identification of the non-archaeological samples of local origin, directly on or around the bog (bog and bog margins). Fig. 7.3b summarises the identification of the wood remains found in archaeological features.

Some of the wood used in archaeological contexts may have had a local origin, but it was assumed that most of the wood will have been brought onto the bog from the nearby dryland and bog margins. Large quantities of wood were necessary for the archaeological structures and it is unlikely that those quantities would have been found on the bog itself.

The results indicate a marked difference between the marginal forest and bog trees, and the utilised woodlands. In the marginal forest, alder was by far the most common species, while hazel hardly occurred. In contrast, hazel was the most frequent species found in trackways, platforms and hurdles (Fig. 7.3b). This difference must be a consequence of the moisture situation locally. On the bog and its margins, wet conditions prevailed, stimulating the growth of alder trees, while drier conditions on slopes leading to the bog and/or dryland were more suitable for hazel growth. Most of the hazel found in the bog must have been transported to the bog from elsewhere. It is difficult to locate exactly where the hazel could have grown, but it is unlikely that the distance was great and it is possible to envisage hazel growth on most of the slopes leading towards the bog.

In total, seventeen wood species were identified in Derryville Bog (see List 1 in additional files). However, only seven occurred regularly, namely, alder, birch, hazel, ash, oak, willow and apple type. The main wood species in archaeological and non-archaeological contexts were the same, and there were few differences between the environmental samples and the archaeological samples, the biggest difference being the use of hazel in archaeological structures against a backdrop of very little hazel in the immediate vicinity. Nevertheless, most of the wood found in archaeological contexts had a local origin, either from the bog margins or the woodlands upslope leading to the bog. This means that, as suggested in the introduction, the wood collected from Derryville Bog does not give much information on the situation in the hinterland.

Charcoal

Fewer species were identified in the charcoal samples (Figs 7.3c and 7.3d). The samples came from archaeological sites, mostly small burnt branches or pieces of roundwood. No chips or pieces from bigger stems were found. The charcoal was in good condition, generally having none or few insect channels, indicating the use of fresh wood. There were no indications for the use of stored wood, gathering of twigs and branches or the use of discarded building material.

The age of the branches was rather young and averaged around twelve years, but some hazel samples derived from trees as old as forty years.

Hazel accounted for the majority of the charcoal. This was somewhat surprising as ash and oak are traditionally better represented and hazel was not part of the local (bog) vegetation. The quantity of hazel charcoal indicates that hazel was readily available. The natural bog vegetation and its immediate margins did not include much hazel, which most likely grew on drier ground, upslope. This would imply that most of the wood used for fires, as well as for the trackways, was collected on the slopes surrounding the bog and not sampled in the immediate surroundings or in the marginal forest. The quantity of willow used for firing was low, compared with the results from archaeological wood and the non-archaeological wood (see Figs 7.3a and 7.3b). It is likely that willow was avoided as a fuel.

The twelve charcoal species identified included the wild cherry type, holly and sloe/blackthorn. It is unlikely that these species grew on the bog proper; they more likely grew in the marginal forest. Yew occurred regularly and was more common in non-archaeological samples. It was probably taken from the bog itself, considering that several yew trunks were found in the bog (see Chapter 3). However, it is unlikely that yew was widespread in the area. Birch formed only a minor part of the charcoal.

The small quantities of elm in both the wood and charcoal samples may have originated in drier land, further away. It cannot be excluded that some of the oak was also of similar origin. However, finds of mature ash and oak stumps in the western area leading towards the bog indicated that enough dry periods existed to enable ash

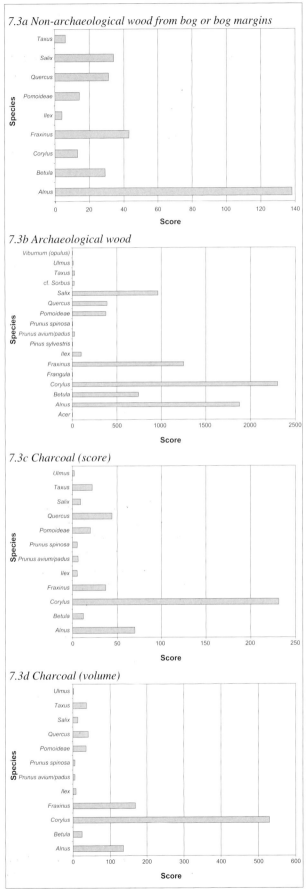

Fig. 7.3 Wood and charcoal identification.

and oak trees to germinate and establish themselves locally, and most wood probably came from these types of forests. It is unlikely that elm grew here.

Conclusion

The results of the wood and charcoal analyses point to a similar origin for most of the material found within archaeological contexts. Most obvious is the dominance of hazel in both groups, indicating the transportation of hazel to the bog through the marginal areas. The high values for alder in the wood samples indicate that trees were available in the bog marginal areas too. Both the wood and charcoal samples were dominated by just seven species. Most of them, except for hazel, would have been found on the bog margins, with willow and birch possibly derived from the bog itself. It is also likely that bog yews were added to the fires.

The western and eastern sides of the bog

Wood

Archaeological remains were found on both the western and eastern sides of the bog, mostly in the marginal areas on the fringes. The quantity of wood remains analysed allows comparison of the wood usage on both sides.

The most conspicuous difference between the western and eastern sides was the quantity of alder and hazel used (Figs 7.4 and 7.5). On the eastern side, alder was the dominant wood species in the archaeological wood samples. In fact, the picture closely resembled the natural woodland situation. The eastern side, therefore, may represent a fairly undisturbed marginal woodland unaffected by intensive human interference.

The woodland cover on the western side, in contrast, included much less alder. This may be due to the fact that the slope on the western side was much steeper than on the eastern side and thus provided less space for alder, with its preference for wet conditions. However, it may also be argued that the bog marginal area on the western side was more intensively exploited. This latter argument, which is supported by the greater use of hazel in the area, is considered the most likely explanation.

In bog marginal forests, hazel only occurs in low quantities because although hazel grows like a weed in Ireland, it does not like the wet conditions that occur in bog marginal areas. The hazel used in the archaeological sites in Derryville probably grew on drier ground somewhat away from the wet margins. The presence of hazel in the western sites is interpreted, therefore, as a sign of locally managed woodland, with the hazel deriving most likely from clearance. The low values for alder and high values for hazel on the western side strongly suggest that most habitation around Derryville Bog was concentrated in this area and that prehistoric man frequently visited the bog margins.

Charcoal

Most of the charcoal samples identified from Derryville Bog came from *fulachta fiadh* dated to the Bronze Age. Loose charcoal spreads from Killoran 18 were also attributed to the Bronze Age. Later samples, such as the post-medieval sample from Killoran 223 (consisting only of oak), were left out of consideration. The charcoal results, therefore, only give information on the Bronze Age stages of Derryville Bog.

The general picture was similar to that of the wood results presented above, with hazel dominating the western side and values for alder much higher on the eastern side. However, the role of hazel in the west was much more obvious in the charcoal samples than in the wood samples (Fig. 7.6). Almost two-thirds of the charcoal found in the western area (Figs 7.6a and 7.6c) consisted only of hazel, which, followed by ash, was clearly the most used firewood in this area. However, the whole range of wood from marginal woodland, including willow, was represented to some degree, with alder present in very low amounts.

On the eastern side (Figs 7.6b and 7.6d), the amount of alder was almost equal to that of hazel and, together, they formed approximately two-thirds of the material. Willow, on the other hand, was scarcely used. Oak was more common than on the western side.

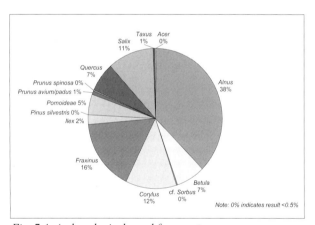

Fig. 7.4 Archaeological wood from eastern area.

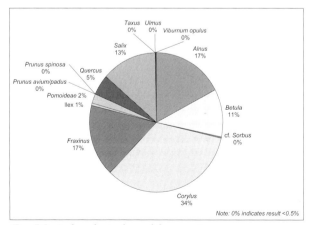

Fig. 7.5 Archaeological wood from western area.

On the western side, the quantity of ash indicated that it was the second choice firewood after hazel. If hazel was left out of the results because of its dominance (Fig. 7.7), this picture was even more evident. In the west (Figs 7.7a and 7.7c), almost all firewood consisted of ash and hazel, the rest being of minor importance.

It can be excluded that the charcoal was the result of burnt houses because it is unlikely that houses were built close to the bog margins, considering the wet circumstances. Houses were more likely built on the dryland. Nor was most of the charcoal the result of burnt vegetation because most samples were taken from within *fulachta fiadh*, where wood or charcoal was used to heat the stones. It is possible that part of the charcoal found in Killoran 18 was derived from local trees, such as alder, yew, birch and willow. Unlike the charcoal samples analysed, the charcoal particles found in pollen profiles indicated an in-wash of charcoal and could thus be connected with the burning of the local vegetation and subsequent erosion.

The charcoal results from the western side did not reflect a marginal woodland but an open environment, with hazel and ash growing somewhat further away on the dryland. The fact that some elm was found within the charcoal samples indicates that at least some material came from the hinterland. Elm may have been used elsewhere, given that pollen diagrams consistently show the presence of elm trees (see Chapter 6). The low values for oak suggest the absence of much oak in the area, at least during the Bronze Age. Yew probably grew on the bog. The remainder of the trees must have grown locally.

On the eastern side, the charcoal results mirror the non-archaeological environmental wood samples, especially with regard to the quantities of alder and ash. Oak was more common in the east than in the west, but was never the favoured firewood within a sample, given the prevalence of alder. Oak was probably growing in the local marginal forest. This was also noted by surveys of this area.

It may be concluded that the prehistoric inhabitants of the area did not, on the whole, gather firewood at the bog margins but brought it to the sites from further afield. On the western side, hazel was the dominant wood species, with a little ash and elm also used. On the eastern side, alder was the preferred species. Most of the wood was fresh and was not, therefore, stored wood or wood picked up from the forest floor. The charcoal results on the western side suggest a rather open situation. The wood was brought to the bog margins from elsewhere, with some local trees being harvested.

Conclusions

The wood and charcoal results clearly show that hazel was well represented at all stages in the area, in natural stands as well as in managed woodlands, the latter especially in the west. Alder was notably dominant in the east.

The marked difference in the presence of alder in the west and east may be due to the much steeper slope on the western side, which would have provided less space for alder, with its preference for wet conditions. Also, Casparie (Chapter 3) identified much more water dynamics on the western side of the bog, which could have prevented the growth of alder trees there. Another factor to consider is the more intensive exploitation or clearance of the bog marginal area in the west.

The occurrence of a mixture of species in wood and charcoal samples indicates that, although the woodlands were certainly managed on the western side, the pressure on the woodlands surrounding the bog was not too great. Trees were at least able to sprout and fill gaps in the marginal areas, maybe during years when the bog was very wet and uninviting and therefore not visited very much. A period of ten to thirty years would suffice for a simple secondary forest to develop. Certainly, in the time after a bog burst, the nutrient-rich situation would have yielded very favourable conditions for quick tree growth.

Woodland management

Introduction

Woodland management forms an integral part of modern society. Since medieval times woodlands have been managed following strict rules. The need for timber necessitated long-term management in which certain trees were allowed to grow for a prolonged period in sheltered conditions. The underwood, providing rods and sails for a variety of purposes, was perhaps still more important. This was cut at regular intervals to provide long, straight poles of vigorous growth and identical form. Apart from woodlands, hedges were needed to keep animals out or in. They were essential, especially where both agriculture and animal husbandry occurred, as will have been the situation in most small prehistoric communities. To avoid animals eating young shoots growing from underwood, it was necessary to fence off specific areas. The trees were not only important for their wood but also other parts, such as bark, leaves for fodder and nuts, were vital supplements to the prehistoric diet and life.

While the management of medieval woodlands is to some extent historically documented, the archaeological record is much more difficult to assess. The amount of prehistoric wood samples analysed is generally too small to allow statistical comparisons. Moreover, it is not always clear how the difference between managed and natural situations can be observed in the wood itself. There are many factors involved, such as local situation regarding nutrients and competitors, climatic conditions and indeed human impact. Also, the prehistoric way of managing woodlands may very well differ from the modern situation.

In this section, particular attention will be given to hazel, because the wood record indicates that the use of hazel increases over time and may reflect woodland management. Hurdles, generally manufactured of hazel, will be dealt with first.

Hurdles

Hurdles are generally made of long, straight rods of about 3m in length. In Derryville Bog, these were primarily of hazel. In managed situations, where stools are cut at regular intervals in a procedure called coppicing, hazel and willow yield such rods in about four years, while it takes birch seven years and oak eleven years (Coles *et al.* 1978). The traditional hazel cycle is often quoted as seven years (Edlin 1973), but there are indications that this has varied greatly over the last few centuries and has tended towards longer cycles in recent years (Rackham 1980; Morgan 1988). According to Lindsay (1974), the optimum cycle for oak was twenty-five years, with the purpose of providing bark for tanning. Ash, too, was often cut at an interval of twenty-five years. Also, alder can produce strong growth after coppicing. After removal of all wood from a stool, new shoots arise on its sides, therefore, the lower part of alder rods often include a thickened part, in the form of an elbow (Coles *et al.* 1978). Long, thin and straight grown rods also occur naturally in close, young, secondary woodlands of birch and ash etc. (Hayen 1997). Hurdles are regularly found in archaeological contexts, not only in Ireland but also in England and Continental Europe, where they are used for many purposes.

On the Continent, modern hurdles are traditionally made in springtime, between April and May, when the wood is soft and pliable and no leaves are present (Tuinzing 1988). Because of the cold, the material is too hard and brittle to weave during winter periods. Also, humidity is an important factor. These hurdles are mostly made of willow. It is not clear whether the situation was different in prehistoric times, using hazel instead of willow. Experiments carried out by Coles and Darrah (1977) indicate that it is quite possible to make hurdles in a (managed) forest situation using freshly cut hazel rods in November. According to Morgan (1988), the making of hurdles was traditionally a winter activity. One could suggest that most hurdles were manufactured in or close to settlements, where the material was stored during the cold period. In contrast to modern practices, prehistoric hurdles were made of complete stems.

The easiest way to obtain material from coppiced stands of hazel is the complete clearance of stools, or clear-felling. The practice of clear-felling would leave a crop of stems of uniform age but of varying diameter because of varying growth rates of the stems (Rackham 1977).

Draw-felling is the selection of suitable rods and poles from each stool, leaving other stems to continue their growth. The result is material of uniform size but varying age. This method is less practical because more stools are needed to obtain the material and the working area around and in a stool is limited because of the remaining rods.

Rackham (*ibid.*) introduced the term 'topping for rods' where leading shoots of rods were cut away, resulting in laterally grown shoots and an injured stump. Normally, a hazel stem grows straight for one or two

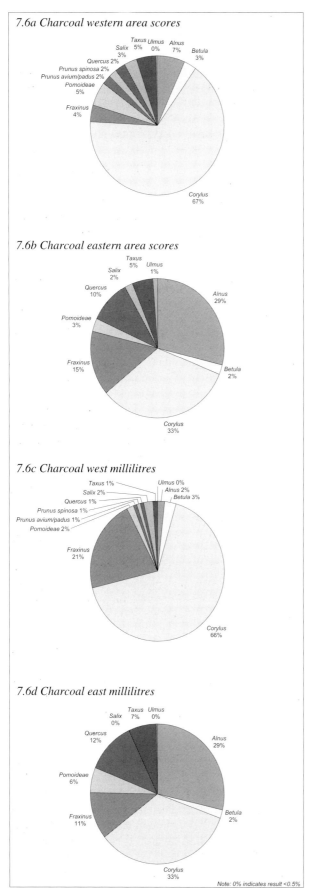

Fig. 7.6 Comparison of charcoal from eastern and western areas.

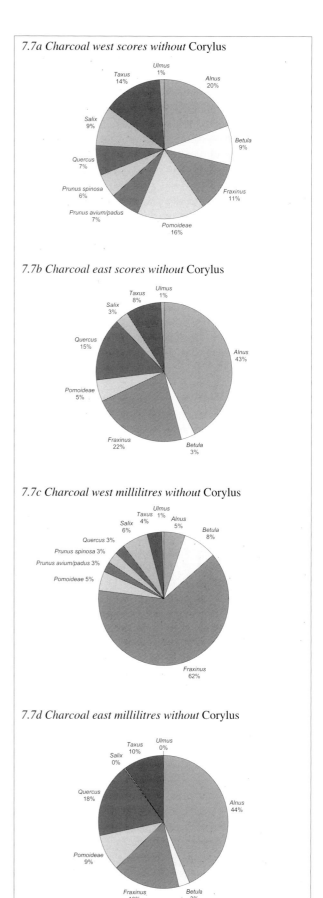

Fig. 7.7 Comparison of charcoal from eastern and western areas with hazel omitted.

years and then develops a kink at the end of a years growth. This kink is sometimes exaggerated by an injury to the stem during its lifetime. Apparently, the tops were harvested after some years but growth of the remaining rods continued for a few years afterwards. This procedure results in a recognisable injury and, moreover, leaves marks in the formation of annual rings.

The age and diameter of individual rods and sails from prehistoric hurdles can possibly give information as to whether one of the above mentioned strategies were involved. In the Somerset Levels, Morgan (1988) has investigated Neolithic and Bronze Age hurdles, mostly made of hazel. She concluded that many variations existed between the tracks as well as through time, which may or may not have resulted from human preference and locally available material. Draw-felling was, according to Morgan, the most likely method applied to provide the wood for the trackways. The quantity of rods required, the fast growth rate and size uniformity pointed to extensive cutting from stools which had been cut over before. Hurdling is said to last about five years when submerged in fresh water, and eight to ten years when used upright (*ibid.*).

In Derryville Bog, wood found in archaeological contexts forms only part of the human activity in the area. Although much of the wood represents material from the marginal forest, it is clear that hurdles must represent an organised situation. It is unlikely that unmanaged wood yields enough regular, long, uniform and knot-free rods for the fabrication of hurdles. The presence of a hurdle must therefore reflect regular manufacturing of hurdles on the dryland. All the hurdles from Derryville Bog were found in Killoran, crossing wet areas. Because of their flexible and closely woven structure, hurdles provide a very effective way of crossing a wet area by spreading the weight of traffic, indicating that local people knew their environment.

The hurdle rods analysed from Derryville Bog were generally taken from a part of a panel where all rods were represented. This means that the maximum diameters noted for these samples do not always indicate the width of the base of the rods where they were attached to the stool.

The four trackways that utilised hurdles were Killoran 75, Killoran 312, Killoran 314 and Killoran 315, all dating to between 400 and 5 BC.

Killoran 75

Killoran 75, dated to 385–50 BC, apparently included two hurdles, one overlying the other. The lower hurdle was broken and consisted primarily of hazel in bad condition. Most of the wood was felled after the growth season. Other species included were some birch, willow and elm. The latter must have derived from the dryland. The birch rods were between seventeen and twenty-two years old. Hazel, on the other hand, was much younger. The hurdle was made of wood between three and eight years old, with notable peaks at four and five years (Fig. 7.8). Although the quantity was rather low, it cannot be

excluded that these pieces result from clear-felling. It is also clear, considering the age pattern of birch that this material came from a different source, probably local. This could mean that the hurdle was manufactured close to the site, incorporating local birch shoots. The elm rod in the hurdle was three years old and could, therefore, have come from the same area as the hazel.

The 7.5m long hurdle on top of the broken one was the most conspicuous aspect of this site. This hurdle was made of birch and ash but primarily hazel. Two pieces of ash were incorporated as sails. Their ages were twenty-two and twenty-nine years. One rod was of birch, which was three years old, while one sail was thirty years old. The hazel sails varied widely in age, between four and twenty-seven years, with a peak at seventeen years (Fig. 7.9). Although size is an important factor in choosing the material for a hurdle, the size varied considerably from 0.5cm to 5cm in diameter. The sails varied from 3cm to 5cm in diameter and the rods were 1cm to 3cm in diameter. The rods varied in age between two and fourteen years (Fig. 7.10) but were primarily six years old, with notable drops at four and seven years. The clustering of the ages seems to indicate clear-felling, rather than draw-felling, following a six year cycle.

The difference between rods and sails is apparent in Fig. 7.11, where one cluster of small and young pieces represent primarily rods and a group of larger pieces of older age represent sails.

It is interesting to compare the age distribution of the other components of this site with those of the hurdle. The three important components were birch, ash and willow. They represent brushwood and roundwoods. Without following the detailed archaeological phases, and excluding pieces older than forty years, some trends are obvious (Figs 7.12, 7.13 and 7.14). Both birch and willow peak at six years, comparable to the peak of six years in the hazel rods. Notable drops are found at nine years. This would almost suggest felling activity six years prior to the construction of the site. The curves for ash are very irregular, concentrating between nine and twenty-two years, but extending well beyond this range. As the numbers are rather small, it is not clear whether the peaks at nine, twelve, fourteen, seventeen and twenty years are indicative of a rotation cycle or mere coincidence.

The material for the hurdles of Killoran 75 incorporated, considering the presence of elm, material from dryland sources, but the hurdles were made locally, using material found close at hand. The source material was managed woodland, consisting primarily of hazel, which was cut at regular intervals prior to the manufacture of the hurdles. The age curves for birch and willow differ from those of ash, indicating a different source. The age curve for hazel, however, is not dissimilar; perhaps the origin of the hazel was not far away from the marginal forest. There are indications that forest management was not restricted to hazel but included marginal forest trees, including birch and willow.

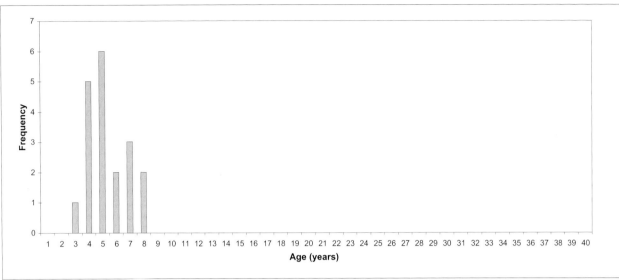

Fig. 7.8 Killoran 75 broken hurdle, Corylus *ages.*

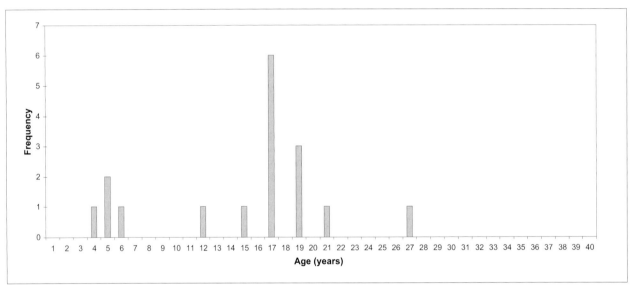

Fig. 7.9 Killoran 75 hurdle, Corylus *sails ages.*

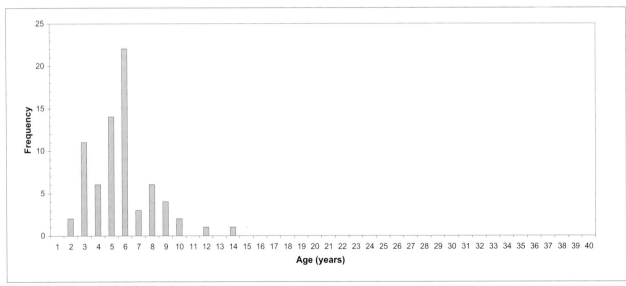

Fig. 7.10 Killoran 75 hurdle, Corylus *rods ages.*

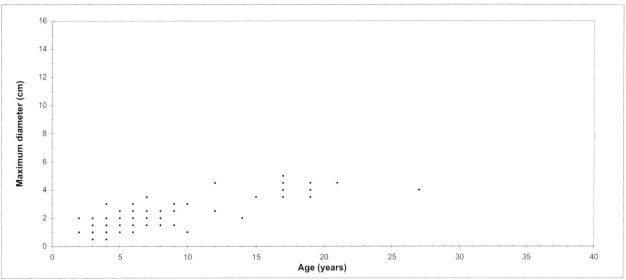

Fig. 7.11 Killoran 75 hurdle, Corylus *age versus diameter maximum.*

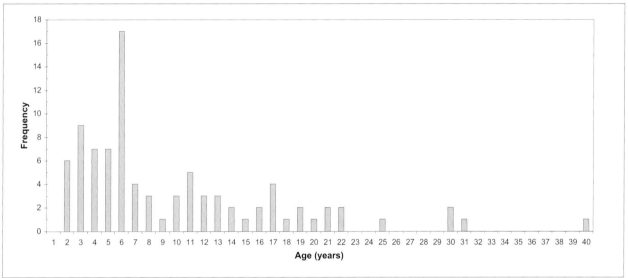

Fig. 7.12 Killoran 75, Betula *ages.*

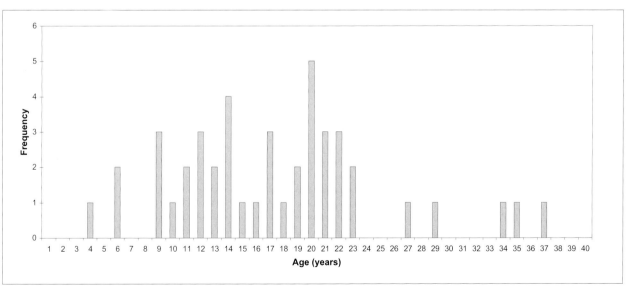

Fig. 7.13 Killoran 75, Fraxinus *ages.*

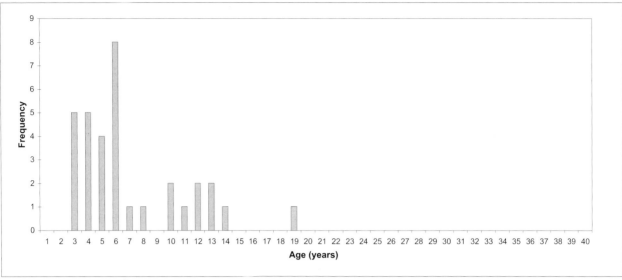

Fig. 7.14 Killoran 75, Salix ages.

Killoran 312

Killoran 312, dated to 385–50 BC, incorporated one hurdle made of double sails with interwoven rods consisting of hazel with a large proportion of ash. There is a close relationship between the age and size of the hazel components of the hurdle (Fig. 7.15). The hurdle (Context 2) consisted of sails between five and sixteen years old, primarily varying between eight and eleven years, but the rods strongly peak between five and seven years (Fig. 7.16), with few older than eight years. Although draw-felling seems more likely (because of the spread in years), it may be that the rods derive from clear-felling because of the peak between five and seven years. The clear-felling method does not exclude the presence of younger rods. Moreover, the sampling done in the field can result in material including lower, medium and upper parts of rods that do not always show the same amount of annual rings. Lastly, it is known that diffuse-porous species such as hazel, perhaps depending on their position in a stool, can skip the formation of a new annual ring, because they can still use the previous years' vessels.

The hurdle was laid down on regular brushwood (Context 3), which, surprisingly, did not deviate much in age from the hurdle (Fig. 7.17). The peak is clearly at six years. No wood was older than seventeen years. In contrast, the overlying brushwood (Context 1), forming a path towards the hurdle, included wood up to thirty-six years old. The wood peaks at seven years, a very uniform distribution (Fig. 7.18).

The ash used in the hurdle also seems to have been selected primarily for size (Fig. 7.19) and varied in age from four to thirty years. Two concentrations seemed to be present. One concentration consisted of rods between four and thirteen years old; the other group consisted of rods between eighteen and twenty-nine years old. No difference in size between the rods was observed; therefore, draw-felling seems the more likely method. Three sails were nine, eleven and twenty years old, respectively.

The brushwood underlying the hurdle contained ash of various ages, with one apparent peak at nineteen years. The quantities, however, were too small to draw further conclusions. The wood was selected for size. The brushwood forming the pathway to the hurdle included somewhat more pieces, which tended to concentrate on the lower group between five and ten years. There was a notable drop at nine years (Fig. 7.20).

Birch occured regularly in the layer overlying the hurdle. It is interesting to see that the age composition of this material differs from that of hazel and ash (Fig. 7.21). Birch strongly peaks at eleven years old and has a notable drop at thirteen years old. No wood less than six years old or more than twenty-one years old was used. The different age composition indicates a different origin for this material; most likely birch grew close to the site. Considering the high proportion of eleven year old wood, it cannot be excluded that birch was also managed to some extent. In size, however, birch is not as rigidly concentrated as hazel and ash.

Apart from concluding that draw-felling is the most likely method applied, the results are very indicative of the use of wood of similar size and age for the hurdle as well as the brushwood. This strongly suggests that this site was constructed on one occasion and that the material was prepared close to the site. Hazel and ash both had the same origin but birch, which was used in the upper layer, was collected from another area.

Killoran 314

Killoran 314, dated to 370–5 BC, incorporated two hurdles, one of which was in a degraded, broken state. Some of the rods of the broken hurdle contained bark beetles indicating that it had been flooded in springtime, before the adult beetles could hatch (see Chapter 8). The hurdle included some birch rods, between seven and twelve years old, felled in Spring and two pieces of alder, ten and twelve years old, respectively. A specific felling season could not be established for the hazel, although the presence of adult bark beetles indicates that the hurdle was made in wintertime or early spring. The rods and sails made of hazel were primarily chosen for their size (Fig. 7.22). The age varies between three and sixteen years old with a notable peak at nine years and a drop at eight years (Fig. 7.23). The quantity, however, was too small to draw any conclusions.

The second hurdle was substantial and was made primarily of hazel, but also included alder, birch and willow. The latter two were young rods, three years old, but the alder fragments varied between seven and twelve years old. Some older pieces (sixteen/twenty-two/twenty-five years old) may represent sails. The alder pieces were probably of local origin, most likely from the marginal forest. The hurdle was in good condition with few insect channels. Most of the wood was felled in wintertime, after November. The close relationship between size and age indicates woodland management (Fig. 7.24). This hurdle seems to have been made from younger material than the broken hurdle (Fig 7.25). In particular, three year old rods seem to have been used. Generally, most of the material was between two and six years old. As the sampling method, in addition to the position of the rod in the stool itself, can result in varying ages, the concentration of ages must reflect a careful selection of rods of the same size and age following clear-felling of an entire crop. The stools yielding the material were probably harvested three years and five/six years prior to the production of the hurdle.

Killoran 315

Killoran 315, dated to 200–170 BC, consisted of one hurdle made of hazel but with one sail made of ash, which was ten years old. Most of the hazel was primarily chosen for size (Fig. 7.26), a logical choice considering the fact that hurdles are made of same-sized rods. The age of the hazel varied between three and twenty-

four years, but a peak occurs at ten years and minimal values are found at nine and thirteen years (Fig. 7.27). No rods younger than three years of age were used. The age distribution is suggestive of a felling period ten years prior to the manufacture of the hurdle. It is not clear how the minimal values should be interpreted, but it is suggested that the year after felling little wood of value was produced. In contrast to other sites, much of the wood from Killoran 315 was felled at the start of the growth season.

Hurdle trends

All of the hurdles in Derryville Bog derive from the western side near the Killoran headland, close to marginal forest and with rather steep access to dryland. The use of hurdles is confined to the period between 400 and 5 BC. Most of the hurdles were laid down on a prepared surface, consisting of brushwood or roundwoods, and continual use of the area is further indicated by the presence of more than one hurdle at two sites. Most hurdles were made of hazel, but other material was also incorporated. At Killoran 75, there are indications of *in situ* manufacturing of hurdles incorporating local material. It is suggested that most of the hazel derived from dryland situations where stools were coppiced at regular intervals. At this stage, draw-felling seems to have been the most likely method of harvesting, although some clear-felling (Killoran 75 and Killoran 314) cannot be excluded. The method of topping cannot be established with the available material, but the regular presence of elbows is indicative of the coppice method. This implies that woodland was fenced off for the deliberate production of underwood at least by the Iron Age, but probably also at earlier times. Most likely this practice was much older. In the Somerset Levels, elaborate coppicing practices were already established in the Neolithic period (Morgan 1988; Rackham 1977).

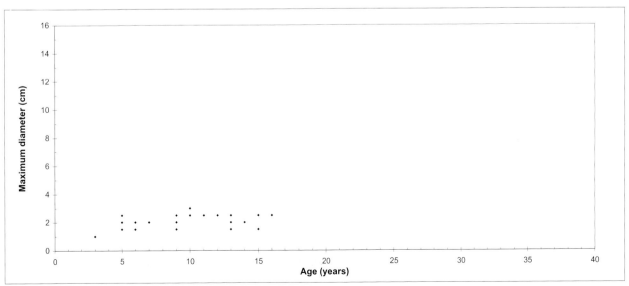

Fig. 7.15 Killoran 312 hurdle, Corylus *age versus diameter maximum.*

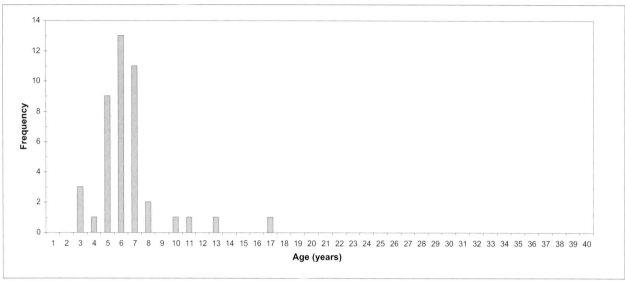

Fig. 7.16 Killoran 312 Context 2, Corylus *rods ages.*

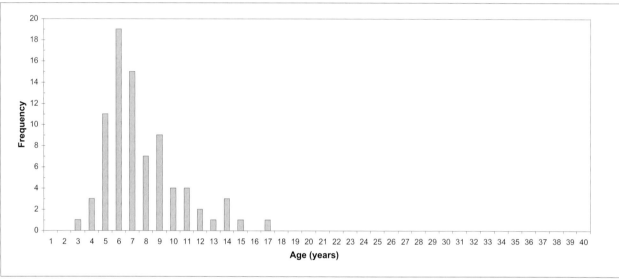

Fig. 7.17 Killoran 312 Context 3, Corylus *ages.*

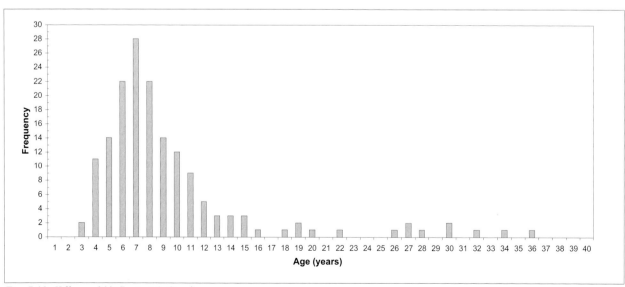

Fig. 7.18 Killoran 312 Context 1, Corylus *ages.*

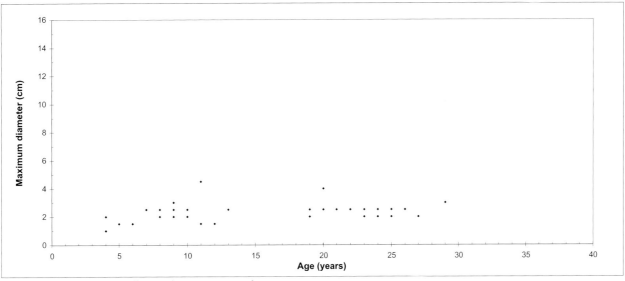

Fig. 7.19 Killoran 312 hurdle, Fraxinus *age versus diameter maximum.*

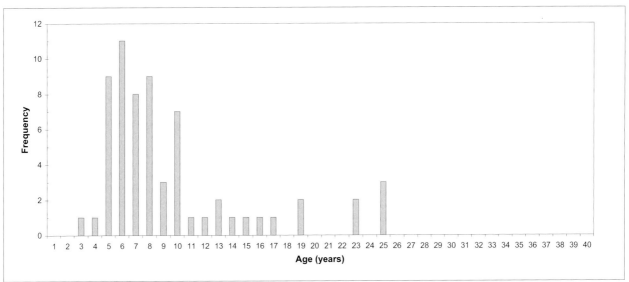

Fig. 7.20 Killoran 312 Context 1, Fraxinus *ages.*

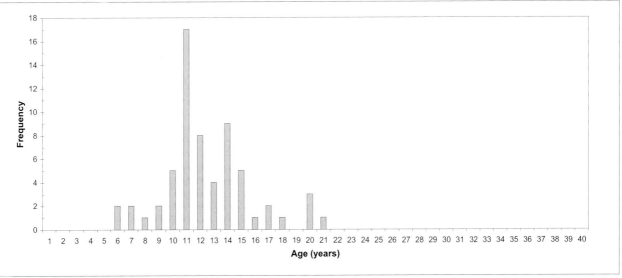

Fig. 7.21 Killoran 312 Context 1, Betula *ages.*

The manufacture of hurdles apparently did not follow strict rules. The majority of the wood for some hurdles was felled in springtime (Killoran 315), others during wintertime (second hurdle at Killoran 314 and Killoran 312), but most hurdles contain a mixture of wood harvested at different times of the year. No strict felling cycle can be established; the predominance of rods of certain age varies from hurdle to hurdle. Peaks of four/five years, five/six years, six years, seven years, nine years and ten years all occur. Although this might seem loosely structured, the presence of the hurdles alone reflects a very organised situation on the dryland.

Other sites with large quantities of Corylus

The following structures, apart from Derryfadda 13, are those where hazel provided the bulk of the material used in the construction. Derryfadda 13 is included here for comparative purposes.

Cooleeny 31—transition period

Hazel was the major constituent of this causeway, occurring in all of the construction phases. The ages of the fragments used showed great variation, which is not indicative of managed woodland (Fig. 7.28). Most wood fell between four and twelve years of age, with a notable peak at eight years and a conspicuous drop at ten years. The generally even distribution resembles, at first glance, a natural situation.

Phase 1, the levelling of the ground surface, included hazel ranging from 1cm to 6cm in diameter at the centre of the causeway. Only a few other species were found here, but some young elm branches indicate a dryland origin. The age distribution (Fig. 7.29) is relatively narrow, between three and seventeen years old, with peaks between seven and twelve years and notable drops at eight and eleven years. Age and size combined reflect a group with similar characteristics, perhaps indicating a similar source (Fig. 7.30). Even when taking this into consideration, a managed background is not immediately clear. On the south side of the causeway, the age distribution was slightly different (Fig. 7.31), with peaks at nine, eleven and thirteen years and drops at ten, twelve and fourteen/fifteen years. Again, the age and size form a similar cluster.

The other phases merely mirror the findings for Phase 1, except that the variation in size is still greater. Only a few fragments are ten years old, in contrast with Phase 1. In Phase 3, peaks occur at seven, eight and fourteen years, respectively, with drops at five, ten and thirteen years. It is not surprising that the surface layer of the causeway was very varied in size and age, because fragments also included pegs, roundwoods and timbers. Here especially, fragments of four and eight years old were found with few pieces of seven and ten years of age (Fig. 7.32). Other fluctuations are probably not significant.

In conclusion, causeway Cooleeny 31 gives few indications for an origin in managed woodland. The age versus size distribution of the fragments indicates a similar source, probably on dryland ridges close to the site (Fig. 7.33). The components of the trackway differed slightly, depending on which part of the hazel shrubs were used. Most age differences can be explained in this way. The only unexplained fact remains the low presence of ten year old fragments. As the excavations only revealed a small part of the surface, it is very likely that such fragments were deposited elsewhere in the track.

Cooleeny 64—Late Bronze Age

This brushwood track consisted of large quantities of *Corylus* of fairly uniform size and age. Most pieces were of less than 3cm in diameter. The brushwood was selected based on size. Their spread in years compared with the narrow size range (Fig. 7.34) is suggestive of draw-felling. The ages varied between two and twenty-four years old, but mostly between six and fourteen years (Fig. 7.35). Peaks occurred at ten and twelve years, which may indicate previous clearances of coppice-stools. The regular growth pattern combined with the straight, regular, knot-free appearance suggests a managed rather than a natural origin.

Derryfadda 13—transition period

Although in the description of the separate sites (see additional files) it is mentioned that the fairly undisturbed marginal forest was used for the construction of this site, for comparison the characteristics of hazel are mentioned here. The wood displayed a large variation in age, between four and sixty years old, but most pieces were less than eighteen years old. Several peaks and drops are visible in Fig. 7.36, giving the ages of the separate samples. Interestingly, the age variation is dissimilar to most other sites. In Fig. 7.37, where size and age are plotted against each other, no selection is obvious.

In summary, it is clear that peaks occur at specific ages, namely five, eight/nine, fourteen, sixteen/seventeen, twenty-five, thirty and thirty-five years. Drops occur at seven, twelve/thirteen, and eighteen/nineteen years. Unfortunately, the numbers are rather small, which makes interpretation hazardous. At this stage it is not clear how to interpret the variations, but it looks as if something occurred on a regular basis, perhaps at intervals of five years. It cannot be excluded that prehistoric man visited the area on an irregular basis and harvested some of the hazel shrubs on the ridges leading to the marginal forest.

A small number of pieces exhibited features characteristic of managed wood. These had a wide first annual ring and were cut after the growth season. The age/size distribution (Fig. 7.38) shows a cluster of items, which indicates selection for size. The ages of these pieces vary between five and thirteen years, with peaks at five, and eight/nine years (Fig. 7.39). The fragments, viewed under the microscope, were like and uniform in appearance.

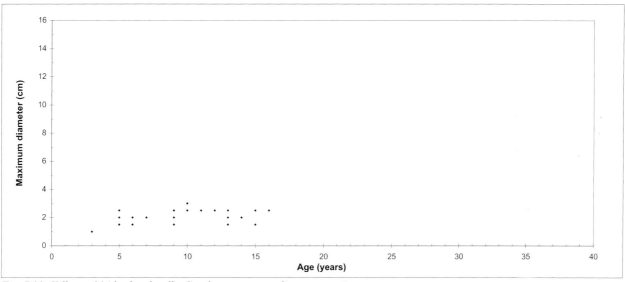

Fig. 7.22 Killoran 314 broken hurdle, Corylus *age versus diameter maximum.*

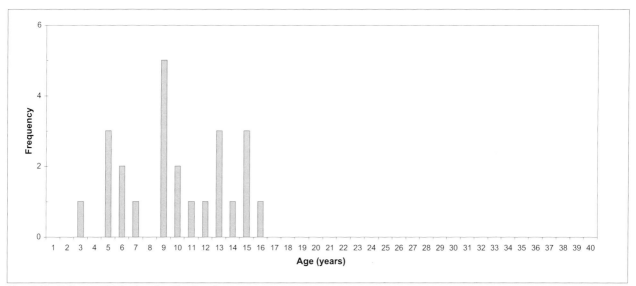

Fig. 7.23 Killoran 314 broken hurdle, Corylus *ages.*

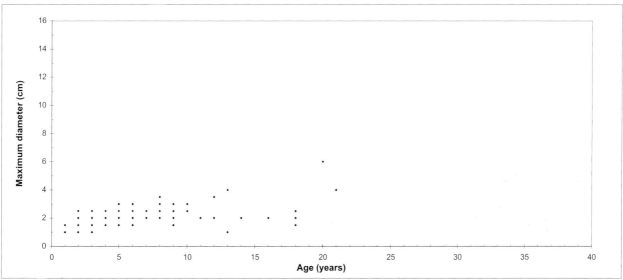

Fig. 7.24 Killoran 314 second hurdle, Corylus *age versus diameter maximum.*

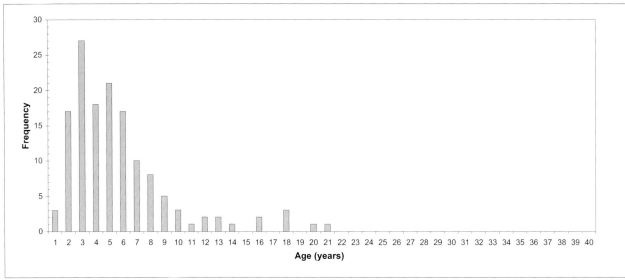

Fig. 7.25 Killoran 314 second hurdle, Corylus *ages.*

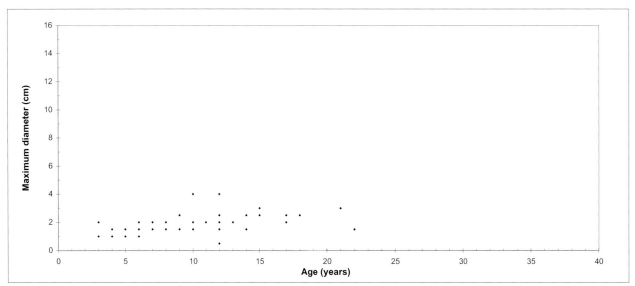

Fig. 7.26 Killoran 315 hurdle, Corylus *age versus diameter maximum.*

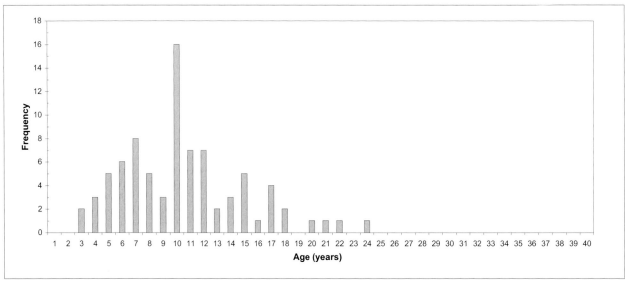

Fig. 7.27 Killoran 315 hurdle, Corylus *ages.*

Derryfadda 204—Early Bronze Age

The brushwood in this site included relatively large amounts of *Corylus* of uniform size and age. The sizes varied from 1–2.5cm and almost all the pieces were between seven and fourteen years old. Only one piece was older (twenty-three years old). All the wood was felled after the growth season in contrast to other wood found in this site. The age patterns are peculiar (Fig. 7.40). In contrast to other sites, there is no wood younger than seven years old present. The peak at seven years and drops at eight and eleven years are very conspicuous. Also noteworthy is the fact that many samples barely added wood during the last six years before felling. This could indicate that competition within a stool inhibited growth.

The results point to the presence of coppiced wood that was harvested regularly. The rods were ready for harvesting, considering their growth pattern, and maybe from an economic point of view should have been felled earlier. The stools were probably clear-felled, considering their close age patterns. Felling occurred seven years before and possibly at three year intervals ten and thirteen years prior to the use of the material. This means that by the Early Bronze Age coppice stools were already available and consequently areas of underwood were managed and fenced off.

Derryfadda 210—transition period

The hazel fragments from this site were mixed pieces of brushwood which measured 2–4.5cm in diameter. At the time of excavation, the regularity in size and form was noticed. The ages of the pieces concentrate between five and sixteen years old (Fig. 7.41). The majority of the pieces were seven to twelve years old, with a few pieces of eleven and thirteen years of age. No wood less than five years old is present. The pieces were selected specifically for size, as is apparent from Fig. 7.42. The results are suggestive of managed wood, but some caution is necessary because of the small numbers. If the material represents managed wood, then draw-felling was the most likely method. The irregularity in age distribution may indicate some form of clearing seven and twelve years before use, at intervals of four or five years.

Killoran 18—Middle Bronze Age

As the building of this causeway was a single event, most of the hazel from this site is likely to be contemporary. The ages vary between one and thirty years of age, but most items fell between three and nineteen years of age (Fig. 7.43). Some fluctuations are apparent, such as peaks at three, nine and fourteen years with smaller ones at five, seven, ten and twelve years. Few pieces of four, six, eight and thirteen years of age were found. Most of the wood was brought to the bog from the eastern side and thus reflects vegetation on this side of the bog. If the variation in age is reflecting felling activities, then prehistoric man must have visited the eastern area regularly. The wood could have been felled periodically at intervals of two and five years.

It is difficult to interpret the variation in ages because the spread in years is considerable. Hazel was used for a variety of purposes within the causeway. The hazel stakes belonging to the initial construction of the site clearly reflect short-term activity. The wood was uniform in size and age, varying between ten and seventeen years. Most stakes were three years old (Fig. 7.44, Field 17). Logically, the stakes were selected for size (Fig. 7.45). Unfortunately, the amount was too small for further conclusions.

A quantity of hazel was analysed from Field 15 and reflects the general pattern (Fig. 7.46). This material too was selected for its size. Although selection was used in choosing and preparing stakes, it is not clear whether Killoran 18 is representative of a managed woodland on the eastern side of Derryville Bog. It seems that a natural environment is more likely.

Killoran 54—medieval period

The catalogue describes this site as consisting almost exclusively of *Corylus* of very uniform size and form, the majority felled after the growth season (Fig. 7.47). The stakes were carefully selected from rods 1.5–2.5cm in diameter. The age of the stakes concentrates between six and fifteen years of age, with the majority consisting of seven year old stakes and a significant drop in stakes of eight years of age (Fig. 7.48). The spread in years is considerable. The origin of these stakes must have been managed woodland, considering their form and size, but the harvesting method was probably draw-felling rather than clear-felling. The peak may indicate that stools were (partially) cleared seven years prior to the use of the wood.

Killoran 230—Middle Bronze Age

Many sites in the Killoran area included large quantities of hazel, which was the most frequently analysed species from this site. The age distribution (Fig. 7.49) is not indicative of managed woodland, as there is considerable variation between three and more than forty years of age. However, there seems to be many fragments of nine and eleven years of age. Minimal values were noted for pieces fifteen years of age. The hazel was apparently selected at random because there is no apparent preference for a specific size (Fig. 7.50). Most of the wood was felled after the growth season; the annual rings being fully formed. The reason for mentioning this site lies in the fact that some fragments show a thickened part in the form of an elbow. This is a characteristic of coppiced wood. However, it cannot be excluded that such characteristics sometimes occur in more natural situations without much human influence.

The only context where management can be proposed was a tightly packed layer of brushwood, including many thin branches with bark. Although the quantity of hazel was low, the ages were concentrated between three and eleven years, peaking at six years of age (Fig. 7.51). The size a uniform 1.5–2.5cm diameter. As other wood species were also present, the use of local brushwood seems likely here also.

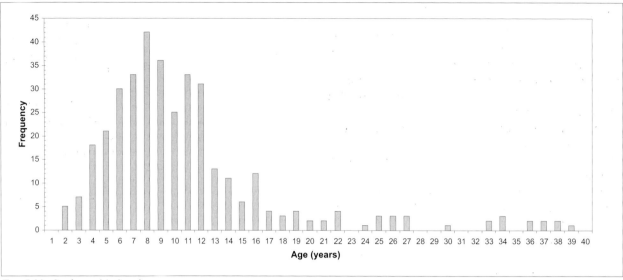

Fig. 7.28 Cooleeny 31, Corylus *ages.*

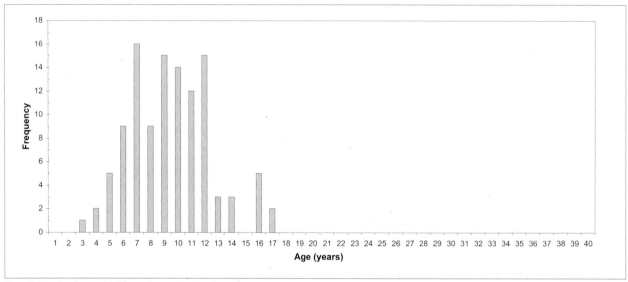

Fig. 7.29 Cooleeny 31 Phase 1 centre, Corylus *ages.*

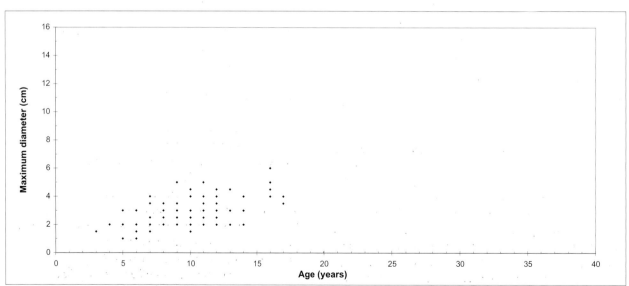

Fig. 7.30 Cooleeny 31 Phase 1, Corylus *age versus diameter maximum.*

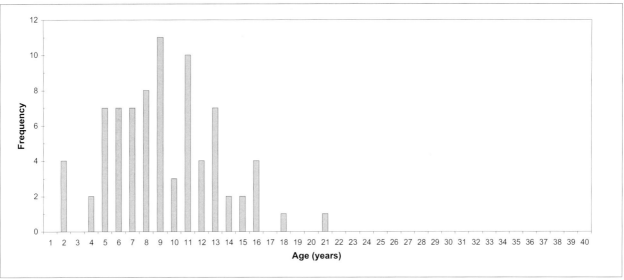

Fig. 7.31 Cooleeny 31 Phase 1 south, Corylus *ages.*

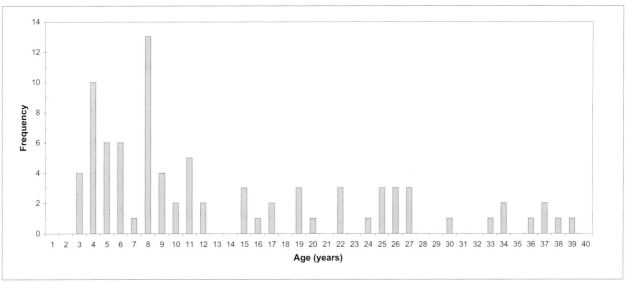

Fig. 7.32 Cooleeny 31 Phase 4, Corylus *ages.*

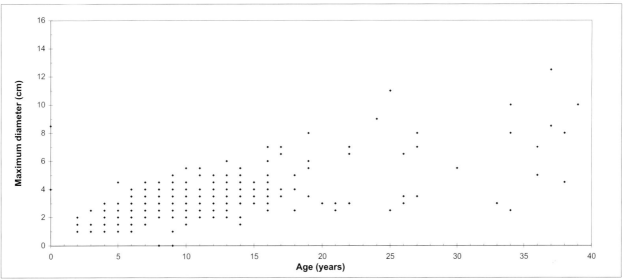

Fig. 7.33 Cooleeny 31, Corylus *age versus diameter maximum.*

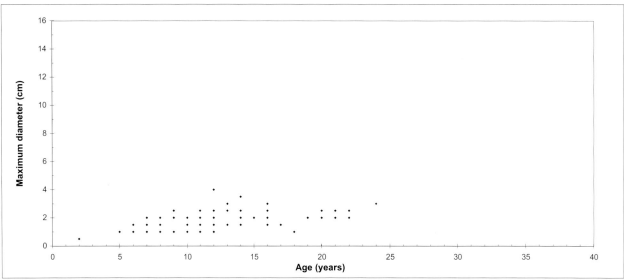

Fig. 7.34 Cooleeny 64, Corylus *age versus diameter maximum.*

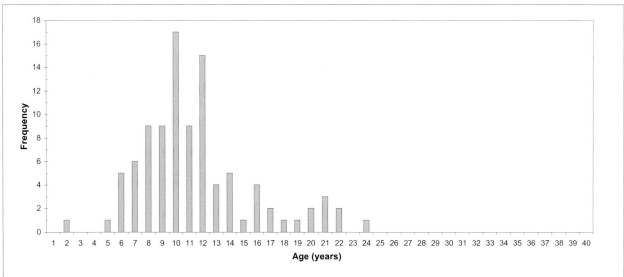

Fig. 7.35 Cooleeny 64, Corylus *ages.*

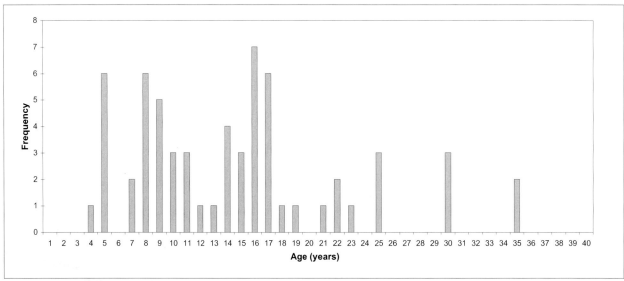

Fig. 7.36 Derryfadda 13, Corylus *ages.*

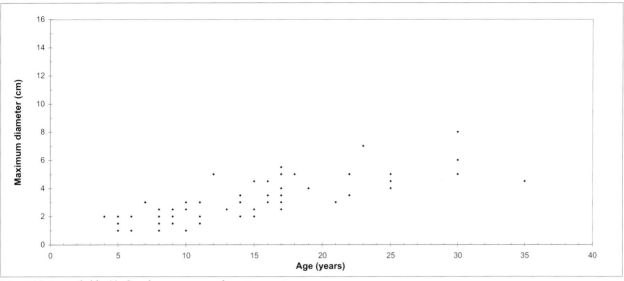

Fig. 7.37 Derryfadda 13, Corylus *age versus diameter maximum.*

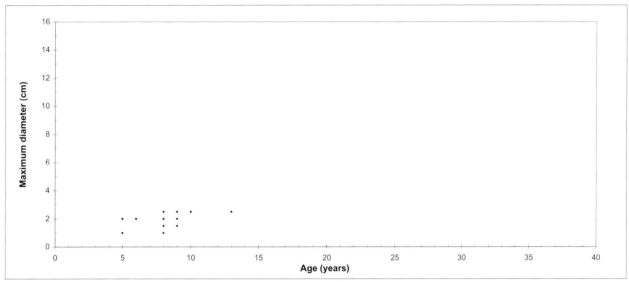

Fig. 7.38 Derryfadda 13, Corylus *first year good growth age versus diameter maximum.*

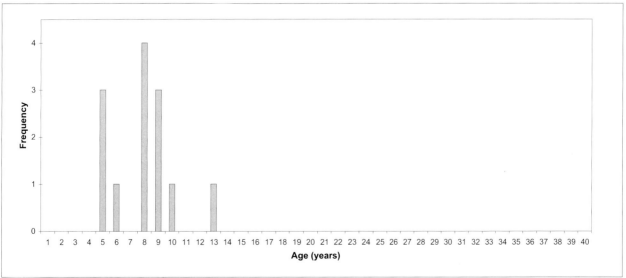

Fig. 7.39 Derryfadda 13, Corylus *first year good growth ages.*

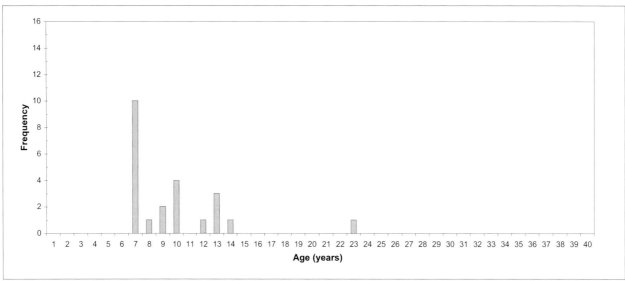

Fig. 7.40 Derryfadda 204, Corylus *ages.*

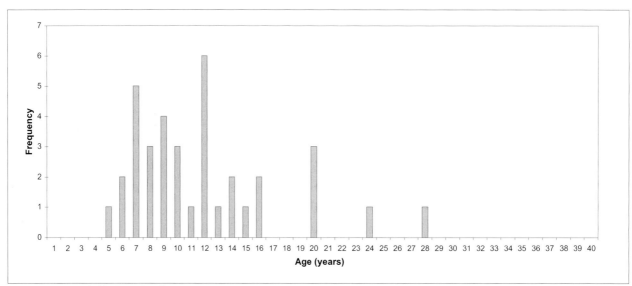

Fig. 7.41 Derryfadda 210, Corylus *ages.*

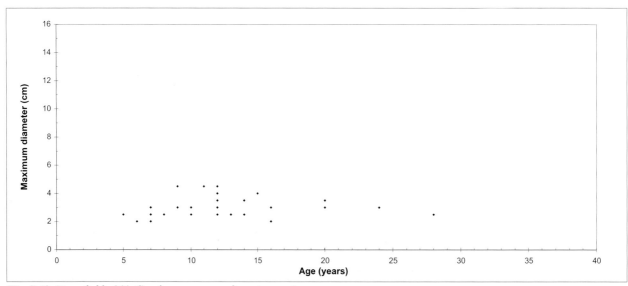

Fig. 7.42 Derryfadda 210, Corylus *age versus diameter maximum.*

Killoran 237—Middle Bronze Age

Most wood analysed from this site was hazel brushwood and, therefore, deserves some attention here. This brushwood derived from a separate cutting and was very similar in appearance. All the wood was felled in springtime, around March. From Fig. 7.52 it is clear that the rod-like pieces were selected based on size. It is tempting to suggest that remnants of rods are represented here, although it cannot be excluded that the pieces were regularly grown brushwood. If this material is indeed the result of managed hazel stools, then draw-felling was the most likely method employed. The ages vary between three and twelve years old, with a single exception of eighteen years of age (Fig. 7.53). In contrast to other hazel found at this site, most of the fragments had a wide first annual ring, which is a characteristic of coppiced wood. Half of the samples had narrow annual rings in cross section. Barely any wood formed, especially during the last two years. This could indicate that rods were in competition with each other and thus grew close together. Although the quantity of samples was rather low, the tendency was towards the use of six year old pieces. Low amounts of four and seven year old fragments were found.

Killoran 241—Middle Bronze Age

Although relatively few hazel items from this site were analysed, the results are mentioned here because of the age pattern (Fig. 7.54). The pieces were similar in size, ranging from 2cm to 4cm in diameter (Fig. 7.55). All of the wood was felled after the growth season and was in rather good condition. The majority of pieces were nine years old and no pieces were below four years or above ten years of age. This is indicative of clear-felling of a hazel stool nine years prior to harvest. The material derived from managed hazel stools on the dryland.

Killoran 253—Late Bronze Age

This wicker-lined trough had much hazel indicative of managed woodland. Part of the lining had collapsed and loose fragments inside the trough were interpreted as rods. In interpreting the site, the fragments were treated separately.

The rods of the wicker-lining were between two and eleven years old, with a notable peak at three years (Fig. 7.56). No rods of seven or eight years of age were present but this may be the result of the limited quantity analysed. In contrast, the sails of the lining peak at six years of age (Fig 7.57). No seven year old wood was present. One sail of eighteen years of age lies far beyond most of the hazel. The rods were selected for their size (Fig. 7.58), which follows a rigid pattern of 1–2cm in diameter.

The rod-like fragments from the trough exhibited distinct wide first annual rings, indicating fast growth. This is the result of coppicing. The age patterns of these pieces (not part of the wicker-lining) peak at five and especially seven years of age, and drop at four years of age (Fig. 7.59). Few pieces extend beyond eight years of age,

but some concentrate between seventeen and twenty-two years of age. These pieces seem to form a separate group of sails (Fig. 7.60). Other loose pieces found in the trough without the wide annual ring but very rod-like in appearance, also concentrate between two and eight years of age, peaking at five years (Fig. 7.61). A second group is found between seventeen and twenty-two years of age, a single piece reaching thirty years of age (Fig. 7.62).

It can be assumed that the trough was made on one occasion and that the rods and sails therefore derive from the same source. Size was the major factor determining whether a piece was used as a sail or rod. The age difference between the two can therefore result from thickness and their relative positions; the three year old rods may represent upper parts and the six year old pieces lower parts of the poles on a stool. However, it cannot be excluded that coppice stands on the dryland included stools which were cut at different intervals. The practice of draw-felling is the most likely explanation. This means that the rods and sails were selected for size from stools, irrespective of their age.

It is puzzling why the age patterns of the pieces in the trough resemble the wicker lining so much. Especially interesting is the peak at seven years of age of the rod-like fragments because this peak is manifestly absent from the wicker lining. It is tempting to suggest that these fragments resulted from another structure belonging to the trough, perhaps a kind of lid, contemporary with the lining and coming from the same source. The material clearly represents coppiced wood with a dryland origin. The rods were of small size and were generally less than ten years old. Sails represent larger pieces and were around twenty years old. The latter material could derive from different stools to the rods, but it cannot be excluded that other methods prevailed in prehistoric times and that the material derived from one stool only.

Killoran 301—Iron Age

The brushwood from this site consisted primarily of hazel and may indicate a managed background. Most fragments were between 1.5cm and 3cm in diameter. The ages varied considerably (Fig. 7.63) between three and twenty-five years of age, but significant peaks were found at six and eight years of age. Although the numbers were relatively low, it may be suggested that, on the basis of the uniform appearance, this brushwood derived from managed stools (Fig. 7.64). Draw-felling must have been the applied method. The stools may have been (partially) cleared six and eight years previously.

Conclusions

Hazel played an important role in the life of prehistoric man around Derryville Bog. This is reflected in the continual use of hazel in the trackways and platforms. The manufacture of hurdles is especially indicative of sophisticated methods, not only of preparing wood, but also of harvesting crops and managing woodlands. As the hurdles all derived from the same area and period, they reflect a well-organised society

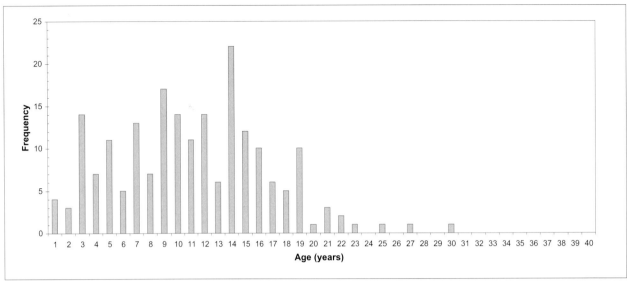

Fig. 7.43 Killoran 18, Corylus ages.

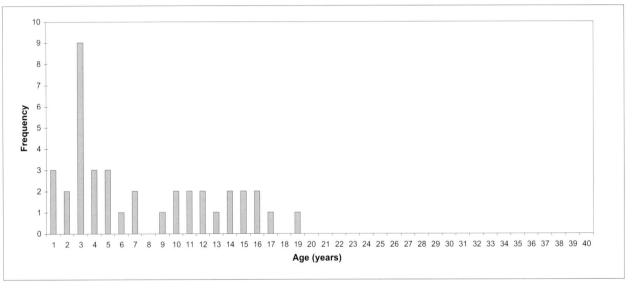

Fig. 7.44 Killoran 18 Field 17, Corylus ages.

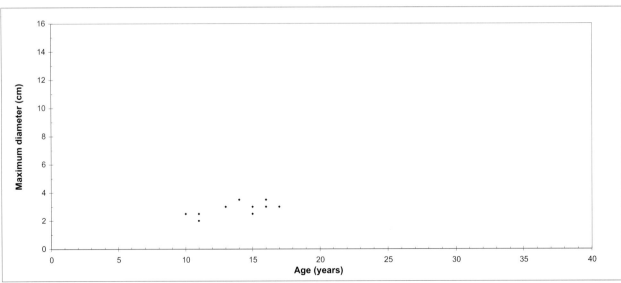

Fig. 7.45 Killoran 18 Field 17, Corylus stakes age versus diameter maximum.

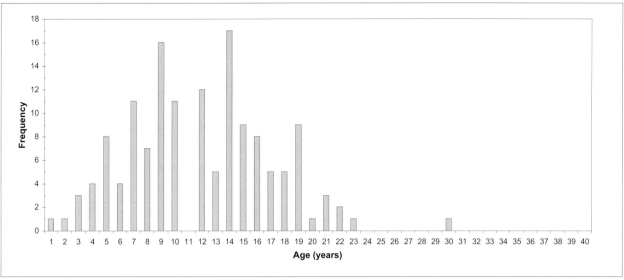

Fig. 7.46 KIlloran 18 Field 15, Corylus *ages.*

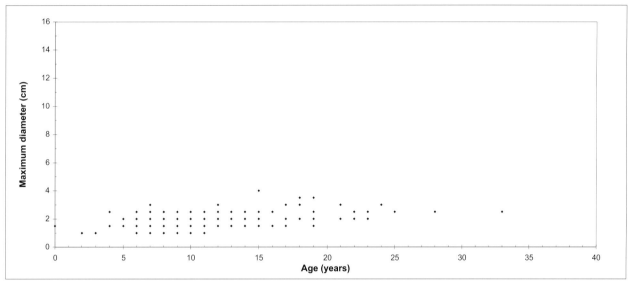

Fig. 7.47 Killoran 54, Corylus *age versus diameter maximum.*

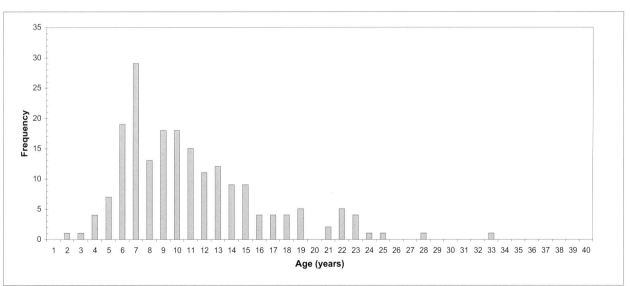

Fig. 7.48 Killoran 54, Corylus *ages.*

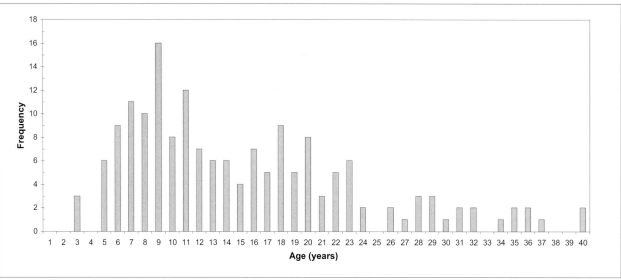

Fig. 7.49 Killoran 230, Corylus *ages.*

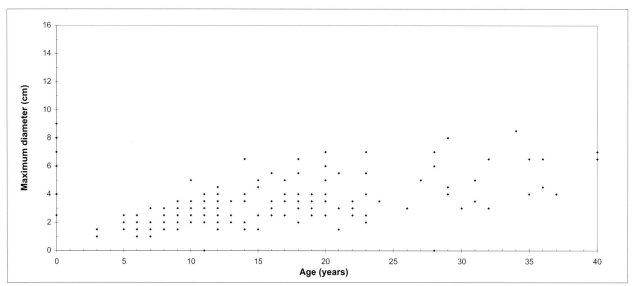

Fig. 7.50 Killoran 230, Corylus *age versus diameter maximum.*

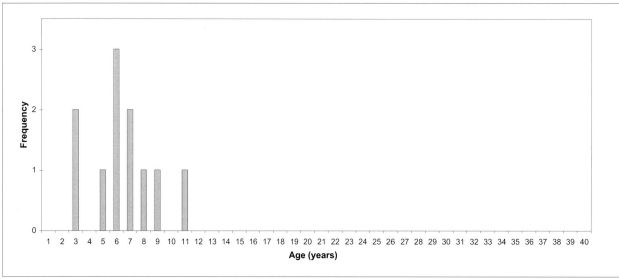

Fig. 7.51 Killoran 230 Context 2, Corylus *ages.*

on the Killoran dryland, with well-managed woodlands of underwood in the Iron Age. Killoran 301 is another example of a site in which managed hazel was used.

The knowledge of woodland management was not confined to the Iron Age. In the same area, the Late Bronze Age wicker-lined trough from Killoran 253 undoubtedly reflects an organised landscape with managed coppice-stools.

Killoran 241, during the Middle Bronze Age in the same area, although consisting of few samples, may reflect the method of clear-felling.

In the Derryfadda area, woodland management involving hazel was being practiced by the Early Bronze Age. It is difficult to trace human influence in the material derived from this area, but it cannot be excluded that Derryfadda 13 and Derryfadda 210, both from the transitional period, include hazel possibly derived from managed situations. Surprisingly, the material from the substantial causeway Cooleeny 31 seems to be of randomly gathered material.

At this stage it is difficult to ascertain felling cycles for the Derryville material because it is not always clear how to interpret the peaks and drops in the ages. The peaks seem to concentrate between five/six and seven years of age, but peaks also occur at eleven and twelve years of age. Even three year old rods were preferred for the wicker lining of the trough at Killoran 253. Drops occur at different times, depending on the sites. Hopefully, future work can explain why hazel from managed situations often shows such a variety in ages.

To conclude, Fig. 7.65 gives the age distribution for all of the hazel sampled from Derryville Bog. In contrast to other species, such as *Alnus* or *Betula*, the ages are very regularly spread and most of them are less than forty years old. Most items are between two and twenty-two years old, with peaks at six and seven years of age. After twelve years, the number of fragments present declines sharply. In the author's opinion, this concentration of

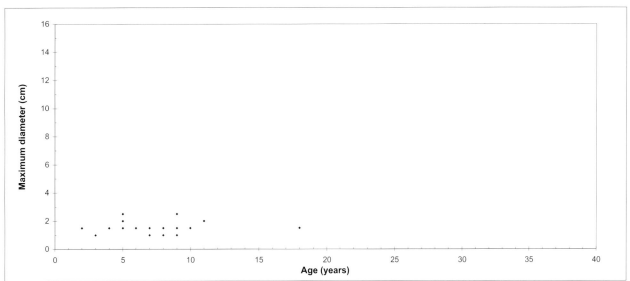

Fig. 7.52 Killoran 237 Context 3, Corylus age versus diameter maximum.

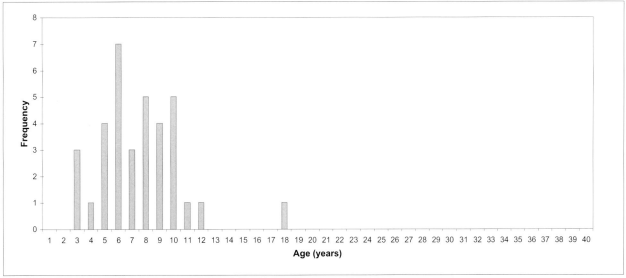

Fig. 7.53 Killoran 237 Context 3, Corylus ages.

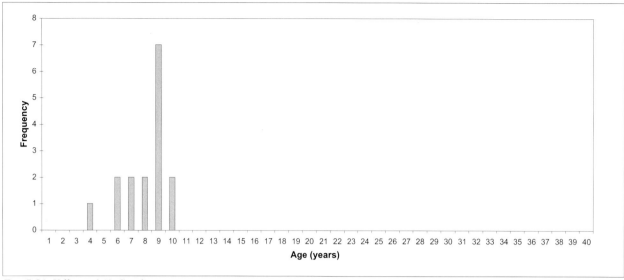

Fig. 7.54 Killoran 241, Corylus *ages.*

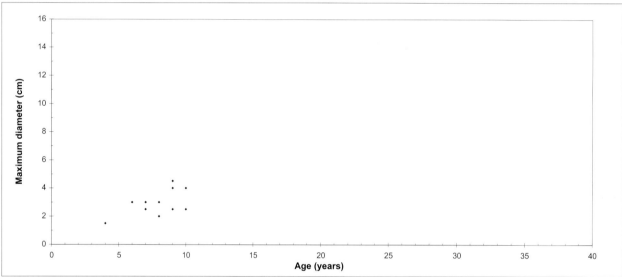

Fig. 7.55 Killoran 241, Corylus *age versus diameter maximum.*

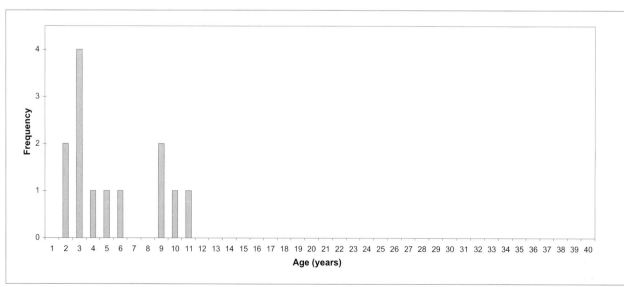

Fig. 7.56 Killoran 253 rods, Corylus *ages.*

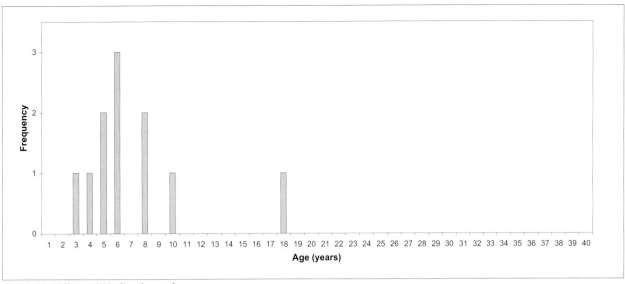

Fig. 7.57 Killoran 253, Corylus *sails ages.*

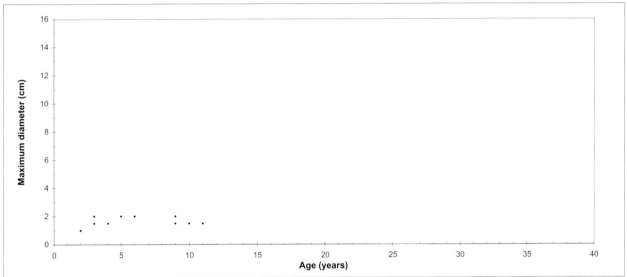

Fig. 7.58 Killoran 253, Corylus *first year good growth ages.*

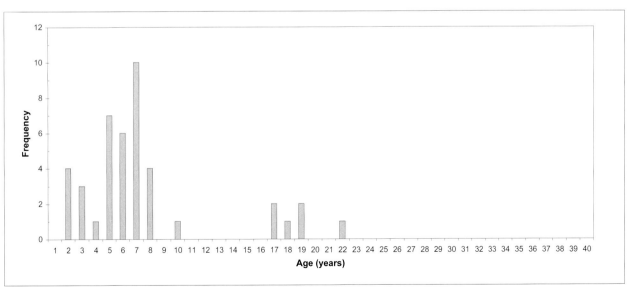

Fig. 7.59 Killoran 253, Corylus *first year good growth ages.*

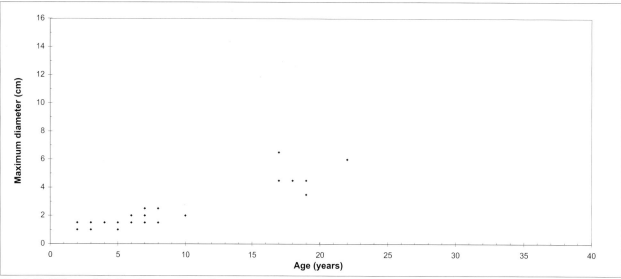

Fig. 7.60 Killoran 253, Corylus *age versus diameter maximum first year good growth.*

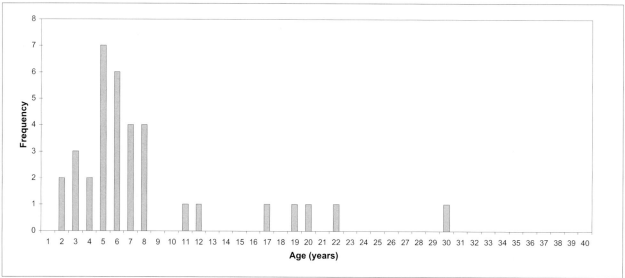

Fig. 7.61 Killoran 253, Corylus *ages other pieces.*

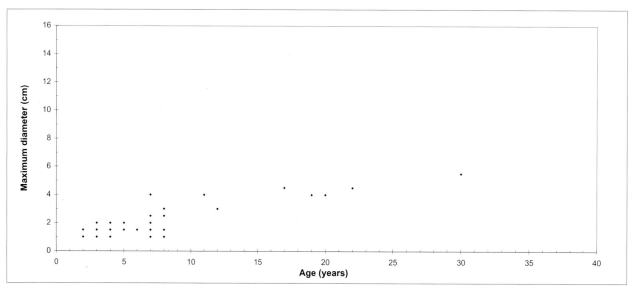

Fig. 7.62 Killoran 253, Corylus *age versus diameter maximum other pieces.*

wood between two and twelve years of age, combined with the regular spread, may indicate a managed situation and not a natural situation. Management could have included the favouring of young, flexible shoots by continual and regular clearing of shrubs or stools. However, this is not always clear from the archaeological and botanical record. This idea can be used as a starting point for further research into the matter.

General trends according to areas

Introduction

From the previous sections, the picture that emerges is that most of the wood used in the archaeological sites in Derryville Bog was gathered locally, in 'brook' forest and marginal woodlands (alder and willow in wet areas, oak and ash on drier grounds or during drier periods), on slopes directly surrounding the sites (hazel) or even from the bog itself (yew and birch). Very little came from the hinterland (elm and oak).

To check whether local topographical conditions promoted certain species, wood identifications were analysed by geographical situation around the bog, independent of their archaeological date. Sites that did not clearly fall into a specific area were omitted (see Appendix 1b, List 2, additional files).

The areas and sites around Derryville Bog defined for this analysis were, from northwest to east, the Killoran 75–240 cluster, the Killoran 69–306 cluster, Cooleeny 31, the Derryfadda cluster and the Killoran 18–305 cluster (it was assumed that causeway Killoran 18 was made from the east towards the west). One site, the brushwood track Cooleeny 64, was looked at separately. This site did not fit clearly within any of the above groupings, but nevertheless must reflect the vegetation on the western margin south of Killoran 69 and Killoran 306, and north of Cooleeny 31. A chart (Fig. 7.66g) representing the results of the natural, non-archaeological wood samples was prepared to facilitate comparison.

Findings

The main observation was again the varying importance of hazel and alder. The most intriguing result was the dominance of hazel in three areas: Killoran cluster 75–240 (Fig. 7.66a), Cooleeny 31 (Fig. 7.66d) and Cooleeny 64 (Fig. 7.66c). From the discussions above (see also Fig. 7.66g), it has become clear that hazel only occurred in low quantities in the local vegetation and did not form part of the local 'brook' forest on the margins of the bog itself, but grew somewhat further away on drier grounds.

The presence of hazel in the Killoran area is interpreted as indicative of local woodland management (especially in Killoran cluster 75–240, where hurdles were made from hazel). This managed woodland might have taken the form of rough coppicing, but also clearance and subsequent expansion of hazel shrubbery.

It is tempting to suggest that the habitation around Derryville Bog was concentrated in the northwestern and southwestern areas, and not elsewhere. This idea might also be supported by the fact that only in these two areas was ash the next most frequently used species after hazel.

The pollen diagrams show that ash was a regular constituent of the vegetation on the dryland, together with oak and elm. Although it is possible that some ash grew in marginal woodlands during drier periods, this does not seem to be the origin of the western material given the absence of oak, which would be expected to occur alongside such growths of ash.

Although Cooleeny 64 (Fig. 7.66c) was only one site, the high value for hazel agreed with the conclusion that hazel was much better represented on the western side. The high values for the crab apple-type at Cooleeny 64 were quite unique. Although it cannot be excluded that crab apple grew locally on the bog, it is more likely that it grew in the marginal woodlands or in more elevated positions on drier ground. The high values for hazel and crab apple, the low values for willow and the absence of birch could mean that the location of the wood supply for Cooleeny 64 was fairly dry at the time of the tracks construction (later Bronze Age).

Values for birch were relatively high in Killoran cluster 75–240 (Fig. 7.66a). This could reflect local stands of this tree, perhaps *Betula pubescens*. The area could have been open and, to a certain extent, dry. Birch is a light-demanding tree that stimulates the growth of herbs and grasses underneath. The sites in the Killoran 75–240 cluster represent a considerable period of time, however, which in the dynamic situation of the bog, must have affected the local situation several times. Noteworthy were the low values for oak and relatively high values for ash.

The high values for willow in Killoran cluster 69–306 (Fig. 7.66b), compared with other areas, could represent a local wet area in which willow abounded, maybe close to mineral-rich water from streamlets seeping from the dryland or a draining system within the bog marginal area. Alder, also, may well have grown locally, sprouting during somewhat drier periods and continuing growth during wetter stages in the 'brook' forest on the margins of the bog. Values for oak were relatively high in this cluster.

The chart for natural wood from non-archaeological sites (Fig. 7.66g) closely resembles that of the Derryfadda cluster. The Derryfadda cluster, therefore, may represent a fairly undisturbed marginal woodland situation. The regular occurrence of crab apple, wild cherry and holly suggest a rather open, light growth. Oak could have grown in this woodland too, but could also very well have grown on the surrounding dryland or on the bog itself during drier periods. The name Derryfadda points to the presence of oak forests in historic times, indicating good local conditions for oak growth. In the bog, several finds of mature oak trees point to dry periods of considerable length, enabling oak trees to establish themselves on the peat.

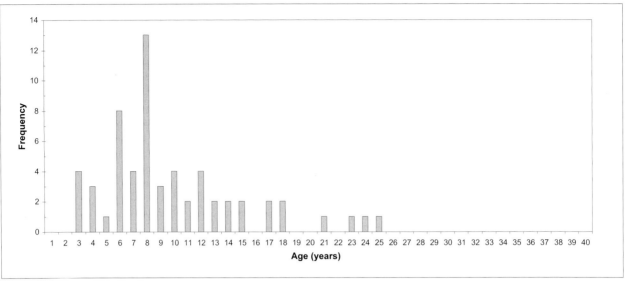

Fig. 7.63 Killoran 301, Corylus *ages.*

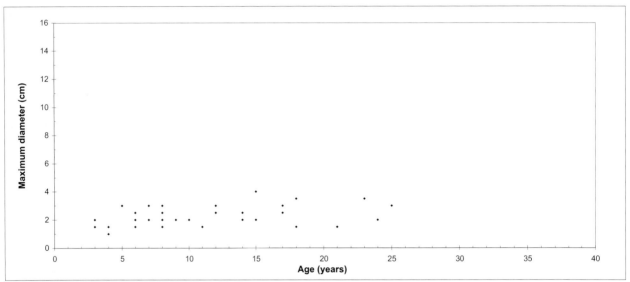

Fig. 7.64 Killoran 301, Corylus *age versus diameter maximum other pieces.*

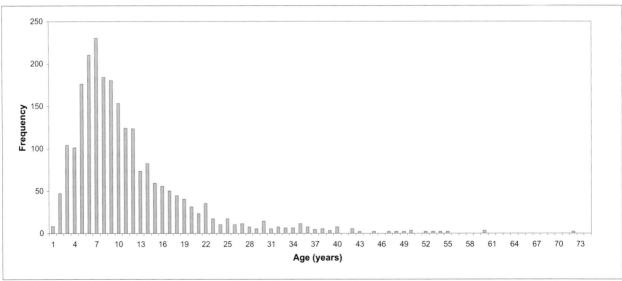

Fig. 7.65 All sites, Corylus *ages.*

General trends in the pollen sequence

Introduction

The pollen record (Chapter 6) shows alternating periods of clearance and regeneration, giving information on the dryland sequence. The analysed wood was compared with these pollen zones to check whether the local/marginal woodlands behaved in the same way. Around 600 BC, a lot of woodland development and archaeological activity occurred in connection with bog burst D. Therefore, the zones were adjusted in order to avoid the activity being ascribed to two different zones.

There was scarcely any wood dating from the period after 80 BC, therefore, no sampling was done in Zones VIII and VIIii. Zone IX (after AD 700) was only represented by two sites, both stake rows, maybe boundary markers rather than trackways, consisting almost exclusively of hazel. The results of Zone IX, therefore, did not receive extra attention (see Fig. 7.67i, List 3 in Appendix 1b, additional files).

Comparing pollen with wood

It is certainly worthwhile to compare the results of the pollen analysis with those of the wood analysis. However, such comparison can be problematic for a number of reasons.

Firstly, pollen dispersal is determined by the dispersing agent (wind, insects and water), while the presence of wood in the bog, except for the natural *in situ* vegetation, is exclusively determined by human action. Differences between the pollen record and the wood analysis may therefore result from the fact that an insect-pollinated species such as holly, crab apple or willow, is generally under-represented in the pollen rain, while a wind-pollinated species such as alder, hazel or Scots pine is over-represented in the pollen rain. The behaviour of the insect-pollinated species is generally difficult to trace in pollen diagrams, except when species grow very close to the sampling sites.

Secondly, pollen zones comprise periods of 300 years, each long enough for many generations of secondary woods to develop. Certainly after a bog burst, the resulting nutrient-rich environment would create favourable circumstances for the development of secondary woodlands on the affected margins within ten to thirty years. The resolution may thus be too coarse to detect any comparable trends, especially in the dynamic environment of Derryville Bog. Most archaeological sites in the bog marginal areas were, moreover, only used for a short period, mostly less than five years, and the wood was mostly felled at the moment of construction. In this way, the results of pollen analysis and wood identification might seem to contradict each other, while in fact they are the result of the same event. For example, when low ash values in the pollen record are interpreted as a period of clearance, the wood record for the same period might record much ash usage at archaeological sites.

Thirdly, clearance to obtain trees for domestic purposes and construction use (building of houses and barns, fuel for cremations or food preparation, kitchen utensils, furniture etc.) does not show up in the pollen record from the wetland area because people did not live on the bog itself. When excluding certain species and concentrating on dryland species, the signal might be regional and not capture the local events on the slopes surrounding the bog, where people lived in their huts and accessed the marginal bog. Some measures to manage woodlands (leaf foddering) could prevent the flowering of trees, thus resulting in low pollen values. On the other hand, activities such as coppicing could stimulate flowering.

A minor problem is presented by the limitations of the methods themselves, especially regarding identifications. As mentioned before, it is not always possible to identify wood to genus level. One of the most problematic species in Ireland is the Pomoideae-type of wood, which includes crab apple, hawthorn, rowan and whitebeam. In pollen analysis too, it is not always possible to distinguish the pollen grains of *Sorbus*-type from those of other Rosaceae. This means that the Pomoideae-type identified in the wood record could partly correspond with *Sorbus*-type pollen. Also, some other taxa, such as yew and poplar, produce little pollen and their pollen is often lost during the preparation of the samples.

Wood sequence following the pollen zones

The pollen sequence is presented in Chapter 6. Zone VI is divided here into Zone VIi and VIii.

Zone I: pre-1910 BC *(Fig. 7.67a)*

Three archaeological sites could be ascribed to this zone. The wood results reflect a mixture of dryland and wetland trees. On the dryland, a mixed forest of oak and ash was found, while in the wetland area, an alder carr woodland exists along the border of a fen. The data closely resembles the non-archaeological wood pattern (Fig. 7.67j), and the environment is therefore interpreted as a fairly natural situation without much human interference. Alder is the most common species in this zone, and alder and ash together form two-thirds of all the wood. Yew is best represented in this zone too and the only pine identified comes from this zone. Values for hazel are low. The results agree with those of the pollen analysis.

Zone II: 1910–1640 BC *(Fig. 7.67b)*

Only one site could be ascribed to this zone and few items were analysed. The results therefore only give some indications. There is a strong drop in alder and a complete disappearance of holly. In contrast, ash, birch and, notably, hazel have increased, ash showing the highest values in this zone. Together, alder and ash still represent approximately two-thirds of all the wood. Oak is

7.66a Killoran cluster 75 & 240

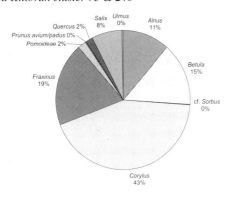

7.66b Killoran cluster 69 & 306

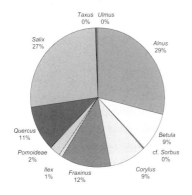

7.66c Cooleeny 64 area

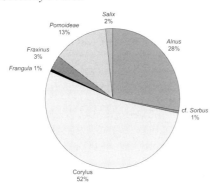

7.66d Cooleeny cluster 31

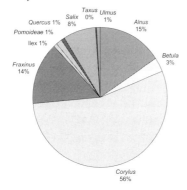

7.66e Derryfadda cluster

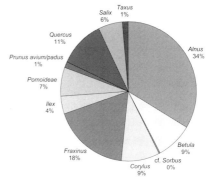

7.66f Killoran cluster 18 & 305

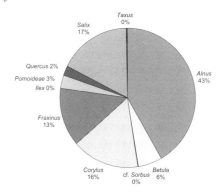

7.66g Environmental samples, total results

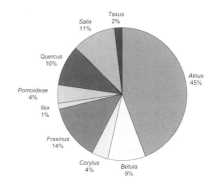

Note: 0% indicates result <0.5%

Fig. 7.66 *General trends according to areas.*

well represented. The reduction of alder may have been caused by local changes following the transition from fen peat to raised bog peat around 1800 BC (see Chapter 3). Although the quantities analysed are quite low, there are no disagreements with the pollen results.

Zone III: 1640–1310 BC *(Fig. 7.67c)*

Six sites fall within this zone and a reasonable amount of wood was analysed. There is a notable rise in the values of alder compared with the former zone. This expansion of alder may be related to a bog burst, but the low number of identifications in Zone II may be distorting the picture. If the wood results for Zone II are omitted, a gradual decline of alder results. Willow is very well represented and must represent locally wet areas. Hazel shows a further increase. Otherwise, this zone is not unlike Zone I, suggesting that a certain amount of seldom accessed marginal woodland remains. There is a marked reduction in the use of oak and ash. The pollen results show the same trends as the wood analysis, although the wood results do not reflect the intense clearance evident in the pollen record. However, it is clear that the use of oak and ash is lower than in Zone I or Zone II and that primarily wetland wood species are being used, indicating a well-developed marginal forest.

Zone IV: 1310–1000 BC *(Fig. 7.67d)*

Wood from ten sites was analysed from this zone and shows a markedly different picture than the former three zones. There is less indication of the presence of a well-developed marginal forest, particularly when the results are compared with the environmental samples. Values for alder and ash are low, and both now represent only one-third of all the wood, instead of the two-thirds in Zones I–III. Hazel, in contrast, has expanded considerably. The low values for alder indicate that the marginal forests were intensively used. This picture becomes even more noteworthy when it is considered that this zone includes the period after bog burst B of 1250 BC, which must have resulted in a considerable expansion of the local marginal border capable of bearing trees. The increase in hazel and the drop in alder suggest an open marginal situation in which a regular transport of hazel (and maybe oak) from more elevated positions to the archaeological sites begins. The pollen data indicate a period of regeneration for the dryland resulting from reduced activity and a serious expansion of ash favoured by previous clearance of woodlands. This pattern is not obvious in the wood data, which mostly represent the local situation around Derryville Bog.

Zone V: 1000–700 BC *(Fig. 7.67e)*

Three sites are represented in this zone. The wood identifications are limited, so the results are only indicative. The zone includes Killoran 69, a complicated trackway with respect to wood preservation. The oak from this

trackway, in particular, seemed to be in very bad condition, indicating that the trackway was in use for a considerable period of time or, more likely, suffered exposure to air resulting from a bog burst. The wood results from this zone are markedly different to those from all other zones. Not only are values for oak dramatically high, but values for hazel are conspicuously low. After Zone V, oak is never important again, with values of less than 5%. Ash is very well represented and occurs in almost equal numbers to oak. Yew, virtually absent from Zones II–IV, makes an appearance. The data for the few sites available might suggest that the area was not frequented at the time, but surprisingly this did not result in a re-establishment of alder locally. It might be that local conditions were too dry for the germination of alder, but then the lack of habitation is puzzling. The pollen record indicates clearance and continued, albeit variable, activity, with a re-establishment of oak and ash. Also, a minimal value for hazel is noticed. This phenomenon has been recognised in other pollen diagrams in Western Europe and has been interpreted as a drop in pollen production due to a climate change that resulted in a rise in the water table and an extension of fens and bogs (van Geel *et al.* 1996; Caseldine *et al.*, this volume). It is not clear whether such a change occurred in Derryville Bog. The local water dynamics indicate a precarious water balance resulting in many bog bursts long before this zone. Nevertheless, the low hazel values in the pollen and wood records remain an interesting point for discussion. The pollen record agrees very well with the wood results with respect to the low hazel values and relatively high ash and oak values.

Zone VIi: 700–400 BC *(Fig. 7.67f)*

As mentioned above, Zone VI was divided into two subzones because of the intensive activity related to bog burst D of *c*. 600 BC. Seven sites fall into this subzone. Ash and oak show decreased values compared with Zone V, while hazel recovers to some extent, as does alder. Willow is very well represented and derives from local sources on the western side of the bog, close to Killoran 248, indicating locally wet situations. Alder represents only one-third of all the wood identified. The pollen record points to restricted open areas with an otherwise regenerating woodland and scrub situation, comparable to Zone IV, with reduced values for ash and oak. The wood results do not resemble Zone IV to a great degree, although the trends for ash and oak are the same. However, the pollen and wood results do not disagree.

Zone VIii: 400–200 BC *(Fig. 7.67g)*

Seven sites are also represented in this zone, indicating activity in the west as well as the east. In the western part of the bog, several hurdles derive from this period, explaining the high values for hazel. This must point to managed forest (see above). Alder shows the lowest value in this zone, pointing to the minimal occurrence

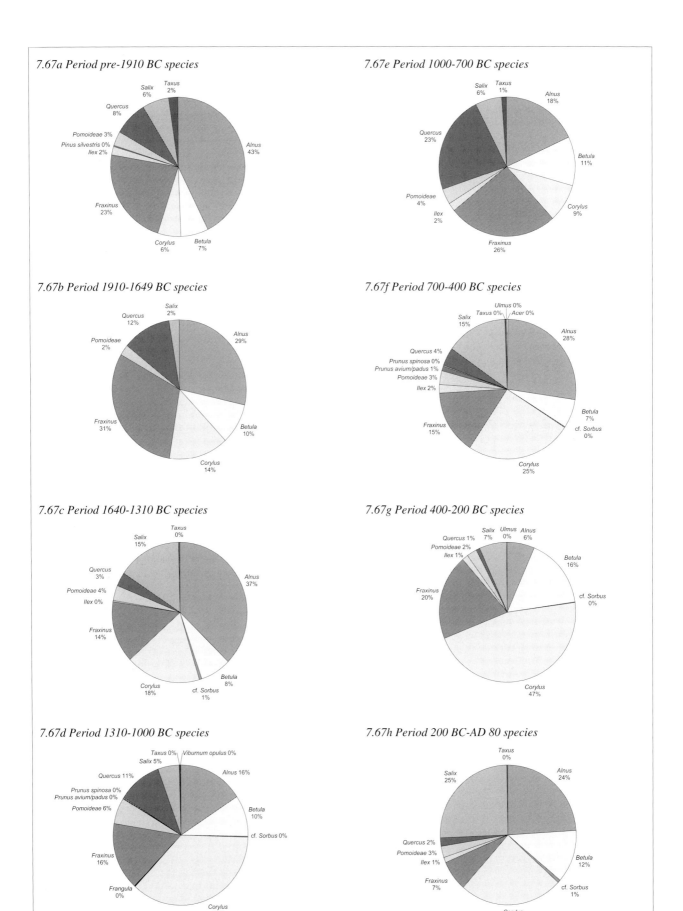

Fig. 7.67 *Zone/percentage all species. Continued next page.*

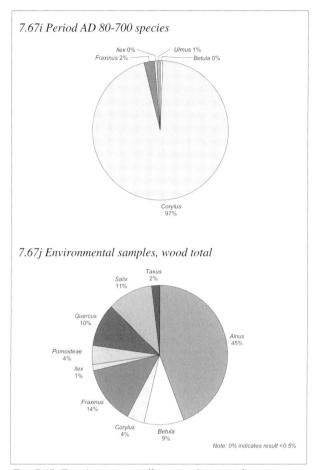

7.67i *Period AD 80-700 species*

Ilex 0% Ulmus 1%
Fraxinus 2% Betula 0%

Corylus
97%

7.67j *Environmental samples, wood total*

Taxus
2%
Salix
11%
Quercus
10%
Alnus
45%
Pomoideae
4%
Ilex
1%
Fraxinus
14%
Corylus
4% Betula
9%

Note: 0% indicates result <0.5%

Fig. 7.67 Zone/percentage all species (continued).

of marginal alder forest. There is a further decline of oak, which virtually disappears, while ash, on the other hand, expands. The high values for birch must be explained by the local expansion of the tree, maybe directly following the locally drier conditions after the bog burst at the start of this zone. According to the pollen record, this zone represents a period of regeneration with low-level activity, comparable to Zone IV. Alder, according to the pollen record, is continually present locally. Towards the end of the zone, charcoal is washed into the sediment on the western side, representing small-scale erosion. Again, it is difficult to compare the pollen and wood results. The low levels for alder wood can partly be explained by the fact that most of the samples derive from the western side of Derryville Bog, where the steep slopes do not facilitate the presence of an extensive alder forest during this stage, especially not during periods with a rising water table. The low levels for oak in the wood record are not mirrored in the pollen record. The intensive use for ash may be in agreement with the pollen results, but then again, the intensive use of hazel for hurdles can only be explained by a high degree of woodland management, which may be difficult to detect in the pollen record. Taken as a whole, therefore, the pollen record and the wood record in this zone disagree.

Zone VII: 200 BC–AD 80 *(Fig. 7.67h)*

Only two sites are represented in this zone, but the numbers are comparable to those for Zones I and IV. Most noteworthy are the reduced levels for ash and the increase in willow and alder. The lower values for hazel can be explained by the absence of hurdles, but values are comparable to those for Zone VIi. The high values for willow point to increased local wetness. Compared to Zone VIi, the marginal western alder forest has to some degree re-established itself. The low values for ash, the lowest in the wood record, could point to the absence of ash at least from the marginal areas and probably its minimal occurrence on the dryland. The pollen record indicates a period of clearance, with silt in-wash from a well-defined area in the west at the start of the zone. It points to an initial reduction of oak and the presence of damp pasture. The low values for oak and ash are found in both the pollen and wood records which could point to woodland clearance. Given that only two sites are represented in the wood record, there are no firm indications for clearance close to the bog border.

Conclusions

In general, the pattern produced by the pollen record is mirrored in the wood record, although the wood record lacks firm indications for clearance compared to the pollen. The number of sites analysed clearly influences the comparability of the methods, which is further complicated by the nature of the archaeological sites and the geography.

Following the timescale presented for the pollen zones, the wood identifications present the following picture:

Species such as birch and willow mostly represent the local trees close to the archaeological sites and may indicate dry and wet circumstances, respectively. Except for Zone VIii, alder is always present in high values and must have been abundantly present in most periods.

There are strong indications for the increased use of (managed) hazel over time and a gradual reduction in the use of alder, which might reflect the intensive use of the bog marginal areas.

Generally over time, the use of oak decreases, but there is a marked rise in oak, accompanied by a conspicuous low value for hazel in Zone V. This zone deviates from the others and might reflect a change in wood usage.

Zones I, II and III probably represent the presence of rather natural woodlands in the environment of Derryville Bog. Afterwards, the woodlands adopt a different appearance, resulting to a great degree from the intensive manipulation of hazel.

The quantities for ash vary considerably. It cannot be excluded that part of the ash found in archaeological sites is of dryland origin. This species might therefore be most suitable for comparing wood results with pollen results from the dryland area. Generally, however, the dryland forest cover of oak, ash and elm represented in the pollen record was hardly used in the bog marginal areas.

Archaeological chronology

In the preceding section the wood identifications were ordered following the pollen zones and according to areas. Here, the identifications follow the archaeological periods, from the Early Bronze Age until the medieval/post-medieval period. As the last period only includes two sites that consist almost totally of hazel, this period is not included. Sites without dates are also excluded. The results are presented in Figs 7.68a–e.

The results show a changing environment. The most noteworthy aspect is the decrease of alder over time. In the Early Bronze Age (Fig. 7.68a), alder represents more than 40% of the wood; after this period it declines to less than 20% in the Iron Age (Fig. 7.68e). Hazel, on the other hand, is not important in the Early Bronze Age, but increases regularly in the periods afterwards.

In the Early Bronze Age, ash and yew are better represented than in later periods. Although the number of sites represented in the Early Bronze Age is relatively low, it looks as if the woodland in this period differs from later periods and reflects the natural woodland situation in the area (see above). The situation in the Early Bronze Age is very similar to the total results of the environmental samples (Fig. 7.68f), except for lower values for ash.

The low values for ash could indicate that this tree was more frequent in dryland areas, but it is more likely that, locally, ash was less common. This can be explained by the increasing acidity of the environment caused by the peat-forming *Sphagnum*. The area of available dryland diminished, however, because of the expansion of the raised bog.

The later periods show increased use of hazel (see above) and a more dynamic woodland situation. Although alder declines, it must have been an important element in the local marginal woodland during all periods, because it always forms a vital part of the wood used.

Another aspect indicated in the diagrams is the preference for oak in the later Bronze Age when the numbers of oak are equal to those of alder. In later periods, oak is less important. This does not necessarily indicate that oak was unimportant in the area during other periods; it is very possible that oak was used elsewhere for other purposes.

The fluctuating values for willow could be interpreted as being the result of changing local circumstances, especially, regarding water level and water quality (as demonstrated by Killoran 248). Whenever willow percentages are high, such as the Middle Bronze Age and especially the Iron Age, high water tables and/or influx of mineral-rich water from the dryland is suggested.

Over time, the importance of hazel in the bog area increases, while local vegetation, including alder, is used less. To bring material from the dryland to the bog would have been difficult as marginal forest generally included much roots and undergrowth. Therefore, increased use of hazel could mean that the bog marginal areas were less well established in later periods, especially in the Killoran area.

The provenance and quality of the wood remains

The wood remains found in Derryville Bog came from two sources: trees directly bordering the access areas and trees growing on the dryland (i.e. locations that were effectively 'non-wet' or 'non-bog') further away from the bog. Prehistoric man knew his environment and hence must have been efficient; implying that the effort employed to make trackways and platforms was as minimal as possible. Therefore, it was assumed that most of the prehistoric wood found in Derryville Bog was of local origin, coming from either the marginal woodlands or from the hilly slopes directly surrounding the area. This idea was supported by the fact that the bulk of the wood remains represented only a few wood species. The natural woodland vegetation in the wider surroundings on the dryland is better represented in the pollen diagrams (see Chapter 6).

The wood identifications showed a dominance of alder and low values for hazel in the non-archaeological wood remains deriving from bog depths. The importance of alder and the low values for hazel were mirrored in the wood samples of Early Bronze Age date. In contrast, archaeological wood remains produced a much higher proportion of hazel at the expense of alder in later periods. It is suggested that the archaeological wood samples reflect the composition or nature of the woodland in the immediate vicinity of the sites. Wherever high values for alder were found, they derived from wet marginal areas. When high values for hazel were apparent, more open and/or drier conditions prevailed. The latter may have been caused by human interference.

The presence of hurdles indicates a clear use of woodland management in the Iron Age. The use of hurdles also indicates that prehistoric people were well adapted to their environment. No strict rules for the felling period were apparent , but hazel as a wood species was generally preferred.

Indications for Early Bronze Age woodland management were noted (Derryfadda 204). In contrast, the substantial trackway Cooleeny 31 seems to have been made from unmanaged wood, using a complete hazel woodland in the vicinity of the site.

A very curious phenomenon in Derryville Bog was the overall poor quality of the wood, compared with the wood collected from other sites in Ireland and elsewhere. Most of the wood was rotten to a certain degree and contained many fungal hyphae. The wood often showed insect channels with frass. Caseldine *et al.* (Chapter 6) noted the absence of *Sphagnum papillosum* and high values for the rhizopod *Hyalosphenia subflava*, an indication of disturbed raised bog. The combination of these three observations led to the conclusion that Derryville Bog was a very dynamic environment. The dynamics can be ascribed to the drastic lowering of the water table occasioned by many bog bursts. Apparently, this resulted

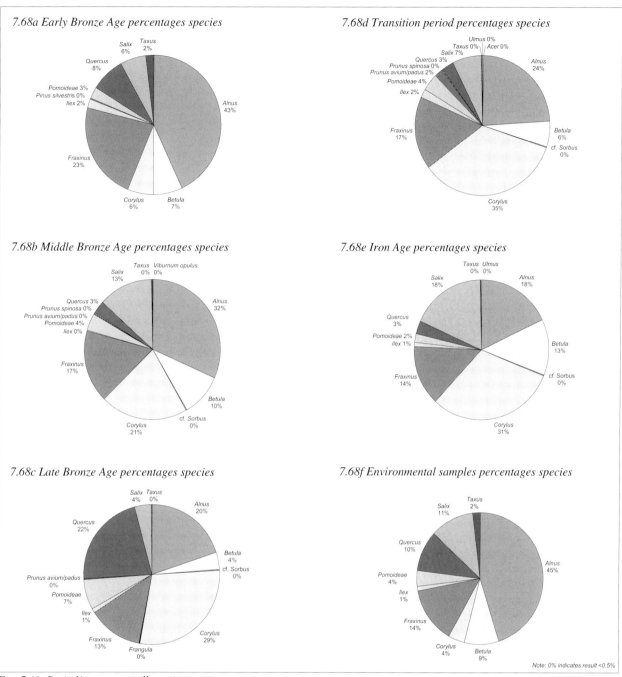

Fig. 7.68 Period/percentage all species.

in thorough aeration of the top layer of peat, resulting in the degradation of the wood. Many trackways were made after a bog burst and later overgrown with a layer of peat of some 20cm. Generally, it took ten years for the water table to re-establish itself to some extent after a bog burst. This short period was apparently sufficient to cause important aeration of the top layer of the peat, including the trackway—a depth of some 50cm. Subsequent bog bursts resulted in renewed lowering of the water table, which must have exacerbated the degradation processes affecting the wood in the bog.

In many sites, environmental wood indicates the presence of trees locally. Most of these are situated in the marginal areas surrounding the bog and they often represent old forests. The finds of large tree stumps on the slopes leading to the bog were indicative of the presence of such forests. However, most of the sites are not made of timbers. For the construction of the trackways and platforms slender, manageable brushwood and round-woods were preferred. Exceptions to this general practice are the oak planks from Killoran 18, Cooleeny 22 and Derryfadda 23.

8. Coleoptera

Eileen Reilly

Background

Introduction

This chapter looks at samples taken for insect analysis at various sites throughout Derryville Bog. The analysis of insect remains, particularly Coleoptera (beetles) which will be the main focus of this chapter, and their use in environmental reconstruction is relatively new in Ireland but has had a long and distinguished history in Britain and parts of Europe.

In order to recover insect remains a certain degree of waterlogging and organic build-up must be present. Obviously, a wetland setting has both in abundance and, in addition, the interference of humans creates artificial habitat niches that insects exploit and from which they are subsequently recovered.

The archaeological structures sampled range in age from the Bronze Age to the Iron Age, however, the time periods covered, in particular by column sampling, encompassed a much wider time frame. It was hoped that by varying the sampling strategy, both micro and macro-environmental changes would be highlighted.

Sampling strategy

Vertical columns of samples in one area were taken to present a picture of changes through time in the Coleoptera and whether the presence of an archaeological structure changed that faunal variety. These profiles can pick up natural occurrences such as flooding episodes, bog bursts and dry episodes.

Site-specific sampling involved a block of peat taken underneath a structure, at the same level as the structure (usually incorporating some wood from the structure) and immediately above a structure. This type of sampling is used to create a picture of the immediate environment often helping to understand why a structure was sited where it was, the community of species which exploited the site (including importations) and the changes that occurred in that environment when the structure went out of use.

Spot sampling, or subjective sampling, is also very important where something of interest is uncovered but the area is not conducive to profile sampling or site-specific sampling, and where they can be closely linked, both stratigraphically and through dating, to a particular structure or activity. This has certainly proved to be the case with samples from Derryfadda 23 which have produced some of the most significant finds of this project and have important implications for the history of the Irish forest fauna relict.

The results will be looked at in terms of the local environment, drawing on information from clusters of structures sampled on different sides of the bog. They will also be examined in terms of overall landscape changes in three areas: woodland, wetland and dryland usage—pasture and cultivation.

Methodology

The samples ranged in size from 1 litre to 6 litres (all samples over 3 litres were sub-sampled) and were processed at the Killoran House facility during the excavation season. They were processed using the paraffin flotation method outlined by Coope and Osborne (1967) and expanded by Kenward (1980). The flots were then sorted in alcohol using a low-powered binocular microscope and the extracted remains identified using the usual range of keys (see bibliography) and the Gorham and Girling Coleoptera Collections at Birmingham University (with help from Dr. David Smith). One species (*Rhyncolus ater* Linn.) was sent to Dr Nicki Whitehouse, Department of Archaeology and Prehistory, Sheffield University (now of Queen's University, Belfast) for confirmation of identification. The species lists (Tables 1–10, additional files) are given in taxonomic order according to the revised lists of British (Kloet and Hincks 1977) and Irish Coleoptera (Anderson *et al.* 1997).

Explanatory note on the habitat data and the figures

The specific habitat information given in Tables 1–10 (see additional files), is adapted from Robinson (1991) and every sample taken has been analysed according to this information. The key is as follows:

A: aquatic; B: bankside/waters edge; C: carrion; D: disturbed or bare ground; F: foul (dung); G: grassland; M: marsh (fen/bog); T: terrestrial, occurring in a variety of habitats; V: decaying plant matter and W: woodland or trees. The specific habitats are then generalised into general habitats and presented in bar charts for each site (Figs 8.1–8.4 and 8.6–8.12) throughout the text. These are (i) aquatic; (ii) dung/rotting vegetation; (iii) dead wood; (iv) marsh/bog/aquatic plants; (v) trees/carr woodland and (vi) terrestrial/pasture. Species are assigned to the most appropriate general grouping, i.e. if a species normally occurs in a marsh environment then it is assigned to the marsh/bog/aquatic plants group. Habitat data was gleaned from BUGS Ecology database (Buckland *et al.* 1996) and various keys and written sources (see bibliography).

A number of references will be made to the status of species as they are recorded in the *British Red Data Books: 2. Insects* (Shirt 1987). This Red Data Book (RDB) is a catalogue of all the rare, vulnerable and extinct insect species in Britain. The list of Coleoptera in this catalogue has since been updated by Hyman (1992; 1994), with additional notes on current distribution and known ecology. Various categories are given including Notable B (a species found in a restricted number of locations in Britain), Rare, Local (restricted in its choice of habitat) etc. This kind of catalogue is not available for beetles in Ireland, so variations may occur in the status of certain species in Britain and Ireland (M. Morris, pers. comm.). Where Irish data is available for the status and distribution of species today, they will be referred to. However, for any species that is not on the current Irish list (Anderson *et al.* 1997), Hyman's (1992; 1994) study is the best method of gauging their current distribution and status in our nearest neighbour.

The local environment

Eastern bog margin

Column samples—Neolithic to Iron Age

A number of column samples were taken through structures in this southeastern cluster covering long periods of time. A vertical column in the drain face on the east side of Field 46 to mineral soil, incorporating Derryfadda 13a, was taken. This site has a date of 767–412 BC but the column may start in late Neolithic levels. This sample was looked at from the earliest layers to the most recent (bog surface) and the habitat data is presented in Fig. 8.1. The species list is in Appendix 1A, Table 1 of the additional files.

The transition layer between the original mineral soil layer and the development of the fen peat (124.48m OD) above had no discernible insect remains of any type, indicating a low organic content and a relatively dry layer compared to those above.

Moving up through the column, a more typical fen fauna was identified (sample 6). The aquatics, *Hydroporus angustatus* and *Graphodytes* sp., indicate peaty pools of acid water, while *Ochthebius* sp. and *Hydreana* sp. are indicative of stands of fresher water. *Paracymus scutellaris* and *Hydrobius fuscipes* are found in shallow acid waters. All these species taken together indicate the beginning of the development of the fen.

Phylopertha horticola, however, is indicative of the nearby drier land, possibly pasture land. *P. horticola* is a pest on many types of bushes and at the roots of grasses, clover and cereals. It usually infests poor quality pasture, often on slopes of hilly areas where there is high rainfall. As this species is present in most of the samples from this site it can be assumed that it was fairly common throughout the local pasture land and that there was a constant presence of such land nearby. The numbers of individuals overall is low and this is probably due to the high silt content. This layer seems to represent the interface between the original drier land, at this point, with its sparse woodland and the inundation of that land, which resulted in the inception of the fen proper in this area.

A maturing of the fen is represented from (sample 5) 124.83–124.98m OD by an increase in insect numbers and variety, probably Early to Middle Bronze Age in date. The wood remains are most likely roots and indicate a bog-edge forest that is fairly well represented in the species present. However, also well represented are

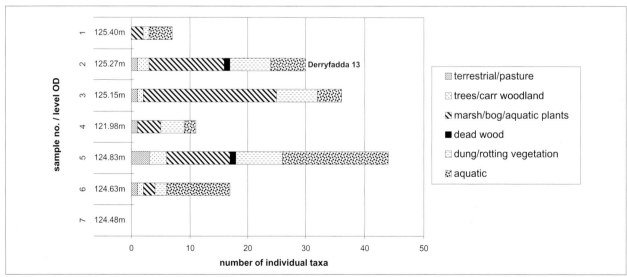

Fig. 8.1 Habitat data from column of samples taken through Derryfadda 13a, Field 46.

pool/bank-side species and drier, pasture species. A small number of species occur in rotting and decaying vegetation but may also indicate dung, carrion and other fouler habitats. These include *Cercyon melanocephalus*, *Megasternum obscurum* and *Staphilinus* sp. In particular, *C. melanocephalus* is found in the dung of large herbivores (i.e. sheep, cow and horse) and less frequently in other decaying matter.

Of the plant feeders, both wetland and drier land species are indicated. *Phylopertha horticola* occurs again, while *Agrypnus murinus* is often found under stones in fields and gardens rich in humus. It has been noted as a pest on various cultivated plants, particularly vegetables. *Gastrophysa viridula* feeds on members of the dock family, a classic waste/cultivated ground plant species. These species and the dung feeders may have been washed down into the fen from nearby grassland during a flooding episode. This level has a higher number of aquatics than any other sample in this sequence, which may be indicating the same thing or a rise in the water table. *Rhinoncus perpendicularis* is also common on various members of the dock family in both moist and dry habitats. *Plateumaris sericea* occurs on great reedmace, burreed and yellow iris, species typical of fen or marshland.

The most significant finding from this level is *Dirhagus pygmaeus* (Plate 8.1), a species listed as rare in the RDB and is absent from the current Coleoptera list for Ireland. It is an inhabitant of woodland and copses of old deciduous trees, including birch, alder, oak and hazel. The larval stage probably develops in dead wood and has been found in old oak and beech stumps (Hyman 1992). Its absence from the Irish list is indicative of two things. The general clearance of deciduous woodland throughout the country from earliest times has contributed to this and probably many other species of beetle disappearing from

the Irish record. Also, detailed studies of the pockets of ancient woodland which still exist have not been carried out and could possibly produce examples of some of these species, albeit in drastically reduced numbers. Its presence at this level is the clearest indicator of stands of ancient woodland on the eastern margin.

A dramatic difference in species numbers and variety is noted at 124.98m OD (sample 4, Fig. 8.1). Only one true aquatic species occurred, pointing to a drier phase in this area for a period of time. From comparative peat morphology and testate amoebae data, a bog burst dated to 1250 BC (bog burst B) has been identified in this area (see Chapters 3 and 6). This would explain the dramatic drop in the water table and, consequently, a radical reduction in water species. *Phylopertha horticola*, the grassland beetle is represented by four examples, while wetland species are all secondary indicators, i.e. the ground beetle *Pterostichus diligens* and the rove beetles *Lathrobium* sp. and *Stenus* sp. that thrive in moist biotopes. *Pterostichus gracilis* (RDB status: notable) occurs in wet vegetated soil and is therefore more eurytropic than many species of its genus. The phase would appear to last for up to four hundred years. A return to normal fen conditions in noted from 125.15m OD (sample 3) until the building of Derryfadda 13a sometime between 767 and 412 BC.

The ground beetles *Elaphrus cupreus*, *Pterostichus nigrita* and *Dromius* sp. are all species of carr and wet deciduous woodland. There are two examples of *Plateumaris discolor*, which feeds at the roots of cottongrass, a raised bog species, and among *Sphagnum* but also occurs on sedges. There are a number of dryland/pasture indicators in this sample also. As well as *Phylopertha horticola*, *Dascillus cervinus* is found which feeds on flowers and shrubs, generally on dryland. The weevil *Alophus triguttatus* is found on a number of species such as hemp agrimony, dandelion and common comfrey. These plants occur on the margins of wet areas often in damp grasslands or flooded pasture and on waste ground. The presence of these beetles in the sample could be indicative of flooding of the nearby dryland (contributed to by the removal of tree cover, for example) and its subsequent draining into the basin in which the bog was developing.

The structure occurred at 125.27m OD in this column (sample 2) and a number of roundwoods were examined as well as the peat surrounding the wood at this level. No direct woodland or dead wood feeders were recovered. Along with the usual species of ground beetle, *Agathidium rotundatum* is associated with the fungi of various trees while *Bryaxis* sp. is found in rotting wood mould, under bark and in leaf litter. Aquatic species such as *Agabus paludosus* and *Hydraena* sp. often indicate running water or fresh water, which may indicate the presence of a nearby ditch or stream bringing fresh water down from higher ground. Nearby dryland or pasture is also indicated.

Plate 8.1 Dirhagus pygmaeus.

Occasionally, animals may have strayed into the bog edge forest to graze but, in general, the numbers of dung indicators on the trackway is small and any found are more likely to be casualties from nearby pasture.

About 0.15m of peat above the structure was examined and proved quite unproductive. All insect remains recovered were indicative of raised bog. Species such as *Pterostichus minor* and *Hydroporus* sp. are typical of wetland from raised bog to fen. *Cyphon* sp. generally occurs in wetland areas from base-rich fens to acidic raised bogs. However, the dung beetle *Aphodius fimetarius* is found in hay refuse, deer and cow dung and, again, is an indicator of nearby dryland or pasture.

Two small column samples were taken in Field 50 from just under the platform Derryfadda 6 (dated to 380–5 BC) to mineral soil. The platform had been built in marginal forest peats in an area of root systems, directly on top of the eastern landfall of the trackway Derryfadda 215, dated to 457±9 BC. The column also

incorporated Derryfadda 216, a *fulacht fiadh*, constructed on a natural layer of fen peat and dated to 1400–990 BC. One column was located at the northern end of the platform in the drain face on the west side of Field 50, the other in the middle of the platform to the south in the same drain face. The results are presented in Figs 8.2 and 8.3 but will be looked at together. The species list is presented in Table 2 of the additional files, with the results from both columns combined together. These columns represented the easternmost bog marginal environment examined and, although not as long as at Derryfadda 13a, cover a period from the Early Bronze Age to the early Iron Age.

The number of insects recovered from just above mineral soil (samples 22 and 26, approximately 125.03/125.08–125.13/125.28m OD) was very small. A small number of unidentifiable beetle remains and fly puparia were present. This layer was similar to the basal fen layer at Derryfadda 13a and appeared to be nutrient-poor or leached out, militating against the preservation of insect remains.

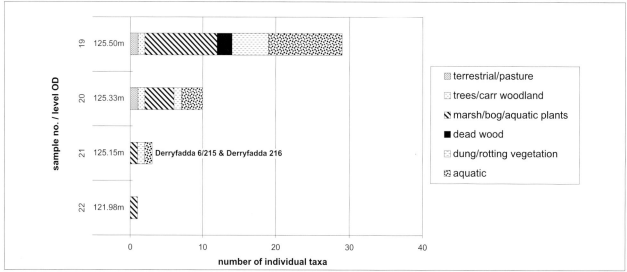

Fig. 8.2 Habitat data from Derryfada 216 and Derryfada 216, Field 50. Northern column.

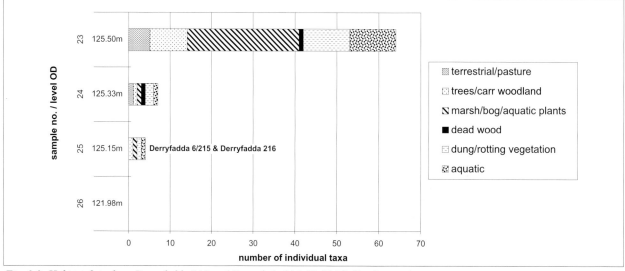

Fig. 8.3 Habitat data from Derryfadda 215 and Derryfada 216, Field 50. Southern column.

Unsurprisingly, the number of insects recovered from the *fulacht fiadh* ash and stone layer (sample 21 and 25) was also small. *Limnebius* sp., *Cyphon* sp. and a possible *Hydropous* sp. all indicate stagnant water but were in a very poor state of preservation. Their presence in this layer was most likely due to waterlogging of the *fulacht fiadh* by a rising water table or natural flooding from the upland.

A layer of dark brown silty carr woodland peat that developed across the top of Derryfadda 216 (samples 20 and 24, 125.33–125.50m OD) after it went out of use proved to be very interesting. From the northern end of the site, where the ash layer was thinner, the fauna is more indicative of fen with standing water. *Pterostichus nigrita*, *Agonum* sp., *Hydroporus* sp., *Cyphon* sp. and *Hydraena* sp. are all typical of this environment. *Dascillus cervinus* and *Phylodrepa* sp. are more typical of dryland or pasture. *Leiosoma deflexum* feeds on wood anemone, marsh marigold and buttercup and so can occur in both wet and dry places. The dryland species are unsurprising as this site is extremely close to upland.

However, from the middle of the site (the second column), the fauna is more indicative of a forest floor (sample 24). The ground beetle *Abax parallelpipedus* is a pronounced forest species, preferring shaded rather moist habitats (Lindroth 1969). *Leiosoma oblongulum*, a species not on the current Irish list and classified as notable B in the RDB, is found in broad-leaved woodland in leaf litter, damp moss and is also associated with the buttercup family in grasslands (Hyman 1992).

The most significant find is the Scolytid *Tomicus minor/piniperda* (Plate 8.2). Only the head was found and on the basis of this alone it was not possible to separate the species. However, while both are found in the same habitat, they attack standing, weakened or recently fallen conifers. *T. minor* is extremely rare while *T. piniperda* is more common. *T. piniperda* was found at Thorne Moors (Buckland 1979) but *T. minor* has never been found in an archaeological context. It is confined to Scotland and the native Scots pine belt, and in one pocket of natural coniferous woodland

in Dorset, in southern England. *T. piniperda* appears to have adapted to the more recent species of conifers such as Norway and Sitka spruce, introduced to Britain and Ireland by the Forestry Commissions, and is therefore more widespread. In this context, however, they could only have been attacking Scots pine, which was native to Ireland at this time. Scots pine is represented throughout the pollen profiles, albeit in small numbers. Casparie has noted that relics of the Boreal pine forests occur but usually at the base of fen peat, however, in a number of locations, small stands of usually badly grown pine existed on top of the fen peat (see Chapter 3).

This small number of species produces an interesting picture of the bog-edge at this point, as a mixed woodland is clearly established in this area but may be very localised. It could be that the ash, stone and charcoal provided an artificially dry base for this section of the forest, which can also be seen from the western bog margin at Killoran 240.

On top of this layer, a loose woody peat, red brown in colour developed in which Derryfadda 215 and, subsequently, Derryfadda 6 was constructed. From the northern column (sample 19, Fig. 8.2), some of the substructural brushwood was included for analysis. Taking both samples together, this layer proved to be the richest in terms of insect remains (sample 23, Fig. 8.3). This is unsurprising given that it represents a mature fen peat and also the organic build-up around the substructure of the platform. However, in terms of habitat diversity the picture presented is fairly predictable—fen with carr woodland. The picture is perhaps clearer than before with not only ground species indicating the presence of alder/willow carr in particular, but also a number of wood and plant feeding species indicating the same thing.

While the actual composition of the fauna is unremarkable, the contrast in terms of diversity and number between it and other layers, both here and in other parts of the eastern margin, is notable. The presence of the platform has added a range of decomposer species while fen itself has provided the standard range of wetland

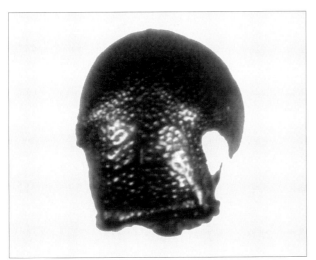

Plate 8.2 Tomicus minor/piniperda.

Plate 8.3 Hydroporus melenarius.

indicators. However, no species specifically associated with dung, animal hides or cadavers was recovered from the trackway substructure.

Of the ground species, *Leistus* sp. is a woodland genus but is also hygrophilous and can be found in swampy woodland. *Agonum fuliginosum* is found in alder carr, willow thickets and moist deciduous/mixed woodlands, as is *A. obscurum*. *Agonum gracile* is found in very wet places in sedge fens and among *Sphagnum*. A great number of species testify to water, particularly stagnant or acid water with *Sphagnum* and other detritus.

The weevil *Dorytomus taeniatus* occurs on varieties of willow, and *Phyllobius* sp. occurs in many trees species, including alder. One fragment of an anobid (wood-worm beetle) was also found.

Site-specific and spot samples—Bronze Age

Samples were taken from the trackway Deryyfadda 23 at two points and are presented in stratigraphic order in Fig. 8.4. The species list is presented in Table 3 of the additional files. Samples were taken immediately above, at and below the structure in an extension cutting in Field 40. Two spot samples were taken through a section of a thoroughly rotted (but once very substantial) timber (Timber 9) in the main cutting in Field 41. The site has two dendrochronological (felling) dates of 1606±9 BC and 1590±9 BC from Timbers 1 and 4, respectively. A pollen diagram and a study of the testate amoebae from this site are presented in Chapter 6.

Studies of the peat morphology and the palaeohydrology of the bog in this area show that raised bog growth began in the vicinity of Derryfadda 23 c. 1800 BC, 400 years prior to raised bog growth 400m north at Killoran 18. The majority of species from below the structure strongly indicate a very wet raised bog environment (sample 12). *Hydroporus melanarius* (Plate 8.3) occurs in peat mosses and six other examples of *Hydroporus* sp.

also occur, indicating acidic water and *Sphagnum* pools. The genus *Enochrus* sp., of which one example occurs, are increasingly rare, their status these days ranging in Britain from rare to occasional or very local (Hyman 1994). They are strong indicators of acidic conditions.

The weevil *Micrelus ericae* is a true raised bog species as it occurs on heather and heath plant species and, combined with the water beetles above, shows an area of mature, wet raised bog upon which the trackway was built. A well-developed hummock and hollow system of bog growth was identified in this area by Casparie (see Chapter 3).

Samples taken through the very rotted wood of the structure (samples 8 and 9; 124.16m OD) shows the usual range of wetland indicators as well as species specifically indicating raised bog. However, the range of woodland indicators is the most significant aspect of the trackway samples.

The most important find was twenty-seven examples of *Prostomis mandibularis* (identified immediately by Dr. David Smith from photographs in Buckland 1979), an extremely rare species now extinct in Britain and Ireland and confined to a small number of areas in Europe (Plates 8.4 and 8.5). From archaeological contexts it has only been found twice before, at Thorne Moors and at the Sweet Track on the Somerset Levels. It is very much a creature of primary, undisturbed natural forest and is now restricted to the few areas in Central Europe where sub-primary forest remains (Horion 1960; Palm 1959). It was, however, predicted by Horion (1960) that this species would disappear altogether from Central Europe as, even in these sub-primary forests, tidier forestry practices were removing its habitat (it is primarily recorded from damp, rotten oak and pine on forest floors). It has a holarctic distribution, occurring in the southern part of Sweden, localities in Denmark, parts of Germany, also parts of Portugal, southern France, Sardinia and parts of Italy (Fig. 8.5).

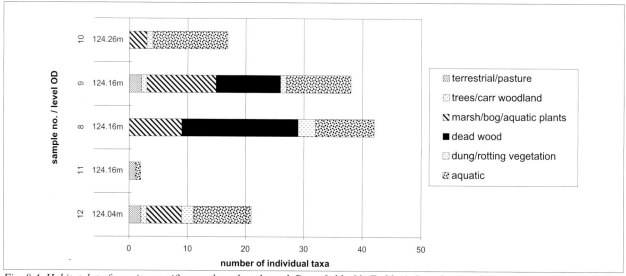

Fig. 8.4 Habitat data from site specific samples taken through Derryfadda 23, Field 41. Samples 8 and 9 are spot samples taken from Timber 9, Field 40.

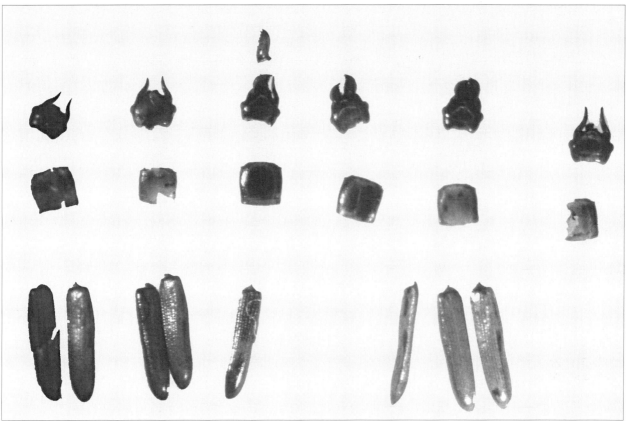

Plate 8.4 Prostomis mandibularis.

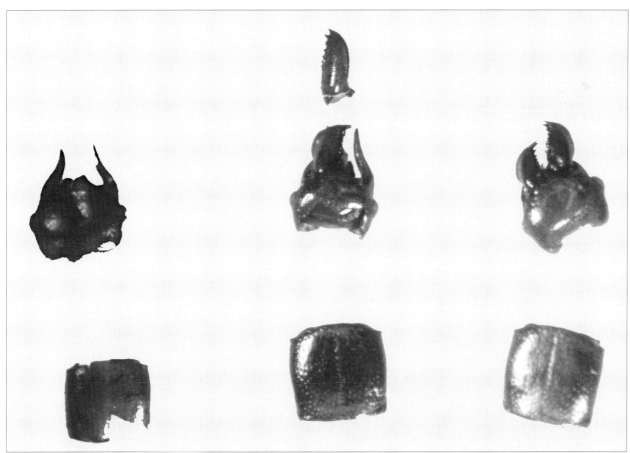

Plate 8.5 Prostomis mandibularis *(close up of head and thorax)*.

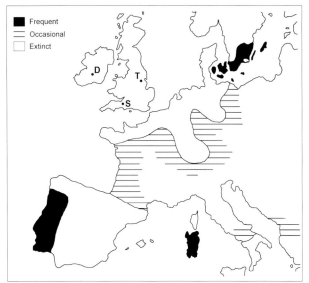

Fig. 8.5 Present distribution of Prostomis mandibularis in Western Europe (adapted from Buckland 1979). D=Derryville, S=Somerset Levels and T=Thorne Moors.

Added to that are two more species that do not occur on the Irish list of Coleoptera, one of whom is listed as endangered in the RDB. *Teredus cylindricus* has only been recorded from Sherwood Forest and Windsor Forest (Donisthorpe 1939) in Britain. On the Continent, it is distributed unevenly occurring on some Swedish islands, in southern Germany and Austria, with some old records from the Lower Rhineland, Thuringia and Bavaria (Horion 1951), and is considered a primary forest species. From archaeological contexts it has been found at Thorne Moors (Buckland 1979) and Runnymede Bridge (Robinson 1991). It is generally found in old beech and oak that has been infested with wood-boring beetles and in the nests of ants (particularly *Lasius brunneus*) in old trees. It appears to be a predator to ants and, interestingly, the most common anthropod at trackway level, apart from beetles, were ants, with twenty-nine heads recovered. One example of *Phloeophagus lignarius* was found which, although not uncommon in Britain, does not appear on the Irish list. It occurs in the sapwood of hardwood tree species and was also found at Runnymede Bridge (Robinson 1991).

Two other species found here that could not be taken beyond genus level were *Rhizophagus* sp. and *Cerylon* sp. Nevertheless, their presence amplifies the information from the previous species. *Rhizophagus* sp. is found under bark of all types of trees from pine to deciduous and also in sap and in tree fungi. *Cerylon* sp. is generally found under the bark of standing and fallen dead trees including oak, elm, beech, birch and lime. They are also found in the galleries left by other wood-borers and, in general, are native to areas of primary woodland.

Most of these species are indicators of primary or fairly unmodified secondary (sub-primary) forest cover,

which must have existed on the fringes of the fen and raised bog. *Prostomis mandibularis* probably represents the last stages of the decay of wood, while the other species are found in drier wood.

The pollen diagram at this point (DV III) shows falls in tree pollen before the building of Derryfadda 23, probably corresponding to the removal of trees for construction of this and other trackways. However, plenty of primary woodland would seem to remain, which would have supported these beetle species, including birch, pine, elm and oak. The number of beetles from above the trackway (sample 10) is quite small but all indicate the continued development of the raised bog in this area after the trackway had gone out of use. Indeed, the area appears to be even wetter than before with true aquatic species making up the majority of individuals present, i.e. *Hydroporus melanarius*, *Enochrus* sp. and *Cyphon* sp. Only one anomaly appears, *Agriotes* sp., which occurs at the roots of grassland species. This was probably an accidental casualty from the nearby dryland.

Site-specific samples—Iron Age

Derryfadda 9, a platform radiocarbon dated to 395–180 BC, was also sampled for site-specific environmental information (Fig. 8.6). The species list is presented in Table 4 of the additional files. Casparie identified a bog-edge forest below the site and it appears to have been constructed within this (see Chapter 3).

Approximately 0.15m of peat below the site was examined but did not reach mineral soil (sample 614). It was very poor in terms of remains, indicating standing water and nearby grassland.

The platform itself produced slightly higher numbers but did not significantly increase or add to the species diversity (sample 613). The presence of *Dorytomus taeniatus*, common on willow, could be explained by local standing trees. Many water species were present and it has been suggested that the site was submerged in water for a time. These include three 'rarities' from an Irish perspective—*Agabus striolatus*, a species of relict fen carr, *Graphodytes* sp. and *Limnebius* sp. A number of species that occur on decaying vegetation were found, which probably reflect the accumulated vegetation on the platform. Although hazelnut shells were found at trackway level and below it, no specific indicator of hazel was found.

The layer above the platform was the most productive, although the numbers were still small compared to other sites. The majority of species were marsh/aquatic plants species, however, two species reflect the nearby upland, *Aphodius* sp. and *Phylopertha horticola*. The aquatic element appears to increase from below the site to above the site. This seems to be the case for most of the sites on the eastern margin and attests to the increased wetness overall of this area during this period. Nothing in the assemblages gave any clear indication as to the use of the platform.

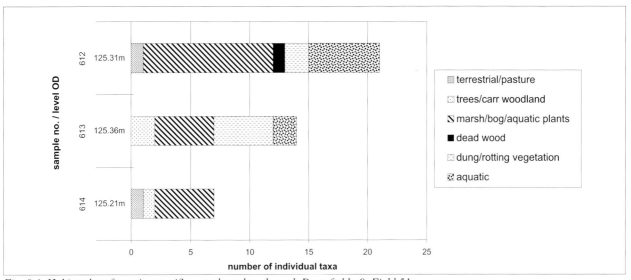

Fig. 8.6 Habitat data from site specific samples taken through Derryfadda 9. Field 51.

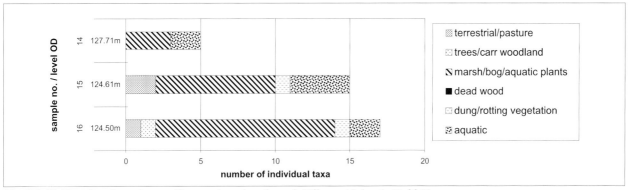

Fig. 8.7 Habitat data from site specific samples taken through Killoran 18 (east), Field 39.

Site-specific samples from the eastern end of Killoran 18—Bronze Age

These were taken from Killoran 18, the stone causeway, at Field 39 towards the eastern end of the trackway. A wood sample at this point produced a dendrochronological date of 1440±9 BC. A pollen diagram was also produced for this site (see Chapter 6). Killoran 18, although built not long after Derryfadda 23, was built in fen peat on top of a ridge which runs east–west across the whole basin. Peat morphology studies would appear to indicate that the building of the trackway itself profoundly changed the environment of the bog, precipitating the inception of raised bog over the whole length of the trackway (see Casparie, Chapter 3). The habitat data is presented in Fig. 8.7. The species list is in Table 5 of the additional files.

Below the trackway (sample 16, 124.5m OD), the majority of species show a typical fen environment. *Dyschirius globosus* is eurytropic on all types of wet ground. *Othius* sp. is a predator on small Carabidae and ants, and is found in mosses and leaves. A build-up of rotting plant remains would account for *Philonthus* sp. and *Bryaxis* sp., while *Brachygluta* sp. occurs in mosses and in decaying wood. One example of *Micrelus ericae* indi-

cates the presence of raised bog plant species nearby. These species are reasonably well represented in the pollen zone T18ii, below the trackway construction level.

At trackway level (sample 15, 124.61m OD), a portion of a roundwood with many insect channels visible was taken as well as the surrounding peat. Although a lot of frass (white coating from larval channels) was recovered, no full individuals were found. Indeed, the channels may well have been the result of ant damage that would have provided prey for *Othius* sp. and *Quedius* sp. found in the previous sample. No clear woodland/dead wood indicators were recovered.

The fauna from the peat layer surrounding the wood yielded a typical range of raised bog species including *Plateumaris discolor,* which feeds at the roots of cotton grass and *Sphagnum*.

The layer from immediately above the trackway yielded poor numbers but all of the species indicate a raised bog environment (sample 14). *Plateumaris sericea* feeds on great reedmace and bur-reed, which are common near stands of water. Samples of almost pure cotton grass, as this one was have, unfortunately, proven to be very unproductive. The pollen diagram shows an increase of sedge pollen (undifferentiated) throughout zone T18iii, which indicates increased wetness throughout this period.

Samples from the opposite drain face were looked at as part of the preliminary survey by the IAWU in 1995 (IAWU 1996a). The two samples, one from the trackway and one from immediately above the trackway, produced remarkably similar results both in terms of assemblage size and environmental indicators. Indeed, a species tentatively identified in those samples as *Rhynchites* sp. (without the use of a comparative collection) has since been correctly identified as *Micrelus ericae*, the heather beetle. Its host plants, as noted in sample 16, are well represented in the pollen diagram. So, the lack of wood indicators holds true for these samples also.

The stark contrast between the trackway fauna of Derryfadda 23 and Killoran 18 would seem to suggest that many of the species found at Derryfadda 23 had already invaded the wood before it was brought out on to the bog. This idea will be explored more fully in the discussion below.

Western bog margin

Column samples from the western landfall of Killoran 18 and associated sites—Bronze Age to the early historic period.

Killoran 18 proved quite unproductive in terms of insect remains towards its eastern end. However, the nature of the peat in the middle of the bog, both at trackway level and above, was not conducive to good samples. The peat on the western margins proved to be more productive, particularly the fen peat. A column sample was taken in

Field 17 from 124.14m OD (*c.* 0.2m below trackway level) to 126.087m OD (1.46m above the structure into raised bog). A sequence of samples for pollen and testate amoebae analysis was taken at the same location. A spot sample from a yew stump in Field 14 was placed stratigraphically in the column, as it added extra information to the site profile. Dendrochronology dates for Killoran 18 include 1542±9 BC (IAWU 1995). The species list is set out in Table 6 of the additional files and the habitat data is in Fig. 8.8.

The basal layer of light grey/brown peat had only two species present (sample 6510, 124.14–124.27m OD), *Limnobaris pilistrata*, which is found on reeds and rushes, and *Chrysomela* sp. There is only one member of this genus listed for Ireland, *C. aenea*, which is found on alder, but it is recorded only rarely in England (Maynard 1994). However, there are three other members of this genus in Britain found mainly on willow and poplar. All but the poplar species (Harde 1984) are rare or have not been recorded for a long time in Britain and are considered to be ancient broad-leaved woodland species (Hyman 1992; 1994).

The roots of the yew stump in Field 14 were used to support the substructure of Killoran 18 in this field (sample 6937, 124.2m OD). The assemblage from it is small but interesting in its composition. There were no aquatics and very few marsh/wetland indicators except *Cyphon* sp. and *Bagous* sp., an increasingly rare genus of wetland plant feeders. *Agonum fuliginosum* is found in wet carr woodland and *Rhynchaenus* sp. is a leaf miner in various deciduous and carr woodland tree species, not usually found in

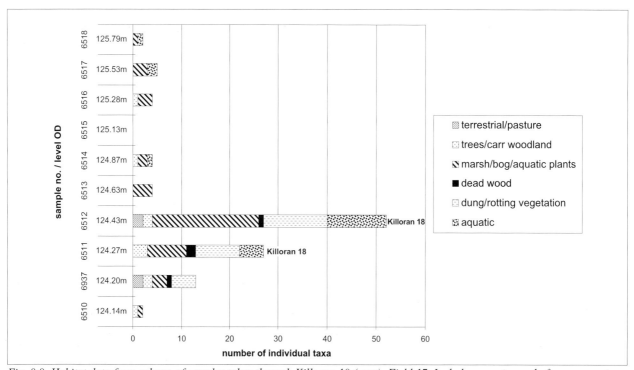

Fig. 8.8 Habitat data from column of samples taken through Killoran 18 (west), Field 17. Includes a spot sample from a yew stump from Field 14 (sample 6937).

385 and 50 BC. The *Menyanthes* pool, which formed in this area as a result of discharges from the upland Killoran bog system, is dated to *c.* 600 BC. Peat formed to a depth of 0.8m between mineral soil and the building of Killoran 234. Due to the two discharge systems operating in this northwestern margin, peat formation occurred quite rapidly. It is difficult to say, therefore, how many years 0.8m of peat growth represents but a safe assumption is that it started probably somewhere in the Middle to Late Bronze Age. The species list is set out in Table 9 of the additional files and the habitat data is in Fig. 8.11.

The layer above mineral soil was dominated by beetles indicating woodland, probably carr. A lot of naturally occurring wood was found at this level (sample 6020, 125.56–125.76m OD). The ground beetle *Pterostichus niger* is a pronouncedly woodland species, while the water beetles *Suphrodytes dorsalis* and *Agabus melanarius* are common in freshwater forest pools with leaf detritus. Other species indicate a waterside environment, such as *Dryops* sp. and *Alophus triggutatus,* which is found on common comfrey. The Scolytid *Hylesinus oleiperda* is found in thin twigs and branches of ash (Alexander 1994). It has not been recorded from Ireland before, is very local in Britain (Duffy 1953) and is considered rare on the Continent (Talhouk 1969; Lekander *et al.* 1977).

The water beetles present in the layer above (sample 6021) indicates a degree of stagnation of the water present. *Paracymus scutellaris* is tolerant of shallow acid waters, and *Hydrobius fuscipes* and *Limnebius* sp. are found in vegetated pools and ditches, usually stagnant. Also, marsh and fen plants such as reeds and sedges are indicated by the presence of the leaf beetle *Donacia* sp.

and the weevil *Erirhinus acridulus*. However, some wood is indicated by the presence of *Cerylon histeroides*, which is found under bark in rotting wood, and one example of *Tomicus minor/piniperda* (found at Derryfadda 6), a pine indicator (Alexander 1994).

In the level above this, a change is noted in the form of an increase in freshwater run-off (sample 6022, 125.96–126.16m OD). In particular, the water beetle *Ochthebius minimus* is found in freshwater or indeed running water and *Agabus bipustulatus* is found in freshwater pools. A build-up of rotting vegetation and presence of dung is indicated by *Onthophilus striatus, Megasternum obscurum* and *Aphodius equestris* (a forest species, Landin 1961; Jessop 1986). Two interesting species present were *Melanotus* cf. *erythropus* and *Denticollis linearis*. These beetles are found in rotting wood feeding on other insects and their waste, particularly Scolytidae (Alexander 1994).

A well-developed woody fen peat is noted in the next level (sample 6023, 126.16–126.36m OD). Here there is a slight increase in freshwater or running water indicators such as *Ochthebius minimus*, however, they are certainly balanced by stagnant water indicators and species found in *Sphagnum* pools and detritus pools with increased acidity. The fen environment is clearly indicated by these and the weevil *Limnobaris pilistrata*, which is found on rushes and reeds. However, carr woodland is more clearly illustrated by the presence of woodland taxa such as *Bembidion harpaloides, Agonum assimile, A. fuliginosum* and *Pterostichus niger*. The anobidae *Grynobius planus* is found, as are the secondary wood feeders *Denticollis linearis* and *Melanotus* cf. *erythropus*. Again, a number of

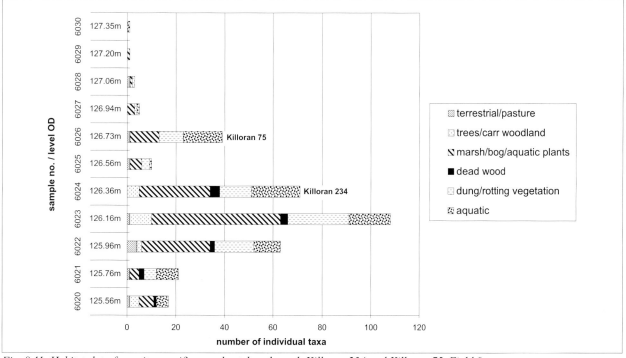

Fig. 8.11 Habitat data from site-specific samples taken through Killoran 234 and Killoran 75, Field 9.

dung beetles were found including *Aphodius lapponum*, found in sheep and deer dung in grassland and open ground, and *Geotrupes vernalis*, found in sheep and fox dung in grassland and moorland (RDB status: notable B, Hyman 1992). Other open ground species include *Phyllobius* sp., found on trees, grasses and nettles, and *Dascillus cervinus*, found on flowers and bushes. The presence of nearby open ground, possibly pasture, is a constant picture throughout these levels and mirror findings from the column at Killoran 240. It is likely that the presence of these species at this level relates to the beginnings of pollen zone DV V (1000–600 BC). This shows a marked drop in all tree species on the western margin (and throughout the bog area) and an increase in plantain (see Chapter 6).

The most numerous single genus is *Cyphon* sp., included as a fen/marsh indicator here as the larvae requires water to develop (Harde 1984). However, the adults are found in many situations, including under leaf litter, reed litter and in detritus pools, and would be common in a bog marginal forest environment.

The trackway Killoran 234 occurs at the next level but is in a very poor state and indeed the range of species present differs very little with the two levels below it (sample 6024, 126.36–126.56m OD). There are a high number of freshwater/running water indicators, in particular, a number of species which indicate freshwater or forest pools. Continuity from previous levels is evident from both wood feeders and dung/rotting vegetation species. Another dung species, *Aphodius foetens*, is present, found usually in deer and cow dung (Jessop 1986). While the leaf beetle *Prasocuris phelandrii* is found on brooklime, a waterside species (also found at Back Lane, Dublin, Reilly 1997).

The layer of *Scheuchzeria*-dominated pool peat shows a dramatic decrease in species and numbers typical of the poor faunal nature of this kind of peat (sample 6025). Interestingly, no water beetles are present except *Paracymus scutellaris*, which is highly tolerant of acid water. Other plant feeding species such as *Donacia* sp. and *Plateumaris discolor/sericea* are found on various waterside and raised bog species including *Sphagnum* and cotton grass.

The hurdle of trackway Killoran 75 is laid in this deep layer of *Scheuchzeria* peat at a height of 126.73–126.94m OD (sample 6026). Indeed, it would appear to have been laid at the deepest point of the pool and it seems evident that it could not have been used for a long time. It does not contribute significantly to the fauna itself, i.e. no wood feeders or even secondary feeders are recovered except *Atomaria* sp., a fungal feeder on wood. The main result of its presence is to increase the number of water beetles and detritus species compared to the pure layers of *Menyanthes* peat above and below. This may be because many of these beetles require some terrestrial element, whether it be at larval stage or otherwise, to survive, in particular, *Coelostoma orbiculare* and *Limnebius* sp. (Friday 1988). Detritus collecting on the surface of the track may have provided a habitat for taxa such as *Megasternum obscurum* and *Hydrobius fuscipes*. Many more aquatic plant feeders are present such as *Erirhinus* sp., *Limnobaris pilistrata* on reeds and rushes, and *Micrelus ericae*, on heather.

The hurdle was then enveloped by the *Scheuchzeria* and again a total drop in faunal numbers and diversity results (sample 6027). However, two species recorded here were found nowhere else in Derryville. *Sphaeridium scarabaeoides* is found in cow dung and *Chaetarthria seminulum* is found under moss and *Sphagnum* in bogs. *S. scarabaeoides* is a reasonably widespread species today but *C. seminulum* is considered quite rare. It is not clear why they should be found here and nowhere else.

From 127.06m OD to the top of the sequence, the peat is dominated by poorly humified cotton grass (samples 6028, 6029 and 6030). This is the start of the true raised bog in this area. However, one interesting finding is *Bembidion lampros* (6028), a ground beetle usually found in cultivated or waste ground. It is obviously a casualty from nearby dryland and from its level would seem to occur towards the end of pollen zone DV VII (200 BC–80 AD). This pollen zone records a major inwash of silt from the dryland corresponding to significant woodland removal in the specific catchment areas of the western bog margin (see Chapter 6).

Site-specific samples from southwestern Cooleeny cluster—Early Iron Age

Three samples were taken from the surface of causeway Cooleeny 31 in Field 19. Samples from within the causeway and beneath it were not possible due to the nature of the excavation and the height of the water table at this point. The primary aim of taking samples was to see if the beetles could shed any light on the use of the causeway, i.e. if dung beetles were present it might indicate animals and cart transport. Cooleeny 31 has been dated to 790–380 BC and 778–423 BC, and its massive construction and weight would appear to have contributed to the Cooleeny bog burst dated to *c.* 600 BC (bog burst D). The list of species is presented in Table 10 of the additional files and the habitat data in Fig. 8.12.

In findings not unlike Corlea 9, County Longford (Reilly 1996), the assemblage better reflects the surrounding environment in which the causeway is lying than the actual causeway itself. This is due mainly to compression of peat from above onto and between the timbers of the causeway. Although there is definite evidence for a build-up of organic matter on the causeway surface, only one true dung beetle is found—*Aphodius prodromus*, found mainly in deer and horse dung (rarely in cow dung). This would not be enough to offer any definite conclusions as to causeway usage. Two species most likely directly related to the causeway are *Silpha atrata*, which is predatory on snails under mosses and bark, and the bark beetle *Xyloterus signatus* (not recorded from Ireland before). This beetle is classified notable B status in Britain and is

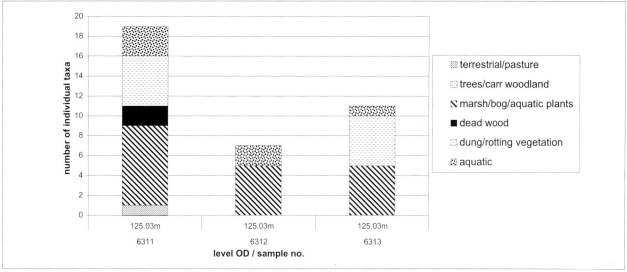

Fig. 8.12 Habitat data from samples taken across the surface of Cooleeny 31, Field 19.

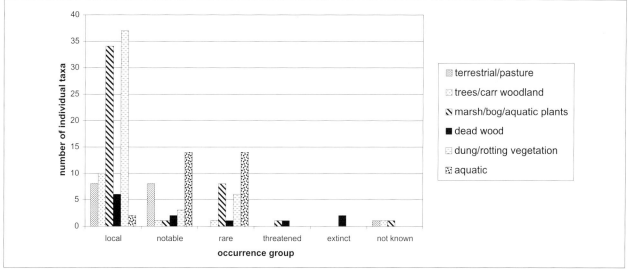

Fig. 8.13 Total number of individual taxa from selected occurrence groups from all samples, Derryville Bog.

found in ancient broad-leaved woodland in dead wood of oak and birch (it has also been found in ash, alder and beech, Hyman 1992). Like a number of other bark beetles from the western bog margin, this species has adapted to woodland management techniques and is found in newly felled trees in broad-leaved woodland plantations.

Landscape changes

Introduction

The information from the insect remains can be used to reconstruct macro-environmental changes and to identify relict landscapes. Three broad areas will be looked at with variations between the eastern and western margins highlighted, if any.

Taking all the species and genera together, and graphing their current status in Britain and Ireland (Fig. 8.13), a clear picture of these relict landscapes can be identified.

The vast majority of species from the 'notable B' to 'extinct' groups are related to woodland and wetland, particularly fen. The 'local' group adds carr woodland and dung to this picture. This is mainly due to the fact that species of dung beetle have become localised due to fragmentation of habitats and changes in agricultural practices; the same is true of carr woodland. A large number of species relating to wetland also appear here due to increased pressure from commercial peat production and agriculture on the few remaining natural wetland habitats in both Ireland and Britain.

Woodland

Some notable species found are now either rare or extinct in Ireland (Table 8.1). The finding of *Prostomis mandibularis* is particularly interesting as it gives us a very clear picture of the surrounding primary or sub-primary woodland during the Bronze Age, woodland that has all but disappeared from Ireland. It would seem that

P. mandibularis can also shed light on the construction of Derryfadda 23. The larger timbers had mortice holes and it was thought that they may have been reused. The location of Derryfadda 23, built on raised bog away from marginal forest, means that this species would not have invaded the wood after it was laid down in the bog as it is found only in forests, usually in the last stages of decay in damp wood. As Buckland noted in his study of Thorne Moors, "having evolved in a world carpeted with continuous forest and immediate proximity of suitable hosts for oviposition, many woodland insects tend to have very low dispersal potential" (1979, 144). This implies that the wood was already infested before being brought out to the construction site. It seems that it was not felled and brought to the bog immediately, but rather lay either on the forest floor or in proximity to it for long enough to be invaded by this species. The same can be said for *Teredus cylindricus* and probably for *Phloeophagus lignarius*, given the similar context in which they were found.

Dirhagus pygmaeus, however, although it does not appear on the Irish list, is found in carr woodland tree species such as alder and birch. The level from which it came at site Derryfadda 13a represented a maturing of the fen in this area prior to 1250 BC. Therefore, this species was recovered in a natural setting rather than from wood used in the construction of a site.

Similarly, *Tomicus minor/piniperda*, was found in the relatively dry layer above the *fulacht fiadh* Derryfadda 216, but before trackway Derryfadda 215 was constructed. Although the dry and well humified nature of this layer was not conducive to good preservation, the species found pointed to a covering of mixed woodland.

Apart from *P. lignarius* and *T. piniperda*, the species above are all considered primary woodland species. They are all rare, extinct or threatened in Britain (status in Ireland unknown) and have not adapted well to forest management practices or fragmentation of forest cover. These species were all recovered from the eastern margin from Middle Bronze Age contexts, and although similar periods were examined from the western margin, no species of equivalent rarity were found, except possibly *Rhyncolus ater*. Pollen diagrams show that clearance started in small amounts at first from 2300 BC onwards, with more widespread clearance from 1640 BC until about 1310 BC. The building of Derryfadda 23 falls into this time period. It would seem that enough damage had already taken place in the habitat of these species as they are not recovered from any site or layer later than this period. The earliest human activity may, therefore, have been on the eastern margin, however, the more widespread clearance and settlement on the bog margins would seem to have been concentrated on the western margin. A huge number of dryland sites of Bronze Age date have been excavated in this area compared to the relative paucity of dryland sites from the east.

This western settlement bias can also be seen from the number of beetle species found on the western margins that would not be considered particularly sensitive to woodland management. Although almost all do not appear on the Irish list and are of notable status in Britain, in general species such as *Xyloterus signatus*, *Hylesinus olieperda* and *Xyleborus dispar* are found in thin dead branches of trees such as ash, alder, hazel etc., and are well suited to coppiced woodland.

Species/Genus	Occurrence Status U.K./Irl.	General Habitat	Context in which species recovered
Dirhagus pygmaeus	rare/NL	dead wood	Derryfadda 13 natural
Prostomis mandibularis	extinct/NL	dead rotting wood	Derryfadda 23 structural
Teredus cylindricus	threatened/NL	dead wood	Derryfadda 23 structural
Rhyncolus ater	rare/NL	dead wood	Killoran 18 (west) structural, Killoran 240 oak stump, natural
Phloeophagus lignarius	local/NL	damp rotting wood	Derryfadda 23 structural
Hylesinus oleiperda	local/NL	dead wood	Killoran 234/75 natural
Xyloterus signatus	notable B/NL	dead wood	Cooleeny 31 structural?
Xyleborus dispar	notable B/NL	dead wood	Killoran 314 structural
Tomicus minor/ piniperda	rare/common/ NL/common	dead wood	Killoran 234/75 natural, Derryfadda 215/216 natural
Rhynchaenus sp.	varied status/ varied status	trees/carr woodland (leaf miner)	Killoran 243 structural, Killoran 18 (west) natural/structural
Tachinus rufipennis	rare/NL	trees/carr woodland	Killoran 240 (trough) natural

Table 8.1 Woodland species, occurrence status and context. (NL=Not Listed)

These species are found from Bronze Age through to Iron Age levels and, where found in structural wood, can provide anecdotal evidence for determining when a site was built. *X. dispar* was recovered in large numbers from a piece of hazel rod in the hurdle of Killoran 314. The area was very wet at the time it was laid down (370–5 BC) and the hurdle became waterlogged reasonably quickly after being laid, as five layers in total were used to compensate for the waterlogging.

This species, along with many of the other Scolytidae found at Derryville, emerge as adults in spring. They disperse and burrow into new wood, preferably newly fallen or cut, and lay eggs. The imagines are developed by early autumn and live inside the wood throughout their immature stage feeding on ambrosia fungus that develops inside their boreholes. This generally requires high humidity to develop. The new adults emerge the following spring. This life cycle takes one year to complete and would indicate that the wood was felled for this hurdle in spring, with the hurdle subsequently being submerged in water, killing the almost fully developed adults sometime in late winter or early spring.

All the Scolytidae species recovered, regardless of their ability to cope with woodland management, have suffered from severe fragmentation of their habitats. It may be that all of the species listed in Table 8.1 are still found in Ireland today and due to a lack of detailed studies of pockets of ancient woodland have been overlooked. It is more likely, however, that sustained clearance and manipulation of woodland from Neolithic times onward, and particularly during the seventeenth to neneteenth centuries, has meant the destruction of most of its ancient forest insect fauna also.

Wetlands

A number of aquatic genera found at Derryville are considered rare or vulnerable for much the same reason. The general use of midland raised bogs for commercial peat cutting has resulted in the destruction of many species natural habitats. Hammond noted that "undisturbed fens are rare and can only be found in a few counties in Ireland. Owing to their small size their representation on the map is not possible, even though there continued existence … is under threat from agricultural and urban pressures" (1981, 26). The study by Foster *et al.* (1992) of 289 water beetle assemblages collected since 1983 produced data on the frequency of occurrence of 165 different species of water beetle. It showed that some species of *Limnebius* and *Enochrus* are restricted to only one site each, and in a review of ten different water habitats, most of these species are restricted to base-rich fens, acid-rich fens occasionally receiving base-enriched water and acid bogs undiluted by base-rich water. Although they may be common in these relict areas, genera such as *Enochrus*, *Graphodytes* and *Limnebius* are extremely localised and could be threatened by further shrinkage of these habitats. It is worth noting that no modern example

of *Agabus striolatus* (Derryfadda 9) was found in any of these assemblages.

The water beetles can be useful in reconstructing the changing hydrology of the bog in different locations (Figs 8.14 and 8.15). This is particularly useful where long sequences were sampled but this was not always possible. It must be noted, however, that the information is quite subjective as some water beetle species could be placed in multiple categories.

The purpose of these illustrations is to give a general idea of the sensitivity of the water beetle population to changes in hydrology.

Fig. 8.14, from the western margin, demonstrates the changing water regime in the vicinity of Killoran 234 and Killoran 75. The continuous run-off from the Killoran upland is well illustrated by the presence of species indicating flowing water. The freshwater element is largely illustrated by beetles occurring in freshwater forest pools, particularly *Agabus bipustulatus*, *A. melanarius* and *Suphrodytes dorsalis*. These fresher pools of water would have been short-lived as leaf fall increased stagnation and eventually lead to the formation of peaty hollows.

The increasing aquatic presence led eventually to the formation of the *Menyanthes* pool, which is represented by a complete drop in species presence. This is mainly due to the nutrient-poor nature of pure pool peats of this type. The presence of Killoran 75 brings the population up again temporarily, but this is short-lived and as the whole area is subsumed by raised bog, the numbers of water beetles (and all other categories) drops, those species present being indicative of acidic conditions. This information ties in with findings from the peat morphology (see Chapter 3) and the testate amoebae (see Chapter 6).

On the eastern bog margin, Fig. 8.15 illustrates the changing bog hydrology in the vicinity of Derryfadda 13a. The earliest layers illustrate the developing fen with the flowing water species indicating the nearby discharge system. The bog burst of 1250 BC (bog burst B) is clearly shown at sample 4, with a complete drop in all water species noted. The peat above this shows a forest floor fauna, indicating the drop in the water table and relatively dry nature of this area after the bog burst. A gradual recovery is seen as an increase in water beetle species by the construction of Derryfadda 13, however, a drop is again seen after this time. This could be caused by the influence of the Cooleeny bog burst of *c.* 600 BC (bog burst D); however, the exact date of the construction of Derryfadda 13a would be needed before this could be confirmed. The other reason for a drop in species is the development of raised bog and certainly the layers above the site show the beginnings of raised bog growth.

There is a general trend in all of the site sequences showing an increase in aquatic and raised bog beetle species in the later layers. This corresponds with well-established views on the nature of environmental change at macro-level throughout the Bronze Age and early Iron

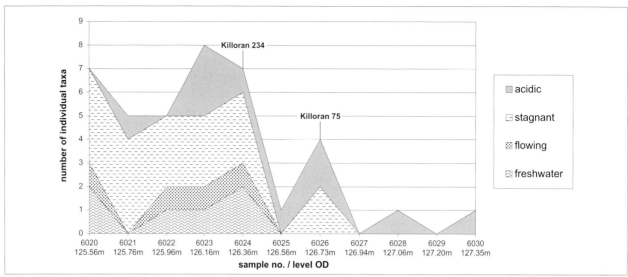

Fig. 8.14 Hydrology of Derryville Bog using water beetle species as indicators, as seen at column through Killoran 234 and Killoran 75, Field 9.

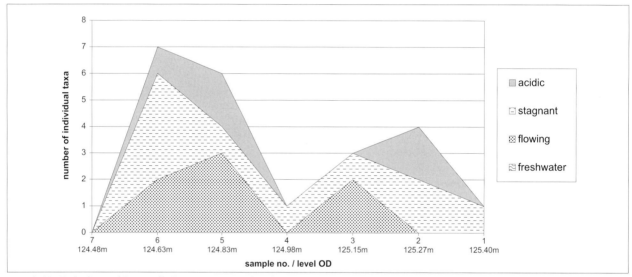

Fig. 8.15 Hydrology of Derryville Bog using water beetle species as indicators, as seen at column through Derryfadda 14, Field 46.

Age and the detailed peat morphological and testate amoebae studies of the bog. It is believed that climatic deterioration occurred during the Late Bronze Age and again at the end of the Iron Age, although it must be noted that widespread forest clearance on the nearby upland would have increased surface run-off in the immediate bog environs. This would have amplified increasingly wet conditions and encouraged raised bog development.

A small number of species found are now restricted to a more northerly range in Britain, for example, *Erirhinus acridulus*. This species was found in Windsor Forest Park by Donisthorpe (1939) but it is generally restricted to the northern Scottish islands such as the Outer Hebrides and Shetland. Lindroth (1973) noted it from between N and 69N in Norway but not above the treeline, and also from western Iceland. Its restricted occurrence may be due to the general shrinking of natural wetlands in Ireland and else-where, but may also be due to general climatic warming.

In general, however, unlike findings from the Somerset Levels (see Girling's various papers in the *Somerset Levels Papers* 1977–85), no beetles found at Derryville can be described as truly thermophilous (i.e. presence determined by temperature), either in terms of warmer temperatures or colder temperatures. Most of the extinct, rare and localised species found would seem to be restricted by habitat destruction or alteration.

Dryland: pasture and cultivation

The illustration of the changing nature of the surrounding upland is well demonstrated by the pollen data. However, the beetles have also shed some light on this. The wood-land indicators have been discussed but many species that feed on grassland, meadow and disturbed ground plant species are found in the samples. Also, a large number of different dung beetle species, which may originate from animals grazing in bog marginal forest or as casualties

Species/Genus	Occurrence Status	General Habitat	Context in which species recovered
Bembidion lampros	common	cultivated/waste ground	above Killoran 75, raised bog sample
Calathus fuscipes	common	dry grassland/cultivated	above mineral soil Killoran 243
Pterostichus melanarius	local	grassland/disturbed ground	peat below Derryfadda 215
Chaetocnema concinna	common	various weed species	structural level Killoran 18 (west), level below Killoran 69, Cooleeny 32
Gastrophysa viridula	not known	various weed species	fen/alder carr level Derryfadda 13a column
Rhinoncus perpendicularis	widespread	various weed species	same as above
Alophus triguttatus	notable B	various grassland herbs	structural level Killoran 18 (west), mature fen levels Killoran 234/77 mature fen levels Killoran 234/77
Leiosoma oblongulum	notable B	buttercup family	peat below Derryfadda 215 & above Derryfadda 217
Agrypnus murinus	widespread	roots of grassland	fen/alder carr level Derryfadda 13a level
Geotrupes vernalis	notable B	dung in moor/grassland	trough of Killoran 240, Killoran 234 level and peat in level below
Onthophilus striatus	common	cow & deer dung etc.	fen/alder carr below Killoran 234
Cercyon sp.	varied status	various dung/rotting vegetation	structural level Killoran 243 amongst others
Megasternum obscurum	widespread		Field 17/23 trackway level Killoran 18, Killoran 243 various levels, below Killoran 69 etc.
Phylopertha horticola	widespread	roots of poor/wet grassland	almost every level Derryfadda 13a, Derryfadda 215/216 & Derryfadda 9. Different levels Killoran 240/245
Aphodius equestris/ foetens/lapponum/icterius	all very local	horse/cow/deer dung etc.	Killoran 240/243, Killoran 234/75, Cooleeny 31, Derryfadda 215/216, Derryfadda 13a

Table 8.2 Upland, pasture/cultivation indicators, occurrence status and context (selected species only).

from nearby pasture, are found. There are also a very small number of beetles associated today with cultivated ground. Although no firm conclusions can be made for the presence of cultivation, taken together with data from the pollen diagrams and dryland excavation evidence, a tentative picture can be drawn for particular parts of the bog margin. The information is summarised in Table 8.2.

The presence of pasture/open grassland is indicated by particular species from all levels within the columns. On the east, this is best represented by the beetle *Phylopertha horticola*, which occurs at almost all levels at Derryfadda 13a, Derryfadda 6 , Derryfadda 215 and Derryfadda 216, and below Derryfadda 9. It is mostly found on poor pasture, particularly that which occurs at the edge of rivers and lakes prone to flooding. This type of land must have been quite common at the edge of the developing fen, particularly in areas with steep slopes running down to the basin. Other species, such as *Agrypnus murinus* and *Gastrophysa viridula*, were found in Bronze Age levels at Derryfadda 13a. *Agrypnus muri-*

nus feeds at the roots of grassland and *Gastrophysa viridula* feeds on disturbed ground weeds.

Equally, from the peat level below Derryfadda 215 (constructed in 457±9 BC) the ground beetle *Pterostichus melanarius* occurs, found in grassland and disturbed ground. A very small number of dung beetles were recovered from the eastern margin samples. Although open ground and poor pasture or water meadow was probably present here, the number of indicators is quite small adding to the picture that the eastern margin had a lower level and variety of settlement activity.

Ground beetles, cultivated/disturbed ground weed feeders and dung beetles are present from the earliest levels on the western margin. Some of the dung beetles present, e.g. *Aphodius equestris*, are forest species, however, it is very likely that the bog margins were frequented by grazing animals. Indeed, some trackway levels have dung beetles present, which may represent use by animals, but could also indicate rotting vegetation build-up on the trackway surface.

Weed and disturbed ground feeders are present at many levels, most notably *Chatocnema concinna*, *Alophus triguttatus* and various *Sitona* spp. They most likely arrived in fen and raised bog levels as casualties from upland through surface run-off. The ground beetle *Bembidion lampros* is a species of cultivated ground and, as mentioned above, its finding in later levels at Killoran 75 corresponds with a period of forest clearance.

Equally, *Calathus fuscipes* from the trough of Killoran 240, is predominantly found on cultivated ground. The period around and after the construction of the *fulacht fiadh*, the Middle Bronze Age, saw widespread forest clearance and undoubtedly some cultiva-tion, as has been indicated by findings in the pollen record. The trackway level at Killoran 18 (west) also recorded a number of open ground species adding to the picture of increased dryland activity around the Middle Bronze Age.

The lull in clearance and settlement activity post-1310 BC, as indicated by the pollen record, is not as clear in the faunal record. However, there is an increase in dung and general disturbed ground species around the construction of Killoran 234. Built between 775–415 BC, this would seem to correspond to the inter-face of pollen zones DV V and DV VI, which shows clearance and continued, if variable, activity.

9. Wetland structures: typologies and parallels

Sarah Cross May, Cara Murray, John Ó Néill and Paul Stevens

Classification

Introduction

This chapter introduces and defines the terminology used in the catalogues of excavated material and in the subsequent discussions of the archaeology of Derryville Bog. In the absence of a common and consistent set of classifications for wetland material, we have attempted to simplify and clarify those already in existence. For example, several terms, such as road, track, trackway and path, have been used to classify structures crossing part or all of a bog. In Ireland, the word togher, an anglicised form of the Irish word *tóchar*, is often used to describe both trackways and causeways, although the word strictly means a causeway. As this word has gained currency as a generic term for both causeways and trackways, it was not useful to redefine it here.

The terminology and typologies used for the initial survey of Derryville Bog by the IAWU (Moloney *et al.* 1993a; 1993b; 1995; IAWU 1996a) were based on the then Office of Public Works classifications and described the wide range of human activities identifiable in wetlands. As constructional details of many structures were not apparent without excavation, it was presumed at the outset that crossmatches could be made with the published site typologies (e.g. Raftery 1996). Unexcavated sites, preserved outside the area of development, are listed in their original classifications in the additional files.

Examination of the excavation results and palaeoenvironmental investigations of structures in the study area of the bog resulted in the incorporation of additional aspects to regroup and reclassify the wetland material. Survey classifications were changed or modified based on the increased level of definition in morphology and function. Where appropriate, survey classifications, such as stake row, were upheld. Elsewhere, survey classifications, such as platform, were used with tightened definitions. Additional classifications included *fulachta fiadh*, burnt mounds, archaeological wood, charcoal spread, bank and hut site. One important group of survey classifications—togher, puddle togher and road—were re-evaluated and grouped into causeways and trackways.

In total, thirty-four trackways, six *fulachta fiadh*, six platforms, three stake rows, two causeways, one burnt mound, a possible hut site, a charcoal spread and a bank feature were excavated in Derryville Bog. A further eleven sites produced evidence for humanly altered wood but no structure, and some twenty-eight sites were examined for environmental purposes, producing no evidence of archaeological activity. The excavations and field walking on the dryland areas west of Derryville Bog produced thirty-three sites. There were two settlement sites, two cremation cemeteries, four individual cremations, eighteen *fulachta fiadh*, two burnt mounds and six possible *fulachta fiadh*, plus a number of other sites producing evidence of archaeological activity. The classifications and definitions used for the dryland material are already in existence and are widely used, so further clarification was unnecessary.

Trackway

The classification trackway is a morphological interpretation based on detailed investigation by excavation of individual structures. It represents a structure that did not bridge a bog system or cross from one dryland margin to another and that was over twice as long as broad. The necessity for short, wide trackways in a marginal area may suggest a number of functions. One possibility, drove roads, is not supported by the condition of the surfaces or the microfaunal evidence. Other domestic and non-domestic functions could have been fulfilled by the trackways and these are discussed in Chapter 15, with reference to social and economic implications.

Causeway

The classification causeway is a morphological interpretation based on detailed investigation by full excavation of individual structures. It represents a structure that bridged a bog system, crossing from one dryland margin to another. Two structures in Derryville Bog were classified as causeways: Killoran 18 and Cooleeny 31.

Platform

The term platform was used when a structure did not appear to provide access from a dry area to a wet area. It was classified as a site less than twice as long as broad, distinguishing these sites from short trackways in areas of very wet peat that were intended to facilitate passage across that area.

Fulacht fiadh

The classification *fulacht fiadh* (plural *fulachta fiadh*) is commonly used in Ireland to describe a burnt mound of fire-cracked stone and charcoal with an associated pit or trough. Such sites are not unique to wetlands and have been the topic of much analysis both in Ireland and abroad (Buckley 1990b; Barfield and Hodder 1987; Ó Drisceoil 1988). Outside Ireland, the term burnt mound is used, but this term was used in the Lisheen project for a separate site type (see below). *Fulachta fiadh* are usually sub-circular in shape, forming a horseshoe shape around the trough, and often have associated platforms and hearths, although this is not a defining factor. All are located in areas where the trough can naturally fill by water percolation or from a spring or river. They have been interpreted as cooking sites, saunas and industrial sites; however, their primary purpose was to boil water. They are largely confined to the Bronze Age (Brindley *et al.* 1989–90).

Burnt mound

Burnt mounds are differentiated from *fulachta fiadh* by the absence of a trough or pit. This classification was applied to a minority subset of sites within the larger group of *fulachta fiadh*. However, burnt mounds were grouped with *fulachta fiadh* for functional analysis and dating, and represent a variant of the same class of site.

Stake row

A stake row was defined in excavation as an alignment of upright wood of small diameter. The alignment could be a single, double or multiple row. In none of the cases in Derryville Bog was there firm evidence to prove that woven panels were inserted between the uprights. The function of such sites appeared to be primarily one of demarcation.

Archaeological wood

The classification archaeological wood was applied to worked wood showing no obvious structure or pattern but representing human activity in the bog. While some of this evidence could represent the remains of destroyed or disturbed structures, the variation in the range of low-level indicators of a human presence suggest that the activities taking place in the bog did not always require a structure to define them.

Other classifications

Three further site classifications were used for evidence of human activity in Derryville Bog. The classification hut (Killoran 66) referred to an enclosure of structural wooden uprights capable of supporting a roof and constituting a small dwelling structure. A number of other sites, such as Killoran 18, produced evidence of human activity pre- or post-dating the structure. In other cases, human activity took the form of areas of charcoal classified as charcoal spreads, as at Derryfadda 6, Killoran 18 and Killoran 223. A field bank feature, classified as a bank, was also recorded at Killoran 224. Dating evidence identified that these activities took place in the Neolithic and continued up to the present day.

Wider parallels

Introduction

This section places the Derryville Bog material in the wider context of wetland archaeology. It was not possible to furnish firm stylistic and typological distinctions that could provide a basis for morphological dating criteria for these structures. However, structural comparisons with other material were possible for most of the wetland material.

Constructional parallels

The typological classification of structures in Derryville Bog can be compared and contrasted with other classifications of wetland material from Ireland and Europe, notably the classification of track construction types carried out for most of the Irish material by Raftery (1996, 211–30) and for Continental material by Hayen (1957; 1987) and Casparie (1987). Analysis of tracks built using stone has largely been omitted from such assessments, so a cursory analysis is outlined here. Other wetland site types can also be compared with the broad range of archaeological material recovered from bogs, although a less comprehensive overview of other wetland structures exists. Although conventional typological comparisons deal with changes through time, this has proved impossible for the Derryville material, as many structural types are used intermittently throughout the prehistoric and historic periods.

Trackways/causeways

Trackways and causeways are the most frequently occurring structural types in Irish bogs. Thirty-four trackways and two causeways, dating from the Early Bronze Age to the Iron Age, were excavated in Derryville Bog. A further body of material that included medieval trackways was preserved to the north of the study area. Trackways and causeways were defined on a functional basis, as outlined above. This distinction has been made elsewhere (Raftery 1990) but has not been applied in analytical comparisons. Therefore, for analytical comparisons, trackways and causeways are taken as a single grouping. An alternative approach subdivides trackways by carrying capacity, to assess scale and function.

Most of the trackways and causeways excavated in Derryville Bog were not constructed in a single style for their full length often changed with the terrain. Where a constructional typology is discussed below, it is the most characteristic of the individual structure. Given the range of basic options—brushwood, roundwood, timber and stone—it would appear that differentiation on the basis of function and carrying capacity will become more important as a larger corpus of these structures is published. However, constructional comparisons could be made for almost all the structures, which are grouped here by the construction types defined by Raftery (1996).

Corduroy roads

Five of the trackways and one of the causeways recorded in Derryville Bog were similar in construction to the corduroy roads discussed by Raftery (*ibid.*, 218–23) and the *Pfahlwege* discussed by Hayen (1957). This type of construction, using transverse wood, either unsupported or with a longitudinal substructure, is extremely common in all periods and all areas. Pegs and uprights were also a feature of these tracks.

In Derryville Bog, the corduroy roads are Derryfadda 13, Cooleeny 31, Derryfadda 203, Derryfadda 215, Killoran 226 Structure 2, Killoran 243 and Killoran 314, which date from the Middle Bronze Age to the Iron Age. The trackways were 7–246m long and 1–8m wide. Few of the trackways were pegged at regular intervals and longitudinal supports were rare, other than on Killoran 314, which also incorporated a hurdle panel at one end. The causeway Cooleeny 31 (Plate 9.1), dating to around 600 BC, could be paralleled in width to Togher B at Leigh/Longfordpass South, County Tipperary (Rynne 1965), which is a mere 2km to the south of Cooleeny 31 and bridges the same chain of bogs.

Plate 9.1 Cooleeny 31. Bronze Age-Iron Age corduroy road.

This type of construction is known from as early as 2850–2470 BC in Ireland at Cloonbony, County Longford (Raftery 1996, 220) and at a similar date in the Netherlands at Nieuw Dordrecht (XXI (Bou) (Casparie 1982). From this date onwards, it is part of a building tradition that includes wide walking surfaces pegged through a longitudinal substructure or overlying runners. Planks, as well as whole roundwoods, are used as part of this tradition, although the shorter trackways of similar construction at Derryville Bog are a sharp reminder that the structures with wide walking surfaces also vary considerably in their intended function.

Hurdles

Four trackways in Derryville Bog dating to the Iron Age made use of hurdles as one of the main components. Tracks using hurdles correspond to the hurdle tracks construction typology described by Raftery (1996, 213–5). The excavated sites incorporating hurdles as a main feature were Killoran 75, Killoran 312, Killoran 314 and Killoran 315.

At Killoran 75 (Plate 9.2), the remarkable 7m-long hurdle panel appears to have been constructed specifically for the track. It provided a stable working surface at the end of the track, where it met a pool. In all other cases, the hurdles were probably used expediently to deal with the particularly wet nature of the fen at the time. At Killoran 312, the hurdle was placed on a very firm base, which weakened and damaged it. At Killoran 314, the hurdles were part of the substructure of the track. At Derryfadda 315, the hurdle was only one of a number of different attempts to stabilise a pool in the marginal woodland.

Hurdles have been compared on the basis of their weave patterns, their woodland management patterns and their construction patterns (Raftery 1990; 1996, 178, 213–5). On all of these points of comparison, the Derryville hurdles fit well with the Irish pattern, which is best shown in the material from Annaghbeg, Corlea and Derryoghil, County Longford (*ibid.*, 201–2). The hurdles from Derryville Bog are Iron Age, while those from the Mountdillon Bogs in County Longford are Bronze Age, which underlines the stability of construction techniques across time. The size range of the Derryville hurdles is similar to that of the larger hurdles in Mountdillon. The simple weave pattern is also similar. Occasional departures from the over-under pattern allowed length extension through the addition of new rods. The double sails from Killoran 312 show stylistic variation within a group (Coles 1987, 155). While the wood from the Derryville hurdles all seems to have been managed to some extent (see Chapter 7), the largest panel at Annaghbeg 2 was cut from naturally growing hazel (Raftery 1996, 441–2). Perhaps most importantly, hurdles from both sets form part of a trackway rather than being the basis of the track, suggesting that it was the structural qualities of hurdles that were important rather than the ease of construction using prefabricated panels.

Plate 9.2 Killoran 75. Iron Age hurdle track.

While the making of woven panels was widespread in European prehistory, their use in wetland trackways is more idiosyncratic. Irish hurdles seem to have been used in a different manner to those in Britain and the Netherlands (hurdles have not been recorded in the rest of Europe (*ibid.*, 214). Thus, all of the Derryville hurdles were found at the ends of tracks, and some of the Mountdillon hurdles were used in particularly wet spots in the centre of a route (*ibid.*, 126, 149)). This is the main distinction between Irish hurdles and those found in Britain and the Netherlands, where hurdles were used in continuous trackways as the main walking surface (Coles and Orme 1977b; 1977c; Coles *et al.* 1982). The example from Bourtanger Moor, Drenthe, the Netherlands, forms a broad and sturdy surface that may be associated with the exploitation of bog iron (Casparie 1986, 192–210) and is a considerably more intensive use of woven panels than is seen in the Irish evidence. The Derryville Bog hurdles played a role as platforms within trackways rather than routes within a landscape.

Roundwood paths

Six of the trackways excavated in Derryville Bog correspond to the roundwood path construction type described by Raftery (1996) and use longitudinal roundwood as the main feature of their walking surface. Sites falling into this type were Derryfadda 206 (Plate 9.3), Killoran 57,

Killoran 69, Killoran 230, Killoran 301 and Killoran 306. These structures dated from the Late Bronze Age to the Iron Age and measured between 1m and 2m in average walking surface width. Some structures contained a mixture of brushwood and roundwood, whilst others incorporated planks and pegs.

This construction type differs from the brushwood track in the size of longitudinals used. At Derryoghil, County Longford (*ibid.*) and Drenthe, the Netherlands (Casparie 1984), a distinct group of tracks with two or three roundwoods forming part of the walking surface was recorded. A similar type of structure was found at an Early Bronze Age site in Bramcote Green, England (Thomas and Rackham 1996). This group strongly resembles structures with longitudinal planks as part of the walking surface and fits into the tradition of narrow tracks, which is discussed below.

Brushwood tracks

Three of the trackways excavated in Derryville Bog could be classified as brushwood tracks according to the typology described by Raftery (1996). They are Cooleeny 64 (Plate 9.4), Derryfadda 201 and Killoran 248, which dated from the Middle Bronze Age to the Iron Age. These track-

Plate 9.3 Derryfadda 206. Bronze Age–Iron Age roundwood path.

Plate 9.4 Cooleeny 64. Middle to Late Bronze Age brushwood track.

ways used longitudinal brushwood as the main feature of the walking surface, which was 1.5–2m wide. Other features incorporated were pegs (Killoran 248), roundwood (Cooleeny 64 and Killoran 306) and timber and roundwood (Killoran 248). Similar trackways include Killoran 241, which began as a longitudinal brushwood deposit and ended as an irregular dump of wood, and Killoran 226 Structure 1, which contained longitudinal brushwood. The western end of Killoran 18 also utilised longitudinal brushwood with occasional stones.

In Ireland, the earliest structures of this nature date to the Neolithic, and similar structures are found on the Continent at the same time (*ibid.*, 211). However, structures with narrow brushwood walking surfaces have been dated as late as the medieval period (Breen 1988, 333). This type of construction appears to be a quite simple concept and is part of the wider tradition including plank paths, roundwood paths and some of the stone tracks.

Plank paths

Two trackways were identified as the plank path type described by Raftery (1996). These Middle Bronze Age trackways, Derryfadda 23 (Plate 9.5) and Cooleeny 22, utilised single longitudinal timbers as the main feature of their walking surface. Neither trackway could be shown

to have extended into the margins of the bog. This is an unusual feature, as the majority of trackways in Derryville Bog were laid across an area of the margins.

Both tracks were constructed with single-plank walking surfaces, occasionally held in place by pegs and supported by transverses spaced along their lengths. This type is referred to by Hayen (1957) as *Bohlensteg* and is directly paralleled at the roughly contemporary Curraghmore 16, County Offaly (Moloney *et al.* 1995). A similar track was recorded at Monavullagh, County Kildare (FitzGerald 1898) and a late example, dating to around AD 665, was recorded in the Lemanaghan Bogs, County Offaly (Bermingham 1997). The unusual Derrindiff 1 trackway, County Longford (Moloney *et al.,* 1993a, 33–41), while having many of the properties of Hayen's *Bohlensteg*, is supported by angled pegs in a manner more reminiscent of tracks such as the Sweet Track, which are referred to by Meddens (1996, 332) as 'cradle-supported tracks.' Other variations on the single-plank walking surface are known from the Netherlands (Casparie 1984).

Single-plank tracks, cradle-supported tracks and multi-planked tracks such as Corlona, County Leitrim (Tohall *et al.* 1955) all fit into a tradition that includes those tracks that were essentially only wide enough for single file, one-way traffic. Other tracks of this nature, including some of Raftery's brushwood paths and roundwood paths, can then be grouped into a set of 'single-file tracks' for discussion.

Plate 9.5 Derryfadda 23. Middle Bronze Age plank path.

Miscellaneous trackways

Seven trackways in Derryville Bog did not conform to a coherent layout or were in very poor condition. Derryfadda 42 and Derryfadda 44 had both been destroyed by milling, while Killoran 234 may have had longitudinal roundwoods as a walking surface but had been dismantled in antiquity. Three other trackways, Derryfadda 204, Derryfadda 207 and Killoran 305, were irregularly constructed to simply fill in a particularly wet area.

Stone roads or gravel tracks

No formal classification exists for trackways or causeways constructed using inorganic materials. However, these have been referred to in excavations and surveys as stone roads, gravel tracks, toghers and stone paths. It is therefore possible to compare such sites on the basis of constructional detail.

The use of stone and other inorganic material in trackways and causeways was evident in three structures in Derryville Bog: Derryfadda 17 (Plate 9.6), Derryfadda 311 and Killoran 18, all dating to the Middle and Late Bronze Age. All were distinct in being constructed from the dryland across fen peat during a period when the bog surface was drier, following a lowering of the water table or following a drying of the surface (see Chapter 15). These structures all incorporate additional elements but are largely constructed using stone cobbles, flags or boulders.

Early Irish excavations of stone tracks include sites at Clonsast, County Offaly (Price 1945); Littleton Bog, County Tipperary (Rynne 1965); Timoney, County Tipperary (Lucas 1975; 1977); and Baltigeer, County Meath (Macalister 1932). Recent studies have focused on survey with only limited excavation and dating. Tracks built of stone were recorded during surveys of County Longford and the Blackwater midlands bog system (Moloney *et al.* 1993a; 1993b; 1995), and on Valentia Island, County Kerry (Mitchell 1989, 64–5, 75; O'Sullivan and Sheehan 1996, 21–3).

The Derryville Bog tracks varied in size but were largely small-scale, ranging from 15–500m in length and 1.1–1.62m in width, comparable to the prehistoric stone tracks on Valentia Island. In contrast, medieval stone causeways found in Ireland are longer, wider and more substantial constructions. The 3m-wide St Broghan's Road, Clonsast, dating to AD 700–860, was constructed over 2 miles (Price 1945). Large, multi-layered causeways were also built using successive layers of stone, gravel or timber, as in Bloomhill Bog, County Offaly (Breen 1986; 1987; Moloney *et al.* 1995) and Lemanaghan Bog, County Offaly, where successive resurfacing occurred over 500 years or more between AD 644 and 662 and AD 1151 and 1167 (O'Carroll 1997). This pattern has also been noted in Continental Europe (Casparie 1987). Large prehistoric examples are known from Ireland: two tracks in Baltigeer, County Meath, multi-layered constructions of timber and

soil, were given an Iron Age date by the excavator (Macalister 1932). Similar structures are also known from Kvorning and Krogsbølle, Denmark (Schou Jørgensen 1996), and Groos Heinss, Germany (Hayen 1970).

Other inorganic materials, including pebbles, cobbles, gravel and deposits of soil, were also utilised on stone trackways and causeways, as seen at Killoran 18. Deposits of sand were noted at Cooleeny 31. The quarrying and transportation of large stone would require a considerable investment in time. The use of pebbles, cobbles, gravel and soil may have provided some of the advantages of using quarried stone, while being less labour intensive.

Wood was also frequently used in conjunction with stone, often as a foundation, as at Killoran 18, where brushwood was used as a foundation for the stone surface. Brushwood foundations were common in many other stone tracks, as in causeway XXVII (Bou) in the Netherlands (Casparie 1987, 40–1). Elsewhere, other materials have been used, such as sods of turf at XXVII (Bou) (Casparie 1987) or straw at Baltigeer and Clonfert 2, County Galway (Moloney *et al.* 1995). At Killoran 18, the main structural component changed from largely stone in the east to a completely wooden construction in the west, where wetter conditions were encountered.

Plate 9.6 Derryfadda 17. Middle to Late Bronze Age stone trackway.

Stone as a building material in bogs has the advantage of durability and can provide a long-lasting, dry footing if the terrain below is firm. However, it also has disadvantages, and a heavy stone paving will not easily hold together in wet terrain. Stone was regularly used in large medieval causeways and trackways, but the assumption that stone tracks solely date to the medieval period may simply be a product of insufficient dating evidence. The use of stone in this context appears to be related more to the nature of the surface terrain and hydrology. Additional work on the peat morphology of stone-built trackways and causeways would provide an opportunity for further comparison.

Carrying capacity

The broader groupings of trackways and causeways can be subdivided on the basis of their width, based on a functional assessment of carrying capacity. As outlined above, a cross section of the different constructional types fits into a group of structures that, by their very nature, only anticipate people using the track alone or in single file. The implicit function of these sites does not anticipate the facility to allow two-way traffic. The fact that none of the trackways or causeways appears to have been deliberately widened as a response to congestion suggests that there was no significant deviation from the intended function.

In the case of roundwood paths, definite elements of the plank paths can be identified, such as Derryoghil 7, County Longford (Raftery 1996, 127–9), which had transverses and grouped longitudinals that mirrored the single-plank style construction. At Bramcote Green, England, a roundwood path appears to have been replaced by a plank path (Thomas and Rackham 1996). The cradle-supported tracks could also be derived from within a similar tradition and show that it is quite likely that no type should be seen as having primacy.

The date range of this group of construction types ranges from the 3807 BC Sweet Track, England (Hillam *et al.* 1990) to the AD 665 Lemanaghan, County Offaly, structure (Bermingham 1997), and it has the lowest possible carrying capacity of any type of functional walking surface. As such, a number of points can be made about the implications of the use of such narrow tracks.

The economical use of time and materials must be one of the main reasons for the recurrence of this type of structure. The same amount of materials used in a narrow track can cover a much greater distance than in a short, wide track.

In particularly wet areas, such as fen, where the pedestrian is often unable to see the ground surface, a narrow track laid within reed lawns is less visible than one that is a greater intrusion on its environment. During hunting or fowling in fens, this would afford people more potential as a hiding place.

It would also be of importance if the structure were an element of a ritual landscape where ceremony and the action of the individual had great weight. Narrow structures are also easier to hide from those unaware of their presence, highlighting the defensive nature of these structures, which can be easily hidden and can also act as a bottleneck against unwanted intruders.

Platforms

Six excavated sites in Derryville Bog produced structures classed as platforms. These were Derryfadda 6, Derryfadda 9 (Plate 9.7), Derryfadda 210, Derryfadda 211, Derryfadda 214, Derryfadda 218 and Killoran 237, which dated from the Early Bronze Age to the Iron Age. The platforms were constructed of brushwood and roundwoods, with some pegging to retain the structure, as on Killoran 237a. There was no evidence of stone deposits. None of the platforms produced evidence of enclosing fences or palisades and most appear to have been constructed within the marginal forest.

Typically, the Derryville platforms were 5–10m in length and width, encompassing an average area of around 32m². Platforms present on *fulachta fiadh* measured around 12m² on average. These figures would suggest that around two to three people could work comfortably at the *fulachta fiadh* and perhaps four to six on the platforms.

Platforms, therefore, appear to be of a lower order than the proto-crannógs and wetland settlements excavated at the likes of Cullyhanna, County Armagh (Hodges 1958), Clonfinlough, County Offaly (Moloney *et al.* 1993b) and Killymoon, County Tyrone (Hurl 1996). Detailed survey of the large number of artificial islands around Lough Gara has also shown that large and small sites form part of the overall lakeside settlement pattern (Fredengren 1998). Work at Knocknalappa, County Clare (O'Sullivan 1997), has also identified platforms and two *fulachta fiadh* of Bronze Age date on the lakeside margins, the smallest platform (Site 3) being around 20–30m². It is possible that some of the more substantial platforms identified by the survey of Cooleeny townland could provide evidence of habitation.

The difference in order of magnitude between settlement sites like Clonfinlough and the Derryville Bog platforms may lie in the nature of the available resources. The lakeside settlements could exploit a larger resource base than the fen platforms, allowing for a more permanent settlement of the actual area. Platforms of various sizes were identified by the survey of Cooleeny townland and these have been dated to the latter half of the first millennium BC and represent more substantial fen activity. These may produce a better parallel for the Derryville platforms.

Three of the six Derryville platforms produced artefacts (see Chapter 12), the highest correlation for any of the wetland site types. Of the range of artefacts found— coarse pottery at Derryfadda 218, a spear shaft at Killoran 237 and two wooden shafts at Derryfadda 9— only the spear shaft can be paralleled on a trackway, Derryfadda 13, which is 1,000 years later in date.

Plate 9.7 Derryfadda 9. Iron Age platform.

This similarity between artefacts found on site provides a further parallel between lakeside and fen platforms. The wooden shafts found on the Iron Age platform Derryfadda 9 and a number recovered from crannógs in Scotland (Munro 1882) suggest that their function may be connected with the wetland environment in which the sites were located.

The microfaunal evidence shows a proliferation of marsh and aquatic insects at the level of two platforms, Derryfadda 6 and Derryfadda 9. Species associated with dead wood were markedly absent, as were species associated with dung or cadavers.

Platforms had a range of domestic, social or industrial functions that cannot be directly reconstructed from the finds or the environment. They may have served as bases for hunting or wild-fowling in the fens and marginal forest or as stockpiles for harvested reeds or rushes. Equally, they may have been constructed as isolated preserves for certain social or ritual customs.

Stake rows

Three stake rows, Derryfadda 19, Derryfadda 209 and Killoran 54, were excavated in Derryville Bog; a further five were recorded outside the study area during the initial survey. The excavated stake rows varied in length from 38m to 406m and dated to the Late Bronze Age and the medieval period.

A number of stake rows have been identified in Irish wetland surveys (Moloney *et al.* 1993a; 1993b; 1995; IAWU 1996a; Raftery 1996; Lawless 1992). Descriptions of a number of others are recorded in the Ordnance

Survey memoirs, such as at Lisconnan and Kilraghts, County Antrim (Day and McWilliams 1993). In Britain, a number of post alignments are known, such as at Flag Fen (Pryor and Taylor 1992) and Runnymede Bridge (Needham 1991).

Of those recorded, a wide diversity exists in terms of length and function. Interpretation of function is also varied and largely dependent on context. The 800m-long post alignment at Flag Fen seems to have been associated with ritual deposition (Pryor and Taylor 1992). Some of the stake rows at Bofeenaun and Cloghbrackfar, County Mayo, have been interpreted as deer traps (Lawless 1992). Stake rows are generally interpreted as boundary markers or defining terrain, as is the case with Killoran 54 and Derryfadda 209, where they delimited a dangerous wet area. The primary phase of Killoran 18 was a double stake row across a large portion of the fen. The purpose of the stakes appeared to have been to provide markers for the construction of the second construction phase and may have been in use as an active pathway during the interval. At Derryfadda 17, the stake row was added after the initial construction and acted as a marker to the increasingly overgrown site.

Huts

One site, Killoran 66, was classified as a hut built on the surface of the bog. This site was identified and sampled during the initial survey and was dated to the early medieval period. The site was destroyed by peat milling before full excavation could take place. It was recorded as a 5.6m x 4m sub-rectangular setting of large oak posts.

Although the contemporary bog surface had been destroyed, reconstruction of the original deposition level of the structure suggested that posts had been inserted to a depth of around 1.2m. This would appear to allow for a roofed structure rather than an open enclosure. In the Glastonbury and Meare Lake Villages, England, house construction took place after the transition to ombrotrophic *Sphagnum* peats (Bulleid and Gray 1948). There are descriptions of huts built on the surfaces of bogs from the seventeenth century onwards (Feehan and O'Donovan 1996, 60).

Fulachta fiadh, burnt mounds and pits

General

Eighteen *fulachta fiadh* and ten burnt mounds (seven classified from the survey) were identified within the study area. Six of the *fulachta fiadh* and four of the burnt mounds were recorded in Derryville Bog; the majority were recorded on what is now dryland, west and north–west of Derryville Bog, but what would have been marsh or blanket bog in the prehistoric period. Only one *fulacht fiadh*, Killoran 253, was actually constructed on the contemporary surface of the bog.

A number of theories have been suggested for the function of these sites, including cooking, bathing and industrial processes. Mythology and the historical literature have also contributed to the argument (Barfield and Hodder 1987, 90; Lucas 1965; Ó Drisceoil 1988; 1990; 1991; O'Kelly 1954; Roberts 1998).

The site types *fulacht fiadh* and burnt mound appear to be mainly a final Neolithic/Bronze Age phenomenon. Similar sites, such as boiling pits, roasting pits and furnace pits, dating to the medieval period were also noted in the study area. However, the prehistoric types are usually quite distinct. Both are discussed below.

The main distinction between *fulachta fiadh* and burnt mounds is the presence of a trough or boiling pit in the former (Brindley *et al.* 1989–90). However, as some burnt mounds produce evidence of metalworking waste by-products and other evidence suggestive of a completely different function, a prehistoric date should possibly be considered as the more useful definition as a *fulacht fiadh*. This argument is reviewed below.

Structural variation

The classification *fulacht fiadh* was assigned to sites where both a trough and a spread of heat-shattered stone and charcoal were present. This term is regularly used in Ireland, while the term burnt mound is used in Britain. The latter term, however, covers a broad spectrum of activities and chronological periods, whereas *fulacht fiadh* refers to a specific type of site. Where a trough was not identified on a site, it was simply labelled as a burnt mound for convenience.

Fourteen percent of the excavated sites in the study area had no trough (and so were designated burnt mounds). However, this may reflect the multifunctional aspect of *fulachta fiadh*, in that, on some sites, the use of a subsurface trough was not required. Of the preserved pre-bog (and demonstrably prehistoric) sites, only 10% had no trough. Many of the sites classified as burnt mounds were noted during the survey.

Within the archaeological record, *fulachta fiadh* have considerable variation in size and form, both in the surviving form of the mound and type of trough in use on the site. It is interesting to note that only one of the ten *fulachta fiadh* and burnt mounds preserved below Derryville Bog had a mound over 0.5m in height. The classic description of *fulachta fiadh*/burnt mounds as having horseshoe- or kidney-shaped mounds (e.g. O'Kelly 1954) was not borne out by the evidence from Derryville Bog, where this shape was represented by only 10% of the fully preserved pre-bog sites and none of the dryland sites.

Low, irregular spreads of burnt material were more typical of the Derryville Bog sites and many of the sites excavated on the surrounding dryland. Although it is often assumed that low spreads of material are the remains of the base of a levelled mound, there was no evidence of large spreads of burnt stone in topsoil around these sites, and it is possible that the low mounds are the actual extent of the original dump of firing debris.

In looking at *fulachta fiadh* as a group, the trough appears to be the morphological variable most susceptible to typological analysis. Troughs vary in size, shape and capacity, as well as in lining. The range of trough types present around Derryville Bog can be paralleled on sites outside the study area. The presence of hearths and platforms has also been noted on a limited number of sites, mainly among wetland examples.

The main distinction between the *fulachta fiadh* excavated in Derryville Bog and those excavated on the dryland derive from the increased preservation of organic remains, particularly wood, on the wetland sites.

Troughs

The majority of sites recorded during this study accumulated around a single trough; however, multiple troughs were recorded at Killoran 17 and Killoran 19. Multiple troughs were also noted at Ballylin 2, County Limerick (Brindley *et al.* 1989–90, 27), and Tintagh, County Roscommon (Opie 1997). Backfilled troughs were re-dug and smaller versions inserted, as at Killoran 27 and Derryfadda 216, or re-lined, as at Killoran 240 (Plate 9.8). The reverse may also apply of course, but no evidence would remain if the trough was enlarged. At Killoran 19, three re-cuts were noted around a widened spring. A spring was also utilised at Killoran 304, allowing a clean water supply at the fen edge.

Circular, sub-circular, square, rectangular, oval and oblong troughs were all recorded. The shape varied from site to site, but generally the troughs had regular, straight or concave sides and flat bases. The size of the troughs also varied considerably, from a maximum length of

Plate 9.8 Killoran 240. Middle Bronze Age fulacht fiadh.

3.45m at Killoran 27 to a minimum of 1.2m at Killoran 19 (trough C10), with an average length of 1.87m. Depth varied from 1.22m at Killoran 27 to 0.12m at Killoran 22 Structure 2, with an average depth of 0.43m. An analysis of the dated troughs showed some typological changes through time. Sites from the Early Bronze Age showed a clear preference for circular or subcircular troughs, those from the Middle Bronze Age tended to be square or sub-rectangular and the Late Bronze Age troughs were rectangular, although a large oval example (Killoran 27), with a later re-cut rectangular trough, proved to be an exception. There was also an oval trough at Killoran 5 dating to the end of the Bronze Age.

As there is a large number of excavated, but unpublished, *fulachta fiadh*, establishing the chronological or functional significance of the troughs at a national level is not yet possible.

The capacity of the troughs varied from a maximum of 1,745 litres (Derryfadda 216) to a minimum of 195 litres (Killoran 22, Structure 2), with an average of 983 litres. The capacity of all the troughs was calculated using the maximum inner dimensions, although the actual capacity of each trough would have varied according to the height of the water table. Within the study area, trough capacity varied across the time framework and the geographical distribution of the sites, suggesting that sites were constructed in various sizes throughout time. Deeper examples may have been built to access a low water table.

Troughs were constructed to allow fresh water to percolate through the sides and base, demonstrating a good level of local knowledge and, in some cases, a

high level of skill in identifying suitable spots where the optimum conditions existed. This objective was achieved by locating the sites near the edge of a bog, as at Derryfadda 216 and Killoran 1, 21, 14, 22, 17, 240 and 265; on the bog surface, as at Killoran 253; or within a marshy area, as at Killoran 26. Killoran 19 and Killoran 304 were located directly over springs. In the case of Killoran 19, several attempts were made to access the spring. At Killoran 27, a very large pit was dug into the limestone to access the water table. This site was reused and suggested that the water table in this area had fallen. Killoran 26, dated to the same period and located in the same area, showed evidence of a burnt lining, suggesting that here too the water table had dropped significantly.

Most of the troughs had been deliberately backfilled with firing debris, possibly to prevent animals drinking and then fouling the water, or merely falling into the open troughs. This backfilling may be a product of the seasonal use of the sites. In the cases of Killoran 240, Killoran 253 and Killoran 265, peat had accumulated in the troughs, suggesting they were abandoned rather than retired until the next season. This may have occurred due to the shifts in the hydrology that characterised this area of the bog.

An analysis of several of the pre-bog *fulachta fiadh* suggested a large number of reuses for these sites. The trough from one site, Killoran 265, showed up to 615 episodes of reuse with a capacity of 1,100 litres. Thus, these sites were a regularly revisited feature of the landscape. Killoran 240, which was reused over 3,000 times, should probably be seen as contemporary with the set-tlement site Killoran 8. As some sites (including Killoran 240) were prematurely abandoned, due to inundation from the bog, it was not possible to determine whether the size of trough was related to the length of use. However, on the dryland, this did not appear to be the case.

Lining

Many *fulachta fiadh* contain evidence for a trough lining. Some troughs have a thin clay lining, as at Killoran 17 and Killoran 304; Lack West, County Mayo (Buckley and Lawless 1987) and Kilnacarriag, County Wicklow (Hayden 1994, 80). Others have complex timber constructions, such as the mortised plank box construction at Killoran 240. Evidence for a plank lining was also found at Derryfadda 216, Killoran 26, Killoran 27 and possibly Killoran 22 Structure 1. Other examples include Rathmore, County Kerry; Catstown, County Kilkenny and Ballynoe West 1 and 2, County Cork (Brindley *et al.* 1989–90, 26–7). A wicker-panelled trough was recorded at Killoran 253 and a carved oak trough was recorded at Killoran 265. Other examples of carved trunks were excavated at Clashroe, County Cork (Hurley 1987) and Curraghtarsna 25, County Tipperary, interpreted as a reused log boat (Buckley 1985).

Peat or moss was used to line a trough at Killoran 253, the moss being packed against the side of the wicker panel, up against the peat. Analysis showed it to be from raised bog, outside the locality. As the trough at Killoran 253 had been sunk through the fen peat into the boulder clay below, the peat plugging the sides would have acted as a filter to water that leaked in through the sides. At Killoran 19, a plug of peat had been inserted into a spring that had been widened for the trough.

Elsewhere, stone flags are commonly used as trough linings. There was some evidence to suggest the presence of a stone lining in a trough at the largely destroyed site Killoran 19. The remainder of the sites excavated did not use stone. This may in part be due to the poor availability of good sandstone flag stones in a predominantly limestone area; however, sandstone was widely used in the boiling process and sandstone flags were found on the causeway Killoran 18. It can only be assumed that wood was the preferable and easier option in this landscape.

Unlined troughs predominated on the dryland sites. While the conditions were not conducive to the preservation of organics on most of these sites, it must be assumed that in some cases there was no lining. Portable and removal troughs of leather, wicker or wood may have been used at these locations. However, it cannot be ruled out that the pit was sufficiently clean to be used unlined. Unlined troughs are known from Ballylin 2, County Limerick (Brindley *et al.* 1989–90, 27), and Derry 1 and 2, County Laois (Duffy 1996).

Hearths

An informal hearth consisting of a concentration of charcoal, sometimes in a small depression, was noted at Derryfadda 216, Killoran 5, Killoran 29 and Killoran 304. Elsewhere, formals hearths have been noted, such as at Clashroe, County Cork (Hurley 1987, 95–105). The construction of a hearth may not have been necessary after the first use, as the mound would have provided an ideal location for the fire.

Platforms

Platforms were a feature of the wetter sites in Derryville Bog. At Killoran 253, a stone-built platform was constructed under the mound, providing a dry working platform. In the case of Killoran 265, a wooden platform was thrown down to extend the dry working surface of the mound. A glacial moraine at Killoran 304 provided a dry surface.

Firing debris

Most of the fully excavated sites from the study area contained a majority component of fire-cracked sandstone, measuring less than 0.1m in diameter. This suggested that the primary function of the sites was to boil water. The use of sandstone over limestone has a practical consideration in that sandstone cobbles can be heated and cooled around five times before they are unusable (Buckley 1990a). Limestone shatters violently on contact with hot water (becoming calcium hydroxide) and also leaves a scum in the water and so is unsuitable for cooking. Nevertheless, limestone was used on sites from the study area, as evidenced by the content of each mound. In some cases, up to 50% of the mound was limestone (Killoran 25 and Killoran 27).

The ratio of sandstone to limestone at some of the sites was significant, in that there was a decrease from 98% sandstone: 2% limestone in the trough (where the stones from the last firing were preserved) to 90% sandstone: 10% limestone in the mound. This indicated that the sandstone was probably used five times more often than the limestone (which is likely to have been used only once). In the case of Killoran 265, the ratios of sandstone to limestone in the trough and mound were much closer. It appears that, on this site, with its long carved trough, the presence of calcium hydroxide in the water did not affect the trough's function, as it did on other sites. It may be that sites with these long, narrow, carved troughs represent a significant functional sub-group of *fulachta fiadh*.

Burnt mounds

Burnt mounds were found in isolation, i.e. without a trough. At Killoran 25, a very large spread of material, 30m in length by 23m in width, was excavated, but no trough was revealed. The absence of a trough was also noted at Killoran 13 and Killoran 316. It is possible that

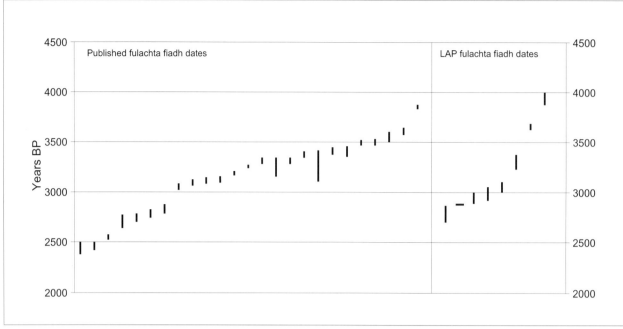

Fig. 9.1 Dates from fulachta fiadh *in Britain and Ireland compared with those from the study area.*

these sites represent a separate functional or cultural activity. However, the similarity and chronological relationship with *fulachta fiadh* make it likely that they were the same. The use of a portable trough has already been mentioned, and it should perhaps be considered here as well. The presence of fire-cracked stone must indicate contact with water. It would seem unlikely that a trough would be located away from the mound, but this is also a possibility. All three sites were located close to *fulachta fiadh*.

The absence of a buried sod below the burnt mound was noted on all the excavated sites. This may be the result of bioturbation making the sod indistinguishable from the mound material, or it may be that de-sodding took place, and therefore site preparation, implying knowledge of the size of the mound in advance. However, it seems unlikely that an area would be de-sodded for a random spread of waste stone.

Dating

Eight *fulachta fiadh* sites in Derryville Bog and the surrounding dryland were dated by radiocarbon analysis or dendrochronology. The date range fits into the general pattern of sites from Ireland and Britain. The oldest site, Killoran 17, is in the transition to the Bronze Age and slightly earlier than the national trend (see Fig. 9.1). Generally, however, the dates from the study area tend to cluster closer to the later Bronze Age, with the youngest site, Killoran 26, not earlier than 795 BC. Of the *fulachta fiadh* dated, four were from the dryland, 25% of all potential dryland *fulachta fiadh*. This small proportion of dryland sites reflects the wider pattern, with an even distribution throughout the Bronze Age, compared with those from Derryville Bog, which concentrated in the later period.

Finds and function

The majority of *fulachta fiadh* excavated in Britain and Ireland have been absent of finds (Barfield and Hodder 1987; Buckley 1985). Only two sites from the study area produced finds, with a further three producing animal bone. A saddle quern was retrieved from beside Killoran 240 and a small chert scraper was retrieved from Killoran 1. The animal bones were recovered from Killoran 5, Killoran 22 Structure 1 and Killoran 27. These consisted of long bones and teeth from sheep. No bone was recovered from the pre-bog sites, and it is perhaps surprising that bone was recovered from the dryland sites at all.

The primary function of *fulachta fiadh* was to procure clean, fresh drinking water and we should perhaps not lose sight of the fact that their probable main purpose was as shallow wells, when located close to settlement. The purpose of boiling water in the trough cannot yet be adequately explained.

The combination of measures to ensure clean water, by lining the troughs and by using sandstone, and the presence of finds associated with food preparation, and also the animal bone (scant as it is), suggest that cooking was the main activity carried out at such sites.

In the Bronze Age, perhaps we should perhaps look on this more as a preservative rather than a feasting process. If we consider that pastoralism was a dominant economic element during much of the time when *fulachta fiadh* were in use, it is most likely that they were used to boil up large haunches as part of a curing process. The apparent cessation in the use of *fulachta fiadh* in the Iron Age may coincide with the appearance or dominance of a dairy economy.

It has been suggested that bathing may be another function of such sites. The size and shape of troughs would not preclude this. The comfort of such an activity would require the prior removal of cracked stone at the base of the trough. In at least three cases, at Killoran 240, 265 and 304, a deposit of stone was found in the primary fill, presumably representing the last boiling stones. These sites were inundated by peat and, therefore, this fill was preserved. Elsewhere, the troughs were either naturally or deliberately backfilled by mound material. The presence of the fill would tend to preclude bathing in these three cases. The use of *fulachta fiadh* for industrial processes involving clean boiling water, sheep and hides cannot be ruled out.

In summary, while cooking or curing haunches was undoubtedly a major function of *fulachta fiadh*, several other functional subgroups appear to exist. Firstly, there are the burnt mounds where a subsurface trough is absent. The long carved trough and limestone-rich mound at Killoran 265 appears to represent a second subgroup. The chronology of these subdivisions has yet to be fully explored and a larger data set may, as with trough classifications, produce some important results.

Medieval pits

A number of oval or circular pits similar in form to the troughs from *fulachta fiadh* were dated to the medieval period. One site, Killoran 23, was remarkably similar in form to a trough but lacked a burnt mound. The pit was circular and filled with a series of layers of fire-cracked sandstone and charcoal. Like *fulachta fiadh*, this pit was deliberately sited to allow for water percolation. The site, in fact, was located around the same peaty hollow as the *fulachta fiadh* sites Killoran 12, 13 and 14, and probably fulfilled the same function in providing boiling water.

Other examples of medieval pits include the unusually large pit at Killoran 16, which was completely backfilled with natural sand. Pits likely to be associated with roasting were found at Killoran 3, 11, 15 and 16. All produced evidence of *in situ* burning, charcoal and fire-reddened sides. Some produced evidence of large stones. These oval pits were of similar size and were interpreted as roasting pits.

Conclusions

The variation in site types in Derryville Bog demonstrates the range and extensive nature of human exploitation of wetlands in the past. From the broad range of material identified by excavation and survey, the most prevalent site type was the trackway. A large variation exists in the classification of trackways and causeways, largely due to the incomplete nature of many investigations and the limited understanding of the nature of individual structures. A large number of terms are used to describe trackways and causeways based on either structural definition or vague generic terminology. We have attempted to simplify the process of classification for ease of analysis and comparison. The net result is a set of classifications that can be applied to all similar excavated structures occurring in wetlands. This enables a comparison of factors other than just constructional detail and provides a basis for meaningful interpretation of function.

The number and varied nature of our material is both surprising and significant. Many studies of wetland material have focused on larger structures, which are unrepresentative of the general picture that emerged from Derryville Bog. Many of the structures excavated in Derryville Bog were in poor condition, were often small-scale and were located in marginal areas of bog. The potential for similar large clusters of material going unexamined is therefore great.

Typological distinctions between wetland material cannot be broken down into discrete archaeological or cultural periods. The same structural designs have been in existence for over 4000 years or more, and, although patterns of use may vary, the same fundamental constructional options apply to each generation. A variation in constructional typology may be more a product of response to local bog terrain. Other factors influencing constructional design may be of a functional nature—narrow tracks for single-file traffic, for example. The nature of a structure will also be influenced by the type of person building it, their knowledge of the local terrain and their own requirements from a structure. A local builder will have far greater knowledge than an outsider. These issues are discussed in greater detail in the chapters that follow.

10. Catalogue of wetland sites

Sarah Cross May, Cara Murray, John Ó Néill and Paul Stevens

Introduction

Each individual area of the bog that was assigned a site number and excavated during the course of the project is listed below.

The entries begin with a summary in the following format: site/structure type; date (see Chapters 17 and 18 for a full list of dates); dimensions; main structural components (including main wood species present); predominant cutting season for wood; location/environmental setting; phases identified during excavation; site code; excavation licence numbers; Irish National Grid References (NGR) and level relative to OD (Malin Head, County Donegal). This is followed by the name(s) of the excavator(s).

Cooleeny 22 (Fig. 10.1; Plate 14.2)

Trackway; 1517±9 BC; 160m x 0.32m x 0.06m; planks and roundwood pegs (*Quercus* and *Salix*); raised bog environment; construction, use and destruction phases; site code DER22; licence 96E202 extension; NGR E222349 N166081–E222481 N166056; 124.30m OD (lowest walking surface).

Cara Murray

This track was located in the south-central portion of the study area, south of the multiphase stone causeway Killoran 18. As excavated, this track was composed of three basic elements: longitudinal planks forming the walking surface, transverse timbers on which some of these planks rested and upright pegs, which were used to hold some of the horizontal timbers in place. The site was traced for 160m of its length, east–west across part of the Middle Bronze Age portion of raised bog. Mineral inclusions at the level of the track indicated that the site did continue westwards beyond its last recorded sighting. The condition and nature of the site varied in each drain face, and the site was not visible in all section faces. Portions of the western, central and eastern sections of the site were excavated.

The longitudinal walking surface was constructed of split planks, of which 44% were radial split, 12% were tangential split, 12% were half split and 32% were of other split types. These planks ranged from 2.36–2.62m in length, with most of the planks being 0.52–1.83m long, having been only partially exposed or cut by the drains. The planks were 0.19–0.28m wide and formed a very narrow walking surface suitable for precarious single file pedestrian use. Some of the planks were degraded and did not show any trace of surviving tool marks, while others were completely rotten and had been infested by ants.

Along the excavated length of the track, mortises were present on only three of the longitudinal timbers. One of the longitudinal planks had a mortise that had split due to the displacement of the two pegs which it contained. In all

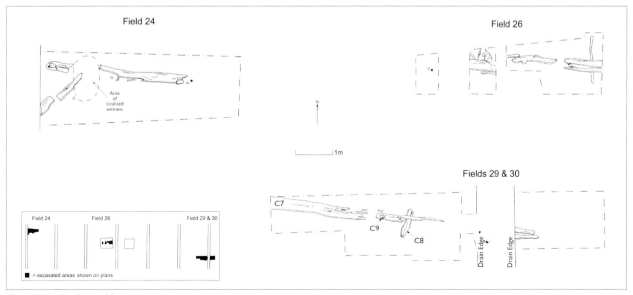

Fig. 10.1 Cooleeny 22.

instances, the ends of the timbers and the mortises were in a very degraded condition. In only a few instances were the longitudinal timbers set on transverse timbers. One such timber was a degraded willow branch, the only non-oak element of the track, and another transverse had a degraded mortise that was secured by a peg that stood 0.20m above the level of the walking surface of the track. Another peg set into a longitudinal plank also stood 0.22m above the walking surface. The pegs were all trimmed, radial splits.

There was no evidence of repair on the site. The second phase related to the destruction of the site, represented by the displacement of a portion of one of the longitudinal planks, which had a damaged mortise securing either end. The plank was radially split and 3.91m in overall length. This was an area of localised wetness and the displacement was probably caused by breakage through use and flooding.

Discussion

The variety of planks used illustrates the woodworking skills of the builders. Some planks were beautifully worked and squared off, while crude tangential splits indicate the simple functionality of the trackways construction. The height of the pegs would have formed substantial obstacles in the use of this very narrow track, and they were probably deliberately set to this height to indicate the walking surface during the winter months, when the water level rose.

This site was set within poorly humified raised bog containing cotton grass and sedges. The track was located roughly 0.90m above the level of the wood-rich fen peat and 0.20m below the transition to moderately humified fen peat. The track post-dated the other single-plank track, Derryfadda 23, by 55–73 years. Both single-plank tracks were almost identical in form and construction.

Cooleeny 31 (Fig. 10.2; Plates 9.1 and 15.5)

Causeway; 778–423 BC; 246m x 8m x 0.5m; roundwood, timbers, brushwood, pegs and sand (*Corylus, Alnus, Fraxinus, Salix, Betula, Pyrus/Malus, Quercus, Ulmus* and *Taxus*); cutting season autumn (65%); fen, raised bog and marginal forest environment; construction, use and destruction phases; site code DER31; licence 97E160; NGR E222247 N165725–E222492 N165572 (excavation E222277 N165705); 125.03m OD (surface).

Sarah Cross May

This is a substantial composite causeway running northwest–southeast from one glacial ridge to another and bridging the mouth of Cooleeny Bog. It is one of only two sites in Derryville that could be used for crossing the bog. One cutting measuring 15m x 8m was excavated on the southern flank of the western ridge, about 60m from its western landfall. As it lay outside the impact zone of the development, the rest of the site remained intact.

Phases of construction

The causeway was constructed in four phases. In Phase 1, brushwood dumped into depressions evened out the general route. This levelling covered an area up to 8m wide. In Phase 2, a triple row of pegs marking a regular 3m width was laid to designate the specific route. A layer of sand was laid between the two outer posts in Phase 3. This was level, both along the length and across the width of the structure. Further brushwood was also laid in soft sections. The surface of the causeway was constructed as Phase 4. It was a mixture of transverse planks and roundwood. One strip of planks and roundwood was laid in the south and then overlapped by another strip laid from the north. It was pegged at both sides and where the two strips joined in the middle. These pegs survived for up to 0.15m above the present surface, but they had clearly rotted during use and could have stood up to 0.50m above the original surface. This surface was 4.5m wide at its narrowest point. While it was level overall, there were substantial gaps and dips between the pieces of wood.

Environment

The causeway crossed Derryville Bog at its southernmost extent—the mouth of Cooleeny Bog. It crossed the southern flank of a ridge on the west and met another ridge to the south of the Derryfadda peninsula. The structure was excavated on the southern flank of the western ridge, about 60m from its landfall, in well-humified fen, amongst marginal forest. At this point, the site was on the northern edge of a large stand of marginal alder stretching up to 50m southwards. It was bounded to the north by a moderately large pool of rannock rush. About 95m west of this point, the causeway met raised bog. It ran across the raised bog for about 50m and then across fen and marginal forest for about another 20m before it hit the southeastern ridge.

There was a very large erosion gully about 20m north of the causeway, approximately 45m from its eastern terminus. This and other erosion gullies through Cooleeny indicate a bog burst heading southward. Bog burst B had affected the area a few centuries (c. 1250 BC) before construction, destabilising the surface. The unstable drainage of this area and the uneven nature of the raised bog required a very firm structure to make crossing possible, even at this very narrow point. Unfortunately, the massive structure, including sand, dammed the outflow of water created by the earlier bog burst and caused another, more massive, bog burst (bog burst D). This would have occurred within a year of the construction of the track and damaged the site beyond repair. The effects of the bog burst can be seen in the line of the track as it crosses the fen peat bordering the southwest ridge.

Relationship between the nature of the bog and the track

This causeway was considerably more substantial and complex than any other structure in the study area. While some of this must relate to its function, some of it relates to the unstable nature of the bog at this point. The technical difficulty of crossing this environment is indicated by the choice to run the track off the southern flank of the western ridge rather than off the eastern point, which would have made a shorter crossing. The ridge itself would have provided a break in the drainage, so the causeway did not meet the main flow until it left the raised bog on the eastern side. This point, where it was crossing fen and approaching dryland again, would have been the weakest section of the track, and it was at this point that the track broke under the pressure of the water flow.

Function

While the top wood was rotted, there was no sign of trample damage or wheel ruts. This may be partly due to the bog burst. The site would have been exposed long past its usefulness as a causeway because of the drop in the water table. Nonetheless, there is nothing to indicate that this track was used for wheeled transport, despite its initial resemblance to the corduroy roads of Central Europe. The analysis of beetle remains (see Chapter 8) shows that what few dung beetles were present fed on the dung of wild mammals such as deer, rather than cattle or sheep. The track seems not to have been used for wheeled vehicles or for droving. The strength of the structure may be

explained by environmental conditions, but its width must have been related to function.

Perhaps the width of the track was deliberately ostentatious, relating to the many over-large structures found in Ireland during the transition between the Bronze Age and the Iron Age. Sites such as the Black Pig's dike and the oak trackway Corlea 1 in County Longford (Raftery 1996) seem to have been built as expressions of power as well as for whatever functions they performed.

The construction of this track in this environment indicates less knowledge of the bog as an ecosystem than is demonstrated by other structures in the study area (cf. Casparie, page 39). This may suggest that its designers were not local. This causeway was set to mark and maintain an important routeway that was required to cross the bog rather than go around it to the south. This was important enough to command a good deal of labour as well as resources and suggests that the site should be seen as an indicator of centralising regional power bases in the area in the later prehistoric period.

Artefact 97E160:31:1

The artefact 97E160:31:1 was found lying horizontally in the lowest layers of brushwood. It is piece of roundwood that has been turned into a small rod (20mm in diameter) with a rounded end. Marks from lathe turning are evident

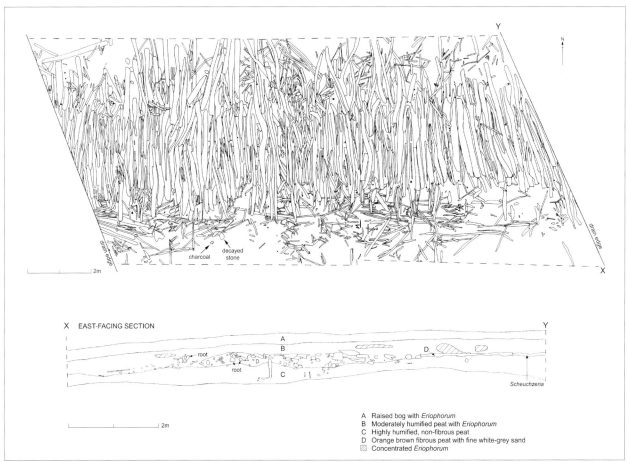

Fig. 10.2 Cooleeny 31.

under low magnification. One end is carved round and smoothed; the other was broken in antiquity. There is nothing in its form that would indicate function. Its position in the foundation layer of the track suggests it was deposited as scrap. However, it is the earliest evidence for lathe turning as a woodworking technique in Ireland.

Cooleeny 44

Trackway; probably Bronze Age; 11.2m x 0.8m; raised bog environment; brushwood; construction phase; site code DER44; licences 94E106 extension and 96E202; NGR E222352 N165627; 124.18m OD.

IAWU and Cara Murray

This was a milled-out brushwood trackway traced for over 11m. The brushwood was 8–18mm in diameter. The trackway was badly damaged, so little of the wood can be said to have been truly *in situ*. The site was situated in *Sphagnum* peat. There were no clear indications of the nature of the structure, although the small diameters of the surviving brushwood suggest that it may have incorporated hurdle panels. If the surviving width of the track approximates the original, it would suggest a narrow brushwood trackway. Although undated, the level of this trackway, south of Cooleeny 31, suggests that it is Bronze Age in date.

Cooleeny 64 (Fig. 10.3; Plate 9.4)

Trackway; 1420–1020 BC; 22.30m x 1.35m x 0.25m; roundwood and brushwood (*Corylus*, *Alnus*, *Pyrus/Malus*, *Fraxinus*, *Salix*, *Frangula* and cf. *Sorbus*); cutting season autumn (51%); marginal forest environment; construction, extension and destruction phases; site code DER64; licence 96E202 extension; NGR E222568 N165914; 124.41m OD (construction level).

Cara Murray

The brushwood track Cooleeny 64 was located on the southwestern side of the study area. The site was excavated at its landfall, in the margins of the marginal forest, where the track ran around the edge of an alder root system located on the southwestern side of the site. The track was constructed of small densely packed roundwood and brushwood. The timbers were set longitudinally with a series of associated transverse timbers. These transverses would have provided greater stability and prevented the longitudinal timbers from either sinking or breaking. The wood was 10–40mm in diameter and was very straight and regular. Thirty-one pieces of alder root were associated with this deposit. Two larger lengths of alder and alder buckthorn roundwood and some stray timbers on the western limit of the site were used to extend the use of the site. These two deposits of wood were separated by about 50mm of silty brown peat. To the south, flooding had displaced a number of timbers that originally formed

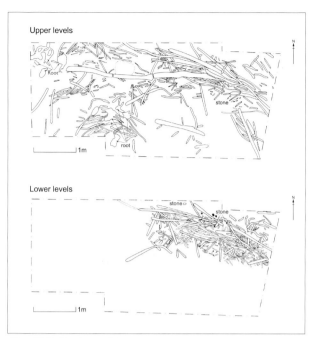

Fig. 10.3 Cooleeny 64.

part of the track. Tree roots were subsequently able to grow in this environment by using the track as anchorage, even as the area became increasingly wet and hypnoid mosses began to invade the site.

Discussion

The brushwood track was constructed at the transition between raised bog and fen peat during the Late Bronze Age, after bog burst B of *c*. 1250 BC. As a result of this event, the water table was lowered by up to one metre, making accessible areas that had previously not been in use. The track was constructed as a brushwood structure, supplemented by large roundwood as the area became increasingly wet. These lengths of roundwood would have formed a precarious walking surface, but the extension of the use of the track was obviously of greater significance than ease of access or comfort. This track was the only site found in this southern area, on the western side of the bog.

Cooleeny 325 (Fig. 10.4)

Archaeological wood; mid-first millennium BC; 3.40m x 1m x 0.05m; brushwood (*Ilex* and *Fraxinus*); cutting season autumn (100%); raised bog environment; deposition phase; site code DER325; licence 97E160; NGR E222279 N165711; 125.28m OD (level of artefact).

Sarah Cross May

This was a small irregular scatter (3.4m x 1m) of brushwood containing an artefact. The material was found in raised bog about 0.25m above Killoran 31. Its distance from the edge of the bog suggested that it was laid to fill a wet hollow in the raised bog where work was already taking place. The

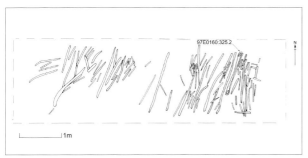

Fig. 10.4 Cooleeny 325.

strong presence of *Ilex* at the site is puzzling. This is an unusual species to use in such a casual construction.

The artefact 97E160:325:1 was found amongst the brushwood. It is a piece of wood that had been carved along its length to produce expanded terminals, which had been perforated. There are tool marks along its length. One terminal is intact and one was damaged in antiquity. While its form is similar to a yoke for carrying, it seems too slight for this purpose. Its position suggests deposition as scrap.

Derryfadda 2

Archaeological wood; no date; 0.47m x 0.12m; roundwood (*Alnus*); raised bog environment; deposition phase; site code DER2; licences 94E106 extension and 96E202; NGR E222944 N166380; 125.27m OD.

IAWU and Cara Murray

This site produced an *ex situ* alder roundwood with a chisel point, located in a BnM drain face. The wood was in very poor condition and had been damaged by milling. There was no evidence of an associated structure at the site.

Derryfadda 6 (Figs 10.5–10.7)

Platform; 380–5 BC; 8m x 7m; brushwood, roundwood and timbers (*Alnus, Fraxinus, Corylus, Betula, Salix,* cf. *Sorbus, Alnus viridus, Ilex* and *Quercus*); marginal forest environment; construction phase; site code DER6; licence 96E202; NGR E222779 N166007; 126.05m OD (working surface).

Cara Murray and John Ó Néill

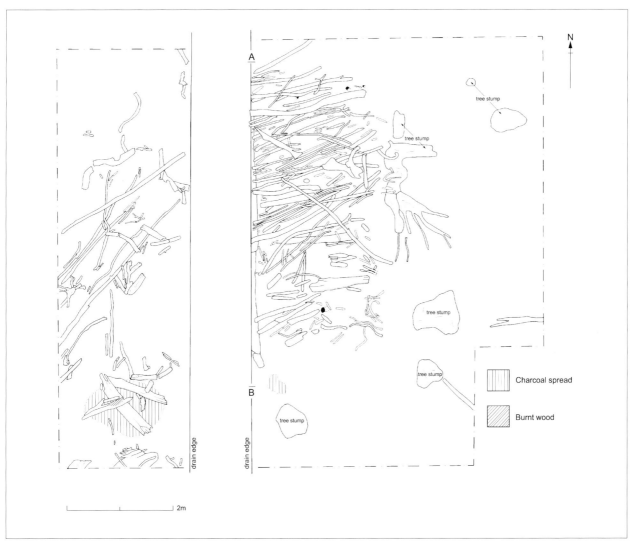

Fig. 10.5 Derryfadda 6. For section A-B see Fig. 10.7.

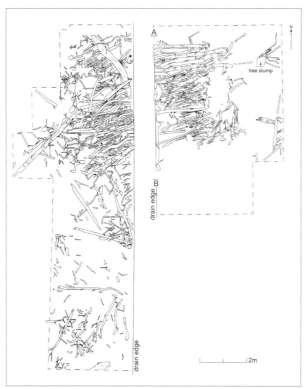

Fig. 10.6 Derryfadda 6 (right) and 215 (left). For section A-B see Fig. 10.7.

This structure was located on the northern side of an underlying glacial ridge on the eastern side of the study area. It measured 5.75m north–south by 3.9m east-west and overlay *fulacht fiadh* Derryfadda 216. The platform had been built in marginal forest peats in an area of root systems, directly on top of the eastern landfall of trackway Derryfadda 215.

It was mainly constructed with roundwoods and brushwood, orientated southwest–northeast. Occasional fragments of cleft wood were also used. There was no apparent orientation to this structure and its main purpose seemed to be to extend the dry area that the *fulacht fiadh* provided. The wood survived in very good condition. A number of burnt oak timbers were found to the south of the site, along with patches of charcoal. A BnM drain had destroyed the central portion of the site.

Discussion

This structure was built to provide a platform at the edge of the navigable area of marginal forest on the top of a gradual slope down into the Derryville Bog system. The wood used in the structure survived in good condition, suggesting that it was in use for a short period of time. The burnt oaks and charcoal found at a slightly higher level than the structure are probably associated with scrub clearance in the area after the structure had gone out of use.

Derryfadda 9 (Fig. 10.8; Plate 9.7)

Platform; 395–180 BC; 8.2m x 3m; brushwood and roundwood (*Betula*, *Fraxinus*, *Alnus*, *Salix*, *Ilex*, *Pyrus/Malus*, cf. *Sorbus* and *Corylus*); cutting season mixed; fen environment; construction and extension phases; site code DER9; licence 96E202; NGR E222766 N165901; 125.12m OD (surface).

Cara Murray and John Ó Néill

This structure was an 8.2m x 3m-wide platform built on the southern side of an underlying glacial ridge on the eastern side of the bog. The site in which the structure was located appeared to have been very wet at the time of construction, with marginal forest root systems surviving below the western side of the structure but only present at the same level on its western side. Peat rich in sedges and bog bean indicated that the area below and to the north of the site was prone to flooding and retaining standing water. These extremely wet conditions created the need for a platform and contributed to the high level of preservation on the site.

Initial use of the site began with a linear deposit of brushwood and roundwood, 4m long and up to 1.5m in width. The timbers used in this part of the structure included alder, holly, ash and willow, felled in winter, spring and summer. It appears that this was a structural element intended to support one end of the main surface of timbers, deposited on top to the south and west. Two pommel-headed batons of cherry wood (artefacts 96E202:9:1 and 96E202:9:2) were found amongst the lower level of wood. Both were about 0.35m long and had a broken notch at one end and a bound area near the other end, which had been trimmed to leave a mushroom-shaped head.

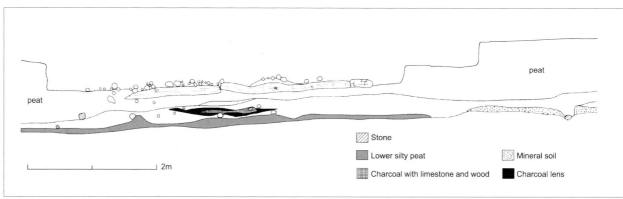

Fig. 10.7 Derryfadda 6 (above) and Derryfadda 216 (below).

The upper level of wood contained a mixture of birch, ash, alder, willow, holly, crab apple, rowan and hazel, felled over all four seasons of the year. These timbers had been laid mostly northwest–southeast to create a walking surface. One or two small stones had been carried in and deposited within the structure, which may have been partially disturbed by the high level of water over the site. Some heavier trunks were laid across the top of the site, either to keep the transverses in position or as wood felled in the area but not used. Substantial wood chips recovered from the site showed that tree-felling took place at the site. Three different sets of axe blade signatures were found on the worked ends recovered from the site.

Discussion

This structure was a platform or possibly a short trackway at the edge of the marginal forest. While the upper surface resembled a trackway in form, the orientation of the site would not have promoted easy access to either end, but more likely to a central part of the site. The function of the two artefacts found at the north of the site is not known and few parallels have been found. The wood chips indicate that tree-felling was taking place at the site, and the pattern of the axe signatures indicates that three axes were in use. The absence of a pattern in the recorded cutting seasons suggests that either the site was in use all year round and accumulated over a period of time or was constructed of stockpiled or waste timber collected from elsewhere.

Derryfadda 13 (Fig. 10.9)

Trackway; 767–412 BC; 24.2m x 3m x 0.70m; roundwood, brushwood and timbers (*Alnus, Fraxinus, Corylus, Pyrus/Malus, Betula, Salix, Prunus avium, Quercus, Acer, Prunus spinosa,* cf. *Sorbus* and *Taxus*); cutting season autumn (48%); pre-bog surface and marginal forest environment; construction and extension phases; This site was recorded in the field as DER13a and the landfall as DER13b. The site had also been disturbed by a BnM drainage pipe and material east of this was recorded during the initial survey as DER12; Site code DER13; licence 96E202; NGR E222691 N165749; 125.40m OD (landfall).

Cara Murray

This track was constructed with its eastern limit on the mineral soil of the southeastern glacial ridge, and extended west into the marginal forest. During excavation, the full extent of the landfall could not be traced as it was covered by a BnM peat stockpile. As excavated, the site ran for a total length of 24.20m and varied in width from 3m at its western limit to 10.8m at its eastern excavated limit on the dryland margin, which was its widest point. The structure consisted of a dense deposit of brushwood and roundwood, between 0.20m and 0.70m deep, which was denser on the eastern, dryland side of the site and on the northern side. Due to the size of the structure and the density of archaeological wood, it was not feasible to excavate the entire substructure. Instead, a 15–20% spatial subsample was taken from five areas, which were fully sampled and consisted of two 3m x 1m and three 2m x 1m areas (Boxes I–V).

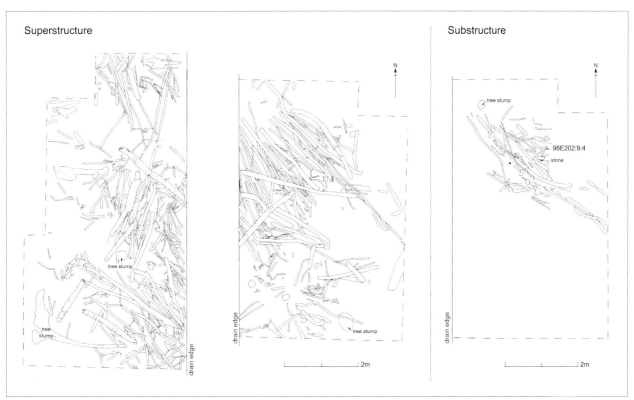

Fig. 10.8 Derryfadda 9. Superstructure and substructure.

Fig. 10.9 Derryfadda 13. Superstructure and substructure.

The eastern end of the site was constructed on glacial till, the western end of the landfall on about 20mm of silty fen peat, with the substructure embedded in boulder clay and moderately humified forest fen peat. The amount of mineral inclusions within the peat decreased from east to west, although silting also occurred within the main body of the structure as a result of proximity to the dryland and associated in-washing. Trees within the marginal forest were growing in this area and the structure was built around these root systems. The higher acidity of the roots caused localised degradation to the wood of the structure, creating apparent voids where the site had rotted away.

The substructure was constructed as a dense foundation of irregularly deposited brushwood, which had a common northeast–southwest trend on its northern side, within Boxes II, III and IV. The substructure was denser on the northern side of the site within Boxes II and III, although the nature of the material was the same. This was recorded in the field as a basal substructure.

Above this, the main body of the structure was irregularly laid, oriented approximately northwest–southeast and consisted of a series of five irregularly placed deposits of roundwood. On the eastern side, two deposits were oriented north–south and northwest–southeast, while the other deposits were oriented northeast–southwest and east–west. This variation in orientation verifies that these deposits were set between the root systems of the marginal forest. To the west, there was less evidence of tree root disturbance, probably because the increasing depth of peat prevented the penetration of roots into the mineral soil. On the western side, where the site ran into the marginal forest, it had two structural levels. The upper level, which measured 13.4m x 9.6m, was a 0.20m deep deposit of roundwood and brushwood that extended beyond the limits of a 4.0m x 3.4m substructure that was also around 0.20m deep. Tree roots caused further disruption by embedding themselves in the structure, crushing and disturbing it.

Two artefacts were uncovered from this site. The first of these, uncovered among the substructural wood during the IAWU survey, was a yew spear shaft (94E106:2) that had been dressed along its entire surface. The second artefact was uncovered within the substructure of Box III and consisted of a wooden shaft (96E202:13:1), 322mm long and 49mm in diameter that had been broken at one end and trimmed at the other. This artefact has been interpreted as a possible spoke or handle.

Discussion

The excavation revealed that this track had been constructed by depositing extensive dumps of brushwood and occasional roundwood across parts of the eastern area of the site to form a foundation. The overlying roundwood was more cohesively structured to form a walking surface. However, within all of this material, the functionality of the structure was more significant than its form of construction, which was a response to the environment. The occurrence of two damaged artefacts within the structure reinforces this idea.

Environmental evidence suggests that the site was constructed prior to bog burst D of *c.* 600 BC. The density of archaeological wood used on the glacial till supports the idea that substantial deposits were necessary to access this area. The continued significance of this area as a route in the Iron Age was such that the construction of substantial, irregular structures were still of value.

Derryfadda 15 (Fig. 10.10)

Archaeological wood; *c.* 1550 BC; 2.72m x 1.83m x 0.14m; roundwood and brushwood; marginal forest environment; construction phase; site code DER15; licence 96E202; NGR E222687 N165776; 124.13m OD.

Cara Murray

This site was located about 25m north of Derryfadda 13, on the western side of the southeastern underlying glacial ridge. It was within dark brown *Sphagnum* peat, which was almost entirely humified. The structure consisted of a deposit of loosely arranged lengths of roundwood and brushwood, oriented approximately east–west, that had been cut by a BnM drain. As such, the structure survived for 2.72m of its length and was 1.33m in width. To the east, there was a smaller deposit of wood, oriented northwest-southeast, which was 2.28m long and 0.70m in width. All of this material appeared to have been deposited to infill an area of localised wetness within the marginal forest. Judging by the location of this material, 0.45m below stake row Derryfadda 209 (990–770 BC), it can be estimated that Derryfadda 15 was deposited about 650 years earlier, around 1550 BC, within the Middle Bronze Age.

Derryfadda 17 (Fig. 10.11; Plate 9.6)

Trackway; 1315–980 BC; 52m x 1.63m x 0.26m; stone, stakes, timbers and brushwood (*Quercus*, *Alnus*, *Pyrus/Malus*, *Fraxinus* and *Salix*); pre-bog surface, fen and raised bog environment; construction and repair phases; This site was also recorded during the initial survey as DER16; Site code DER17; licence 96E202 extension; NGR E222628 N165992; 125.40m OD (landfall).

Cara Murray

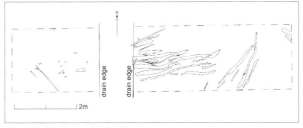

Fig. 10.10 Derryfadda 15.

This trackway was located in the southeastern complex, immediately west of the dryland headland of Derryfadda. It was traced for 52m of its length from landfall in mineral soil at the western end of the site, through the fen and out into raised bog. The excavation of the track was concentrated at its eastern end, but sections also recorded on the western side, where some split planks overlay the main stone body of the site. At its most easterly sighting, in a BnM field drain, the site was located on mineral soil; on its northern side, the site rested on 0.15m of fen peat. At its landfall, it survived as a 2.45m-wide deposit, which was up to 0.20m in depth. The track was composed of a compacted deposit of small stones, some of which were in decayed stone and mineral soil. A small deposit of oxidised peat, immediately below the level of the site, indicated the construction level of the structure. Above the site, well-humified fen peat, wood-rich fen and *Sphagnum* had developed. The trackway could not be traced beyond its excavated eastern limit.

Construction phases

As excavated, the site was constructed of eight separate deposits of stone. Stone did not occur in a continuous deposit but was used where it was needed to create a walking surface. The main portion of each context occurred in the centre of the deposit and was dispersed towards the edges. At these peripheral areas, a single stone often represented the deposit. Although the stone deposits did not merge, traces of decayed stone and gritty inclusions within the peat between the deposits indicated the walking surface. These stone deposits were used to infill pools within the hummock and hollow system of the raised bog.

The deposits ranged in size from 6–1.05m long, 1.63–0.25m wide and 0.30–0.03m in depth. The stones within all of these contexts were predominantly small pieces of sandstone and limestone, some of which had decayed. In only one of the easterly deposits was any variation to this general trend found. Here, cobbles, boulders and flagstones were used, although the majority of stones were predominantly small. Dopplerite formed between some of the stones, indicating the extent to which the stones had moved during the use of the track. The most westerly stone sighting consisted of

a single flagstone, 0.54m x 0.26m x 0.09m, which had partially decayed.

Within a year or two of the construction of the site, the use of the site was extended by the addition of small split timbers and wood-chips. Three of the stone deposits were supplemented with small amounts of oak, willow and burnt ash split timbers from 0.20m to 0.41m in length. A large amount of small pieces of split oak, ash and alder were also used. One of the split oak pieces was charred on its underside. Two pieces of alder stem base or root fragments were also charred and some very small charcoal fragments of birch and alder were associated with one of the deposits.

On the southern side of the most westerly stone, three stakes were set in a linear arrangement, 1.12m to 1.18m apart and 0.65–0.89m in length. Two of the stakes were trimmed, radially split timbers of oak. The third was an alder roundwood that had been cut at the beginning of the growing season to a wedge end and then had the side branches removed. It was this piece that was used to produce the radiocarbon date for the site. Six additional split stakes were recorded from the section face of the drain in the original survey and were recorded as DER16. Trimmed, split oak stakes were also found in association with three of the other stone deposits, within the main body of the site. They were set approximately 7.50m and 12m apart. None of the stakes were used to anchor material in place. They were all of oak, two were radially split and one was tangentially split and trimmed.

Within a couple of seasons of the construction of the site, the water table had risen and these stakes were used as markers to indicate/demarcate the line and safe route of the track. Although the site was not found to continue west of its last excavated sighting, an artefact was uncovered during the clearance of the field drain to the west. This artefact, 96E202:17:1, consisted of an alder bucket fragment 139mm in maximum length, *c.* 85mm in maximum width and 8–21mm in thickness. It had a 5.6mm thick croze for the insertion of a base plate.

The local environment

This site was constructed after bog burst B of *c.* 1250 BC. As a result of this event, the water table was lowered by up to one metre, making areas that had previously not been

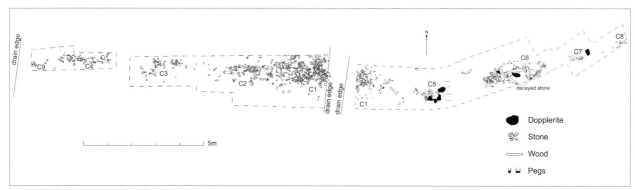

Fig. 10.11 Derryfadda 17. Eastern and western sections.

in use accessible and allowing what would otherwise seem to be an illogical and ineffective construction method.

The excavated portion of the site was set in raised bog. On its eastern side, the site was associated predominantly with cotton grass and *Sphagnum* pool peat, representing the hummock and hollow system within the raised bog. This peat was moderately to well humified, with patches of sedge and heather that had hardly decomposed at all. Other plant elements noted were areas of reeds, sedges and mosses. In places, the site was associated with slightly more compact peat with mineral inclusions. This material could not be followed as an archaeological feature. The stones were used to infill the hollows of the raised bog, as the hummocks provided a sufficiently solid walking surface.

The track provided a crude stepping-stone type method for crossing this area of the raised bog. This would appear to be an unusual method of construction, as the stones would have simply sunk into the pools. However, the existence of the site indicates its successful use. It appears to have only been used for one or two seasons before it was necessary to supplement the site with additional wood. Some of the material used to supplement the site may have been taken from deliberate clearance of the marginal area. This root material and rotten willow was probably associated with the original clearance of the site in the immediate vicinity. The rotten willow may have resulted from the desiccation caused by the lowering of the water table subsequent to the bog burst.

Derryfadda 19

Stake row; AD 640–890; 100m x 0.04m; posts (*Corylus* and *Betula*); cutting season winter (2 of 3); raised bog environment; construction phase; site code DER19; licence 96E202; NGR E222695 N165823–E222670 N165925; 125.27m OD (top of post).

Cara Murray

This stake row was located 50m north of Derryfadda 15. It was approximately 100m in length and has been described as a stake row, although it only consisted of

five stakes whose association was tentative given the distances (15–70m) between them. The length of the stakes varied considerably: Stakes B (birch), D, C (hazel) and A (hazel) measured 0.16m, 0.30m, 0.45m and 1.88m, respectively. The diameters ranged from 35–40mm. No archaeological wood was associated with any of the stakes.

Discussion

Timber from this site dated to AD 640–890; the complex stake row Killoran 54 dated to AD 668–884. While the dates are similar, the nature of the sites could not be more different. They do, however, occur along the same northwest-southeast orientation, at the edge of a small underlying rise in the mineral soil, which would probably have produced a slightly drier area in the overlying material.

Derryfadda 23 (Fig. 10.12; Plate 9.5)

Trackway; 1590±9 BC; 34.50m x 0.5m x 0.12m; planks, roundwood and pegs (*Quercus* and *Fraxinus*); raised bog environment; construction and use phases; site code DER23; licence 96E202; NGR E222600 N165763; 123.97m OD (lowest level).

Cara Murray

This single-plank track was oriented approximately east–west and ran for a distance of approximately 34.5m across raised bog, about 40m west of the southeastern underlying glacial ridge. No other sites were found in close association. The majority of the track was constructed of large split planks, set end to end, with each end of the planks sitting on a smaller transverse half-split length of roundwood. Much of the wood was in poor condition and had dried out during the use of the site. The trackway may have been exposed for ten to fifteen years, during which time disturbance or slippage of the timbers also occurred.

Nine longitudinal timbers, seven transverse timbers, six pegs and an amount of disturbed or displaced timber fragments were recorded from the superstructure. Four of the longitudinal timbers ranged in length from 2.4m to 4.3m. Two of the longitudinal timbers, both of which were located in Field 41, differed from the rest. Timber 9

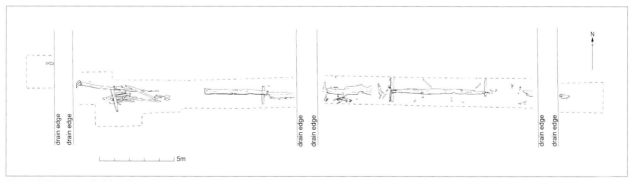

Fig. 10.12 Derryfadda 23.

was completely rotted and was without any cohesive structure or form; it was sampled for beetle analysis. Timber 13 was a reused timber that had been tangentially split. It measured 6.7m in maximum length and 0.35m in maximum width. Both of the ends were very degraded and rested on transverse timbers.

Four mortises were cut into Timber 13. Mortises B and D, on the northern side of the timber, were 2.28m apart and 2.44m and 4.80m, respectively, from the western end of the timber. Mortises A and C were located on the southern side of the timber and were 2.06m apart, and 2.06m and 4.26m, respectively, from the western end of the timber. Both Mortise D and Mortise B were damaged and only remained in part. The transverse timbers were predominantly half-split with two radial splits and varied in condition. They ranged from 0.45m to 1.75m in length and 80mm to 190mm in width, and were set 0.68m to 6.30m apart. The pegs associated with this site were irregularly set and, due to subsequent displacement, were not related to any of the mortised timbers. The pegs were set 0.20mm to 44mm apart and ranged in diameter from 50mm to 60mm and from 0.76m to 0.99m in length.

A substructure was only present along one section of the site, to infill an area of localised wetness. It measured 4.60m x 2m and was 0.40m in depth, and was comprised of sixteen timbers and some smaller fragments of wood. The substructure was constructed as a series of roughly east–west oriented timbers, 1–1.63m long, 0.19–0.24m wide and 0.05–0.12m thick, with trimmed radial and tangential splits. These timbers were overlaid by a single transverse plank, 1.82m long, 0.22m wide and 75mm in thickness. This was a trimmed radial split with a mortise on its southern end and damage on its eastern end.

Discussion

This single-plank track has been dendrochronologically dated (Q9370 and Q9369) to give the best estimate felling date range of between 1590±9 BC and 1606±9 BC or later. Examination of the peat profile revealed that the site was constructed above a gully in the mineral soil at 122m OD. Due to a superficial desiccation around 1600 BC, the subsoil was reflected in the shrinkage of the bog surface, and this area was probably used as an accessible route at that time. However, as the area became increasingly wet, it was necessary to construct the trackway. The testate amoebae analysis indicated that the area dried out again shortly after the construction of the track, verified by the degraded nature of the wood, which suggested that the track was in use for ten to fifteen years. Some of the very rotten wood (e.g. Timber 9) decomposed prior to its use in the structure, and had been infested by beetles that were primary forest indicators. Samples were taken from this site for pollen analysis, and the relational peat profile was also examined in detail.

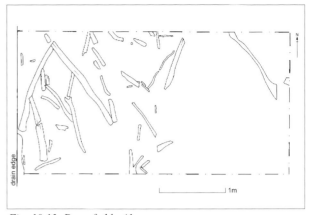

Fig. 10.13 Derryfadda 41.

Two single-plank tracks of this type were constructed during the Middle Bronze Age within the southern discharge zone of the raised bog. The other track, Cooleeny 22, post-dated this site.

Derryfadda 41 (Fig. 10.13)

Archaeological wood; probably Middle/Late Bronze Age; 2.40m x 2m x 0.15m; roundwood and brushwood; fen environment; deposition and disturbance phases; site code DER41; licence 96E202; NGR E222680 N165749; 124.08m OD.

Cara Murray

This site consisted of randomly deposited pieces of roundwood and brushwood, which were laid down in extremely wet conditions within fen peat. There was no cohesive form to this material, which was well spaced and of mixed orientation. However, some of the roundwood was worked. The site can be securely fixed 0.66m above the Early Bronze Age track Derryfadda 204, suggesting that this material can be roughly attributed to the Middle/Late Bronze Age.

Derryfadda 42

Archaeological wood; probably 515–365 BC; 1.12m x 0.8m; brushwood and a peg; marginal forest environment; construction phase; site code DER42; licence 96E202; NGR E222677 N165753; 125.08m OD.

Cara Murray

This site was comprised of the remains of a destroyed structure, which may have been a trackway. The surviving fragments were horizontal lengths of brushwood and one upright peg measuring 0.14m in length. The brushwood was 20–30mm in diameter. Re-investigation did not reveal any further evidence concerning the nature of the structure. Although no date was obtained for this structure, on the basis of its level within the peat, it may be roughly contemporary with Derryfadda 210.

Derryfadda 201 (Fig. 10.14)

Trackway; 390–190 BC; 4.80m x 2.50m; roundwood and brushwood (*Fraxinus, Alnus, Pyrus/Malus, Ilex, Salix, Betula* and cf. *Sorbus*); cutting season autumn (68%); marginal forest environment; construction and use phases; Related to the non-archaeological site DER92; Site code DER201; licence 96E202; NGR E222727 N165762; 125.25m OD (lowest level).

Cara Murray

This trackway was constructed of irregularly laid and densely packed brushwood. The wood was very straight and regular, measuring 0.30–1.70m in length and 20–40mm in diameter. A BnM drainage pipe had disturbed material to the northwest of the structure. It did not appear to have significantly disturbed the site, however. As it survived, the main extent of the structure measured 5.02m northeast–southwest and 3m northwest–southeast, with some associated wood on the northern side of the site that may have been disturbed either in antiquity or by the concrete drain pipe.

Discussion

The site was located 0.18m above the mineral soil on a small 0.30m high knoll on the western side of the southeastern underlying glacial ridge. Part of the eastern side rested on the fine brown grey clay with sand and fine gravel inclusions that formed the knoll, while the remainder rested on *Sphagnum* peat with forest debris inclusions. A large birch root system bounded the eastern side of the structure. The location of this structure, at this high level on the glacial ridge, seems to indicate the use of this area prior

to the collapse of the marginal forest. Large non-archaeological alder related to the non-archaeological site DER92 overlay the site, which represented this tree collapse as the raised bog encroached on the area. This encroachment must have begun towards the end of the Iron Age and continued to develop into the Early Christian period.

Derryfadda 203 (Fig. 10.15)

Trackway; 385–165 BC; 15m x 3m; brushwood and roundwood (*Alnus, Betula, Fraxinus, Corylus, Salix, Pyrus/Malus, Ilex* and *Quercus*); cutting season autumn (46%); marginal forest environment; construction and extension phases; site code DER203; licence 96E202; NGR E222759 N165928; 125.51m OD (surface).

Cara Murray and John Ó Néill

This site was a 15m long trackway with a 3m wide walking surface and a number of small deposits of wood at either end of the main structure. The trackway began in the peaty soils of the bog marginal forest on the slope down into the Derryville Bog system. It ran east–west from the southern side of an underlying glacial ridge, out towards the edge of the bog marginal forest. A series of root systems was present below and at the level of the site; one of these was a 0.12m diameter piece of oak that had been felled in summer, along the southern side of the track.

The eastern end of the track was considerably wider than the main body of the structure—up to 6.5m wide. Tree roots at the northern side provided a secure enough footing that the track continued south of the trees until it encountered more roots around 8m further west, where

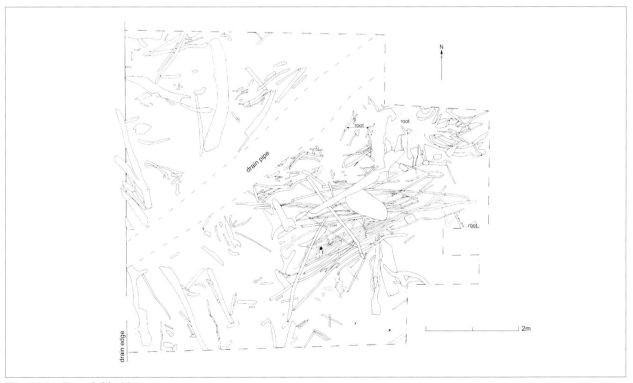

Fig. 10.14 Derryfadda 201.

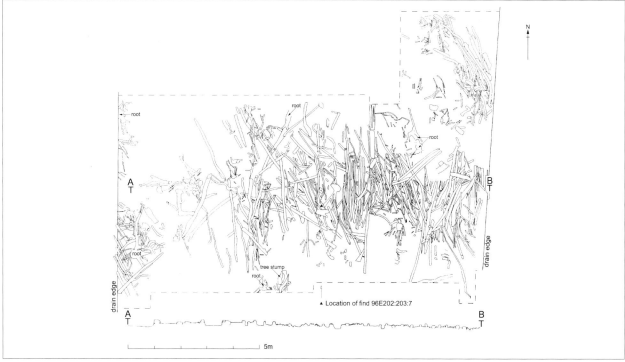

Fig. 10.15 Derryfadda 203.

there were few archaeological timbers. Beyond these roots, a small number of timbers represented the western end of the structure at the interface between the marginal forest and the raised bog. These timbers were aligned northwest-southeast and continued for 3m, their western limit having been destroyed by a BnM drain.

The walking surface was formed by transverse brushwood and roundwood, and the wood species used on the site included alder, birch, ash, hazel, willow, crab apple, holly and oak. A date of 385–165 BC was obtained from these timbers. In a number of places, individual timbers had been displaced well beyond the northern and southern limits of the walking surface. There was a large variation in the cutting season from spring through autumn, with almost half being felled in autumn. A 104mm long notched peg (96E202:203:7) was found among the timbers at the centre of the site. This had one end broken and the other trimmed along the shaft to leave a pronounced head. Most of the wood used on the site was in good condition, although wood on the surface was fairly degraded.

Discussion

This track provided access to an area of the bog marginal forest on the underlying ridge. The felling seasons of the wood appear to indicate that the trackway may have been constructed over spring, summer and autumn, which is supported by the felling of an oak alongside the trackway in summer. Alternatively, the wood may have been stockpiled or reused from elsewhere. The condition of the wood suggests the track may have been in use for a small number of seasons.

Derryfadda 204 (Fig. 10.16)

Trackway; 2120–1755 BC; 4.10m x 2.37m; roundwood and brushwood (*Corylus, Betula, Alnus, Quercus, Fraxinus, Salix, Taxus* and *Pinus sylvestris*); cutting season mixed; fen environment; construction phase; site code DER204; licence 96E202; NGR E222680 N165749; 124.84m (surface).

Cara Murray

Located on the western side of the southeastern underlying glacial ridge, this track was formed by six deposits of roundwood and brushwood, oriented northwest-southeast. Individual timbers ranged from 0.75m to 2.13m in length; many were in fragmentary condition. Surviving woodworking on these timbers was also in poor condition. The timber occurred in well-packed deposits, with irregular gaps between them, which would seem to have been intended to give a surer footing for walking across this part of the bog.

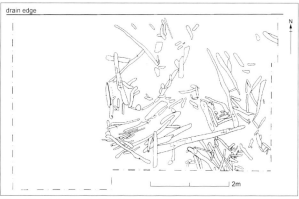

Fig. 10.16 Derryfadda 204.

Discussion

The site was located in very well-humified fen peat containing bog bean, an indicator of extremely wet conditions. Some small fragments of stone were deposited within the site, as part of either its construction or use. The wood identified as forming part of the structure was more mixed than that from the later prehistoric sites. As well as the usual wetland indicators, drier species were also found, along with pine, which appeared to have occurred as stands within the fen environment. All of the wood within this structure was in very poor condition. This was a feature of structures within the fen peat environment, as at the other Early Bronze Age sites, Derryfadda 218 and Derryfadda 207, located in the southeastern area.

Derryfadda 206 (Fig. 10.17; Plate 9.3)

Trackway; 785–375 BC; 5m x 1m; brushwood and roundwood; marginal forest environment; construction phase; site code DER206; licence 96E202; NGR E222779 N166027; 125.40m OD (surface).

Cara Murray and John Ó Néill

This structure ran northwest–southeast for 5m, with a 1m wide walking surface. It was the most northerly of a group of sites on an underlying glacial ridge, on the eastern side of the bog. A large number of alder root systems were present below the site, during and after construction, perhaps indicating the presence of alder carr rather than the more mixed marginal forest that had developed by the time of the construction of Derryfadda 214 and Derryfadda 6.

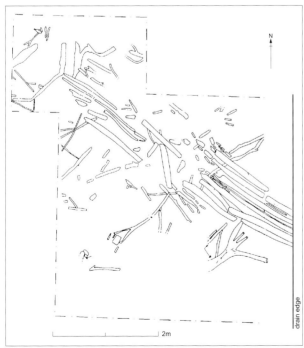

Fig. 10.17 Derryfadda 206.

The walking surface of the structure consisted of roundwood laid longitudinally among alder roots. Only one structural level could be identified. The wood was in poor condition and often broken due to later root disturbance and the differential compaction of the forest peat below the site.

This structure acted as a short trackway within the bog marginal forest, probably alder carr in this case, that had grown around the northern slope of a glacial ridge on the eastern side of the bog.

Derryfadda 207 (Fig. 10.18)

Trackway; 2205–1795 BC; 6.22m x 3.40m; roundwood, brushwood, pegs and timbers (*Alnus, Fraxinus, Quercus, Betula, Pyrus/Malus, Salix, Ilex, Taxus* and *Corylus*); cutting season summer (50%); fen environment; construction phase; site code DER207; licence 96E202; NGR E222677 N165772; 123.87m OD (base).

Cara Murray

This trackway was located on the western side of the southeastern glacial ridge and was oriented approximately northeast–southwest. As excavated, the track measured 6.22m northeast–southwest and 3.40m northwest–southeast. The northern limits of the site could not be traced due to the high level of the water table, which caused continual flooding in the cuttings. These conditions also prohibited the entire excavation of the substructure, and, consequently, it was only possible to excavate and record a 2m x 3.80m subsample of this portion of the site.

The trackway was constructed within the fen environment using non-archaeological wood to form a secure foundation on which to construct the site. As exposed within the subsample area, it was comprised of irregularly laid longitudinal and transverse timbers, which did not form any cohesive pattern and which were interspersed with fragments of brushwood and wood chips between the larger timbers. Above this, the superstructure measured 6.22m long, *c.* 3.40m wide and about 0.15m in depth. The densest section of the track, which formed the 1.96m wide walking surface, was composed of longitudinal roundwood. Some of the timbers at the western limit of the site were roughly set at right angles to the main body of the structure and appeared to have formed its western limit.

Discussion

The site was located within moderately humified fen peat containing forest debris. The wood identified as forming part of the structure was more mixed than that from the later prehistoric sites. As well as the usual wetland indicators, drier species were also found, along with pine, which appears to have occurred as stands within the fen environment. All of the wood within this structure was in very poor condition. This was a feature of structures within the fen peat environment, as at the other Early Bronze Age sites in the area, Derryfadda 218 and Derryfadda 207.

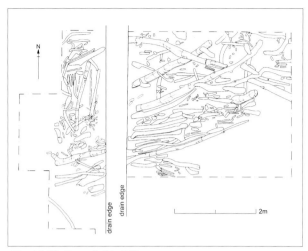

Fig. 10.18 Derryfadda 207.

Derryfadda 208 (Fig. 10.19)

Archaeological wood; 365 BC–AD 5; 5.07m x 1.71m x 0.15m; roundwood and brushwood; marginal forest environment; deposition phase; site code DER208; licence 96E202; NGR E222696 N165728; 124.63m OD (worked wood).

Cara Murray

This material consisted of three worked timbers located amongst roots and fallen wood within the marginal forest on the western side of the southeastern underlying glacial ridge. It consisted of two pieces of roundwood and one piece of brushwood, which had axe marks on one end. All the pieces were *in situ*. They were 0.36–1.71m long and 15–100mm in diameter.

Discussion

The location of this material on the western side of the southeastern glacial ridge was of particular interest, as it indicated the use of the forest without any need for the construction of sites. The finding of archaeological wood in areas difficult to access gives only a partial understanding of local activities and does not illuminate activity where there was no need to supplement the area with archaeological wood. In a sense, the driest areas would have been used as much as possible, with the wet areas only in use when necessary or when conditions made activity feasible. This site, however, is an important instance of obvious activity without the need for a cohesive structure.

Derryfadda 209 (Fig. 10.20; Plate 14.5)

Stake row; 990–770 BC; 38m x 0.36m x 0.45m; stakes (*Alnus, Betula, Fraxinus, Pyrus/Malus, Ilex, Prunus avium, Sorbus* and *Taxus*); cutting season mixed; marginal forest environment; construction phase; DER15 is 0.45m below DER209; Site code DER209; licence 96E202; NGR E222680 N165739–E222691 N165785; 124.58m OD.

Cara Murray

This stake row consisted of thirty-eight stakes, which ran for a total length of 38m. There were a number of associated root systems and some small brushwood, which appeared to be non-archaeological. Located on the steep western side of the southeastern glacial ridge, west of Derryfadda 211, the site was oriented approximately north–south and curved northeast at its northern limit and southeast at its southern limit. As excavated, the thirty-eight stakes were set in single file, 0.70m apart on average, at an angle of about 78° to the field surface. The stakes were about 55mm in diameter and an average of 0.38m in surviving length. Some small pieces of brushwood were laid horizontally on the field surface at the excavated level of the site, but they did not form any cohesive structure and may not have been associated with the site. Two large root systems were associated with the site, but it was not possible to determine whether the stakes in this area abutted or were overgrown by the root systems.

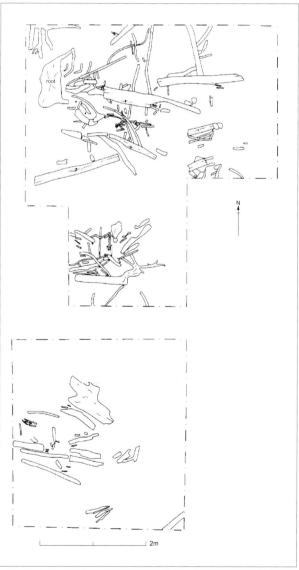

Fig. 10.19 Derryfadda 208.

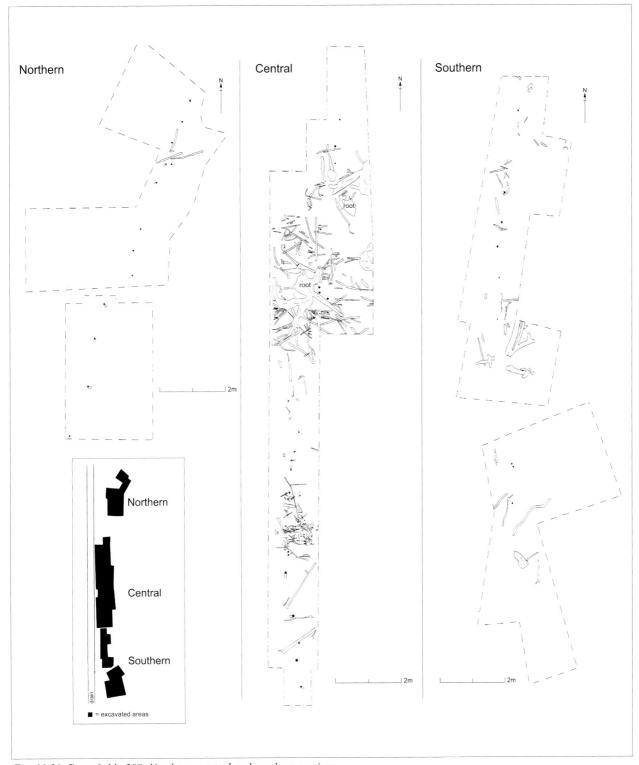

Fig. 10.20 Derryfadda 209. Northern, central and southern sections.

Discussion

It is very difficult to determine whether a site like this was constructed as a single event. One of the stakes survived to its rotting level (124.58m OD; 0.35m long), suggesting that this was the original rotting level of the stake row. The structure would have functioned as a border or fence to demarcate two zones for either safety reasons or use. Dating to the Late Bronze Age, this site represents the earliest indication of territorial demarcation and delineation within the study area. Given its location on the edge of a glacial ridge, it is most likely that the structure separated a safe zone or edge of the marginal forest to the east from an area of fen peat and encroaching raised bog.

Derryfadda 210 (Fig. 10.21)

Trackway; 515–365 BC; 5.96m x 4.02m; brushwood, round-wood and timber (*Betula, Corylus, Alnus, Fraxinus, Salix, Pyrus/Malus, Ilex* and *Prunus avium*); cutting season autumn (60%); marginal forest environment; construction and extension phases; site code DER210; licence 96E202; NGR E222685 N165748; 124.87m OD.

Cara Murray

This site was located on the steep western side of the southeastern glacial ridge, immediately west of the western limit of trackway Derryfadda 13. It had an overall area of 6m east–west by 4.80m north–south, with an insubstantial superstructure approximately 0.10m deep and a substantial broken brushwood substructure 0.20–0.30m deep.

A small deposit of roundwood formed the eastern limit of the site. The main structural component of the track consisted of a dense deposit of brushwood, which survived in a very fragmentary state, having been very badly damaged during the initial use of the site. The brushwood was very regular in size and form, with diameters of 25–45mm and lengths of 0.15–0.60m. The brushwood had two main orientations: northwest–southeast and northeast–southwest. When it was no longer possible to cross this structure, a supplementary deposit of five large pieces of roundwood were placed in irregular positions above the fragmented brushwood to extend the use

of the site. These pieces of roundwood were oriented northeast–southwest, longitudinally across the site, and were 0.20–0.30m apart. The wood was 0.60–4.10m long and 60–180mm in diameter.

Discussion

It was impossible to determine the original form of this structure, as the extensive use of the site had destroyed much of it. However, the mixed orientation of the timbers suggests that the structure was constructed using longitudinal brushwood and roundwood with similar transverse elements to lend stability to the deposit. Although the material was very fragmentary, the destruction caused by its use appeared to be fairly even, as all of the wood was equally badly damaged. This suggests that the structure was either used frequently by heavy traffic or incorrectly constructed for its local environment. This small track can be stratigraphically linked to Derryfadda 13, which occurred on its eastern side, although the association and integration of the sites could not be securely established. It can also be stratigraphically linked to the stake row Derryfadda 209 on its western side, and it immediately overlay the small track Derryfadda 211. It was bounded by natural wood on its northeastern and southern sides, with evidence of very wet peat having occurred on its western side. The peat associated with the eastern end of the site was well-humified, very wet, mossy *Sphagnum*; a similar moderately humified mossy *Sphagnum* peat, with forest debris inclusions, was present on the northwestern side.

Derryfadda 211 (Fig. 10.22)

Trackway; 1265–910 BC; 5.62m x 2.51m x 0.30m; brushwood and roundwood; marginal forest environment; construction phase; site code DER211; licence 96E202; NGR E222685 N165748; 124.68m OD.

Cara Murray

This track consisted of two separate deposits of mixed brushwood and roundwood, set approximately at right angles to each other, in an L-shaped form. The wood within each deposit was laid parallel, and the northern deposit was oriented northeast–southwest, while the southern deposit was oriented northwest–southeast. A brushwood and roundwood substructure was located below the southern deposit. The southern deposit, which measured 3.50m x 1.50m, was more concentrated and compact in construction, with more regular timbers, 1.23–3m long and 48–160mm in diameter. The northern deposit, which measured 3m x 2.4m, was constructed of irregularly laid roundwood and brushwood, 0.30–2.66m long and 63–150mm in diameter. Many of the timbers showed signs of branch trimming. A substructure was unnecessary as a root system underlay this portion of the site.

Fig. 10.21 Derryfadda 210.

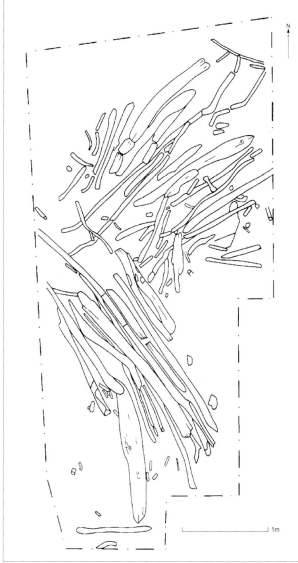

Fig. 10.22 Derryfadda 211.

Discussion

This site was constructed within a pool of reed and bog mosses, and appears to have been overgrown by mosses and cotton grass, about 30cm below the basal level of Derryfadda 210. It would also appear to pre-date many other sites, such as Derryfadda 13 and possibly even Derryfadda 201, associated with this level in the overall stratigraphy of the area. It is more difficult to stratigraphically link the site with sites further afield, although there is a possible association with the material of Derryfadda 213.

Derryfadda 212

Archaeological wood; no date; 1.5m x 0.6m; brushwood (Corylus); raised bog environment; deposition and destruction phases; site code DER212; licence 96E202; NGR E222776 N165890; 125.51m OD.

Cara Murray and John Ó Néill

This site produced two overlapping pieces of hazel brushwood, which showed evidence of felling and branch trimming. No structure could be identified with the timbers, which lay in raised bog peat just to the southeast of Derryfadda 9 on an underlying glacial ridge. An area around the two timbers was examined for other associated evidence, but none was found. As the timbers were exposed on the modern surface of the bog by peat extraction, it is likely that all other traces of a structure had been destroyed.

Discussion

This site provided further evidence for activity on the glacial ridges on the eastern side of the Derryville Bog system. As up to 0.4m of peat had formed within the marginal forest, between the construction of Derryfadda 9 and these two timbers, they are likely to be several centuries later in date.

Derryfadda 213 (Fig. 10.23)

Platform; 1315–915 BC; 4.10m x 3.5m x 0.57m; roundwood and brushwood; fen and marginal forest environment; construction, use and destruction phases; site code DER213; licence 96E202; NGR E222693 N165728; 124.30m OD.

Cara Murray

This site was located in the southeast of Derryfadda on the western side of the southeastern glacial ridge. It was above the non-archaeological site DER90, which produced the dendrochronological date 2718±9 BC. This material was located immediately below the site as a result of bog burst A of 2200 BC. The overall area of the site measured 8.50m (north–south) by 3.80m (east–west). Within this area, the main body of the site consisted of the wooden platform, which measured 4.10m (east–west) by 3.50m (north–south). The timbers were 0.80–4.10m long and 140–40mm in diameter.

It was constructed predominantly of parallel roundwood, which overlay a deposit of irregularly laid brushwood and roundwood, with a mix of forest floor debris and foundation material up to 0.55m deep. To the northeast and southeast, two small irregular deposits of loosely scattered timbers had been displaced from the main area of the site during seasonal flooding episodes, as had a similar deposit of brushwood to the southeast. The peat associated with this site was moderately to poorly humified, with a strong *Sphagnum* presence, indicating that the overall area of the site was very wet.

Discussion

This was one of three Late Bronze Age sites in the area. All of the material was set in wood-rich fen peat, indicative of a wet fen wooded environment. Of these sites, the environmental evidence suggests that the site was constructed prior to bog burst B (1250 BC) and that the stone track Derryfadda 311 was constructed afterwards, when the area had dried out sufficiently.

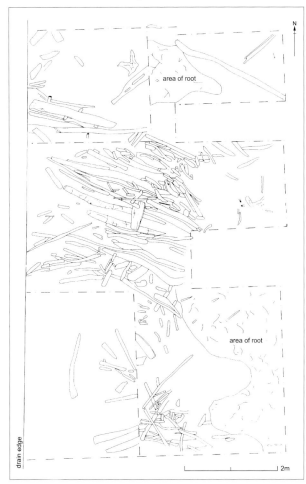

Fig. 10.23 Derryfadda 213.

Derryfadda 214 (Fig. 10.24)

Platform; 390–75 BC; 5m x 3.5m; brushwood and roundwood (*Alnus, Corylus, Betula, Fraxinus* and *Pyrus/Malus*); marginal forest environment; construction phase; site code DER214; licence 96E202; NGR E222776 N166001; 126.10m OD (surface).

Cara Murray and John Ó Néill

This structure measured 5m east–west and 3.5m north–south. It was one of a group of structures overlying a glacial ridge on the eastern side of the bog and was the last to be built, being at a slightly later stage in the development of the area than Derryfadda 6. A large number of root systems below the structure indicated the presence of marginal forest cover and irregularly laid brushwood and roundwood was deposited among these. The area in which this wood was deposited was defined by the surrounding root systems, rather than crossing an area between them.

Only one structural level could be identified and the wood used was in fairly good condition. A BnM drain had removed a substantial portion of the centre of the site.

Discussion

This structure acted as a platform within the bog marginal forest, which had grown around the northern slope of a glacial ridge on the eastern side of the bog. The site was located around 50mm higher up in the peat than Derryfadda 6, which was 4m to the north. The rate of peat growth may indicate a gap of less than a century between the two. As the surviving timbers of the structure were in fairly good condition, the site may not have been exposed for very long.

Derryfadda 215 (Fig. 10.6)

Trackway; 457±9 BC; 17m x 2m; brushwood and roundwood; marginal forest environment; construction phase; related to Derryfadda 6; site code DER215; licence 96E202; NGR E222778 N166013; 125.93m OD (surface).

Cara Murray and John Ó Néill

This structure ran approximately north–south for 17m and was up to 2m wide. It lay among a group of sites overlying a glacial ridge on the eastern side of the bog and directly overlay *fulacht fiadh* Derryfadda 216. A platform, Derryfadda 6, was built directly on top of its southern end. A number of root systems were present below the site, during and after construction, indicating the presence of marginal forest on the ridge.

The walking surface had been built using roundwood and brushwood in various orientations and a number of

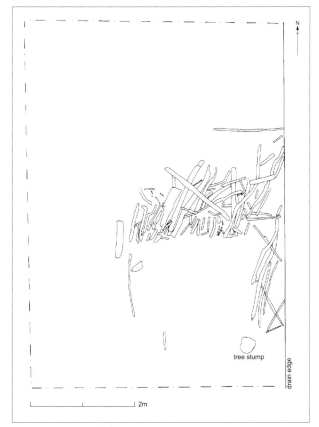

Fig. 10.24 Derryfadda 214.

uprights had been used to hold parts of the structure in place. One of the larger trunks used on the structure provided a felling date of 457±9 BC. The ordering of the wood used on the structure was very random and only one structural level could be identified. Where worked ends survived on the wood, the character of the individual facets was similar to those on Derryfadda 6. Despite this, the wood survived in a fairly poor condition and the site may have been exposed for a substantial period of time.

Discussion

This structure acted as a short trackway within the bog marginal forest, which had grown around the northern slope of a glacial ridge on the eastern side of the bog. It was built around 457±9 BC and was exposed for a substantial period of time, after which platform Derryfadda 6 was built over the southern end. The evidence from the woodworking suggests that the same type of axe was used on timbers from both structures, linking this structure to the Iron Age rather than the Late Bronze Age.

Derryfadda 216 (Figs 10.7, 10.25 and 10.26)

Fulacht fiadh; 1400–990 BC; 19m x 14m; trough type, plank-lined and rectangular (*Quercus*, *Pyrus/Malus*, *Fraxinus* and *Corylus*); pre-bog surface and fen environment; construction, use and reuse phases; site code DER216; licence 96E202; NGR E222777 N166009; 125.53m OD (top of trough).

Cara Murray and John Ó Néill

This *fulacht fiadh*, located overlying a glacial ridge on the eastern side of the bog, consisted of a timber-lined trough, a hearth and a mound of firing debris scattered over a wide area. The trough began as a pit dug into the subsoil and lined with a mixture of split planks and roundwood. During the use of the site, a mound of firing debris accumulated around the north and west of the trough, where fen peat had begun to form over the ridge. The spread of mound material was very irregular and covered an area 14m (north–south) by 19m (east–west). It consisted of heat-shattered sandstone (90%) and limestone (10%) in a charcoal-rich loamy clay matrix; stone accounted for 90% of the mound's volume. In the area west of the trough, an initial deposit of firing debris was overlain by charcoal-rich peat on which a more scattered deposit of firing debris lay. As the peat also contained firing debris, it may be assumed that the lower deposit had sunk in extremely wet peat. There was evidence of silt channels in the top of the peat, and it was directly on top of these that the main body of firing debris accumulated. The mounds of firing debris were scattered, but excavation of the intervening areas did not reveal evidence of further troughs, suggesting that firing debris was deliberately dumped in these areas. There was evidence of a hearth, 1.2m long and 0.8m wide, east of the trough, consisting of a dense deposit of charcoal, 0.25m deep.

The trough

A pit measuring a maximum of 2.4m x 1.75m x 0.6m was dug into the subsoil, oriented east-northeast–west-southwest in a 6m wide depression. It contained the remains of a plank lining on the sides and base. In the corners of the trough, paired upright planks retained planks set along the two longer sides, with occasional roundwood added to the side lining. In two corners, both paired uprights survived, but in the other two, one survived and the other was evidenced by a post hole whose profile matched the upright beside it. No evidence for a lining on the two shorter sides of the trough survived. A number of timbers survived at the base, which appeared to be the remains of a basal lining. The planks were all radially split oaks, apart from one tangentially split ash; some crab apple and hazel roundwood were also used among the side planks. Using dimensions from the footings of the plank lining, the volume of the trough was calculated at 1,764 litres.

After a period in which the trough had become back-filled with firing debris, an east-west oriented pit, 1.84m x 1.24m x 0.35m, was dug in the same place. A small number of large stones were found at the bottom of this pit, indicating its continued use as a trough. This new trough had a capacity of 799 litres.

Technical assessment of the firing debris

The volume of the mound material was calculated on the basis of the various sections excavated through the mound and knowledge of the pre-mound land surface. Adjusting for the relative percentage of stone identified in the mound, the mound consisted of 41.32m³ of micaceous sandstone and 4.59m³ of limestone. The absence of stone in the trough meant that an approximate percentage of 27% was used to suggest a number of firings, this figure is based on the percentage from the two similar sized troughs at Killoran 240 and Killoran 253. The geological proportion of 98% sandstone to 2% limestone was also carried over from those two sites. This gave an estimated volume of 0.467m³ sandstone and 0.01m³ limestone in the original trough and 0.21m³ sandstone and 0.004m³ limestone in the re-cut trough.

As micaceous sandstone can be reused up to five times before being reduced below a rough 50mm limit (the average stone size in the mound), the volume of sandstone in the original trough was divided into the mound volume and multiplied by five (Buckley 1990a, 171). This gave a total of 443 firings of the trough, assuming that the geological ratio of each use oscillated within narrow parameters. Assuming the last 10% of the mound material accumulated from the use of the smaller trough, an adjusted total of 497 firings was calculated. In both cases, the limestone would only have been fired once, but this is an inherent property of the scale-down of the 90%:10% ratio of sandstone:limestone in the mound to 98%:2% in the trough.

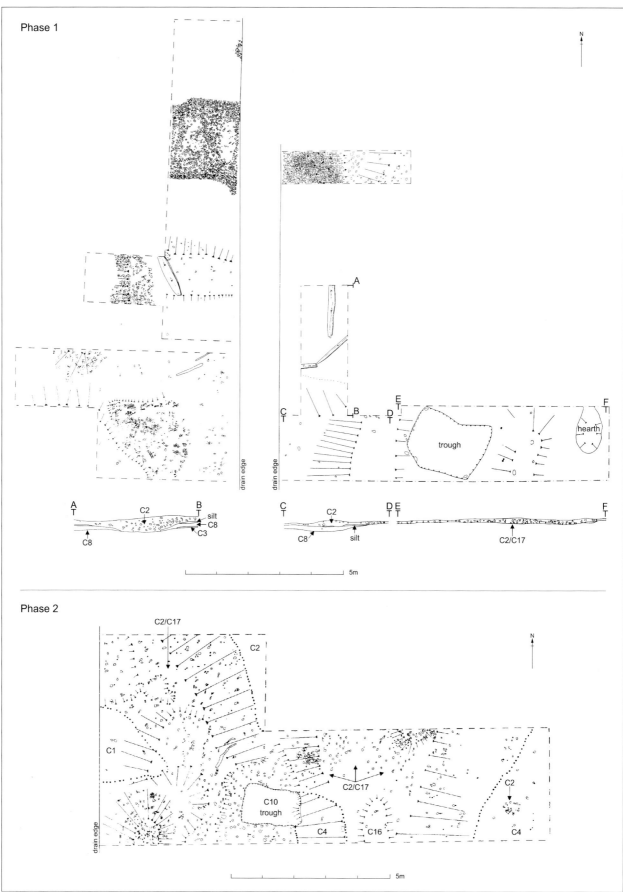

Fig. 10.25 Derryfadda 216, Phases 1 and 2.

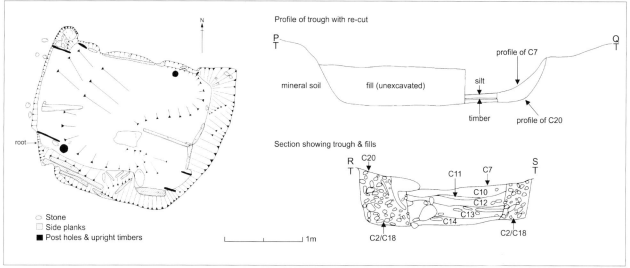

Fig. 10.26 Derryfadda 216, Phases 1 and 2, trough.

To have retrieved the volume of stone used on the site from the local glacial till would have required the extraction of some 103.3m³.

Discussion

The irregular shape of this mound requires some explanation. The location of the trough in a depression would create backfilling problems if firing debris were just scattered around the trough after use. The debris would continually accumulate in the depression and fall into the trough, as eventually happened. As the bog was a matter of metres away to the north and west of the trough, it may be that the firing debris was deliberately dumped in these locations, in much the same way as platforms were added to Killoran 253 and Killoran 265 as they got wetter.

The suggested number of firings of the trough is very much an approximation, but a figure of 400–500 firings would suggest that the site was in use for a number of years, which is likely given the poor condition of the timbers in the trough.

Derryfadda 218 (Fig. 10.27)

Platform; 2290–1935 BC; 4.29m x 4.85m x 0.30m; roundwood and brushwood (*Alnus, Fraxinus, Salix, Quercus, Pyrus/Malus, Corylus, Ilex, Taxus* and *Betula*); cutting season mixed; fen and marginal forest environment; construction, use and destruction phases; site code DER218; licence 96E202 extension; NGR E222776 N166001; 123.36m OD.

Cara Murray

The bog burst

This site was constructed at the top of the fen peat deposit after bog burst A of 2200 BC had reduced the water table levels in the area. The bog burst disrupted the peat and environmental material immediately below the site, where part of the sequential peat development had been removed. These effects can be seen in the associated dendrochronological dates of 3368±9 BC (Q9540) and 2718±9 BC (Q9401).

As a result of its location in an unusually dry environment, this site was in a very poor state of preservation and had suffered more extensive exposure to the elements in comparison with many of the other sites. In general terms, archaeological sites located at the top of fen were more highly degraded.

Phases

Two phases were identified on the site. The first was the construction and use of the site, associated with some fragmentary sherds of pottery. The site had been so badly destroyed that only the base of the structure survived, and this had been cut by a BnM drain. The site measured 4.85m east–west by 4.29m north–south, forming a crudely sub-circular structure. Most of the site survived on the eastern side of the drain, where three deposits of wood formed the structure.

The southern deposit of wood, oriented northwest–southeast, consisted of timbers 0.85–1.27m long and 0.06–0.16m in diameter. On the northern side, a similar concentration of wood was oriented northeast–southwest. This second deposit was less dense, with timbers 0.59–0.99m long and 0.04–0.18m in diameter. Some additional wood, including two larger, north–south oriented timbers and a small amount of east–west oriented brushwood, was located between these two deposits. All of this material was located at the base of the wood-rich fen peat, immediately above the fen peat material. Amorphous peat deposits were located to the east of this material, indicative of increased activity and the use of the area as a walking surface. On the western side of the drain, the site survived as a few pieces of brushwood of alder, ash and hazel. Two small sherds of pottery (96E202:218:1 and 96E202:218:2) were associated with this material.

The second phase was the destruction of the site, the continued development of the mixed marginal forest and the collapse of trees within the local environment. Some of this wood may in fact have fallen from the dryland ridge with the encroachment of the bog into this area, which also subjected the eastern area of the site to greater environmental influences because of the steep slope of the underlying glacial ridge and the closer proximity of the mineral soil. In places, mineral soil was located only 0.40–0.60m below the level of the site. With the growth and development of the drier marginal forest, alder saplings developed in this area and caused a high degree of root disturbance. The site was constructed at the level of the commencement of this mixed marginal forest; thus, the drier conditions that allowed the sites construction also allowed the development of the marginal forest that led to its destruction.

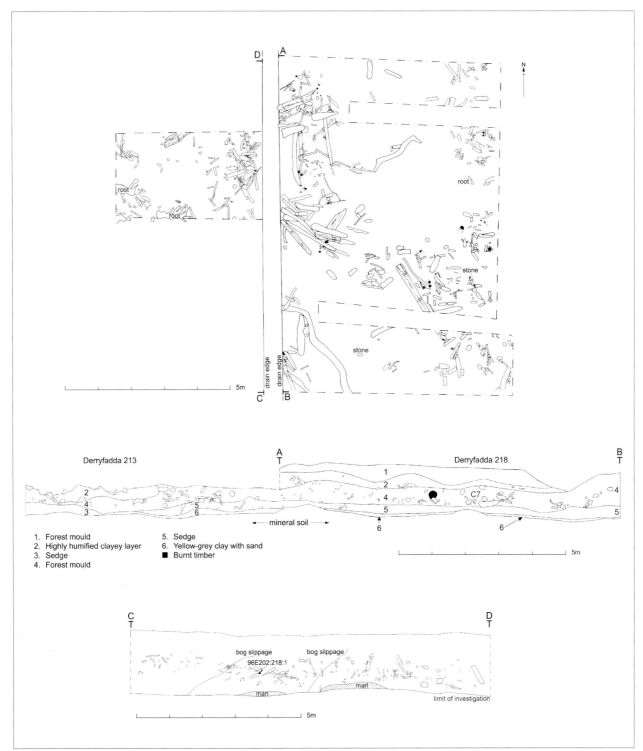

1. Forest mould
2. Highly humified clayey layer
3. Sedge
4. Forest mould
5. Sedge
6. Yellow-grey clay with sand
■ Burnt timber

Fig. 10.27 Derryfadda 218.

Derryfadda 311 (Fig. 10.28)

Trackway; 1450–1030 BC; 15.46m x 0.54m x 0.17m; stone and brushwood (*Alnus* and *Salix*); cutting season mixed; pre-bog surface and marginal forest environment; construction and use phases; site code DER311; licence 96E202 extension; NGR E222688 N165739; 125.17m OD (landfall).

Cara Murray

This site was constructed from the landfall on the western side of the southeastern glacial ridge, across the width of the marginal forest. On its landfall, the track was a 1.42m wide deposit of cobbles and small flagstones, which rested on the mineral soil. These stones comprised 60% cobbles, 30% boulders and 10% small flagstones, of which 80% were limestone (10% of which was decayed) and 20% sandstone. The stone deposit petered out on the mineral soil, where there was no necessity for the track. The main portion of the deposit curved very slightly off the mineral soil in a southwesterly direction. In this area, a deposit of dopplerite, formed of decayed stone and silt with some peat where water had gathered between the stones, overlay the stones in small pockets.

A wooden artefact, 96E202:311:1, was uncovered on the southern side of the site, sitting on a very thin mixed deposit of fen peat and mineral soil. The artefact consisted of a small wooden panel, broken at both ends, with a semicircular notch at one end and a keyhole shaped cut at the other.

The track narrowed as it crossed the marginal forest from 0.82m on its landfall side to 0.54m at its western limit. The nature of the track also varied. Flagstones were used more extensively, increasing to 40%, with cobbles and boulders forming the remaining 60% of the structure. At its western limit, the track was constructed of well-preserved flagstones, which were mostly set with the flat side upwards to form the walking surface. The stones ranged in size from 80mm x 0.16m to 0.40m x 0.23m and were 30–100mm in thickness. They were set within fen peat and utilised the dry root systems of the marginal forest. At the time when the site was in use, the edge of the marginal forest was located

approximately 2m west of the edge of this western limit.

During the use of the site, a small area of brushwood was used to supplement the flagstones on the western side of the site. This deposit measured 1.88m x 0.54m x 0.15m. The wood was 70–310mm long and 20–60mm in diameter. Peat associated with this deposit, and the track in general, was more highly oxidised than the surrounding forest peat as a result of the movement of stones during the use of the site. As such, this brushwood was deposited within a very short period of time after the construction of the site, probably within a season or so.

Discussion

The main structure of the track was laid down as a single event, and there was no evidence of repair or secondary work, in that there was no peat separating any of the deposits, including the brushwood deposit. The increased oxidisation of peat associated with the track was caused by the movement of stones in their use as a walking surface. Alder stools and root systems formed locally drier areas, which were utilised in the line of the track. Only where the brushwood was laid was there any localised wetness. This area became dry as a result of bog burst B of 1250 BC, enabling the construction of the track. In comparison, Derryfadda 213, to the south, was constructed after the bog burst, when the water table had begun to rise again.

Killoran 18 (Figs 10.29–10.39; Plates 4.1 and 14.4)

Causeway; 1440±9 BC; 555m x 1.1m (max 3.3m) x 0.15m (max. 0.35m); cobbles, boulders, stakes, brushwood, roundwood, gravel, clay and planks (*Alnus*, *Salix*, *Corylus*, *Fraxinus*, *Betula*, *Pyrus/Malus*, *Quercus*, *Ilex* and cf. *Sorbus*); cutting season autumn (49%); glacial till, marginal forest, fen and discharge channel environment; pre-construction, construction and repair phases; part of this site was also recorded as DER302, a branch track; site code DER18; licences 96E203, 96E298 and 96E298 extension; NGR E222243 N166233–E222797 N166199; 124.70m OD (eastern landfall).

Tim Coughlan and Paul Stevens

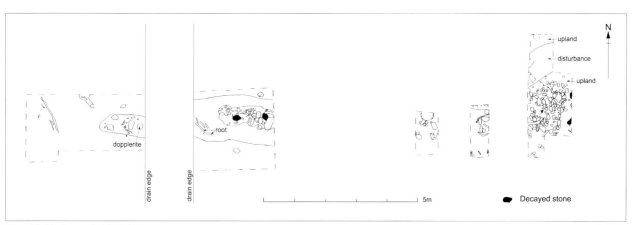

Fig. 10.28 Derryfadda 311.

This site was located in the centre of the study area of Derryville Bog, extending east–west for 555m across the Middle Bronze Age bog to dryland. Archaeological excavation over two seasons revealed the structure to be dated to before 1440±9 BC and exposed for a period of twenty to forty years. The excavation also revealed three phases and considerable variation in the overall building technique over its length. Analysis of the wood revealed a dominance of alder, with high levels of hazel, ash and willow. The majority of the trees were felled in autumn, with some summer and less spring felling (more common in the west).

Environmental context

The nature of Derryville Bog was heavily influenced by the underlying topography of the basin. A large raised plateau across the centre of the bog, tapering to the east, divided the northern bog from the southern and western bogs (Cooleeny in the west). During the Middle Bronze Age, a reed fen covered the central area of the bog, bordered by substantially deeper water to the north and raised bog to the south. Stones were used in the causeway to compress and flatten a matting of the tall reeds, which would otherwise have been impassable. Several glacial moraine ridges extended into the basin from the Derryfadda peninsula. The causeway originated on a glacial ridge in the east, but was also in part constructed from the west. The bog was surrounded by bog marginal forest up to 60m wide. Areas of burning were noted close to the eastern end of the causeway, suggesting a degree of clearance. The central area of the bog drained to the north by means of a 60m wide discharge channel of slow-moving water, located along the western margin of the bog; this was crossed using large deposits of wood. Worsening surface conditions brought about by the presence of this heavy wooden deposit effectively dammed the channel, resulting in the flooding of the entire causeway and the removal of part of the wooden crossing. The effects of the stream overflowing caused the rapid and unusually uniform growth of raised bog over the entire length of the track, rendering it impassable.

Pre-construction

Three isolated features were recorded during the excavation of Killoran 18. The earliest date produced from Derryville Bog of 3339–2924 BC related to an isolated degraded half-split ash timber (T37) located on the eastern side of the study area. This timber was recorded above a small glacial bank and gully feature, located within 0.25m of dark purple amorphous sedge and reed fen peat, indicating fen peat was developing in this area at this stage of the Neolithic. A second tangential split oak timber (T45) was recovered 30m east of T37. This timber was radiocarbon dated to 3020–2613 BC. The timber was located in 20cm of peaty, silty clay with some charcoal flecks, filling shallow hollows in the glacial till. Analysis of this clay revealed it to be naturally formed by fluvial/colluvial in-wash at the margins of the bog. An Early Bronze Age date of 2133–1548 BC was obtained from a burnt spread located 10–30m to the east of the eastern bog margin. The burning may be from an isolated informal hearth or a natural tree fire. There was no further evidence for activity on the ridge at this time.

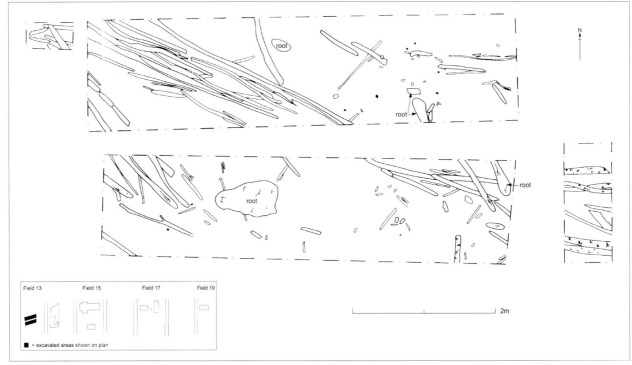

Fig. 10.29 Killoran 18, Field 13. Western terminus.

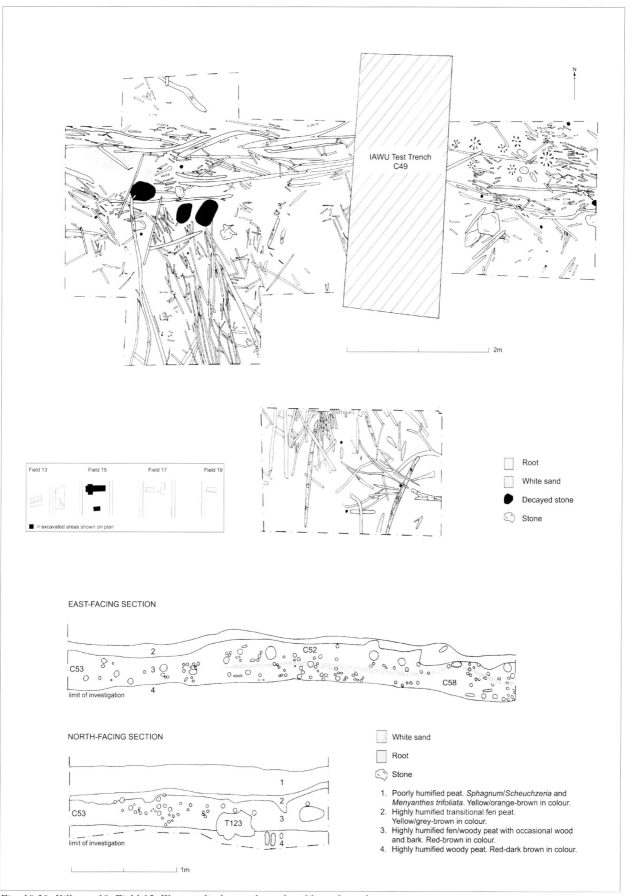

Fig. 10.30 Killoran 18, Field 15. Western discharge channel and branch track.

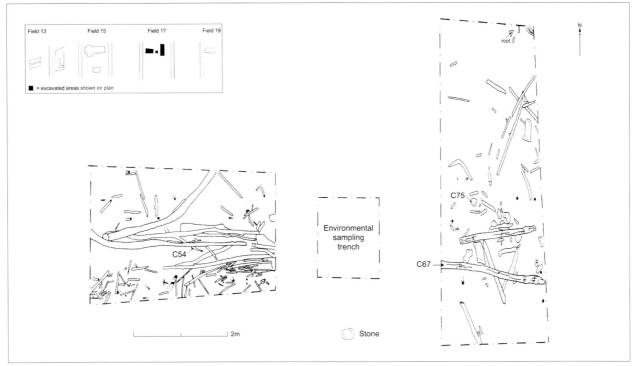

Fig. 10.31 Killoran 18, Field 17, western discharge channel.

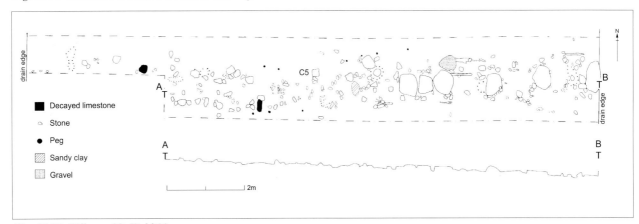

Fig. 10.32 Killoran 18, Field 23.

Construction 1

The line of the causeway was first set out using two parallel rows of stakes as marker posts in the peat. The double stake row was oriented east–west and measured 456–488m long, terminating 10–20m from the eastern landfall and at the main bog discharge channel to the west. The two rows were up to 1.4m apart and contained around 1,400 individual sharpened stakes. The stakes in each row were set on average 0.4m apart and were 60mm in diameter. This phase was radiocarbon dated to between 1745 and 1405 BC.

Construction 2

An irregular paved walking surface was then set onto an already well-worn surface of compressed peat using large and small stone, flagstones and occasional deposits of clay. The stone causeway across the bog was 1.2–1.4m wide and 484–486m long, set within the defined limit of the stakes. This phase of the causeway was dated to 1605–1285 BC from a timber directly below the stone; localised areas of wetter peat resulted in the need for intermittent use of a brushwood and sand foundation for the stones. The brushwood was laid in short stretches, 1.1m long and 1.4m wide, set between the stake rows. In the western trenches, the causeway was excavated as a loose stepping stone arrangement of small angular cobbles. In the central and eastern trenches, the excavation revealed a tightly packed pavement of stones and gravel throughout the structure. Most of the stone used was sandstone (80%); the remainder was limestone (20%), much of which was visibly decayed. The stone was sourced from the surrounding glacial till, which had similar proportions.

The causeway was largely made up of the same constructional elements across the width of the bog, varying

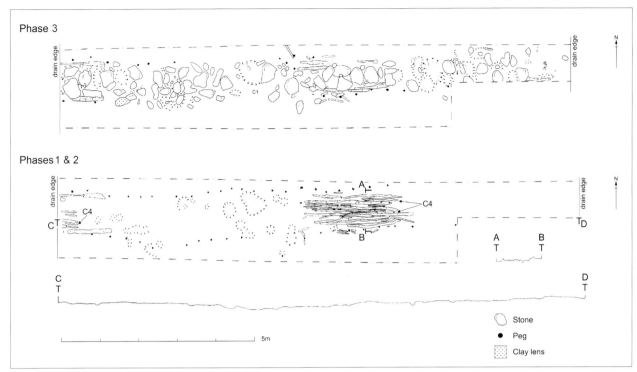

Fig. 10.33 Killoran 18, Field 39, Phases 1, 2 and 3.

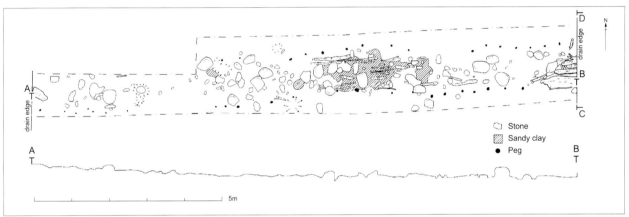

Fig. 10.34 Killoran 18, Field 41. For section C-D see Fig. 10.35.

only in density of stone and stone size. However, both ends of the causeway were notably different in size and were formed in response to the local environment. The western end of the causeway was constructed entirely of a deep deposit of longitudinal roundwood, occasional planks and brushwood, 8m long, up to 0.8–3.3m wide and up to 0.55m in depth. Here, the causeway crossed the much wetter discharge channel, often utilising dry root. The westernmost element of the causeway was a very poorly constructed deposit of brushwood oriented north-west-southeast, then northeast–southwest. This probably represented the remnants of a heavier construction washed away in a violent burst by the channel after construction. There was evidence that the heavy wooden construction blocked the channel and that the resultant flooding eastwards inundated the causeway, creating the uniform growth of raised bog across its length. Several

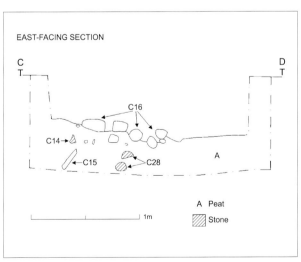

Fig. 10.35 Killoran 18, Field 41. Section C–D. For location of section see Fig. 10.34.

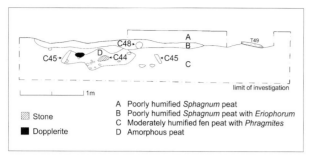

Fig. 10.36 Killoran 18, Field 40, eastern zone.

branch tracks of various lengths, the longest and most substantial (8m x 3m) of which was assigned site code DER302, split off from the main body in the west. These spurs may have represented aborted attempts to find an alternative route. A second, contemporary structure, Killoran 305 (see below), was excavated 10m south of the western end of Killoran 18. This was also a poorly constructed trackway, possibly part of the same process. The eastern terminus was a very well-defined 2.2m wide by 7m long metalled roadway that crossed from the bog onto the glacial till and veered northeast.

Repair

Later intermittent repair to the stone causeway, where elements sunk into softer peat, was also revealed in excavation. Oak planks, some with cut holes or mortises, were laid over transverse roundwood and between piles of stone, and longitudinal planks and roundwood were laid directly over the sunken causeway. This phase was dated to before 1440±9 BC by a plank recovered from the excavation.

Discussion

The features revealed prior to the construction of the Middle Bronze Age causeway provided some of the earliest evidence for human activity in and around Derryville Bog. The two Neolithic timbers may have been contemporary but indicated only in-washed residual material. The construction of the causeway merits more discussion.

Within the first construction phase, several uprights in the stake row were broken and others sealed by the stones of the paving or peat below the level of the stones. This suggests that the uprights were in the bog for some time, possibly up to five months, to allow enough time to rot to the level of the stone deposit. The excellent condition of the majority of the uprights was in stark contrast to the very rotten condition of the horizontal brushwood foundation deposit below the stones. The peat surface must have been very dried out or heavily compressed for the wood to be so damaged. The wood species profiles from the two samples produced a significantly similar pattern, indicating that the wood was sourced from the same forest. The wood was cut in the autumn, summer and spring, the majority in the autumn. Notably, the foundation wood was sourced exclusively in the autumn, implying that the stone phase was constructed at that time.

It is likely, however, that the cutting, sharpening and deposition of approximately 1,300 uprights in the bog only took a week or so. Whereas the collection alone of up to 32m^3 of stone (38,892 kg) would have been a substantial exercise requiring the quarrying by hand of a large area. No such quarry was found in excavation. It would not be unreasonable to assume that the stone phase of construction may have been undertaken over a period of several months or seasons, a significant time period that would have ensured that the stakes were no longer visible by the conclusion of the operation. Significantly, no evidence of dung beetles was found in the causeway, which means the entire operation was carried out by hand, without the aid of animals.

The repair to the causeway occurred intermittently across the length of the track, sealed in raised bog and transitional peat stratigraphically later than the amorphous peat on which the stone surface was laid. This may have been the result of different peat growth in the hollows created by the sunken stones and may not represent a time lag at all between construction, use and repair. However, it is more likely evidence that the structure was used for a considerable length of time and repaired during its use before

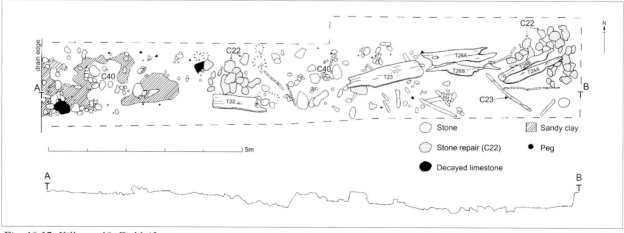

Fig. 10.37 Killoran 18, Field 45.

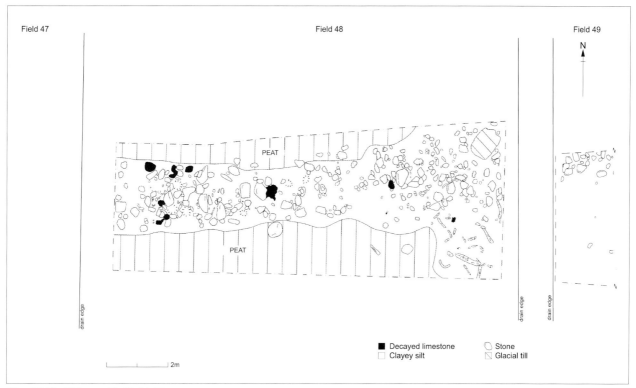

Fig. 10.38 Killoran 18, Field 48.

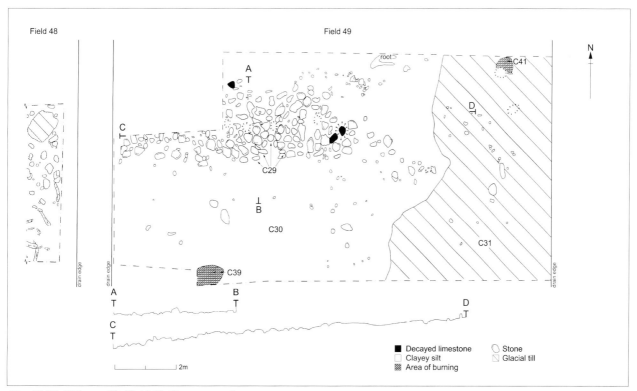

Fig. 10.39 Killoran 18, Field 49. Eastern terminus.

being flooded. This was confirmed by the use of various repair materials, such as piles of stone, planks and round-wood, which implied a deviation from the otherwise organised nature of construction. The wood and stone used for construction appeared to have been stockpiled prior to construction, as there was little deviation from the species and size of wood, or the type of stone used.

The western end of the causeway (wooden section) changed dramatically from a very heavy, multi-layered con-struction to a single layer of brushwood, seemingly dumped with no regularity. This might suggest that the causeway was not completed at this most western point and that the structure was therefore never in fact used. However, it is more likely that the area in question was constructed from the west using smaller stones and available wood; it was then removed or destroyed either naturally or deliberately. The presence of the discharge channel in this area would appear to be strong evidence for a surge of water during a wet period breaking through the dam caused by the structure and carrying all but the lowest material downstream.

Killoran 20

Archaeological wood; 1305–940 BC; 3.58m x 0.35m x 0.07m; roundwood (*Corylus*); cutting season autumn (three out of four); raised bog environment; construction and use phases; also recorded in survey as DER21; site code DER20; licence 96E202 extension; NGR E222670 N166367; 125.15m OD (eastern landfall).

Cara Murray

This site was located in the central area of the TMF footprint, west of the southeastern end of the stake row Killoran 54. The site was located in raised bog and consisted of six isolated pieces of roundwood, with no other archaeological timbers within the immediate area. As it survived, the site measured 3.58m x 0.22–0.35m. It had been bisected by a BnM drain. On the western side, the archaeological wood consisted of three pieces of roundwood, two of which were oriented east-west, overlaying the third, which was oriented northeast-southwest. They were 0.56–0.63m in surviving length and were 50–70mm in diameter. On the eastern side of the drain, there were three roughly parallel pieces of roundwood, oriented east–west, which were 0.69–0.92m+ in length and 50–70mm in diameter.

Discussion

This site was located within poorly to moderately humified raised bog comprised predominantly of bog mosses and cotton grass. The wood was set at an angle, being 70–80mm lower on the eastern side, where tool marks survived. It would appear that the site was constructed on the edge of a pool between two cotton grass hummocks. This site is unusual in that it occurred in isolation to other archaeological activity within the area. It formed one of a complex of sites constructed after bog burst B (*c.* 1250 BC), when this area became accessible for a short period of time.

Killoran 54

Stake row; AD 668–884; 406m x 4.5m; stakes (*Corylus, Fraxinus, Ilex* and *Ulmus*); cutting season autumn (75%); raised bog environment; construction and use phases; site code DER54; licence 96E202 extension; NGR E222409 N166548–E222568 N166345; 125.92m OD.

Cara Murray

This stake row was located in the central portion of the bog, north of causeway Killoran 18 and east of the main concentration of sites in Killoran townland. This unusual complex linear arrangement, which was in places up to four to five stakes wide but elsewhere more irregular, was traced for 406m of its length in a southeasterly direction. The arrangement of stakes varied in each field but, in general, it appeared that the stakes formed two parallel rows with a series of stakes placed in between. Unlike many of the other sites examined within the study area, this site was visible as small stakes projecting from the field surface. In many instances, only the remaining 20–50mm of the stake survived. It was evident that some of the stakes within the structure had been completely milled away.

The number of stakes surviving in each BnM field varied from eight to ninety-one. This should be seen as related to the destruction rather than the construction of the site. In the field with the largest number of surviving stakes, the stakes were set in two distinct concentrations. The more northerly concentration ran across the entire width of the field and consisted of an irregular series of stake rows, one to four stakes wide, 4.15m across. The stakes were structured in two main rows, with peripheral stakes located on either side. The distances between the stakes ranged from 0.16m to 2.0m. Two instances of double stakes were recorded.

Immediately to the south, the second concentration of stakes appeared as two separate stake rows, set in a zigzag-like arrangement at the western end and a more irregular arrangement on the eastern side, which was a maximum of 1.90m wide. Each row was roughly two stakes wide and the stakes were from 0.12m to 0.66m apart. One triple stake arrangement and one double stake arrangement were recorded. All of these stakes were set at varying angles of 5–85° off vertical, with the majority set at 20–45°. There was also a great range in the orientation of the stakes, although the majority of were oriented either northwest or southeast. The stakes ranged from 0.04m to 0.41m in surviving length and from 0.03m to 0.005m in diameter. Roughly 90% were worked to a chisel end, and all were very straight grained and regular in size.

At its maximum, the site was five stakes wide in the centre, tapering off in places to two to three stakes wide. Overall, the staked area was 0.60–3.30m wide. The distances between the stakes ranged from 0.10m to 1.27m. One triple stake arrangement and five instances of double stake arrangements were recorded. Elsewhere, the stakes were all set at varying angles to the field surface, with most set at 20–35°. The stakes ranged from 0.07m to 0.75m in surviving length and from 10mm to 40mm in diameter, with a modal diameter of 0.025m. Roughly 70% were worked to a chisel end and 20% to a wedge end. All were very straight grained and regular in size. One stake appeared to have a coppice heel.

Discussion

As the tops of these stakes were destroyed, the original environment in which they were located has been removed by milling. The field surface surrounding the stakes was comprised, for the most part, of poorly humified *Sphagnum* peat. The site has been dated to AD 668–884 and the surviving field surface dates to *c.* AD 500. No close parallel is known for a contemporary stake row of this complexity.

It has been noted with interest that the site ran roughly parallel to the Kilkenny-Tipperary border to the east. In addition, the underlying topographical survey indicated that the site ran along the line of an underlying ridge, sloping away to the northeast of the line of the site. This ridge would have created different environmental conditions at the level of the ridge that would have continued to be reflected in the area, albeit to a lesser extent, as the peat developed.

Killoran 56

Archaeological wood; no date; 0.36m x 0.40m x 0.05m; brushwood (*Fraxinus* and *Corylus*); raised bog environment; construction and use phases; site code DER56; licence 96E202 extension; NGR E222231 N165716; 124.24m OD.

Cara Murray

This site had been largely destroyed by BnM milling and survived as six isolated brushwood fragments. It was located in raised bog, within the northern discharge zone. No other material was visible in the vicinity. There was no surviving evidence of any woodworking on any of the brushwood. There was a lot of BnM disturbance in this (northern) part of the study area. Drainage channels had been cut east–west across many of the fields to further drain this area, which was one of the wettest areas of the bog.

Killoran 57

Trackway; possibly Middle Bronze Age; 5.5m x 0.37m x 0.12m; roundwood (*Quercus* and *Fraxinus*); raised bog environment; construction and use phases; site code DER57; licence 96E202 extension; NGR E222568 N165914; 123.87m OD.

Cara Murray

This site consisted of two isolated pieces of roundwood that were oriented northeast–southwest. These straight, regular pieces of roundwood were set side by side and were over 5.50m long, as traced. The individual timbers were 90–120mm in diameter. Despite the lack of visible axe marks, the straightness and regularity of these timbers and their narrowness suggested that the wood must have come from a managed environment. There was no environmental wood associated with this material.

Discussion

The site was constructed at the top of a deposit of raised bog that was well humified and almost clay like in consistency. A small pool, represented by a pocket of poorly humified *Sphagnum* peat about 1.50m in diameter, was associated with the site. The site was overlain by poorly humified, well-laminated peat, which was formed of cotton grass, leaves and heather (ling) root systems. These peat types indicate that the site was constructed at a time when the area had become quite dry. The site went out of use during a subsequent period of increased wetness, with periods of drier weather probably representing seasonal changes. The OD levels suggest a Middle Bronze Age date for the site.

Killoran 65

Archaeological wood; no date; 0.22m x 0.036m; brushwood; raised bog environment; site code DER65; licences 94E106 extension and 97E158; NGR E222272 N166573; 126.35m OD.

IAWU and John Ó Néill

This was a single piece of brushwood lying on the field surface. The wood was in very poor condition. One end had been worked into a point, but the wood was too damaged to identify the point type. The wood was situated in the raised bog. This site was in the vicinity of Killoran 66 and may be related to that site.

No structures were identified with this site, either during the original survey or during the 1997 excavations.

Killoran 66 (Fig. 10.40)

Possible hut site; AD 775–887; 5.6m x 4m; brushwood, roundwood and timber (*Fraxinus*, *Salix* and cf. *Sorbus*); raised bog environment; construction phase; site code DER66; licences 94E106 extension and 97E158; NGR E222254 N166590; 126.63m OD.

IAWU and John Ó Néill

This structure consisted of a sub-rectangular setting of uprights, including roundwood, brushwood and half-split posts, measuring 5.6m (east–west) by 4m (north–south). It was bisected by an internal setting of three stakes.

All but the western side was destroyed by peat milling between 1994, when the IAWU first recorded the site, and 1995, when they re-examined the western side and took samples for identification and dating.

The half-split posts ranged from 0.24m to 0.32m in diameter and were from 0.10m to 0.18m in width. They sat at angles of 40–50° to the vertical. Their setting in the ground, however, may have been altered by heavy machinery crossing over the site. The smaller posts ranged in diameter from 60mm to 200mm and sat at a variety of angles; two were almost vertical, while two others had been disturbed and were found lying horizontally.

All the posts had worked ends. The intervals between the post settings varied from 0.2m to 0.5m at either end of the structure. In two instances, pairs of half-splits were positioned immediately adjacent to one another. In the centre of the western side, two smaller posts had been set either side of the line of the larger uprights.

Further inspection of the site was undertaken in 1997, when potential archaeological timbers were identified in the field drain beside the site. Investigation of the area revealed the tips of three worked ends in the milled peat on the field surface, all of which were pencil ends of ash. This confirmed that the last traces of the structure had been destroyed and that the 1997 field surface had been lowered to 126.25m OD. Samples removed from root systems at this level included alder, birch and willow.

Discussion

Investigation of the area of the site between 1994 and 1997 established that a sub-rectangular setting of substantial half-split uprights and roundwood had been present on the site, which had been disturbed by heavy machinery and removed by peat milling. Wood identified on the site included ash, rowan and willow. The large split ash timbers and the sub-rectangular layout resemble the outline of a hut site. The absence of an associated floor level is not surprising, as the worked ends of the timbers were removed in the 0.3m of peat lost between 1994 and 1997. Excavations some 25m to the west identified an Iron Age trackway (Killoran 314) at a level contemporary with the initial surviving remains of this site, indicating that a substantial portion of the site had already been removed by 1994. The position of Bronze Age marginal forest roots at 126.10m OD, a mere 0.4m below the ends of uprights from an eighth or ninth century AD site suggests that as much as twice that amount of peat had already been lost. As such, the posts would have been inserted to a depth of around 1.2m; this would allow for an equivalent standing height, even in raised bog. It would then seem reasonable to conclude that the surviving remains were of the walls of a hut site.

Killoran 69 (Fig. 10.41; Plate 14.6)

Trackway; 838–799 BC; 69.5m x 1.5m x 0.35m; roundwood, brushwood and timbers (*Quercus, Fraxinus, Betula, Corylus, Salix, Alnus, Pyrus/Malus, Ilex* and *Taxus*); cutting season mixed; marginal forest, fen and discharge channel environment; construction phase; site code DER69; licence 96E298 extension; NGR E222203 N166351–E222283 N166350; 126.30m OD.

Paul Stevens

This site was located in the western quarter of the study area. Excavation revealed that the narrow linear trackway was constructed of brushwood and roundwood laid longitudinally, one layer deep. It originated in western

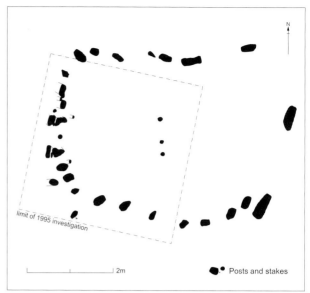

Fig. 10.40 Killoran 66 (after IAWU 1996a).

marginal forest as a narrow scatter of brushwood within a root system and extended northeastwards for 10m. It widened to a maximum width of 2.7m as it turned east–west, then continued east as a less substantial structure of closely packed longitudinal wood, 0.1m in average diameter and 1.1–1.65m in width. At its eastern extent, the trackway was truncated. Analysis of the wood revealed a dominance of oak with high levels of ash. The majority of the trees were felled in autumn, with some felling in winter.

Environmental context

The site was located close to the western bog margin, which, in this area, was undulating, forming small islands of drier ground. Although the trackway did not originate on dry ground, it did utilise an extensive root system within the marginal forest and a fallen oak also used by another trackway (Killoran 306). Killoran 69 crossed wet forest, fen and finally the western bog discharge channel. The end of the trackway was lost due to an erosion gully within this discharge channel. The roundwood from the structure was broken and displaced, with many carried off downstream, suggesting the wood was already waterlogged and brittle. The trackway was then totally submerged by peat.

Discussion

The wood from the trackway was from at least two sources: the majority was short lengths of small, mostly rotten oak quarter splits and ash significantly worn from long exposure (perhaps up to five years); the remainder was fresh brushwood from marginal forest rapidly submerged by peat. This could suggest either that some of the wood was reused from another location or that the structure lay exposed for some time and was resurfaced using local wood from marginal forest during its use. The

Coleoptera analysis of the peat below, within and above the track revealed that the local environment prior to construction was very wet fen, which became open water with floating mat vegetation. This implies the wood was indeed reused from elsewhere, although there was evidence that the structure was built shortly after bog burst C lowered the water table in the surrounding area.

The date of 838–799 BC for this structure came from a brushwood within the structure recovered during excavation. Killoran 69 was stratigraphically contemporary with and thought to be a continuation of trackway Killoran 306 (see below). However, the date range for Killoran 306 (756–407 BC) was well outside that of Killoran 69, even allowing for the use of reused material. The age/diameter profile and the species profile of the two structures also confirmed two different constructions, albeit similar designs and locations, travelling in different directions for different destinations. Killoran 69 appeared to have been constructed as far west as the timber platform at Killoran 306 (see below), the root systems acting as a firm footing further west. It did not appear to have been constructed for more than one person and was not built to last any great time. It was probably built to access the raised bog to the east, crossing the carr forest and western discharge channel.

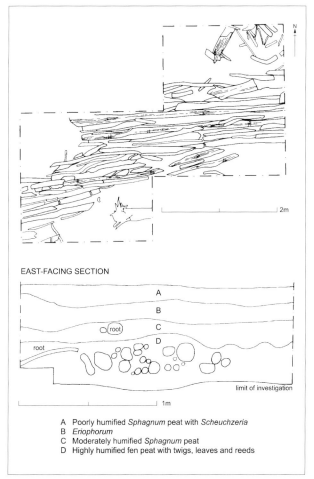

EAST-FACING SECTION

A   Poorly humified *Sphagnum* peat with *Scheuchzeria*
B   *Eriophorum*
C   Moderately humified *Sphagnum* peat
D   Highly humified fen peat with twigs, leaves and reeds

Fig. 10.41 Killoran 69.

Killoran 75 (Figs 10.42, 10.43 and 10.48; Plates 9.2 and 15.3)

Trackway; 385–50 BC; 45m x 2.5m x 0.5m; roundwood, brushwood, hurdle and pegs (*Corylus, Betula, Fraxinus, Salix, Alnus, Quercus, Ulmus* and *Pyrus/Malus*); cutting season autumn (51%); fen and marginal forest environment; construction, extension and repair phases; site code DER75; licences 96E298 extension and 97E160; NGR E222188 N166631; 126.71m OD (hurdle surface).

Paul Stevens and Sarah Cross May

Phases of construction and repair

This was a wooden track leading to a hurdle, which was built in four phases. Phase 1 of the track was of bundles of substantial roundwood (1.5m wide) running in an east–west direction for 30m. The bundles were heavily pegged and very decayed. This phase was followed by two phases of extension. In Phase 2, a hurdle was added to the eastern end of the structure—some years later judging by the relative condition of the wood. In Phase 3, a bundle of brushwood was laid on top, leading to trackway Killoran 312. This could have happened soon after the hurdle was laid. Later again, in Phase 4, more roundwood was added to the top of the structure, perhaps as a repair or a resurfacing. This occurred more towards the eastern terminus.

Environment

This track began on a firm base of alder carr which formed an arc around a pool of bog bean peat, formed in reaction to run-off from the dryland into the bog. The track crossed the centre of this pool and continued to a pool of rannock rush, formed by the combination of the alkaline mineral soil run-off with the acid water flow from the main discharge channel of the bog. This open water pool began just south of the track, near its eastern terminus, and ran about 50m northwards.

The track was designed to deal with these changing peat types. The section on the alder carr was less substantial than the section crossing the bog bean, as the tree roots provided a foundation. The hurdle began at the edge of the pool and was a perfect structure for bearing weight in such wet conditions. There were also alder trees immediately beside the track. These could have acted as anchors and foundations.

The hurdle

The hurdle was 7.2m long and the longest yet recorded in Ireland. Its weave was simple, single rods woven around single sails. The longest rod was 2.75m, and double rods came from overlapping the weave to increase the length. The sails averaged 30mm in diameter, the only double sail providing extra strength to smaller pieces of wood. There were seventeen sails in the hurdle, one of which was marked by two smaller pieces of brushwood. The average distance between sails was 0.33m with a minimum of 0.2m and maximum of 0.63m.

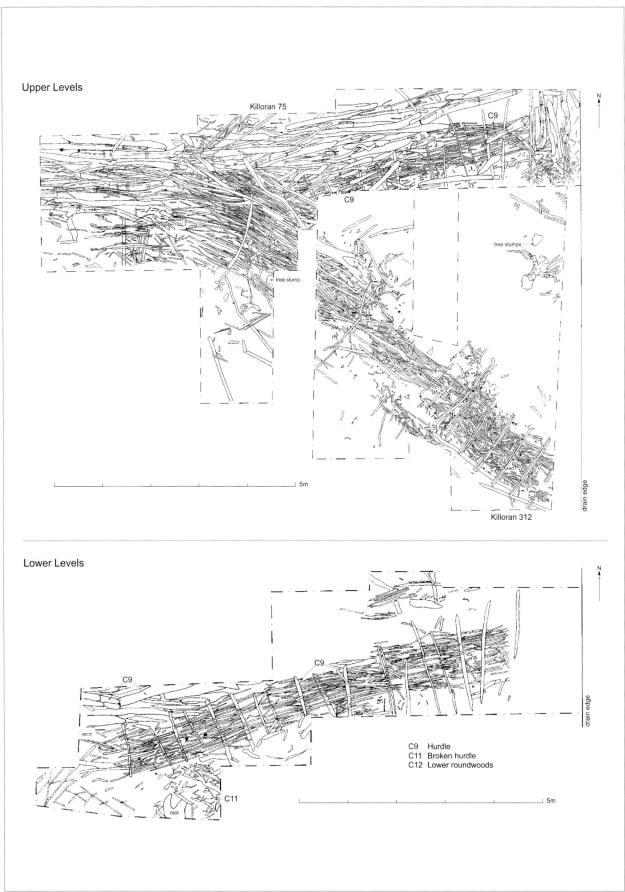

Fig. 10.42 Killoran 75 (upper and lower levels) and Killoran 312, Field 9.

Where the hurdle was covered by brushwood, it was very well preserved. Where it was exposed as a working surface, it was slightly damaged. Where the roundwood used for repair was laid on the exposed section, it was completely destroyed.

Due to the extreme wet conditions, this hurdle could not have been constructed *in situ*. Its construction on dryland could have been part of the regular construction of hurdles for many different uses, like fencing. As the western end of the hurdle was swiftly covered with brushwood, it seems possible that it was not purpose built for the track and could have been originally part of another structure. However, two factors argue against this. Firstly, the hurdle showed little damage in the area protected by the upper layers of brushwood. Secondly, moving such a long structure through alder carr and wet peat must have been difficult. If the full length were not specifically required, it would have been simpler to have chosen a smaller panel from the available stock.

While not part of the initial design of the track, the hurdle served a specific purpose that required its full length. It ran at a slight angle to the rest of the track. Its length was not used to extend the track but to provide a broad platform. From its position on the edge of the *Scheuchzeria* pool, it is likely that it served as a working base that allowed activities at the pool edge to be carried out by a number of people.

Killoran 223

Charcoal spread; medieval/post-medieval; 3.5m x 3.1m x 0.09m; marginal forest environment; site code DER223; licence 96E298 extension; NGR E222094 N166351; 127.60m OD.

Paul Stevens

This site was located in the west of the study area up against the westernmost margin of the present-day bog. It lay immediately north of Killoran 224 and was undoubtedly associated with this structure. It consisted of a large charcoal spread, roughly sub-circular in plan, which was patchy and thinning towards the edges. The structure measured 3.5m east-west by 3.1m north-south, petering out to the east; it was 0.09m deep and overlay silty peat. It was itself overlain by 0.8m of redeposited peat upcast from the excavation of a BnM outer drainage dyke. The site, therefore, probably represents post-medieval field clearance associated with the field boundary Killoran 224. Charcoal analysis revealed the wood to be well-grown oak.

Killoran 224

Bank; medieval/post-medieval; 30m x 2.2m x 0.12m; stones and compressed silty peat; marginal forest environment; construction phase; site code DER224; licence 96E298 extension; NGR E222087 N166348; 127.60m OD (peat surface).

Paul Stevens

This was a linear feature located in the west of the study area in the western margin of the present-day bog, immediately south of Killoran 223. It was composed of small stone boulders within a bank of redeposited blocks of silty peat, compressed to form a regular shallow concave bank. The bank, which was oriented east-west, was traced for 30m into the bog and was sealed by woody peat and a deposit of upcast peat and soil from a BnM drainage dyke.

Killoran 226 Structure I (Fig. 10.44)

Trackway; 450±5 BC and 460±9 BC; 15m x 4m x 0.35m; roundwood, brushwood, timbers and stakes (*Alnus*, *Salix*, *Fraxinus*, *Betula*, *Corylus*, *Quercus*, *Pyrus/Malus*, *Ilex* and *Ulmus*); cutting season autumn (76%); marginal forest environment; construction and repair phases; site code DER226; licence 96E298 extension; NGR E222192 N166344; 126.50m OD.

Paul Stevens

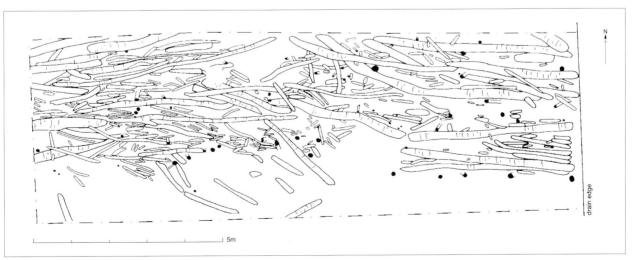

Fig. 10.43 Killoran 75, Field 8. Central zone.

Killoran 226 (Structures 1 and 2) was located in the west of the study area, close to the location of Killoran 69, 248, 301 and 306. The area appeared to represent near continuous use of a route through the marginal forest from the ninth to the second century BC, crossing a localised pool within a shallow north–south oriented trough, with firmer peat to the east. The excavation of this site revealed two distinct trackways—Structures 1 and 2—built on the same alignment but separated by three hundred years: the earlier trackway was constructed and then repaired. The analysis of the wood in Structure 2 revealed a dominance of willow, with high levels of alder and ash in the earlier phase. The majority of the trees were felled in autumn.

Construction

The trackway was heavy and roughly made with large longitudinally placed trunks interspersed with closely spaced longitudinal roundwood and brushwood, 0.12–0.02m in diameter. It was oriented east–west and measured 15m x 4m x 0.35m. The initial phase of construction was dated to 450±5 BC from a large oak trunk recovered from the excavation.

Repair

The trackway was resurfaced within ten years of its construction with a walking surface of closely packed, longitudinal roundwood (average diameter 0.06m) and occasional large split timbers secured by an irregular line of stakes. This construction extended the length and width of the previous structure to 19m x 3–4.4m x 0.16–0.5m. This phase was dated to 460±9 BC.

Environmental context

The undulating relief of the glacial till in this portion of the bog basin resulted in an uneven growth of peat in pools surrounding small islands of drier ground. The structure was constructed across a pool, an area of wetter peat growth, within bog marginal forest. The initial trackway was constructed by laying down heavy timbers over the peat and root systems to form a dry walking surface. A second bog system to the west that invaded the western margin of Derryville Bog induced a very wet area, which was flooded regularly, as evidenced by a build-up of highly humified forest mud. Additional wood was laid down over this mud to form a repaired surface, and several centuries later, a completely separate trackway (Structure 2). The structures were completely submerged by a build-up of laminated peat, deposited in successive flooding events. This was cut through by later erosion gullies. Finally, the hummock and hollow system of a raised bog formed.

Discussion

The initial purpose of this trackway was to form a narrow, single-file walkway across a wet pool in the carr forest; the trackway was later widened. The trackway did not appear to fully extend to the dryland in the west and terminated at a large root system in the east. Both phases were in poor condition on excavation and can therefore be assumed to have been in use as long as they were exposed; they were substantial in size and so built to last. The repair may have been up to twenty-five years later (450±5 BC versus 460±9 BC, although these dates do overlap). A single, unstratified radiocarbon date of 407–175 BC was produced as part of the initial survey.

Killoran 226 Structure 2 (Figs 10.44 and 10.45)

Trackway; 161±9 BC; 40m x 6.8m x 0.6m; roundwood, brushwood, timbers and stakes (*Alnus, Salix, Betula, Fraxinus, Corylus, Quercus, Ilex, Pyrus/Malus* and cf. *Sorbus*); cutting season autumn (49%); marginal forest environment; construction phase; site code DER226; licence 96E298 extension; NGR E222178 N166353; 126.83m OD (landfall).

Paul Stevens

Killoran 226 (Structures 1 and 2) was located in the west of the study area, close to the location of Killoran 69, 248, 301 and 306. The area appeared to represent near continuous use of a route through the marginal forest from the ninth to the second century BC, crossing a localised pool within a shallow north–south oriented trough, with firmer peat to the east. The excavation of this site revealed two distinct trackways—Structures 1 and 2—built on the same alignment but separated by three hundred years: the earlier trackway was constructed and then repaired. The analysis of the wood in Structure 2 revealed a dominance of willow, with high levels of alder and ash in the earlier phase. The majority of the trees were felled in autumn.

Construction

Structure 2 was constructed over the same line as Structure 1 but was substantially larger, measuring 40m x 6.8m x 0.35m. It was constructed on dry ground in a southeast–northwest orientation as a single layer of tightly packed, transverse roundwoods, 0.25–0.17m in diameter. This was laid alongside a small rise in the glacial till, over a lower deposit of transverse brushwood. It was laid directly over Structure 1, separated by a thin layer of mud. The new trackway veered east–west and extended across the line of the earlier track. It widened towards its eastern terminus, which was constructed of diagonal roundwood, brushwood and split timbers decreasing in diameter to the south (0.3–0.07m). The trackway terminated at a root system east of a pool. The trackway was dated, by dendrochronology of an oak plank recovered in the excavation, to 161±9 BC.

Environmental context

The undulating relief of the glacial till in this portion of the bog basin resulted in an uneven growth of peat in

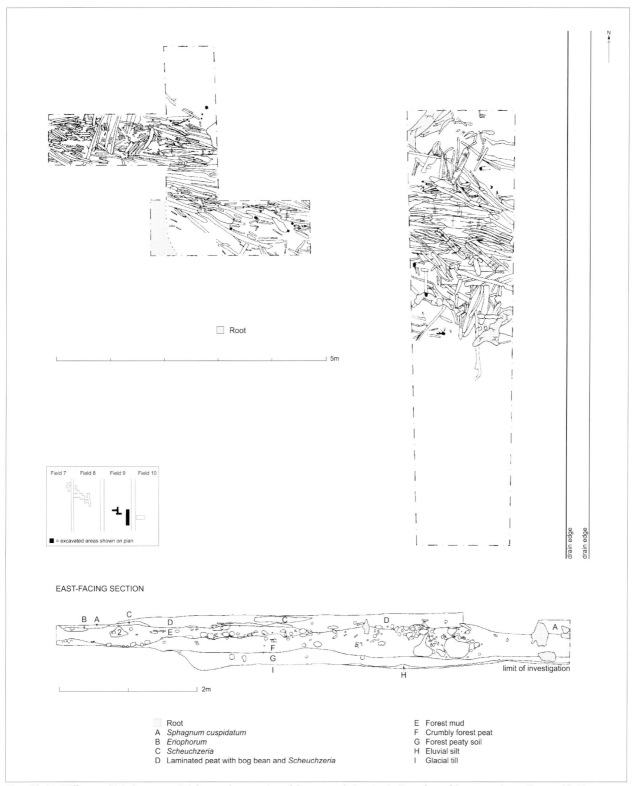

Root

5m

Field 7   Field 8   Field 9   Field 10

■ = excavated areas shown on plan

drain edge
drain edge

EAST-FACING SECTION

limit of investigation

2m

	Root		E	Forest mud
A	*Sphagnum cuspidatum*		F	Crumbly forest peat
B	*Eriophorum*		G	Forest peaty soil
C	*Scheuchzeria*		H	Eluvial silt
D	Laminated peat with bog bean and *Scheuchzeria*		I	Glacial till

Fig. 10.44 KIlloran. 226, Structure 1 (plan and section) and Structure 2 (section). For plan of Structure 2 see Figure 10.45.

pools surrounding small islands of drier ground. The structure was constructed across a pool, an area of wetter peat growth, within bog marginal forest. The initial trackway was constructed by laying down heavy timbers over the peat and root systems to form a dry walking surface. A second bog system to the west that invaded the western margin of Derryville Bog induced a very wet area, which was flooded regularly, as evidenced by a build-up of highly humified forest mud. Additional wood was laid down over this mud to form a repaired surface, and several centuries later, a completely separate trackway (Structure 2). The structures were completely submerged

Eastern terminus

Western terminus

Fig. 10.45 Killoran 226, Structure 2. Eastern and western termini. For section of eastern terminus see Figure 10.44.

by a build-up of laminated peat, deposited in successive flooding events. This was cut through by later erosion gullies. Finally, the hummock and hollow system of a raised bog formed.

Discussion

Two trackways were constructed along the same line separated by three hundred years. It is likely that some visible remains of the earlier trackway were exposed to the later builders. The structure and purpose of the later trackway was quite different to that of the earlier one. It was a substantial construction with a maximum width of 6.8m, one of the widest known Irish trackways after Cooleeny 31 and Derryfadda 13. The western half of the trackway was built in places alongside a shallow bank of glacial till, effectively extending the width of the walking surface. The result was a well-built droveway-type trackway, capable of taking wheeled traffic or cattle, extending from dry ground out 40m across a flooded area to access an exposed root system within the carr forest margin of the bog. It seems unlikely, however, that wheeled traffic would have been driven into the bog.

Both structures were in poor condition on excavation and can therefore be assumed to have been in use as long as they were exposed. Both were substantial in size and so built to last.

Killoran 229

Archaeological wood; 1385–930 BC; 7.5m x 3m x 0.08m; roundwood, gravel and flagstones (*Fraxinus*, *Alnus*, *Salix* and *Taxus*); cutting season autumn (63%); fen and discharge channel environment; deposition phase; site code DER229; licences 96E268 extension and 97E160; NGR E222235 N166672; 125.5m OD.

Paul Stevens and Sarah Cross May

This was a small, sparse scatter of roundwood measuring 7.5m east–west by 3m north–south. Although almost all of the wood uncovered showed signs of working, there was no real indication of a structure. A flagstone was removed during drain clearing, but no further evidence of substantial stone was found. The drain truncated most of the wood, and it is likely that any real structural evidence was removed by BnM draining.

This site would have been very close to the contemporary fen edge. It was underlain in both cuttings by natural silting and erosion deposits due to the development of a discharge channel in the fen. It represents a short episode of working on a wet and unstable surface. The wood and stone present could be a result of the work or the remains of an attempt to clear and stabilise the area to facilitate access. The wood may represent local felling, especially given the yew stump found 10m south of the site on the western edge of Killoran 237.

Killoran 230 (Figs 10.46 and 10.47; Plate 14.7)

Trackway; 1500–1195 BC; 25m x 1.8m x 0.8m; brushwood, roundwood and pegs (*Corylus*, *Fraxinus*, *Betula*, *Alnus*, *Salix*, *Pyrus/Malus*, *Quercus*, *Prunus avium/padus*, *Viburnum* (opulus type), *Prunus spinosa* and cf. *Sorbus*); cutting season autumn (70%); forest fen and discharge channel environment; pre-construction, construction, extension and repair phases; overlay deposits derived from Killoran 265; site code DER230; licence 97E160; NGR E222216 N166658; 125.79m OD (landfall).

Sarah Cross May

Phases of construction and repair

This site was a wooden trackway that was underlain at its western end by pits filled with burnt stone. The trackway was composed of bundles of brushwood and roundwood laid longitudinally and pegged irregularly. Differences in depth and character over the length of the track pointed to many episodes of repair, and six phases of activity were identified on the site. The first two were pre-construction;

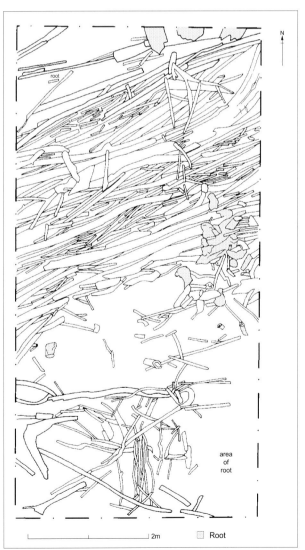

Fig. 10.46 Killoran 230. Central zone.

the remaining four represented the construction, extension and repair of the track.

In Phase 1, two pits were dug in the glacial till. Phase 2 saw these pits filled with material from the neighbouring site Killoran 265, a *fulacht fiadh*. This activity was confined to the westernmost extremity of the site. Phase 3 was the construction of a foundation layer for the track, occasionally including mineral soil. Phase 4 was the laying of a smooth surface. There were two repair phases: Phase 5 was fairly uniform across the length of the track; Phase 6 differed from cutting to cutting and was recorded as sub-phases.

There was no repair of the track at its western landfall, but it was probably extended into this area at the time of the first general repair. The wide range of wood species used in the track also reflected the slow and casual construction process.

The environment and the destination of the track

The track began at the eroding fen edge with stands of marginal alder and was closely bounded by alder trees along its length. It finished at a low ridge in the glacial till that would have created an area of drier peat. Between

the shore and the ridge ran the main outflow channel for the Killoran discharge system of the bog.

The discharge and erosion of this system seems to have fluctuated during the use of the site. The track overlay a silting episode on the eastern side of the discharge channel, indicating a rapid discharge from the dryland. There were reeds and pool mosses overlying this silting episode, indicating a more steady water flow. The extension of the structure westwards shows that the fen was flooding back over the dryland margin during the use of the track.

Artefact 97E160:230:1

A split roundwood with two notches cut into its edge (97E160:230:1) was found in the base of the track at its landfall. It lay horizontally with the rest of the foundation material. There is nothing about the form of the object that would indicate its function. It was probably a component of a more complex object and was deposited as scrap.

Relationship to the fulacht fiadh Killoran 265

The pits pre-dating the track were filled by a broader spread of burnt stone. Stratigraphically, and going by radiocarbon dates, it would seem that the track was only

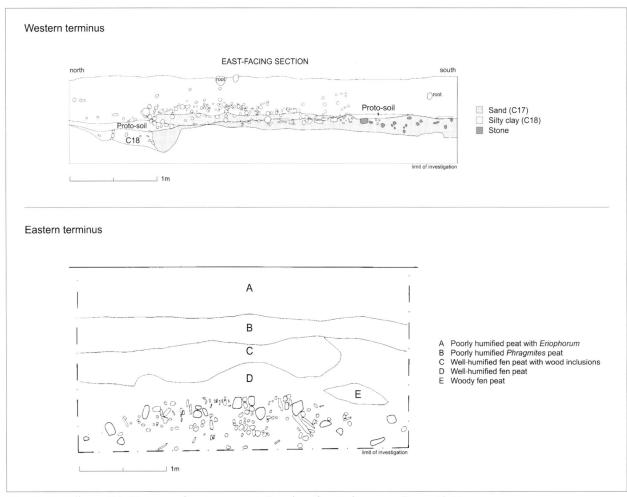

Fig. 10.47 Killoran 230. Western and eastern termini. For plan of central zone see Figure 10.46.

slightly more recent than *fulacht fiadh* Killoran 265, but nothing in the form of the track respected the *fulacht fiadh*. The track passed very close to the *fulacht fiadh* and its wooden platform, but it did not diverge to reach these features. If the *fulacht fiadh* had gone out of use, there is no reason for the track to lead to it, but the complete independence is odd. The people building the track could have used the mound of stone as a staging point in the track, but they did not. Perhaps the *fulacht fiadh* became heavily overgrown due to the high nutrient value of the stone and charcoal, and provided a constraint rather than a conduit for the track. This would mean that the continuity of use was fortuitous rather than deliberate. It would also emphasise the sense of this track as a path through underbrush as well as a structure to reach out into the bog.

Killoran 234 (Fig. 10.48)

Trackway; 795–395 BC; 45m x 2m x 0.25m; brushwood, round-wood and pegs (*Fraxinus, Alnus, Quercus, Salix, Betula* and *Corylus*); cutting season mixed; forest fen environment; construction and abandonment phases; site code DER234; licence 97E160; NGR E222188 N166631; 127.35m OD (landfall).

Sarah Cross May

Structure

This was a disturbed wooden track composed of runners, pegs and some brushwood but no finished surface. While flooding could have damaged the site, it is most likely that the surface was deliberately removed prior to the track being covered in peat. The wood was very poorly preserved, suggesting prolonged exposure before being covered in peat. There was a good deal of variation in the remains—in places, only brushwood and pegs remained; in others, there were clear runners.

Relationship to Killoran 75

The structure lay below Killoran 75. The depth of peat between the two features became deeper as they extended into the bog, as the peat was becoming wetter and perhaps accumulating more quickly. Both sites showed a distinct slope of about a meter over a distance of 40m. At three points, pegs from Killoran 75 intruded on Killoran 234, but there was certainly a lapse of time between the use of the two tracks.

Environmental context

The track was constructed on a slope in which the developing bog was held in pockets by the roots of trees. The large number of stumps and fallen trunks above and beside the track indicated a contemporary marginal forest dominated by alder. A discharge channel, denoted by reeds, may have run along the course of the track. The fen would probably have been less open at this point than at the time of construction of Killoran 75 and Killoran 312.

Abandonment

While the track was built very near to an active run-off zone, it seems unlikely that natural processes would have caused the removal of the upper surface. It could have been deliberately dismantled so that the timbers could be used elsewhere or so that the route created by the track was no longer available. Both scenarios have problems. There are no contemporary tracks in the area where displaced timbers could have been used. Furthermore, in a marginal forest it would be easier to cut new wood than dismantle a track for wood that would be damaged in any case. Nor does there seem to be any good reason to actively change use patterns, there being no strategic advantage. Perhaps this indicates that the use of this track was for something more sensitive than resource procurement.

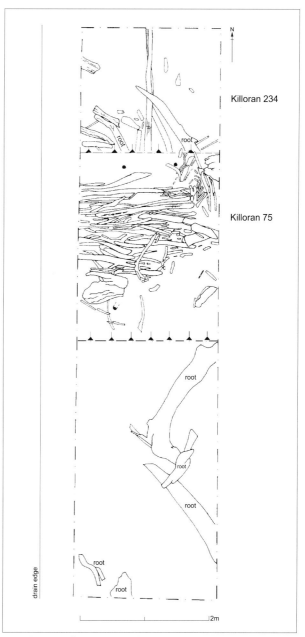

Fig. 10.48 Killoran 234 and Killoran 75. Western terminus.

Killoran 235 (Fig. 10.49)

Trackway; 1212±9 BC; 6.7m x 2m; brushwood, charcoal and stone (*Betula*, *Alnus*, *Salix*, *Quercus* and *Pyrus/Malus*); cutting season spring (67%); marginal forest environment; construction phase; site code DER235; licence 97E158; NGR E222233 N166609; 126.10m OD (surface).

John Ó Néill

This trackway was a 6.7m long by 2m wide dump of burnt stone and charcoal, overlain by an irregular deposit of brushwood, mostly birch, with some alder and willow. The wood was mostly felled in the spring. A best felling date of 1212±9 BC was obtained from a collapsed oak lying directly on top of the brushwood. The upper half of the oak was in very poor condition, suggesting that it collapsed when there was very little peat covering the structure.

The track was aligned east–west within the marginal forest around the western (Killoran) headland of the bog. There was no evidence of heating or burning in the peat below the structure, probably indicating that this material had been brought from a *fulacht fiadh*. The *fulachta fiadh* Killoran 240 and Killoran 265, which are nearby, have produced dates that could place them earlier than Killoran 235.

Discussion

This structure was laid as localised infill within the marginal forest during a period when the water table was recovering from the thirteenth-century BC bog burst (bog burst B), the end of the recovery being marked by the collapse of the oak in 1212±9 BC. The trackway is unusual in its construction, as the use of material robbed from a *fulacht fiadh* mound is not recorded elsewhere in Derryville Bog or further afield. Another unusual aspect of this site is the dominance of wood felled in the spring. This was noted on other Derryville Bog structures but was not common.

Killoran 237 (Fig. 10.50)

Platform; 1685–1400 BC; 10m x 6.5m x 0.4m; brushwood, roundwood and pegs (*Corylus*, *Betula*, *Fraxinus*, *Alnus*, *Salix*, *Pyrus/Malus*, *Quercus* and *Taxus*); cutting season mixed; fen, marginal forest and discharge channel environment; construction and extension phases; site code DER237; licence 97E160; NGR E222233 N166671; 125.33m OD (first phase).

Sarah Cross May

Construction phases

This was a small platform built in two phases. The first phase was an irregularly laid mixture of roundwood and brushwood measuring 6–7m east–west by 6m north–south. It may have been a foundation layer for the platform or it may have been a smaller platform that was repaired and extended. The second phase was a denser layer of irregularly laid roundwood, the final surface of the platform. It measured 8–10m east–west by 6.5m

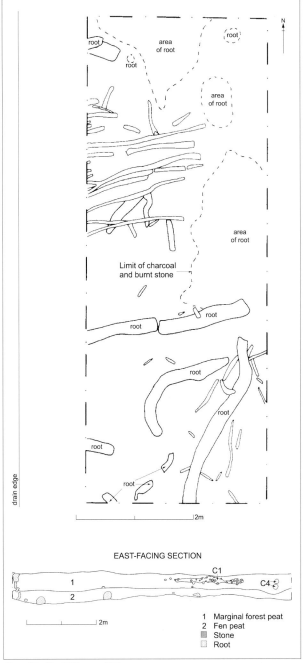

Fig. 10.49 Killoran 235.

north–south. Both layers of wood were held in place with irregularly spaced pegs. The centre of the structure was less organised than its edges.

Microenvironment

This platform bridged a pool in the outflow channel between the actively eroding fen edge and a small ridge of higher, drier land. There was a substantial silting event immediately below the site. These two features mark the beginning of the influence of the Killoran discharge system in this part of the bog. This hydrological shift caused peat to form in a sloping marginal forest. The platform allowed access to areas that had only recently begun to flood.

The same hydrological fluctuations discussed with regard to Killoran 230 are visible here. The moss content of the peat was quite high near the centre of the site, indicating a pool. There was a distinct edge to this mossy peat at the western edge of the structure. Contemporary stumps of alder and oak bounded the pool on the western edge of the platform. The platform did not extend significantly beyond the pool and drier fen peats surrounded it at the time of its construction. A few small stones at the foundation level in the eastern part of the site could be from disturbance to the underlying ridge or part of the silting episode The peat above the site was distinctly more humified, suggesting drier conditions. This increased dryness is also suggested by the many stumps and root systems found about 0.15m above the platform.

Spear shaft (97E160:237:1)

A rounded and pointed stick, 97E160:237:1, was found lying horizontally in the lowest level of the track on its western edge. It is trimmed along its whole length, and the tip is rounded as well as sharpened. The other end was broken in antiquity. It is made from yew and its form and position suggest that it was not simply a structural timber. Comparison with stumps of spear shafts in socketed spearheads, the suitability of yew for spear shafts and wear marks at the appropriate point from the tip all suggest that this was a shaft for a socketed spearhead. The

details and significance of this find are discussed in the catalogue of finds.

Killoran 240 (Figs 10.51–10.53; Plate 9.8)

Fulacht fiadh, trough type, plank-lined and rectangular; pre-1547±9 BC; 19.5m x 19m; *Alnus, Fraxinus, Betula, Quercus, Corylus, Salix* and *Pyrus/Malus*); cutting season mixed; marginal forest environment; construction phase; site code DER240; also recorded as DER244; licence 97E158; NGR E222205 N166572; 126.59m OD (top of trough).

John Ó Néill

Lying just beyond the western margins of the bog in the Bronze Age, this *fulacht fiadh* had been constructed on a slope that carried discharge from the Killoran Bog system down into the Derryville Bog system. Initial use of the site began with the construction of a plank-lined trough (described below), which accumulated a mound of firing debris around it in a rough circle. This eventually became a 19.5m x 19m mound of heat-shattered stones up to 0.95m deep, set in a peaty, charcoal-rich matrix containing sand and silt. This mound flattened towards the western side of the site, and 90% of its volume was stone (90% of which was sandstone and 10% limestone). The charcoal in the mound was identified as hazel (66%), crab apple, oak, alder, yew, wild cherry, blackthorn, elm, ash and willow.

Fig. 10.50 Killoran 237.

A formal hearth was not identified on the site, but a dense deposit of charcoal was recorded directly east of the trough, suggesting that an informal hearth may have been used around the trough.

A saddle quern (97E158:240:1), measuring 235mm x 150mm x 55mm, was found in peat directly west of the site, 0.3m below the level of the trough. This was a horizon in the peat that may be contemporary with the construction and use of the *fulacht fiadh*.

A number of other features were recorded on the site, including an oval pit cut into the subsoil west of the site that contained stones. The pit measured 1.1m (east–west) by 0.85m (north–south) and was up to 0.25m in depth. There was little or no charcoal in the fill of this pit, and no direct relationship could be established between it and the *fulacht fiadh*, but it was covered by trackway Killoran 241. Two other trackways, Killoran 243 and Killoran 314, took advantage of dry areas that the top of the mound presented.

The trough

The trough began as a rectangular pit dug into the subsoil with plank-lined sides. During the phase of use of this lining, the trough would have had a capacity of 1,560 litres. The original lining had been disturbed during subsequent relining but appears to have had timber on all sides. On the northwestern side, a stone had been used to prop the bottom plank into position, while the other end of the same plank had been tenoned and forced into the side of the pit. Two other planks on this side were positioned so that the bottom of each plank was held in place by an overlap, with the top of the plank below. The planks on the opposite side supported each other in a similar fashion, but no evidence survived of how the bottom plank was held in place. A single plank at the southwestern end sat on top of the tenon of the plank below, which held it in position. At the other end, a U-shaped notch in a side plank fitted over a stone in the side of the trough to keep the timber in position. A plank and a number of pegs which were covered over during the relining of the base should probably be viewed as belonging to this original lining. A number of pegs and loose planks associated with this earlier lining had been disturbed, while others fitted into the construction in a way that could not be identified.

Sometime later, the trough was relined with planks. This relining was very well preserved, and, although the lining was structurally weak, the construction technique could be clearly observed. A new lining of mortised and unmortised timbers was added to the base. Lap joints on the ends of the plank on the northwestern side retained the planks at the northeastern and southwestern sides. A lap joint on one end of the plank on the southeastern side held in place the other end of another vertical plank on the southwestern side. The other flat end of the same plank braced the plank at the opposite end. A further plank added on top of the southwestern side was held in

place by two pegs. This was the uppermost plank on this side, but the top of three other sides had a horizontally laid plank added to provide a rim around three-quarters of the trough. These planks had been mortised at both ends, and pegs had been driven down between the planks of the two linings to hold the planks in place. Although none of these pegs were found, the tracks they made survived on the planks below. The plank at the northeastern end had its inner edge carved to fit the edge of the trough.

The planks were a mixture of radially and tangentially split oak and ash timbers. During this phase of use, the trough had a capacity of 1,245 litres. Preserved in the base of the trough was a deposit of stones (98% sandstone and 2% limestone) that appears to be that used during the last time the trough was used to heat water.

The trough had become backfilled with a mixture of charred and uncharred wood in peat rich with charcoal fragments and fragments of fire-cracked stone, which may have been washed down from the mound or broken off the lining. Unburnt ash (87.5%) and hazel were present in this deposit, along with charred hazel (75.8%), ash (20.2%), blackthorn, wild cherry and crab apple. The top of the trough was filled with peat containing charcoal fragments and fragments of wood, including ash, but with alder and birch also present. There was no predominant cutting season for the wood in either deposit.

Technical assessment of the firing debris

The volume of the mound material was calculated on the basis of the various sections excavated through the mound and knowledge of the pre-mound land surface. Adjusting for the relative percentage of stone identified in the mound, the mound consisted of 266.76m^3 of micaceous sandstone and 29.64m^3 of limestone. Some 0.374m^3 of stones was present in the trough, including 0.34m^3 of sandstone and 0.037m^3 of limestone. A volume of 0.47m^3 of stone was proposed for the original trough, which had a greater capacity, giving 0.42m^3 of sandstone and 0.047m^3 of limestone.

As micaceous sandstone can be reused up to five times (*ibid.*), before being reduced below a rough 50mm limit (the average stone size in the mound), the volume of sandstone in the trough was divided into the mound volume and multiplied by five. This gave a total of approximately 3,167 and 3,967 firings for the relined and original trough, respectively, assuming that the geological ratio of each use oscillates within narrow parameters. Multiplying the volume of limestone in the trough by the number of firings produced roughly the same volume as the amount of limestone in the mound, suggesting that it was used only once. This is well below Buckley's figure of six reuses of limestone. This may represent a deliberate attempt to prevent it reducing to calcium hydroxide.

To have retrieved the volume of stone used from the local glacial till would have required the extraction of some 833.63m^3 of stone.

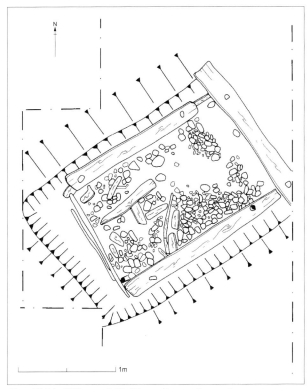

Fig. 10.51 Killoran 240. Plan of central trough.

Discussion

This site was a very well-preserved *fulacht fiadh* that had been sealed by the bog, and, as such, the mound survived intact. The stones used to fire the trough included micaceous sandstone and limestone. The wide range of

wood found as charcoal, including dry ground species such as elm and wet ground species such as alder and willow, appears to derive from various woodland types in the area.

The number of times that the trough was used may seem excessive, but all year round use of the trough would account for the build up of the site in less than ten years; likewise, daily use for six months could be accounted for in less than twenty years. The amount of effort put into the construction and maintenance of the trough would appear to be reflected in this duration of use. In relative terms, this site was in use for five times as long as Killoran 265 and forty times as long as Killoran 253.

Killoran 241 (Figs 10.53 and 10.54)

Trackway; 1547±9 BC; 13m x 1m; brushwood, roundwood (*Alnus, Salix, Fraxinus, Pyrus/Malus, Corylus, Quercus*, cf. *Sorbus, Betula* and *Ilex*); cutting season autumn (58%); marginal forest; construction and extension phases; site code DER241; also recorded in the survey as DER242; licence 97E158; NGR E222194 N166566; 127.37m OD (surface).

John Ó Néill

This was a 13m long trackway. It was less than 1m wide, except where it crossed a wet area on the edge of the *fulacht fiadh* Killoran 240, where it was up to 4m wide and almost 0.5m in depth. The trackway began in the marginal forest peat on the slope down into the Derryville Bog system and extended onto the top of Killoran 240.

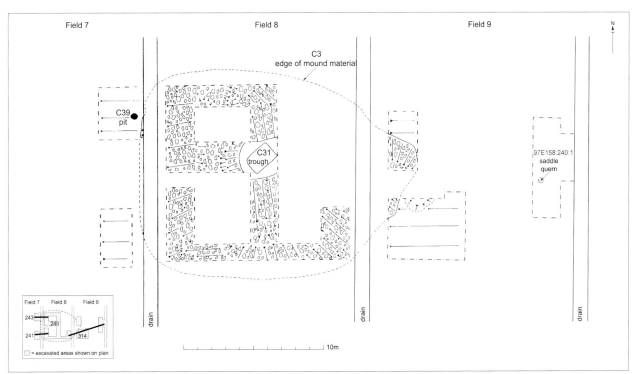

Fig. 10.52 Killoran 240. For plan of trough see Fig. 10.51.

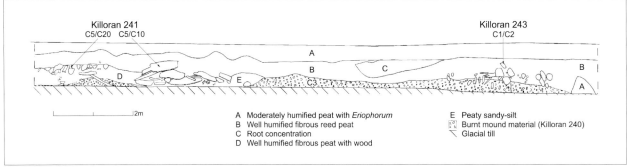

Fig. 10.53 *Killoran 240, Killoran 241 and Killoran 243, Field 7. East facing section.*

The watershed between the Killoran and Derryville Bog systems lay in the area to the east, while the outflow from Killoran occurred to the south. A number of trees had collapsed directly on top of the surface of the trackway, including a number of 100 to 120 year old oaks and a 160 year old ash. One of the oaks produced a date of 1547±9 BC for the last year of growth. Two saplings, an alder (with fifty years of growth rings) and a willow (with less than twenty years of growth rings) indicated that conditions got wetter in the period before the structure was built. The trackway was probably laid down in the twenty years just before the trees collapsed, dating it to the late sixteenth century BC.

The method of construction was not uniform across the length of the track, which began as longitudinally laid alder brushwood, mostly felled in spring. Some possible uprights were present at the eastern end.

An irregular superstructure of brushwood was present where the trackway crossed the depression formed by the western edge of Killoran 240, continuing as an irregularly oriented deposit of brushwood and roundwood. The lower timbers of the track were embedded in the charcoal-rich peat overlying the mound of Killoran 240. A similar pattern was found in the felling season of wood used in different parts of the upper level of the track, with a bias towards the autumn. The wood species used included alder, ash, oak, hazel, willow, crab apple, rowan and holly. There was little variation across different sections of the upper level of the structure.

Discussion

This track provided access to the dry area created by the mound of the *fulacht fiadh* Killoran 240 on the downward slope into the Derryville Bog system. Use of the site may have begun with an initial deposit of alder thrown down as a walking surface. This would have only been wide enough to be used by one person at a time. The wood was mostly felled in the spring, while the pattern of the other material seems to represent additions and extension of the structure in the autumn. The wood used in the structure was in fairly poor condition, suggesting that the trackway may have been in use for a number of seasons.

Killoran 243 (Figs 10.53, 10.55 and 10.56)

Trackway; 979±9 BC; 7m x 3.5m; brushwood and roundwood (*Fraxinus, Alnus, Corylus, Quercus, Ilex, Salix, Pyrus/Malus* and *Taxus*); cutting season autumn (57%); marginal forest environment; construction phase; site code DER243; licence 97E158; NGR E222194 N166575; 127.43m OD (surface).

John Ó Néill

This trackway, which was 7m long and 3.5m wide, was constructed in the Late Bronze Age from the western margin of the bog towards the point where the mound of the *fulacht fiadh* Killoran 240 broke the surface of the peat. The date obtained for the structure, 979±9 BC, was from an oak used in the substructure of the track, which overlay the western margin of the *fulacht fiadh*.

The use of the site began with a dumped mixture of heavy roundwood trunks set as transverses among brushwood and some cleft wood. These were felled in spring, summer and autumn. More roundwood, mostly felled in

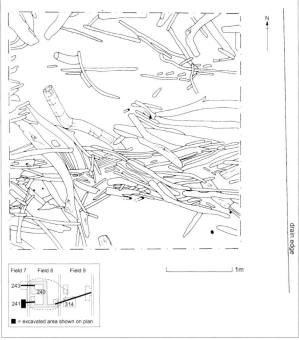

Fig. 10.54 *Killoran 241. For section see Figure 10.53.*

autumn, was added to raise the surface of the structure slightly. Although rotting was noted on the upper timbers the site may not have been open for very long, as this may have resulted from lowering of the water table after the two bog bursts in the ninth and seventh centuries BC.

Discussion

This structure was dumped in the depression formed by the *fulacht fiadh* Killoran 240 and the slope on which it was built. The majority of the wood used in the structure was felled in the autumn and was mostly derived from the marginal forest.

Killoran 248 (Figs 10.57 and 10.58)

Trackway; 394–199 BC; 24m x 2.3m (max. 4m) x 0.25m; brushwood, roundwood and timber (*Salix, Alnus, Fraxinus, Betula, Corylus, Quercus, Pyrus/Malus,* cf. *Sorbus, Ilex* and *Taxus*); cutting season autumn (64%); glacial till and marginal forest environment; construction phase; overlay Killoran 306; site code DER248; licence 96E298 extension; NGR E222189 N166349; 126.90m OD (landfall).

Paul Stevens

This site was located in the west of the study area. Excavation established the site to be a trackway, radiocarbon dated to 394–199 BC. It was oriented east–west and measured 24m long and 2–4m in width. It was constructed of a single layer of longitudinal brushwood and split timbers 0.3m deep, 0.04–0.06m in diameter, except where two extra foundation layers 0.23m deep were required over a pool. The structure originated in the west on a dense floor of root systems on dry ground, and extended east, crossing a deep pool into the carr forest, ending abruptly with a deposit of transverse roundwood close to a root system within the marginal forest. The analysis of the wood revealed a dominance of willow, with high levels of alder. The majority of the trees were felled in autumn and a high proportion in spring.

Environmental context

A shallow, north–south oriented glacial trough within the underlying relief of this area was located in a generally steep sloping western margin of the bog basin. This area remained wet throughout the formation of pre-bog eluvial silt over glacial till and a thick build-up of crumbly forest peat, deposited in a wet brook forest environment. The

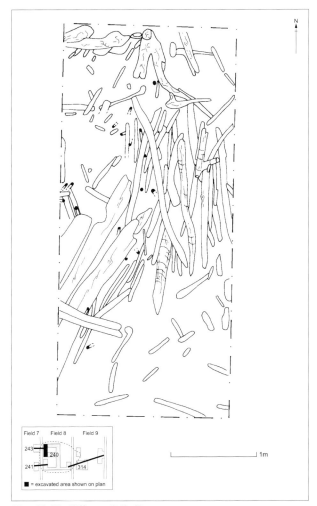

Fig. 10.55 Killoran 243. Eastern zone.

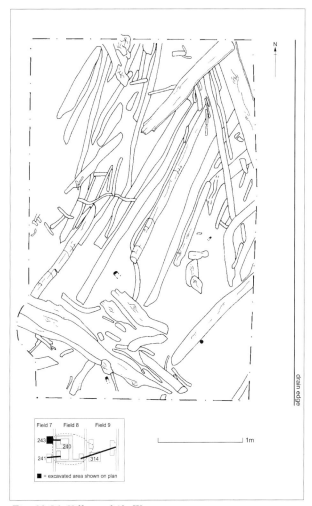

Fig. 10.56 Killoran 243. Western zone.

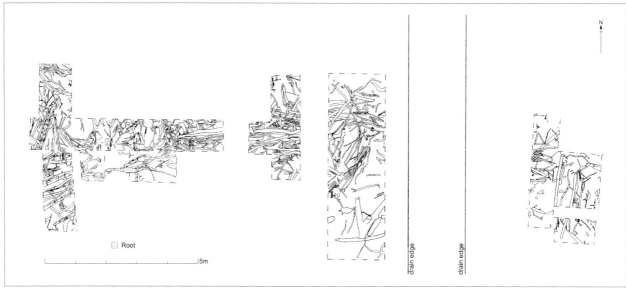

Fig. 10.57 Killoran 248. For section see Fig. 10.58.

structure originated on dryland in the west over a root system and extended east towards the edge of the Iron Age marginal forest, crossing a deep pool. The construction was engulfed by mud from flooding after several years.

Discussion

The very rough construction of this trackway suggests it was built for short-term use. The wood, however, was generally in bad condition from long exposure. Most of the material was felled in the surrounding forest, with other old wood brought in from other sources, possibly other structures. This would explain the dendrochronology date of 641±9 BC or later taken from a split timber recovered from the excavation. The site was located 5–6m north of Killoran 226 and was at a similar level. It seems likely that the structure was using a pre-existing access way into the bog attested to by several previous trackways (Killoran 69, 226 Structure 1, 226 Structure 2, 301 and 306). Although the date range for Killoran 301 was similar (322±9 BC or later) to that for Killoran 248, it occurred at a lower level (126.6m OD versus 126.9m OD). The latter is much closer in level to Killoran 226 Structure 2 (*c*. 161 BC; 126.8m OD).

Killoran 253 (Figs 10.59 and 10.60; Plate 4.2)

Fulacht fiadh, trough type, wicker-lined and rectangular; 1305–940 BC; 10m x 5m; *Corylus, Alnus, Fraxinus, Salix, Betula* and *Pyrus/Malus*; cutting season autumn (87%; platform); marginal forest and fen environment; construction, use and abandonment phases; site code DER253; licence 97E158; NGR E222211 N166765; 125.54m OD (peat below mound).

John Ó Néill

This *fulacht fiadh*, which lay at the northwestern limits of the study area, had been completely buried by the peat. On excavation, it was discovered to have consisted of a wicker-lined trough, a stone and timber platform and a mound of firing debris. It had been built on the surface of the fen, and its eastern side had been disturbed by a BnM drain.

A number of stakes preserved below the mound did not form any coherent pattern, but the lack of evidence of stake holes through the mound material would appear to indicate that their presence is associated with the beginnings or construction of the site, and apparently not its use. The site began as a trough and stone platform (described below). The mound material covered an area 5.5m (north–south) by 6.5m (east–west) and consisted of heat-shattered sandstone (90%) and limestone (10%) in a charcoal-rich loamy clay matrix. The mound material was 90% stone; the wood species identified among the charcoal were ash, willow, alder and hazel. The eastern limit of the mound had been disturbed by a BnM drain, and its highest point was no more than 0.5m above the level at which it had begun to be deposited. The northern edge of the mound had spilled over into the trough and, to the east, the stone and timber platform.

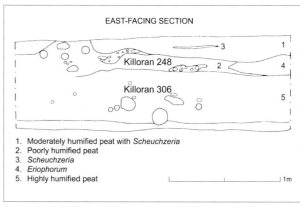

Fig. 10.58 Killoran 248 (upper) and Killoran 306 (lower).

The platform

Sealed below the northeastern corner of the mound was a stone platform that extended for 5m (north–south) by 2.3m (east–west). It consisted of irregular boulders and cobbles of sandstone and, occasionally, limestone. At the time of construction, or perhaps later in the life of the site, timber (alder, ash, hazel, birch, crab apple and willow) was added, mostly wood derived from the marginal forest and mostly (87%) felled in the autumn. Woodchips of alder and birch were found near the platform. Where the mound met the platform, a 0.4m wide setting of stones containing a 0.15m deep deposit of charcoal appeared to represent a hearth. The peat directly below these stones was severely compressed and dried out. The timber of the platform was in poor condition, suggesting that it had lain open for a period of time. However, how long the platform was open after the trough collapsed cannot be ascertained.

The trough

To the west of the platform, and north of the mound, a wicker-lined trough was identified. Cut through peat to the subsoil, the trough must have had original dimensions of around 1.8m x 1.1m x 0.6m, giving it a capacity of 1,188 litres. The weight of the peat on three sides of the trough had led to the collapse of the sides, and, only on the southern portion below the mound, were the original sides intact. The base of the trough had been lined with straight rods, laid transversely. The shorter north side and the northern three-quarters of the longer eastern and western sides survived as piles of wicker wands inside the line where the trough walls had been. On the southern side, an in-out weave around double sails could be identified, with the pattern appearing to continue along the longer sides. The ends of the wands were thrust 0.10m into the peat at the corners, giving the lining some extra stability. All the rods were of coppiced hazel, but no pattern emerged from the cutting season, suggesting that they had been stockpiled over a period of time. The rods used in the base were mostly (61.6%) felled in autumn. *Sphagnum* had been used to plug the sides, and this, along with the presence of an amount of silt in the base of the trough, would point to an attempt to build the trough so it would fill with fresh ground water rather than acidic bog water. The area of the collapsed trough contained some fallen wood and hazel and willow (*Salix caprea*) leaves. The wood used in the trough was very fresh and could not have been exposed for very long, perhaps not even one year.

Technical assessment of the firing debris

The volume of the mound material was calculated on the basis of the various sections excavated through the mound and knowledge of the pre-mound land surface. Adjusting for the relative percentage of stone identified in the mound, the mound consisted of 6.84m³ of micaceous sandstone and 0.76m³ of limestone. Some 0.436m³ of stones was present in the trough, including 0.4268m³ of sandstone and 0.009m³ of limestone.

As micaceous sandstone can be reused up to five times (*ibid.*) before being reduced below a rough 50mm limit (the average stone size in the mound), the volume of sandstone in the trough was divided into the mound volume and multiplied by five. This gave a total of approximately eighty firings, assuming that the geological ratio of each use oscillates within narrow parameters. Multiplying the volume of limestone in the trough by eighty produced roughly the same volume as the amount of limestone in the mound, suggesting that it was used only once. This is well below Buckley's figure of six reuses of limestone. This may represent a deliberate attempt to prevent it reducing to calcium hydroxide.

To have retrieved the volume of stone used from the local glacial till would have required the extraction of some 21.375m³ of stone.

Discussion

The use of the site probably began during the period after the thirteenth century BC bog burst (bog burst B), when the surface of the bog was drier. According to Casparie's calculations, the base of the trough would have been below the water table by around 1100 BC, a time when it was rising quite quickly. It may have increased the water

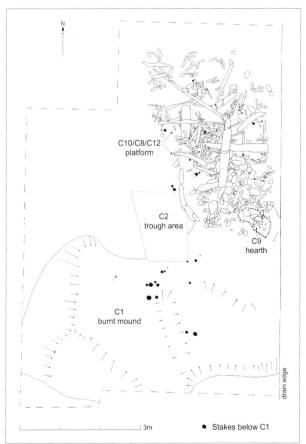

Fig. 10.59 Killoran 253.

Fig. 10.60 Killoran 253. Plan of trough.

content of the peat below the site, leading to the destabilisation and collapse of the trough. Seasonal fluctuations in the water table make it most likely that the site was only used in the spring and autumn, when the water table reached a certain level. The correlation of the autumn felling season for the basal rods of the trough and the platform timbers confirm that this was a time of the year when the site was regularly in use. The deliberate deselection of limestone and the plugging of the sides of the trough suggest an intention to procure fresh water for boiling, without the caustic residue of calcium hydroxide. This would imply that the site could have been used for cooking. If the site was used on a daily basis, the eighty to ninty times it was used can be accounted for in one season.

Killoran 265 (Figs 10.61–10.63; Plate 14.1)

Fulacht fiadh, trough type, hollowed oak, rectangular; 1425–1120 BC; 10.6m x 8.4m; *Pyrus/Malus*, *Fraxinus* and *Alnus*; cutting season spring (73%; platform); pre-bog surface environment; construction, use and extension phases; site code DER265; licence 97E158; NGR E222223 N166679; 125.52m OD (top of trough).

John Ó Néill

This *fulacht fiadh* had been completely buried by the bog and, on excavation, was discovered to have consisted of a carved oak trough, a timber platform and a mound of firing debris. The trough pit had been dug into the subsoil and had been disturbed by a BnM drain, which had removed a 1.5m wide strip through the centre of the site. During the use of the site, a timber platform was added to the eastern side, where the bog had begun to encroach. The mound material covered an area 8.8m (north–south) by 8.4m (east–west) and consisted of heat-shattered sandstone (80%) and limestone (20%) in a charcoal-rich loamy

clay matrix; stone accounted for 90% of the mound's volume. There was no evidence of a hearth at the site.

A number of pits containing burnt stone and charcoal were found below trackway Killoran 230, which passed due south of the site.

The trough (97E158:265:1)

Dug into the subsoil, the trough was housed in a pit measuring 2.8m x 1m x 0.5m. This pit contained a partially hollowed oak trunk (97E158:265:1) that measured 2.7m long and 0.6m in width, which would have reduced the water capacity to around 1,100 litres. The BnM field drain that had bisected the site disturbed around 50% of the southeastern part of this pit. In the intact portion, there was a deposit of stones (60% sandstone and 40% limestone) that seemed to be the remains of the last firing of the trough. The wider end of the trough was in a slightly rotted condition. The top end had been axed, and the underside of the trough had been cleaned of bark and sapwood. The heartwood had been adzed from the cleft face of the trough to a maximum depth of 0.15m. Individual tool marks were not observed on any part of the trough. A technical assessment of the carrying potential and manoeuverability of the trough suggested that it is highly unlikely that this trough was a reused log boat (see Chapter 12).

The platform

A 4m x 3m platform was added to the east of the site, probably while it was still in use, and a date of 1425–1120 BC was obtained from timber used in this platform. The platform was a loose arrangement of timbers used to extend the size of the dry area around the *fulacht fiadh*. The majority of the wood sampled from the platform was identified as crab apple, but some ash and alder was also present. The crab apple was all felled in spring, while the rest of the wood was felled in autumn.

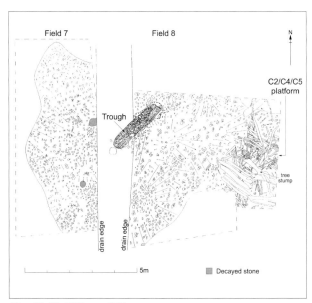

Fig. 10.61 Killoran 265.

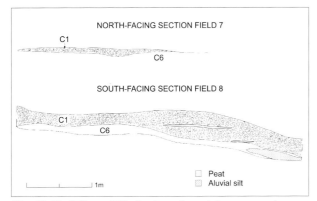

Fig. 10.62 Killoran 265. Sections through burnt mound.

Technical assessment of the firing debris

The volume of the mound material was calculated on the basis of the various sections excavated through the mound and knowledge of the pre-mound land surface. Adjusting for the relative percentage of stone identified in the mound, the mound consisted of 10.37m³ of micaceous sandstone and 2.59m³ of limestone. The volume of stone present in the surviving portion of the trough was 0.07m³, including 0.042m³ of sandstone and 0.028m³ of limestone.

As micaceous sandstone can be reused up to five times before being reduced below a rough 50mm limit (the average stone size in the mound), the volume of sandstone in the trough was divided into the mound volume and multiplied by five (*ibid.*). This gave a total of 615 firings of the trough, assuming that the geological ratio of each use oscillates within narrow parameters. A figure of thirteen reuses of the limestone was calculated on the basis of the sandstone results, compared with Buckley's six reuses, from experimental data. This would suggest that the surviving portion of the trough was not an accurate representation of the overall composition or that there were wide parameters within which the sandstone to limestone ratio oscillated.

To have retrieved the volume of stone used from the local glacial till would have required the extraction of some 32.4m³ of stone.

Discussion

This site differs from the other *fulachta fiadh* in Derryville Bog in its proportion of sandstone to limestone. The number of reuses of the limestone implies that the picture we are gaining from this site is being compromised by the disturbance of half of the trough.

In stratigraphic terms, this site was positioned several metres to the north of Killoran 230, a trackway that appears to have been constructed shortly after the trough went out of use. Two pits below that trackway contained what appeared to be material derived from the mound of firing debris around the trough of Killoran 265. The function of these pits may be connected to the variation in the use of the limestone on the site.

Killoran 301 (Fig. 10.64)

Trackway; 322±9 BC or later; 17.4m x 2.6m x 0.5m; roundwood, brushwood, planks and stakes (*Corylus*, *Salix*, *Pyrus/Malus*, *Betula*, *Alnus*, *Fraxinus* and *Quercus*); cutting season autumn (53%); glacial till and marginal forest environment; construction phase; site code DER301; licence 96E298 extension; NGR E222196 N166368; 126.60m OD (landfall).

Paul Stevens

This site was located in the west of the study area. The excavation established the site to be a trackway constructed of alternating sections of longitudinal roundwood, with diagonal planks and roundwood, measuring 17.4m long and up to 2.8m in width. The roundwood was 0.12–0.02m in diameter. Most of the construction was a single layer, 0.2–0.4m deep, except over a pool where an extra foundation layer was required (0.2m deep). The trackway terminated to the east with a deposit of diagonal planks over a large root system and to the west with a similar diagonal and transverse deposit of planks west of a small dry island and root system. The trackway was radiocarbon dated to 367–195 BC and dated by dendrochronology to 462±9 BC or later and

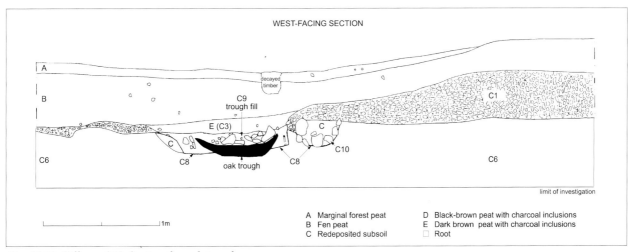

Fig. 10.63 Killoran 265. Section through trough.

322±9 BC or later. The wood within the trackway was mostly in bad condition, suggesting that the structure was exposed for a considerable period. The poor construction, however, suggested it was not meant for long-term use, although the width of the walking surface was large enough for two people to walk in tandem. The analysis of the wood species profile showed a dominance of hazel. The majority of wood was felled in the autumn, but a high proportion was felled in summer and some was felled in spring.

Environment

A shallow glacial trough oriented north–south within the underlying relief of this area was located in a generally steep sloping western margin of the bog basin. This area remained wet throughout the formation of pre-bog eluvial silt over glacial till and a thick build-up of crumbly forest peat, deposited in a wet brook forest environment. The structure originated on dryland in the west over a root system and extended east towards the edge of the Iron Age marginal forest, crossing a deep pool. The construction was engulfed by mud from flooding after several years.

Discussion

The very rough construction of this trackway suggested it was built for short-term use. The wood, however, was generally in bad condition from long exposure. Most of the material had been felled in the surrounding forest, with other old wood brought in from other sources, possibly house structures. The structure was very similar in nature and function to Killoran 226 Structure 1 and Killoran 248.

Killoran 304 (Fig. 10.65)

Fulacht fiadh, trough type, unlined, circular and cut into a spring; 2138–1935 BC; 7m x 5.1m; glacial till and fen environment; construction and use phases; the site was overlain by Killoran 305; site code DER304; licence 96E298 extension; NGR E222223 N166226; 124.74m OD (top of trough).

Paul Stevens

This site consisted of a trough, a low semicircular burnt mound of fire-cracked stone and an area of charcoal-rich clay, possibly representing a hearth. The *fulacht fiadh* was located in the west of the study area, on a gentle undulating slope in the mineral soil, 10m southeast of the western terminus of Killoran 18. Charcoal from the site was radiocarbon dated to 2138–1935 BC. The analysis of the charcoal within the trough revealed that it was composed entirely of branches of hazel from ten to thirty year old trees.

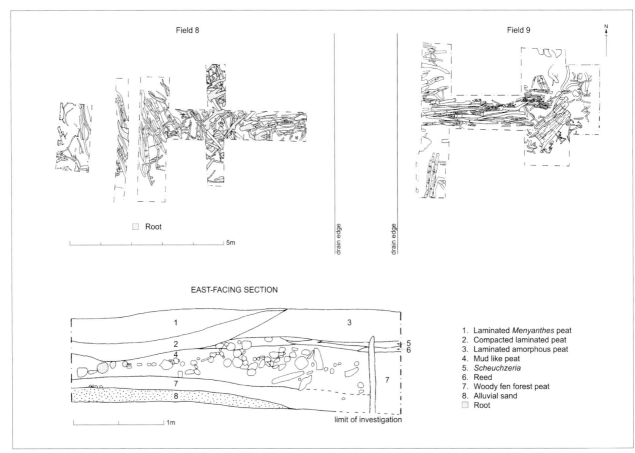

Field 8   Field 9

Root

5m

EAST-FACING SECTION

1. Laminated *Menyanthes* peat
2. Compacted laminated peat
3. Laminated amorphous peat
4. Mud like peat
5. *Scheuchzeria*
6. Reed
7. Woody fen forest peat
8. Alluvial sand
   Root

limit of investigation

1m

Fig. 10.64 Killoran 301.

Environmental context

The site was located on glacial sands and boulder clay cut by the trough. A low glacial moraine located immediately north of the trough may well have been utilised as a dry working platform. Grey eluvial silt formed in patches over the glacial till but was not present under the mound material. Forest fen peat formed over the silt and the site. Over this, a layer of more humified forest mud formed, the result of an initial flooding event, followed by a build-up of laminated peat, indicating very acidic stagnant water and vegetation. This laminated peat was cut through by erosion gullies sometime later due to the saturation of the bog.

The trough

The trough cut through glacial sands, boulder clay and a natural spring. The sub-circular trough measured 1.5m in diameter and was 0.4m in depth. It had a shallow concave profile and a flat base. It was positioned over a natural spring that provided a clean, abundant and effortless water source. The trough contained a thin lining of pinkish-grey silty clay, which did not cover the entire area inside the trough; this may have been a deliberate attempt to line the trough, thereby sealing in the water from the spring, or a natural silting after its use. There were two fills within the trough. The lower fill consisted of small (<0.06m) angular, fire-cracked sandstone with silty sand and charcoal, 0.27m in thickness. The upper fill consisted of large (<0.25m) angular, fire-cracked sandstone with silty clay and charcoal, 0.09m in thickness. It was sealed by a small quantity of small rodent-gnawed hazel nut shells, probably the cache of small rodents rather than human waste.

The burnt mound

The trough was surrounded on three sides by a semicircular, intermittent mound of fire-cracked sandstone measuring 7–5m in diameter; larger stones had slumped into the trough fill. The mound was completely covered by peat.

The mound consisted of fire-cracked sandstone (<0.05m) with silty clay and charcoal. The southern portion of the mound overlay peat, whereas the northern mound overlay a low glacial ridge that extended northwest–southeast across the eastern face of the trough and may have provided a dry working platform.

The hearth

A possible informal hearth site was located to the southeast of the trough and sealed the burnt mound material. This was a sub-rectangular deposit of dark blue clay rich in charcoal inclusions. It was oriented southwest–northeast and was 1.8m long, 1.5m wide and 0.1m in depth.

Discussion

This example of a *fulacht fiadh* was completely covered by peat. The site was obviously located at the spring and utilised the low ridge for a platform. It was not possible to calculate the number of times the trough was used, as the site was truncated to the west by a BnM drain. The mound itself was sparse and intermittent, indicating that the site was used very few times. The upper trough fill contained stones from the boiling of the trough of a larger size than those in the mound or the lower fill, confirming that the stone (micaceous sandstone) was reused possibly up to five times before being discarded, as noted by Buckley (*ibid.*). This suggested that the site was abandoned prematurely. The presence of gnawed nuts indicated an autumn date for abandonment, as no silt occurred between the nuts and the stony fill. The nuts may have been scavenged from the surrounding trees, as the charcoal from the mound was entirely hazel. The earlier stony fill in the trough would suggest that the trough was partially backfilled and not cleaned out prior to its final use; indicating that it would still have worked effectively.

Killoran 305 (Fig. 10.66)

Trackway; 1436–1310 BC; 10.5m x 5.6m x 0.18m; roundwood and brushwood (*Alnus*, *Salix*, *Fraxinus*, *Pyrus/Malus*, *Betula*, *Taxus* and *Corylus*); cutting season autumn (38%); marginal forest environment; construction phase; contemporary with causeway Killoran 18 and overlying part of Killoran 304; site code DER305; licence 96E298 extension; NGR E222224/N166223; 124.90m OD (surface).

Paul Stevens

This site, dated to 1436–1310 BC, was in the west of the study area, south of the western terminus of Killoran 18 and overlying Killoran 304. It was oriented north-northwest–south-southeast. It was a collection of scrappy brushwood deposits laid to form an irregular and incomplete pathway 10.5m x 6.8m x 0.18. The structure originated in the west as a transverse deposit of brushwood overlying fen peat, 2.4m (north–south) by 0.8m (east–west). It partly overlay a small drier area of burnt mound material from Killoran 304 that protruded through the surrounding peat. To the east of this were three intermittent deposits of transverse or longitudinal deposits of brushwood, 1.9–4.5m x 0.7–1.7m.

The structure terminated with a large scatter of brushwood and occasional roundwood laid around a tree stump and root system measuring 4.98m (east–west) by 5.6m (north–south), forming a platform construction. The wood in the structure was in moderate to good condition, suggesting that the structure was covered over by peat in a short time. The wood had been felled in autumn, summer and spring, with marginally more in the autumn than the summer. This pattern was repeated in the cutting season profile for the stake row of Killoran 18, suggesting possible stockpiling. The analysis of the wood species revealed a dominance of alder and high levels of willow.

This structure represented a trackway of very poor construction, although unnecessarily substantial in overall size. The structure fanned out at the eastern terminus around tree stumps within the marginal forest, which suggested the

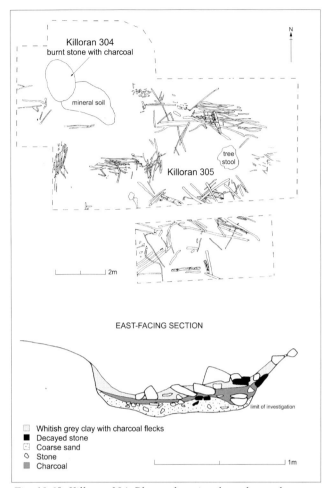

EAST-FACING SECTION

Whitish grey clay with charcoal flecks
Decayed stone
Coarse sand
Stone
Charcoal

Fig. 10.65 Killoran 304. Plan and section through trough.

structure was more for access into the forest for timber than for passage through it. The presence of identical axe mark signatures, as well as a similar date range, suggested this structure was contemporary with Killoran 18, possibly related to the branches phase of that construction, attempting to ford the discharge channel at a drier spot.

Environmental context

Forest fen formed over this area of gently sloping relief, with occasional undulations in which the site was located. Further fen with smaller trees and scrub established over this. The discharge channel for the bog was located just east of the site.

Killoran 306 (Figs 10.58 and 10.67; Plate 4.3)

Trackway; 756–407 BC; 18m x 1.5m x 0.35m; roundwood, brushwood and cobbles (*Fraxinus, Quercus* and *Alnus*); cutting season autumn (83%); glacial till and marginal forest environment; construction phase; site codes DER306 and DER300; DER300 forms the southeastern portion of Killoran 306, which is overlain by Killoran 248; licence 96E298 extension; NGR E222203 N166351; 126.40m OD (landfall).

Paul Stevens

This site was located in the west of the study area. The overall construction measured 18m northwest–southeast and consisted of a single layer of longitudinal roundwood and brushwood, with the addition of extra layers within a deeper pool, up to 0.75m thick. The structure originated in the west, overlying eluvial silt. It consisted of a loose arrangement of mostly longitudinal roundwood, 0.02m in diameter and 1.2m wide, with a single transverse marking the western terminus. The track crossed the deepest peat with several layers of roundwood up to 0.75m thick and 1.3m in width. The central section of the track expanded to 2.35m wide and 0.06m in depth, and included the use of transverse pieces of roundwood.

Several of the structural timbers were charred and small patches of charcoal were noted north of the trackway. The trackway here was bordered to the north by root systems and to the south by a deposit of grey fluvial sandy silt, underlying the trackway (0.11m thick) and overlying peat. The trackway abutted a large fallen oak, 0.8m in diameter, which was partly covered by peat. Several cobble-sized, angular limestone stones were deposited north of the oak, together with occasional roundwood, to form a loose platform. The trackway continued southeast of the platform at a width of 1.5m for 7m, where it petered out at a root system within marginal forest. The wood in this structure had been almost exclusively felled in the autumn. The analysis of the wood species revealed a high proportion of ash and oak with alder, a very small species profile compared with the other sites analysed. This suggested that the structure was built in a short period of time during the autumn.

Environmental context

Eluvial silt, lying over undulating steep sloping glacial till within the bog basin, was sealed by a thick build-up of crumbly forest peat, deposited in a wet brook or carr for-

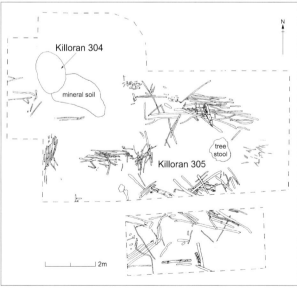

Fig. 10.66 Killoran 305.

est. Over the forest peat was a layer of more humified forest mud, the result of a flooding event, then a build-up of laminated peat, indicating very acidic stagnant water and vegetation. The trackway originated in the west, on the dry glacial till and crossed the early Iron Age marginal carr forest to the platform situated on a bar of dryland acting as a small island. A band of in-washed silt ran northwest by southeast along the southern edge and under the trackway.

Discussion

This trackway originated in the west on dry ground and extended across a shallow, wet, peat-filled hollow in the underlying relief to a root system in the southeast within the marginal forest. The site was stratigraphically under trackway Killoran 248 (394–199 BC) and separated by 0.4m of peat growth. It utilised a large fallen oak tree (dated to 1148±9 BC) with additional limestone cobbles as a staging post or platform. In the northeast corner of this platform, several roundwood lengths extended off from the oak to the northeast, representing part of the earlier trackway Killoran 69 (838–799 BC). The unexpected early date of the oak suggested no peat growth for 600 or so years. This must have been the result of erosion of the peat by a bog burst event. The growth of the oak so close to the bog margin, commencing in 1331 BC, was itself made possible by the lowering of the water table after bog burst B of *c.* 1250 BC. The re-exposure of the oak followed a bog burst (bog burst C) dated to before the construction of Killoran 69 (838–799 BC). Some two hundred years on, trackway Killoran 306 was constructed, with the lowering of the water table from bog burst D in Cooleeny (600±5 BC) once again re-exposing the oak.

The structure was bordered to the south by a layer of grey fluvial silty clay overlying peat, which was abutted by the trackway. This fluvial silt washed in from the Killoran fields to the west, following the lowering of the water table in Derryville Bog after bog burst C. Therefore, the date range for structure Killoran 306/300 (756–407 BC) could be refined to several years after *c.* 600 BC.

The trackway extended northwest–southeast 7m beyond the platform, where it terminated at a root system within marginal forest (labelled as Killoran 300). The trackway, which was generally in moderate to poor condition, would have been exposed for some considerable length of time. It appeared to have been constructed to provide access into the bog rather than across it.

Killoran 312 (Fig. 10.68; Plate 14.3)

Trackway; 385–50 BC; 10m x 1.5m x 0.4m; brushwood, hurdle and pegs (*Corylus, Fraxinus, Betula, Alnus, Salix, Pyrus/Malus, Ulmus*, and cf. *Sorbus*); cutting season autumn (58%); fen and marginal forest environment; construction, repair and extension phases; site code DER312; attached to Killoran 75; licence 97E160; NGR E222223 N166627; 126.78m OD (hurdle surface).

Sarah Cross May

Phases of construction

This was a brushwood track leading to a hurdle, built in three phases. In Phase 1, a regular layer of brushwood measuring 3m x 1m was laid on top of an alder stump. In Phase 2, a hurdle was laid on top of this. In Phase 3, a thick layer of brushwood measuring 2.2m x 1m was laid to connect these features to Killoran 75.

The track and its microenvironment

This track was constructed on the southern edge of the pool of rannock rush described in Killoran 75. There was an inlet of the pool running between the two sites. The first phase of construction was on top of an alder stump, and there was an alder tree between the two sites. The site provided access to and from the southern limb of the arc of alder carr that surrounded the wetter pool peats.

The trackway originally provided a different route to the same place as Killoran 75, but when the two structures were linked, it extended the working space around the edge of the pool. The time frame for this change was probably fairly short, a few years at most. It may even be that the sequence reflected construction. The approach from the alder carr was taken to make extension easier.

The hurdle

This hurdle was smaller than that found at Killoran 75 and was less important to the function of the structure. It measured 3.4m x 0.9m, though the length of the hurdle may have been truncated by a BnM drain. The weave, though damaged, was quite regular—single rods woven around double sails. The individual sails averaged 30mm in diameter, the same as at Killoran 75. The double sails indicated

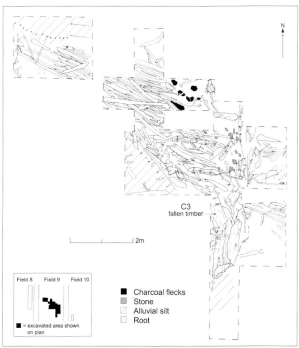

Fig. 10.67 Killoran 306.

a stylistic choice by the hurdle maker, perhaps suggesting a different intended use. The longest rod was 1.75m, and the occasional double and even triple rods appear to have been for overlapping rods rather than a separate pattern. The average distance between sails was 0.38m, the minimum was 0.23m and the maximum was 0.4m.

The positioning of the hurdle on the top of an alder stump and a layer of brushwood took away from its advantages as a flexible surface in wet environments. Placed on a firm footing, the hurdle was badly damaged during use, and most of the rods were broken where they pressed on the sails that held them in place. This was also the only hurdle in Derryville Bog that had any rot. In contrast to the hurdle at Killoran 75, this hurdle my have been reused. While a stable surface was desirable, there is nothing to indicate that this particular hurdle was designed for this use.

Killoran 314 (Fig. 10.69)

Trackway; 370–5 BC; 18m x 3m; roundwood, brushwood and hurdles (*Corylus, Salix, Betula, Alnus, Fraxinus, Pyrus/Malus, Ilex, Taxus, Quercus* and cf. *Sorbus*); cutting season varied by deposit; marginal forest environment; construction and extension phases; site code DER314; licence 97E158; NGR E222205 N166565; 127.25m OD (surface).

John Ó Néill

This 18m long trackway ran east into the bog from the dry area created by the top of a *fulacht fiadh* mound (Killoran 240). Most of its surviving length was excavated, revealing a variation in construction methods. The depth of the site increased from the west to the east, where it ended in a four layer deep platform. The trackway crossed the marginal forest from the top of the mound of Killoran 240 to the edge of the discharge channel.

The upper (walking) surface was present all the way across, from the western landfall on the *fulacht fiadh* mound to the platform at the eastern end of the structure. It was made up of dense bundles of roundwood and brushwood, laid as transverses, which were occasionally pegged down. In the central portion of the trackway, some longitudinal runners were added to give more stability as it crossed the root systems of the marginal forest. These runners were occasionally intermingled with the transverses rather than underlying them. Towards the eastern end of the track, this structure gave way to a system of larger roundwood transverses overlying runners. A wide range of species, including alder, birch, ash, oak, hazel, willow, crab apple, holly, rowan and yew, were used to build the walking surface. Hazel was only present in this deposit at the eastern end, where it was in the form of rods derived from hurdles used in the platform. The timbers used in the trackway had a range of felling seasons, from spring through autumn, with only one timber felled in winter.

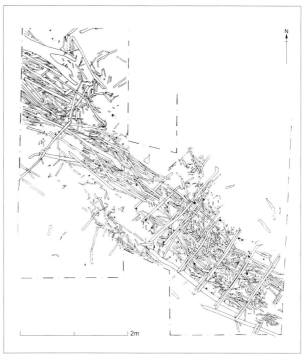

Fig. 10.68 Killoran 312.

The basal level of the platform was made up of a mixture of alder, ash, birch, crab apple, hazel, holly and willow. Some 86% of this wood was felled in the summer or autumn. This deposit was overlain by a hurdle made of two woven panels of hazel rods, mostly felled in winter (63%) or autumn (24%). At this point, the main trackway level was laid across the platform.

A hurdle was then added to the southern end of the trackway level. This contained some birch as well as hazel, all of which was felled in spring, summer and autumn, while the sails were all felled in winter. Galleries in one of the hazel rods contained eighteen adult bark beetles (*Xyloborus dispar*), which are generally found in broad-leaved woodlands from June to August and infest dead or felled wood, grow to adulthood over the winter and then move on to another piece of wood the next spring. As so many adults were found in the galleries, they appear to have been killed in a single event, such as a sudden immersion in water.

Some birch and alder, felled or fallen in winter and spring, overlay the upper hurdle and the trackway level.

Discussion

This trackway is one of a small number of structures from Derryville Bog incorporating hurdle panels. The absence of hazel from the trackway underlying the hurdle is interesting with hazel perhaps being reserved as part of a woodland management strategy aimed at providing rods for wattling. This may also be evident in the apparent stockpiling of the rods for the upper hurdle, during which they were infested by bark beetles, the sails being cut when the hurdle was made.

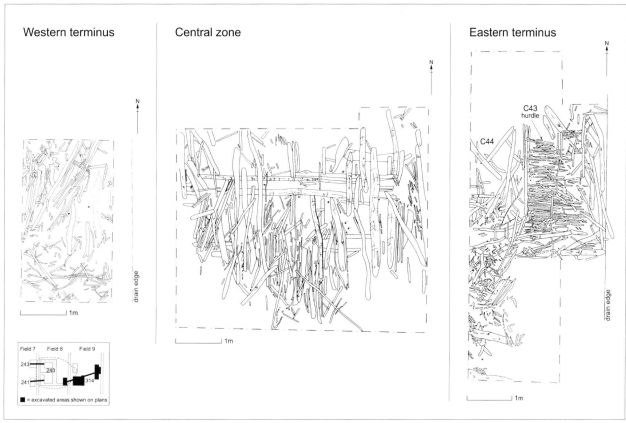

Western terminus | Central zone | Eastern terminus

C43 hurdle
C44

drain edge

Field 7 | Field 8 | Field 9
243
240
241 | 314
■ = excavated areas shown on plans

Fig. 10.69 Killoran 314. Western terminus, central zone and eastern terminus.

There appears to be a pattern to the accumulation of the levels of the structure. The summer/autumn-felled wood at the bottom level of the platform is followed by a hurdle felled in autumn/winter. This is overlain by a trackway of wood felled in spring, summer and autumn that avoids using hazel, and then a hurdle is added in the winter (on the basis of the adult *Xyloborus dispar*). Thus, it may be that the use of the site started as a dump of material in a wet area of the marginal forest that was in use in the summer or autumn and remained in use during the winter, when a hurdle was added. A trackway was built during the course of the next or year or so, during which a hurdle was added to the platform at the eastern, wetter end. As the upper surface wood of the trackway and platform was not very fresh, the site may have remained in use for some time, but a drop in the water table around AD 100–300 may have contributed to the poor condition of the structure.

Killoran 315 (Fig. 10.70; Plate 4.4)

Trackway; 405–180 BC; 7.5m x 4m x 0.3m; brushwood, roundwood, hurdle and pegs (*Corylus, Betula, Salix, Fraxinus* and *Alnus*); cutting season spring (60%); mixed marginal forest and fen environment; construction and repair phases; site code DER315; licence 97E160; NGR E222201 N166651; 126.93m OD (hurdle surface).

Sarah Cross May

Phases of construction

This was a wooden track leading to a hurdle, built in four phases. Phase 1 was a layer of roundwood measuring 4m x 3m laid on top of tree stumps. Since both the stumps and roundwood were birch, this could reflect felling *in situ*. In Phase 2, another layer of mixed wood measuring 5m x 4m was laid on the western edge of this deposit. This probably formed the first surface of the structure. In Phase 3, a hurdle was laid at the northeastern edge, extending the trackway. In Phase 4, another layer of substantial roundwood, measuring 4m x 1.5m, was laid on the western edge of the hurdle. The condition of the hurdle under the roundwood suggested that Phase 4 may have been a slightly later phase of repair.

Environmental position

The structure was built in mixed marginal forest with pools of wet fen. As it overlay substantial stumps in both cuttings, it was clearly built in marginal forest. This appears to be the last of the sloping fen formed from the Killoran discharge system. An environmental cutting taken to the east of the site showed a continuation of the marginal forest but with more reed and cotton grass, indicating wetter conditions. The drain face running through the site showed a small patch of rannock rush, but this only extended about 0.5m into the excavated portion. This is a different situation to Killoran 75 and Killoran 315. The hurdle was certainly in a wet environment, but it was placed firmly on a

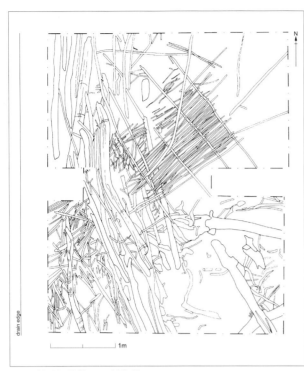

Fig. 10.70 Killoran 315. Eastern terminus.

base of roots and was not providing a flexible surface.

There is the possibility that the lowest level of roundwood represents *in situ* felling. In contrast to most of the brushwood tracks that were excavated, Killoran 315 contained almost no alder. Furthermore, the species composition between each phase was very different. The foundation was almost all birch, and the main surface was three separate bundles, one of willow, one of birch and one of hazel. The hurdle was more purely hazel than the hurdles found at Killoran 75 and Killoran 312, having only one sail of ash. Killoran 315 was also different from Killoran 75 and Killoran 312, and indeed most of the other structures in the bog, in using wood cut in the spring.

The hurdle

This was a simpler and more robust hurdle than the contemporary structures at Killoran 75 and Killoran 312. It was nearly square, measuring 2.5m x 2m. It had a simple, moderately tight weave of single rods over single sails. There were four sails, averaging 40mm in diameter. The longest rod was 2m long, which meant that very few double rods were needed for overlapping. The first space between the sails was 0.33m, the next was 0.4m and the last was 0.5m.

It extended out over wet peat but was centred on a stump. This damaged the hurdle but not as badly as the hurdle at Killoran 312. Most of the hurdle was well preserved. There was nothing to indicate that this hurdle had been reused but neither was there anything to indicate that it had been purpose built.

Killoran 316

Burnt mound; no date; 4.2m x 2.9m; burnt stone and charcoal; pre-bog surface environment; use phase; site code DER316; licence 97E158; NGR E222218 N166552; 125.99m OD (below deposit).

John Ó Néill

Ten metres south of Killoran 240, a deposit of burnt mound material, up to 0.2m deep, was found in an area 4.2m (north–south) by 2.9m (east–west). No other features were associated with this deposit. A number of root systems were present above the site, as well as an extremely rotted oak. The mound included 0.987m^3 of micaceous sandstone and 0.11m^3 of limestone. As no trough was present on the site, no assessment could be made of the number of firings or the relative number of firings of sandstone and limestone.

Discussion

This spread of burnt stones and charcoal derives from the repeated heating of stones on the site. It is possible that the material may have been redeposited from Killoran 240, by natural erosion or by human agencies. Alternatively, some form of portable trough may have been used to collect and heat water.

Killoran 400

Fulacht fiadh, trough type, irregular; Bronze Age; 4.2m x 2.9m; pre-bog surface environment; construction, use and abandonment phases; site code DER400; licence 97E372 extension; NGR E222229 N166600; 125.7m OD (base of mound).

Paul Stevens

This site was revealed during trench cutting for the TMF embankment road. The site was only recorded in section due to health and safety restrictions. It was sealed by peat and consisted of a large burnt mound sealing an irregular trough which cut mineral soil. The site was located on the east-facing slope of shallow sloping glacial till at the contemporary edge of Derryville Bog. It formed part of a cluster of three sites that included Killoran 240 and Killoran 253.

Burnt mound

The burnt mound measured 10.6m east–west and 0.51m in depth and consisted of fire-cracked stone (90% sandstone, 10% limestone), charcoal and sand. The mound overlay glacial till, close to the contemporary edge of the bog margin.

Trough

A roughly U-shaped cut measuring 1.67m wide and 0.3m in depth, representing the trough, was recorded. The cut had a gently sloping eastern side, a steep western side and a regular base. It was filled with burnt mound material of fire-cracked stone, less than 0.3m in diameter, and charcoal-rich silt.

11. Catalogue of dryland sites

Sarah Cross May, Cara Murray, John Ó Néill and Paul Stevens

Introduction

Each individual area of the dryland that was assigned a site number during the course of the project is listed below. The catalogue includes both excavated and unexcavated sites.

The entries begin with a summary in the following format: site type; date(s); dimensions; location/environment; site code; excavation licence numbers; Irish NGR and level relative to OD (Malin Head). This is followed by the name(s) of the excavator(s).

Barnalisheen 1

Burnt mound; Bronze Age; >3m x >0.3m; level glacial plateau; site code BAR1; licence 97E372 extension; NGR E219349 N167618; 125m OD (base of mound).

Paul Stevens

This site was revealed during topsoil stripping for the Carrick Hill borrow area, on level ground south of the glacial knoll of Carrick Hill. It consisted of a small spread of fire-cracked stone continuing east under the baulk and therefore not excavated further. The site may constitute a *fulacht fiadh* or burnt mound and is in a cluster of activity over a very small area within the largely archaeologically sterile area of Carrick Hill.

Barnalisheen 2

Isolated pits; Bronze Age; 2m x 0.7m; level glacial plateau; site code BAR2; licence 97E372 extension; NGR E219364 N167622; 125.417m OD (top of pit).

Paul Stevens

This site was revealed during topsoil stripping for the Carrick Hill borrow area, on level ground south of the glacial knoll of Carrick Hill, south of Barnalisheen 1 and west of Barnalisheen 3. It consisted of two small pits containing burnt mound material of fire-cracked stone and charcoal-rich silt. The pits were oval in plan, with a shallow concave profile, and measured 0.7m in diameter and 0.2m in depth. The site may constitute peripheral activity associated with that at Barnalisheen 1 and 3. It is not a *fulacht fiadh* or burnt mound but may be a potboiler—a smaller version of the burnt mound, usually single phase use and dating to any period. The site was in a cluster of activity over a very small area within the largely archaeologically sterile area of Carrick Hill, which may be contemporary and date to the Bronze Age.

Barnalisheen 3

Isolated pits; Bronze Age; 9m x 3m; level glacial plateau; site code BAR3; licence 97E372 extension; NGR E219369 N167568; 125.417m (top of pit).

Paul Stevens

This site was revealed during topsoil stripping for the Carrick Hill borrow area, on level ground south of the glacial knoll of Carrick Hill, south of Barnalisheen 1 and east of Barnalisheen 2. It consisted of three oblong pits containing burnt mound material of fire-cracked stone and charcoal-rich silt. The pits had irregular, oblong plans and had been partly truncated by machine. They had a shallow concave profile. They were located in a roughly linear (east–west) alignment but were themselves aligned north-south. They measured 1.2m, 1.8m and 2.2m in length, 0.4m; 0.5m and 1m in width and were 0.1m in depth. The site may constitute peripheral activity associated with that at Barnalisheen 1 and 2. The site is not a *fulacht fiadh* or burnt mound but may be agricultural. It is in a cluster of activity over a very small area within the largely archaeologically sterile area of Carrick Hill, which may be contemporary and date to the Bronze Age.

Cooleeny 1

Possible *fulacht fiadh*; Bronze Age; 5m x 5m; break of sloping glacial ridge; site code COO1; licence 97E372 extension; NGR E222135 N165330.

Paul Stevens

This structure was a ploughed-out *fulacht fiadh*, or burnt mound, that was revealed during field walking in a ploughed field west of Derryville Bog and the Cooleeny Complex. The site consisted of a spread of fire-cracked sandstone and some charcoal. It was located on the east-facing break of a slope of glacial till, sloping eastwards to the bog. The area was not to be developed, so the site remained unexcavated.

Killoran 1 (Fig. 11.1)

Fulacht fiadh, square trough, burnt mound and later linear field ditches; Bronze Age; 8m x >7m; chert scraper found; low glacial knoll environment; construction, use, abandonment and later drainage phases; site code KIL01; licence 97E372; NGR E220603 N166853; 127.94m OD (top of trough).

Paul Stevens

This *fulacht fiadh* was uncovered during the stripping of the haul road, a 10m wide construction road. No surface trace of the site was visible prior to excavation. It was located on the break of a slope of a glacial knoll with two linear ditches running parallel (north–south) along the top of the knoll, probably dating to the post-medieval period.

Trough

The trough was a square pit 1.5m long, 1.5m wide and 0.3m in depth. It was filled by fire-cracked stone, charcoal and silty clay.

Burnt mound

The ploughed-out burnt mound was roughly horseshoe shaped and oriented east–west. It measured 8m x >7m x 0.3 m. It consisted of small fire-cracked stones <0.3m in diameter (80% sandstone and 20% limestone), spread out downslope over the trough. A single chert thumbnail scraper (97E372:1:1) was recovered from the mound material. The area stripped for the haul road exposed the majority of the mound, with only a small portion of the southern end remaining under topsoil.

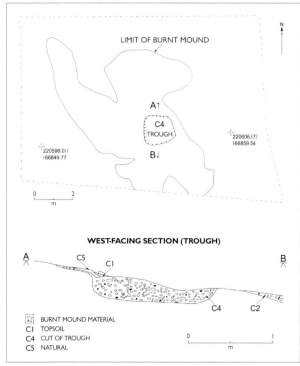

Fig. 11.1 Killoran 1.

Ditches

Two parallel, linear, peat-filled field ditches, probably dating to the post-medieval period, were located on the top of the knoll, 20m east of the *fulacht fiadh*. They were oriented north–south and were *c*. 2m apart. The ditches were not contemporary, but both were in use when they were filled by forest peat growth. The surrounding field boundaries respected the line of the ditches, and one was marked on the first edition OS map (1845).

Discussion

This site represented two distinct phases of use within a broader settlement pattern in the locality. It was located on the western break of the slope of a glacial knoll, overlooking a peat-filled depression to the west and Derryville Bog, north Killoran, to the north. A similar site, Killoran 21, located on a knoll 80m to the southwest, overlooked the same depression.

The burnt mound was partially ploughed away, so no detailed firing calculation was possible; however, the trough capacity was small at 675 litres. Like other dryland *fulachta fiadh* excavated by the Lisheen Archaeological Project, the trough was deliberately backfilled with mound material, suggesting a deliberate decommissioning of the site. The use of sandstone rather than limestone implies that clean water was important and would suggest that cooking was the function of these sites.

The linear ditches were noted on the first edition OS map and appear to be contemporary with the agricultural activity, also noted at Killoran 16 and Killoran 21, associated with the Killoran estate dating to the 1750s (see below).

Killoran 2 (Fig. 11.2)

Peripheral activity; prehistoric; 106m x 10m; linear field ditches and burnt spreads; flint flake found; glacial till and bog environment; construction and abandonment phases; site code KIL02; licence 97E372; NGR E222045 N166427; 128.55m OD (top of burnt spread).

Paul Stevens

Several features were revealed during soil stripping for the TMF berm road at the western edge of Derryville Bog. Killoran 2 was located on shallow sloping glacial till and consisted of three regular, shallow field drains of uncertain date leading up to and beyond the BnM drainage dyke that defined the present-day bog. The ditches were V-shaped in profile and measured over 13m x 0.85–0.6m x 0.3–0.1m. They were filled with dark silty clay. Three ephemeral burnt spreads were also revealed close to the ditches. One burnt spread, measuring 0.6m x 0.3m, close to the northernmost ditch produced a waste flake of pebble flint (97E372:2:1). No other features were noted in the vicinity.

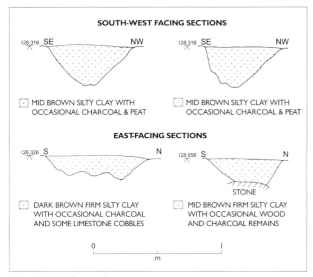

Fig. 11.2 Killoran 2.

The site was located close to the prehistoric activity within Derryville Bog to the east and the settlement site at Killoran 8, to the north. It is possible that the site represents prehistoric cultivation and clearance at the bog side, corresponding to the marginal fen activity of Derryville Bog.

Killoran 3 (Figs 11.3 and 11.4)

Complex; AD 415–630, AD 980–1270 and post-medieval; 35m x 10m; linear ditches, pits and post holes; two chert flakes, a coarse ware pottery sherd and a spud stone found; glacial till environment; ditch phases; site code KIL03; licence 97E036; NGR E22177 N16635; 128m OD (top of burnt pit).

John Ó Néill and Paul Stevens

This site was uncovered during topsoil stripping for the plant site access road on a west-facing slope in otherwise flat ground. It consisted of a number of pits and gullies over an area 35m x 10m.

Parallel linear ditches

In the southern part of the excavated trench, two parallel ditches and a number of other features provided evidence of an Early Christian presence in the area. The two ditches were oriented northeast–southwest and were filled with a mid-brown sandy clay, charcoal and a reddish clay.

The north ditch (C7) was over 7m long and pits appeared to have been dug at either side of its western end, which ran into the baulk. These pits, 1.2m and 2m in diameter, seem to have been contemporary with the ditch. The ditch (C7) varied from 0.9m to 1.35m in width and was up to 0.5m deep, with straight sides and a flat base.

The southern ditch (C8) was parallel to C7 and had a similar shape and fill. It contained some undecorated coarse ware (97E036:3:2), which was found near the base; the pottery had a corky, untempered fabric. The ditch was 5.5m long, 0.8m wide and up to 0.5m in depth. Charcoal from the gully produced a date of AD 415–630.

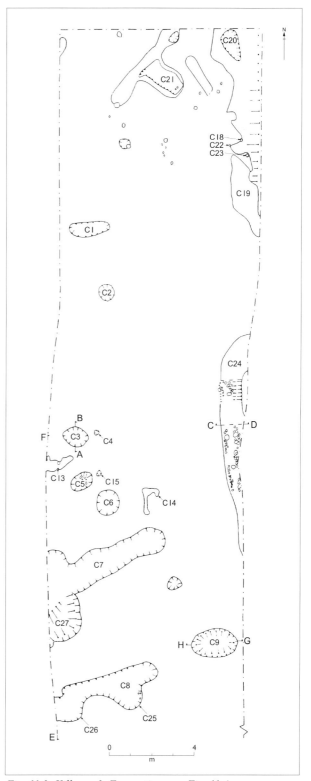

Fig. 11.3 Killoran 3. For sections see Fig. 11.4.

Burnt pit

A burnt pit (C9) was excavated on the eastern side of ditch C8. The pit measured 2m x 1.1m x 0.28m; it was oriented east–west. The steep, straight sides were fire reddened from *in situ* burning. The pit contained several alternating layers of silt and charcoal; the primary fill appeared to be

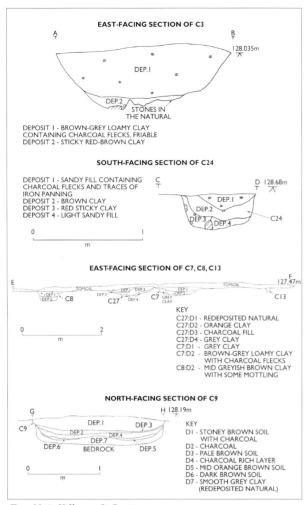

EAST-FACING SECTION OF C3

A · · · · · B

128.035m

DEP.1

DEP.2 STONES IN
THE NATURAL

DEPOSIT 1 - BROWN-GREY LOAMY CLAY
CONTAINING CHARCOAL FLECKS, FRIABLE
DEPOSIT 2 - STICKY RED-BROWN CLAY

SOUTH-FACING SECTION OF C24

DEPOSIT 1 - SANDY FILL CONTAINING   C   D 128.68m
CHARCOAL FLECKS AND TRACES OF
IRON PANNING   DEP.1
DEPOSIT 2 - BROWN CLAY   DEP.2
DEPOSIT 3 - RED STICKY CLAY   DEP.3 DEP.4   C24
DEPOSIT 4 - LIGHT SANDY FILL

0   1

m

EAST-FACING SECTION OF C7, C8, C13

E   F 127.47m
     TOPSOIL   DEP.1 DEP.1   DEP.1   TOPSOIL
  DEP.1   DEP.2
  DEP.2   C8   DEP.2   C27 DEP.4 C7 GREY   C13
                                        CLAY
                                   KEY
0   2   C27:D1 - REDEPOSITED NATURAL
                                   C27:D2 - ORANGE CLAY
       m   C27:D3 - CHARCOAL FILL
                                   C27:D4 - GREY CLAY
                                   C7:D1 - GREY CLAY
                                   C7:D2 - BROWN-GREY LOAMY CLAY
                                        WITH CHARCOAL FLECKS
                                   C8:D2 - MID GREYISH BROWN CLAY
                                        WITH SOME MOTTLING

NORTH-FACING SECTION OF C9

G   H 128.19m
C9   DEP.1   DEP.3   KEY
     DEP.2   DEP.4   D1 - STONEY BROWN SOIL
          DEP.7   WITH CHARCOAL
     DEP.6   BEDROCK   DEP.5   D2 - CHARCOAL
                                D3 - PALE BROWN SOIL
0   1   D4 - CHARCOAL RICH LAYER
                                D5 - MID ORANGE BROWN SOIL
       m   D6 - DARK BROWN SOIL
                                D7 - SMOOTH GREY CLAY
                                     (REDEPOSITED NATURAL)

Fig. 11.4 Killoran 3. Sections.

a burnt lining. Charcoal from the pit was dated to AD 980–1270. The pit formed a rough northwest–southeast alignment with the end of the northern ditch and a number of other pits/post holes that lay in the area immediately north of the ditch (C7); at least one contained what appeared to be packing stones. A sandstone spud stone (97E036:3:1) was recovered from post hole C5.

Post-medieval features

Several features appeared to date to the post-medieval period and represented ephemeral drainage activity. One ditch, C24, running along the eastern limit of the site was a stone-filled field drain. A number of other features along the eastern side of the site, such as burnt spreads and some ground disturbance, were probably also related to a post-medieval field boundary. These post-medieval features were mostly confined to the northern half of the area excavated, which also included a small number of pits and possible post holes.

Discussion

The site appears to represent a number of phases of drainage activity in the Early Christian and post-medieval period, with a phase of activity in the later first millennium

AD that may have been associated with settlement.

The presence of earlier linear ditch features at Killoran 3, and also possibly at Killoran 11 (see below), pre-dating the burnt pit there may relate to a phase of Early Christian land reclamation activity. The site was located 450m from a settlement at Killoran 8, dating to just after this phase, and 750m from a substantial quantity of unexcavated wetland material noted in the survey of Derryville Bog. The Early Christian land reclamation activity may also relate to a phase of activity contemporary with the large enclosure at Killoran 31, dating to the sixth/seventh century AD.

Several other examples of burnt pits were excavated in the area at Killoran 11, Killoran 15 and Killoran 16 (see below). These pits were all similar in size, usually oval in plan, and one, Killoran 11, contained stone; all showed evidence of *in situ* burning. Another example, at Killoran 23, dated to AD 660–880; this site closely resembled a *fulacht fiadh* trough in form and content, but may be part of an Early Christian phase in the area.

Killoran 4 (Fig. 11.5)

Cremation cemetery; Bronze Age; 7m x 5.8m; cremation pits, post holes and pits; crushed coarse ware pot found; brow of glacial ridge environment; construction phase; site code KIL04; licence 97E051; NGR E221331 N166629; 129.149m OD (top of cremation).

Paul Stevens

This site was revealed during monitoring of topsoil stripping for the construction of the waste rock stockpile. Archaeological excavation revealed a cluster of pits, including two containing cremated bone and one with a crushed pot, within an area 7m x 5.8m, roughly aligned northwest–southeast. The site was located along a glacial ridge overlooking a marshy plain to the east and a brook to the west.

Cremations

The cluster produced two cremations. The first cremation was within an oval pit, 0.6m east–west, 0.5m north–south and 0.2m deep, filled with a matrix of charcoal lumps and silty clay containing a small concentration of tiny burnt bone fragments.

The second cremation was contained within an upturned crushed coarse ware pot (97E051:4:1) that was rammed into a V-shaped pit 0.15m in diameter and 0.05m deep. Tiny fragments of crushed cremated bone were contained within the pit within a matrix of silty clay and virtually no charcoal.

Pits

The cremations were located within a loose linear alignment of post holes and pits representing an attempt to mark their location.

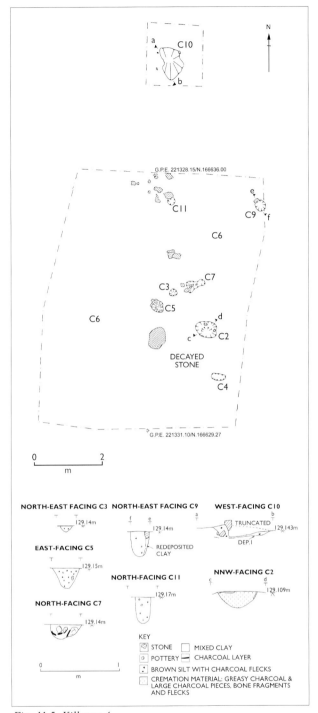

Fig. 11.5 Killoran 4.

Discussion

The coarse ware pot containing the cremation is similar to coarse ware fragments found at the Middle Bronze Age cemetery Killoran 10 and within the Middle Bronze Age settlement Killoran 8. Killoran 4 is also located 45m west of the Middle Bronze Age *fulacht fiadh* Killoran 5 and may be associated with isolated cremations at Killoran 6, a Late Bronze Age site 65m to the south, and Killoran 7, a Bronze Age site 35m to the north, located on the same glacial ridge (see below).

Killoran 5 (Fig. 11.6)

Fulacht fiadh; 1750–1410 BC; 12m x 10m; oval trough, hearth and burnt mound; animal tooth (Ovis) found; glacial knoll environment; construction, use and abandonment phases; site code KIL05; licence 97E372; NGR E221285 N166626; 127.57m OD (top of trough).

Paul Stevens

This *fulacht fiadh* was revealed during topsoil stripping for the waste rock stockpile area. No trace of the site was visible prior to construction. Excavation revealed an area of *in situ* burning or a hearth and a sub-oval trough backfilled with fire-cracked stone and charcoal and partly sealed by a truncated horseshoe-shaped mound of burnt sandstone, silt and charcoal.

Trough

The trough (C5) was an oval pit, oriented northwest–southeast, which measured 1.53m x 1.14m x 0.24m. It was filled with fire-cracked stone, charcoal and silty clay containing a fragment of a sheep/goat molar tooth. An internal stake hole was cut into the eastern wall of the trough, and an associated external stake hole cut the subsoil immediately to the west of the trough. Charcoal from the trough produced a radiocarbon date for the site of 1750–1410 BC.

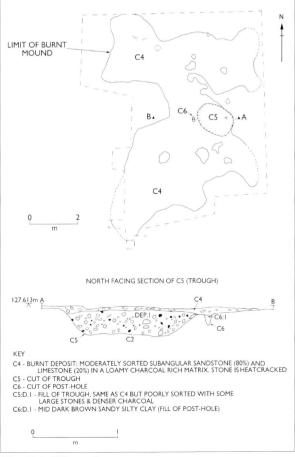

Fig. 11.6 Killoran 5.

Burnt mound

The ploughed-out burnt mound was roughly horseshoe-shaped. It was oriented north–south and measured 8–10m x 7–8m x 0.11m. It contained small fire-cracked stone, <0.15m in diameter, made up of 80% sandstone and 20% limestone and spread out downslope over the trough.

Hearth

The hearth consisted of a heavily burnt area of charcoal-rich natural sand with a fire-reddened surround and measured 2.8m x 1.7m. It was located 4m south of the mound but was probably contemporary. The area was oval in plan and concave in profile, contained within a small hollow. No retaining stone or timber surround was identified.

Discussion

The site, which overlooked a small stream, was located on the western break of the slope of a low plateau on which Killoran 4, Killoran 6 and Killoran 7 were also located. The trough was deliberately backfilled, in common with other dryland sites, and had a capacity of 381 litres. The use of sandstone rather than limestone implies that clean water was important, and this, together with the discovery of the sheep/goat tooth in the trough, as at Killoran 22 and Killoran 27, suggests that the site was used for cooking.

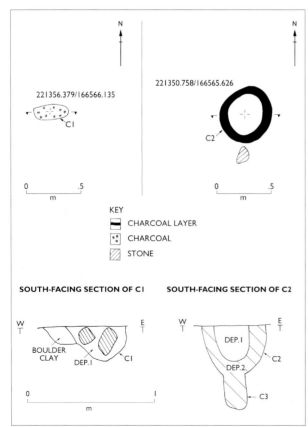

Fig. 11.7 Killoran 6.

Killoran 6 (Fig. 11.7)

Isolated cremation; 1145–900 BC; 10m x 0.5m; cremation pit, post hole and pit; brow of a glacial ridge; demarcation and construction phases; site code KIL06; licence 97E372; NGR E221350 N166565; 129.149m OD (top of cremation).

Paul Stevens

This site was revealed during topsoil stripping for the waste rock stockpile area and a rescue excavation was carried out. It consisted of two isolated pits cut into the subsoil. One pit (C1) contained a token cremation deposit of crushed, burnt bone fragments and charcoal. The charcoal from this pit produced a radiocarbon date of 1145–900 BC. The second pit (C2) appeared to be an associated post hole. C2 was circular in plan and U-shaped in profile, measuring 0.51m in diameter and 0.35m in depth. It cut an earlier post pipe (C3), 0.61m deep and 0.16m in diameter, suggesting that the post marked a burial plot. The site was located at the southern extent of a raised plateau overlooking lowlands to the east. The undated cemetery Killoran 4 and the isolated cremation pit Killoran 7 were also located on this plateau, 65m and 95m to the north, respectively.

Killoran 7

Isolated cremation; Bronze Age; 1m x 0.7m; cremation pit; brow of a glacial ridge; construction and deposition phases; site code KIL07; licence 97E372; NGR E221340 N166664; 128.809m OD (top of cremation).

Paul Stevens

This feature was revealed during topsoil stripping for the waste rock stockpile area. It consisted of an isolated pit cut into the subsoil containing charcoal and a single fragment of burnt bone. The pit had an irregular, oval plan, a U-shaped diagonal profile and measured 0.5m northwest–southeast, 0.2m southwest–northeast and 0.29m deep. It cut an earlier post pipe measuring 0.1m in diameter and 0.41m in depth, suggesting the post marked a burial plot. The site was located at the northern end of a ridge overlooking lowlands to the east. The cemetery Killoran 4 was located 35m to the south, and a second isolated cremation, Killoran 6, was located on the same glacial ridge, 95m to the south.

Killoran 8 (Figs 11.8–11.10; Plate 15.4)

Settlement; 1860–1845 or 1775–1430 BC, AD 685–985; 80m x 50m; houses, post holes, pits and linear ditches; coarse pottery, saddle quern fragment, burnt daub, rubbing stones, hammerstones, a possible whetstone, struck chert and flint were found; low glacial ridge environment; construction, use, reconstruction, abandonment, construction, use and abandonment phases; site code KIL08; licence 97E439; NGR E222097 N166698; 129.6m OD (top of house slot).

John Ó Néill

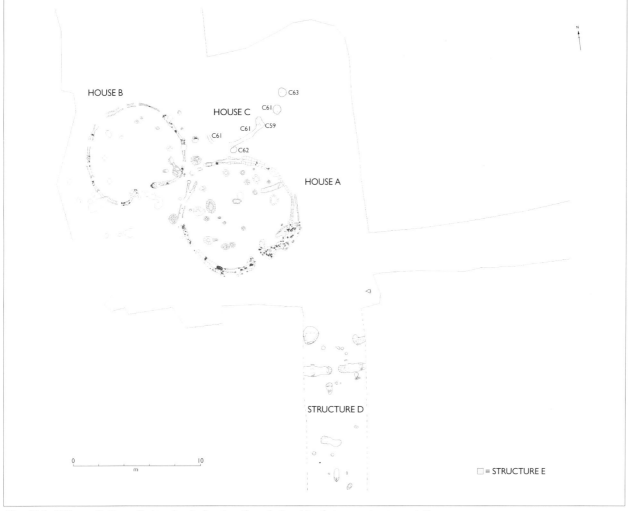

Fig. 11.8 Killoran 8. Overall site plan indicating the relationships between structures A-E.

A number of house sites were discovered during monitoring of topsoil removal for the Killoran northeast peat stockpile and borrow area. The peat stockpile was to be located on a low ridge on the western side of Derryville Bog within 200m of an area in which a series of excavations of *fulachta fiadh* and trackways took place in 1996 and 1997. The wide range of palaeoenvironmental work undertaken in the bog provides a wider context for the site.

After consultation with the Heritage Services, the developer halted topsoil removal in the area around the site and two strips were removed to the east and south of the main features to assess the archaeological potential of these areas. As it had not yet been stripped, the ground north of the site was left untouched. The house sites were uncovered in an area measuring 560m². Two 4m wide extensions were opened for 20m south and 30m east of the main site. No archaeological activity was recorded in the area west of the site during monitoring.

Full excavation of the exposed area took place in the winter of 1997 and identified the presence of three roundhouses (Houses A, B and C), along with a possible fourth structure (Structure D), south of the main site. A possible fifth, earlier, structure (Structure E) was disturbed by the wall slots of Houses A and B. Finds from the site include some sherds of coarse pottery, saddle quern fragments, burnt daub, rubbing stones, a hammerstone, a possible whetstone, some struck chert and some flint (see Chapter 12). The coarse pottery appears to be Bronze Age in date.

House A, House B and House C stood side by side, with the doorways of each affording a view past the wall of the neighbouring structure. It is likely that all three are contemporary.

Modern cultivation marks, lying under a depth of up to 0.2m of ploughsoil, were present across the top of the whole site. In places, the ploughing had gone deep enough to expose the limestone bedrock. There were no intact layers associated with any of the structures, which survived as truncated features dug into the subsoil.

In the two extension trenches, south and east of the site, there was no firm evidence of an enclosure around the site. As none was identified during monitoring to the west of the site, it appears that the settlement was unenclosed. The ground plan of the houses is similar to others from which Bronze Age dates have been obtained.

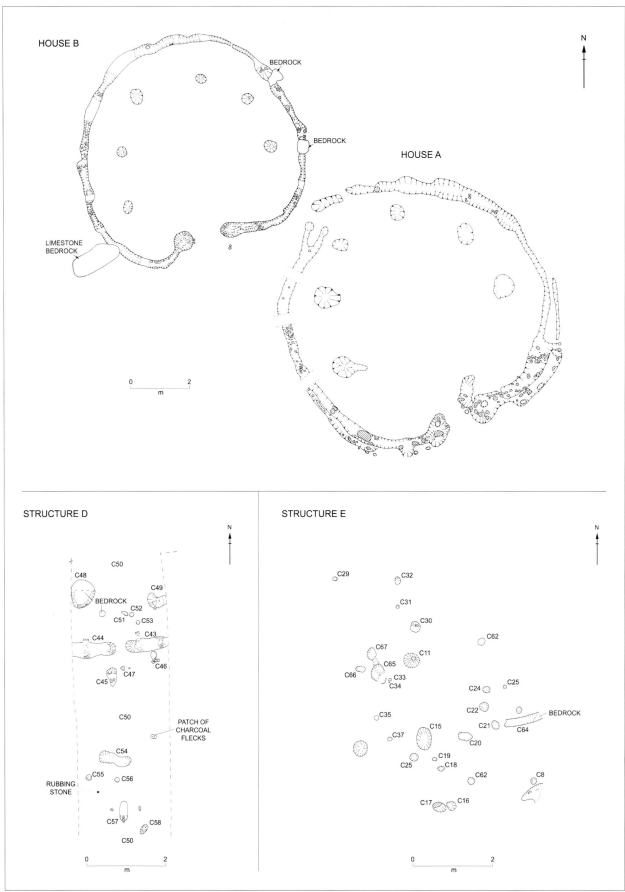

Fig. 11.9 Killoran 8. Plans of structures A, B, D and E.

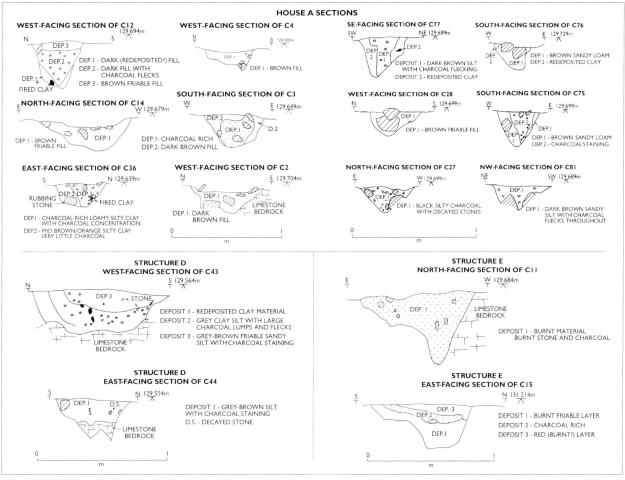

Fig. 11.10 Killoran 8. Sections.

House A

This was the largest house, measuring 9m in diameter. It survived as a circular wall slot with a doorway facing southeast. The wall slot contained occasional stake holes and stone packing, and appears to have contained wattle-built walls. The doorway was marked by two post holes either side of a gap in the wall slot and there was a single stake hole at the eastern side of the gap. An internal circular setting of six post holes appear to have contained the roof supports. There was an additional outer stake setting around the southeastern half of the house. Opposite the doorway was another gap in the wall slot, which may have been a back door. Possible struck quartz (97E439:8:17), a rubbing stone (97E439:8:6), a flint flake (97E439:8:5), saddle quern fragments (97E439:8:22 and 97E439:8:25) and a hammerstone (97E439:8:7) were found in features associated with the house. Numerous fragments of burnt daub were also found (see Chapter 12).

House B

Immediately north of House A was a smaller sub-circular house, approximately 8m in diameter. It had an internal ring of post holes for roof supports, concentric with the outer wall slot. There was a southeast-facing doorway with two stake holes on the western side of the door. Again, the wall slots contained stake holes and stone packing suggesting wattle walls. A number of finds, including some sherds of coarse ware (97E439:8:15 and 97E439:8:24) were associated with the house.

House C

This house lay to the northeast of House A and survived as a doorway facing southeast, part of the western wall slot running away from the door and a number of post holes. It lay very close to a modern tree line and field boundary, which may have removed most of the traces of the house. No other features could be identified with this house.

Structure D

This structure was uncovered 9m south of House A in the southern extension trench to the main site. Features uncovered in this area included an interrupted gully that, on excavation, seemed to contain a doorway of similar dimensions to that of Houses A, B and C, marked by a stake hole on the western side. This doorway faced north towards the other houses. Two post holes were set just inside the doorway, and a gully that may represent the back wall of the structure was found 5m to the south. The nature of this structure is problematic, and it may be that the gully and 'doorway' are part of an enclosure around the main site, although this could not be identified anywhere else.

Structure E

This was a possible post-built structure that was cut by the western side of House A and the eastern side of House B. Three probable structural post holes were identified, along with a large number of other subsidiary features that pre-date the other two houses and seem to form a sub-rectangular pattern. The identification of these features as a structure is somewhat uncertain, however. A chert flake (97E439:8:15) was recovered from a pit, C11, associated with the structure.

Investigations of the northeastern margins of the site, east of House A and north of House C, revealed other features running into areas that were not exposed. As the developer had agreed to fence off the site, no further excavation took place, and these features were covered with plastic, stones and topsoil and left intact.

Killoran 9 (Fig. 11.11)

Road; post-medieval; >30m x >7m; clay and stone layer, cut; brick found; construction phase; brow of glacial ridge; site code KIL09; licence 97E372; NGR E222045 N166427; 124.75m OD (top of road).

Paul Stevens

This site was uncovered during the excavation of a square trench measuring 7.3m x 7m x 0.4–0.9m for the installation of a noise and vibration monitoring station. The site was found to be a linear drove-type road of post-medieval date. It was oriented northeast–southwest, within a shallow cut, and measured over 30m long, over 7m wide and 0.55m in depth. The linear portion was partly revealed in section as a 0.24m deep deposit of brown sandy clay over

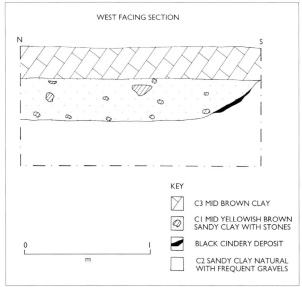

Fig. 11.11 Killoran 9.

yellowish brown sandy clay containing small stones and early post-medieval brick, 0.3m deep, set within a shallow concave cut that was partly burnt and lined with larger cobbles. The line of the feature was traced eastwards for 10m in the section of a service trench for the monitoring station. The road does not appear on the first edition OS map of Tipperary (6" Sheet No. 36), and the field boundaries shown on the map do not respect the line of the road.

Killoran 10 (Fig. 11.12; Plate 15.8)

Cremation cemetery; 1435–1215 BC; 11m x 9m; cremation pits and pits; glacial till, close to bog and stream; demarcation, construction and reuse phases; site code KIL10; licences 97E168 and 97E168 extension; NGR E221792 N166269; 127.63m OD (top of cremation pit).

Paul Stevens

Monitoring of topsoil removal for the construction of an access road to the temporary magazine uncovered a large cluster of small pits that were subsequently established to be a flat cremation cemetery of Middle Bronze Age date. Further non-intrusive excavation was conducted to establish the exact dimensions of the site, for the purpose of preservation. The site was located on a low plateau surrounded on three sides by marshy ground and the Moyne Stream to the south. The Middle Bronze Age stone causeway Killoran 18 lies 400m to the east in Derryville Bog, and the settlement site Killoran 8, as well as the nearby cultivation ditches at Killoran 2, lies 530m to the north.

Excavation was carried out along the line of the access road and revealed a total of seventy pits within an area measuring 11m x 9 m. No visible trace of an enclosing ditch or boundary was noted to the north or south. The pits were clustered with no apparent regularity within a loose east–west alignment, with several inter-cutting one another and three lying outside the cluster to the north. Most were circular in plan and measured 0.75–0.13m (mode 0.3m) in diameter and 0.37–0.05m (mode 0.2m) in depth The cemetery continued east and west beyond the course of the road.

Cremations

Evidence for twenty-eight individual burials was recorded. Twenty-six pits were sealed or partly sealed by a capping of local boulder clay; of these, twenty produced cremated bone and a further eight also contained bone but had no capping. Several cremation pits cut post pipes, including one that still contained an *in situ* post fragment, suggesting the use of wooden marker posts prior to burial.

Post holes/pits

A further twenty-nine pits or post holes without bone were noted. They were filled with a charcoal-rich silty matrix. These fills were analysed for microscopic cremation remains but did not show any evidence of bone.

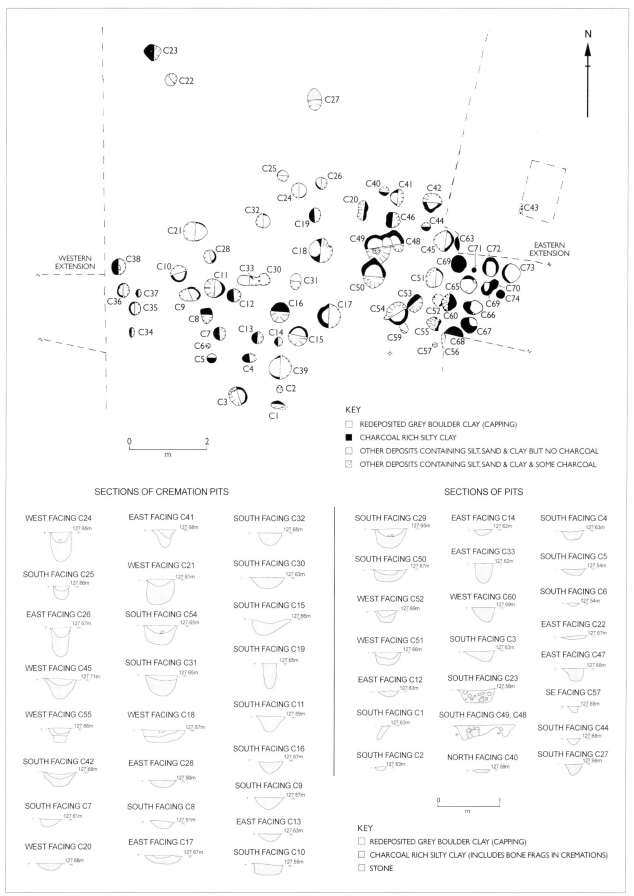

Fig. 11.12 Killoran 10.

A further thirteen pits were uncovered outside the area under development and remained unexcavated; several of these pits were capped, and all were of a similar size to the cremation and other pits.

Discussion

One cremation pit produced a number of fragments of coarse ware pottery (97E168:10:1) similar to that found in the complex of Middle Bronze Age houses at Killoran 8 and also at Killoran 4. It is feasible that this cemetery was contemporary with the settlement Killoran 8, albeit slightly earlier in date, several of the *fulachta fiadh* and the large causeway Killoran 18. Certainly, the dates for these sites represent the most intense phase of use of the landscape throughout the archaeological record, which shows a climax of activity in the Bronze Age. The cemetery is at present the largest known of its type in Ireland.

Killoran 11 (Fig. 11.13)

Roasting pit; medieval/post-medieval; 2.5m x 1.32m; roasting pit, linear ditches and a post hole; an iron object, a coarse ware pot sherd and post-medieval pottery found; sandy glacial ridge; drainage, construction, reuse and abandonment phases; site code KIL11; licence 97E372; NGR E220424 N166774; 131.6m OD (top of pit).

Paul Stevens

This site was uncovered during the monitoring of topsoil stripping for the Killoran east borrow area, a sand and gravel quarry. Excavation revealed a small cluster of features, including a possible medieval roasting pit and a linear field ditch. The site was on a sandy glacial knoll, 12m east of the courtyard of Killoran House, a large manorial house dating to *c*. AD 1783; 70m to the west of a post-medieval pit at Killoran 21; and 170m south of several medieval and post-medieval features at Killoran 16. Topsoil finds from Killoran 11, dating to the post-medieval period, suggested a substantial level of cultivation in the fields around the back of the house. An early medieval date for a large pit and an Iron Age date for a house at Killoran 16 showed an earlier presence in the area.

The site consisted of an oblong pit cutting a large linear field ditch, with an associated post hole. The earliest feature revealed on the site was a linear V-shaped ditch, oriented east–west, measuring over 1.45m in width and 0.6m in depth.

Roasting pit

An oblong pit, 2.5m x 1.32m x 0.3m, cut the linear ditch. It was oriented east–west and was slightly offset from the ditch. The base of the pit contained charred timbers and charcoal. The northern side of the pit was fire reddened and showed *in situ* burning. The upper fill contained a

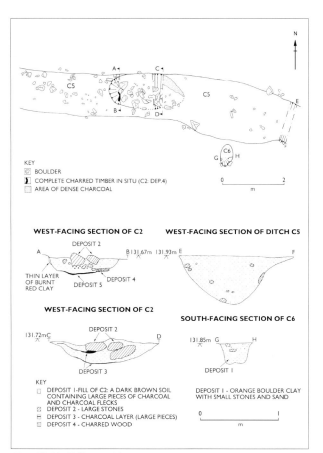

Fig. 11.13 Killoran 11.

large quantity of small and medium-sized limestone boulders within a matrix of silty clay and charcoal. The fill contained two fragments of coarse wheel-turned pottery (97E372:11:1) and a small iron nail point (97E372:11:2). The feature was interpreted as a roasting pit due to the presence of stones, charred wood and *in situ* burning.

Discussion

Several other examples of burnt pits were excavated in the area. Most notable was that at Killoran 3, which was dated to AD 980–1270. Other examples were excavated at Killoran 15 and Killoran 16. These pits were all similar in size and usually oval in plan. Although they did not all contain stone, all showed evidence of *in situ* burning. Another example, at Killoran 23, dated to AD 660–880; this site closely resembled a *fulacht fiadh* trough in form and content, but may be part of an Early Christian phase in the area.

The presence of an earlier linear ditch feature at Killoran 11, pre-dating the pit, may relate to a phase of Early Christian rather than late medieval/post-medieval activity. An early linear feature at Killoran 3, dated to AD 415–630, produced similar coarse pottery to that found in the pit at Killoran 11, suggesting a phase of activity in that period possibly contemporary with the large enclosure at Killoran 31, which was dated to the sixth/seventh century AD.

Killoran 12

Possible *fulacht fiadh*; Bronze Age; destroyed; burnt mound material; sloping glacial till; site code KIL12; licence 97E372 extension; NGR E220220 N166895; approx. 127.5m OD.

Paul Stevens

This site was revealed during tree clearance and removal of a stone boundary wall and ditch to facilitate the installation of an ESB pole. It consisted of a destroyed burnt mound, evidenced in the spoil. The site lay on shallow sloping ground at the southern margin of reclaimed bog and was part of a cluster of *fulachta fiadh* that includes Killoran 13 and Killoran 14, 60m and 160m to the west, respectively.

The site was a 25m (east–west) x 8m (north–south) spoil heap consisting of felled-tree debris, limestone wall blocks and an 8m x 8m x 2.3m pile of fire-cracked stone within a black, charcoal-rich sandy silt matrix. There was no visible trough within the disturbed area. A field boundary ditch, 2m wide and 0.75m in depth, filled with medium-sized boulders, large cobbles and burnt mound material cut the site and may have removed all trace of the trough.

Killoran 13

Burnt mound; Bronze Age; 13m wide; burnt mound material; sloping glacial till; site code KIL13; licence 97E372 extension; NGR E220162 N166918; 125.5m OD (base of mound).

Paul Stevens

This site was revealed during trench cutting for the haul road drainage scheme. A large trench was excavated by machine north of the haul road. The site constituted a burnt mound consisting of a large spread of fire-cracked stone and charcoal. The site was located on the east-facing slope of shallow sloping glacial till and was surrounded to the north and east by bog (drained to the east). It formed part of a cluster of three sites that included Killoran 12, 60m to the east, and Killoran 14, 100m to the west.

The burnt mound was 0.54–1.26m thick and made up largely of fire-cracked stone <0.3m in diameter, overlying yellow sand and gravel glacial till. The mound was sealed by a layer of grey-brown sandy clay to the east of the section; this in turn was sealed by dark grey silt and dark grey silty clay topsoil. No surface trace of the site was visible prior to excavation and no trace of a trough was found. The site continued to the north and south of the drainage trench but was not noted in the topsoil stripping for the haul road.

Killoran 14 (Fig. 11.14)

Fulacht fiadh; Bronze Age; 14.8m x >5.8m; disturbed trough and burnt mound; sloping glacial till; construction, use and abandonment phases; site code KIL14; licence 97E372 extension; NGR E220062 N166934; 125.88m OD (base of burnt mound).

Paul Stevens

This site was revealed during trench cutting for the haul road drainage scheme. A large trench was excavated by machine north of the haul road, with a further area of disturbance found north of the trench. The site was a ploughed-out burnt mound consisting of a large spread of fire-cracked stone and charcoal. It was located on the west-facing slope of shallow sloping glacial till and was surrounded to the north by bog. It formed part of a cluster of three sites that included Killoran 12 and Killoran 13, 160m and 100m to the east, respectively.

Burnt mound

The roughly horseshoe-shaped burnt mound was disturbed in several places by plough furrows and measured 14.8m east–west, over 5.7m north–south and 0.24m in depth. The southern edge of the mound extended into the baulk. No surface trace of the mound was visible prior to excavation and the site was not noted during the monitoring of the haul road, immediately south of the drainage trench. The northwest corner of the mound overlay peat; elsewhere, the site sealed natural sand and gravel glacial till.

Trough

A shallow, roughly circular depression, 3.9m in diameter and 0.33m in depth, may have represented the trough. The cut had gently sloping sides and an irregular base and was filled with burnt mound material of fire-cracked stone <0.3m in diameter and charcoal-rich silt. The estimated capacity of the trough was 12,384 litres, by far the largest capacity of any trough excavated during the project. It is likely that this site represented an informal hearth or the site of a removable trough with a smaller capacity.

Killoran 15 (Fig. 11.15)

Burnt pit; early medieval; 2.46m x 1.5m; burnt pit and oval pit; sloping glacial till; construction and use phases; site code KIL15; licence 97E372 extension; NGR E221173 N166663; 128.32m OD (top of burnt pit).

Paul Stevens

This site was revealed during topsoil stripping for the construction of a compound for the contractor Kvaerner Cementation Ltd. The site was situated at the base of an east-facing glacial ridge. The site consisted of three features—a large burnt pit, a smaller pit and a furrow—spread over an area of some 80m north–south x 55m east–west.

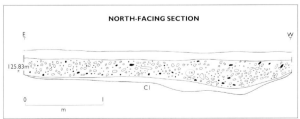

Fig. 11.14 Killoran 14. Section.

Burnt pit

This feature was oval in plan and measured 2.46m east–west, 1.5m north–south and 0.4m in depth. The sides of the pit were substantially reddened and showed *in situ* burning. The pit had near vertical, stepped sides and was filled with several layers of clayey silt, with intermittent layers of charcoal. The feature was disturbed to the northeast by a cultivation furrow.

A second pit was excavated 80m to the east. This consisted of an oblong cut with steep concave sides and measured 0.5m east–west, 0.4m north–south and 0.25m in depth. The feature had several fills of silty clay and charcoal. A re-cut within the centre of the cut measuring 0.3m north–south, 0.15m east–west and 0.25m in depth probably constituted a post hole.

Discussion

Several other examples of burnt pits were excavated in the area. Most notable was that at Killoran 3, which was dated to AD 980–1270. Other examples were excavated at Killoran 11 and Killoran 16. These pits were all similar in size and usually oval in plan, and all showed evidence of *in situ* burning. Another example, at Killoran 23, dated to AD 660–880; this site closely resembled a *fulacht fiadh* trough in form and content, but may be part of an Early Christian phase in the area.

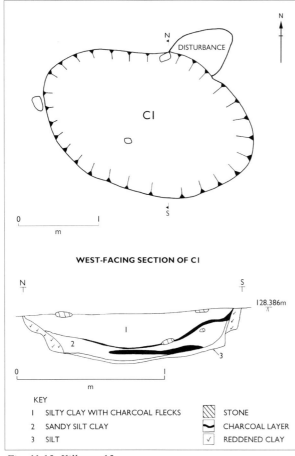

Fig. 11.15 Killoran 15.

Killoran 16 (Figs 11.16 and 11.23; Plate 4.5)

Archaeological complex; 180 BC–AD 425, AD 890–1040 and post-medieval; 60m x 52m; house, large pit, burnt pits, lazy beds and pond; post-medieval iron nails, pottery, brick, clay-pipe fragments, worked stone, bone, a whetstone fragment and a possible rubbing stone found; glacial knoll; occupation, construction, extension and abandonment, pit use and abandonment and cultivation phases; site code KIL16; licence 98E066; NGR E222350 N166900; 131m OD (house).

Cara Murray

This archaeological complex was revealed during topsoil stripping for the Killoran north borrow area, a sand and gravel quarry. The site was located on the brow of a high glacial ridge within an area of undulating land, surrounded on all sides by reclaimed bog. The site lay within a cluster of archaeological activity: Killoran 12 to the west, Killoran 23 to the northwest, Killoran 11 to the south and Killoran 1 and Killoran 21 to the east. The post-medieval manor, Killoran House, was also located nearby. The main concentration of archaeological features was located on top of the eastern side of the glacial knoll. This area (Area 1) measured 20m east–west. Another concentration of material was located to the north of this area.

There were four phases of activity on the site: a house site, a large backfilled pit, sub-rectangular cultivation furrows (lazy beds) and an ornamental pond. Material associated with Phases 1 and 2 was located within Area 1 and related to the later Iron Age/Early Christian period. Phase 3 and 4 material occurred around the site. A series of burnt pits (Phase 3) was undated but may be early medieval. The latest phase of activity (Phase 4) was related to post-medieval agriculture and land improvements, including drainage.

Phase 1: Late Iron Age roundhouse

The house site consisted of a central post, wall posts and a southeast-facing doorway formed by four post holes. Charcoal taken from one of the post holes was dated to 180 BC–AD 425. The house was slightly oblong, measuring 13.22m in minimum diameter and 14.88m in maximum diameter. The partial remains of two shallow wall slots extended southwest from the doorway post holes, tapering to a narrowed shallow end at its northern surviving limit. Within this structure, there was a series of smaller post holes/stake holes, which were located in the eastern area of the structure, west and southwest of the doorway. Many of these were cut at very unusual angles.

Phase 2: Early Christian large pit

The second phase of activity on the site was a large pit located 2.85m north of the northern limit of the house site. It was 3m in diameter and 0.84m in depth, but its function is unknown. The pit appeared to be lined with a thin smear of yellowy orange clay and was filled with natural sand

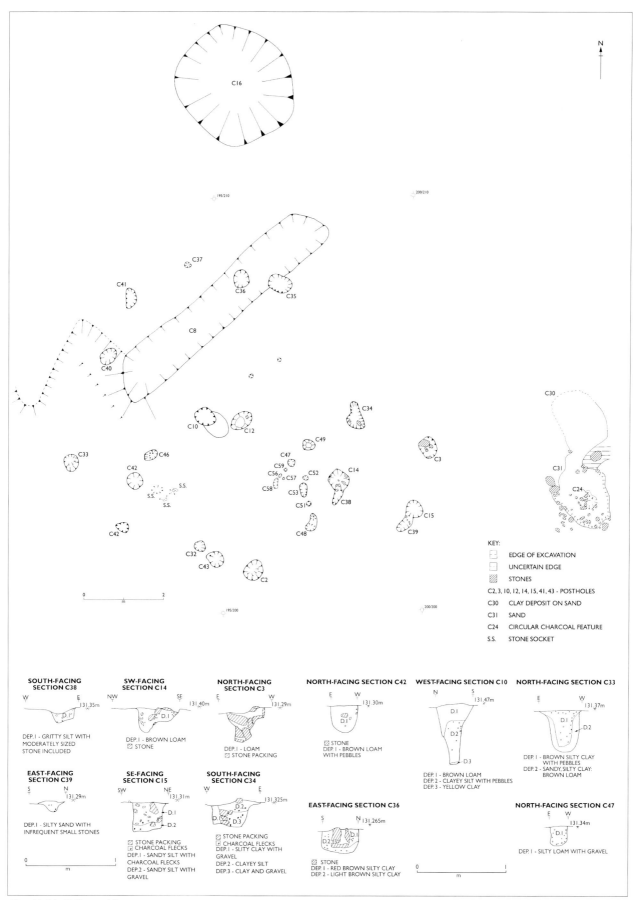

Fig. 11.16 Killoran 16.

and gravels, although it also contained a dark yellowish brown silty clay deposit from which charcoal samples were extracted. This charcoal produced a radiocarbon date of AD 890–1040. The feature was cut by a semicircular arcing gully containing a re-cut post hole. A charcoal axe-marked point was recovered from this post hole.

Phase 3: early medieval burnt pits

The third phase of activity consisted of small, localised areas of burning. Two burnt pits, one cut by a cultivation furrow, were also noted. These pits were oval in plan and had regular, steep sides and were fire reddened from *in situ* burning. The pits contained fills of intermittent layers of silty clay and charcoal. Similar burnt pits were noted at Killoran 3, dated to AD 980–1270; Killoran 11 and Killoran 15.

Phase 4: post-medieval

The final phase was agricultural activity in the form of cultivation furrows and drainage ditches. Numerous northeast–southwest oriented cultivation furrows of various dimensions were interspersed across the entire area; some contained post-medieval brick. These may relate to activity associated with Killoran House after AD 1750.

Downslope of Area 1, to the north and west, the area was much wetter, and a post-medieval artificial pond, 9m x 7.3m, had been dug and surrounded by a deposit of stones. The occurrence of the pond and the discovery of a large stone-lined field drain indicated that this was the result of post-medieval land improvement. Judging by the location of the nearby *fulachta fiadh* sites, the main drainage discharge for this area probably occurred to the west and east.

Killoran 17 (Fig. 11.17)

Fulacht fiadh; 2585–2195 BC; 10.8m x 8m; three inter-cutting circular/sub-oval troughs and a burnt mound; glacial ridge; construction, use, reconstruction and abandonment phases; site code KIL17; licence 97E372 extension; NGR E221210 N166999; 127.47m OD (top of trough 1).

Paul Stevens

This site was revealed during the construction of an access road within the magazine. It was located at the base of a gentle south-facing slope, overlooked by higher ground to the south and surrounded to the north and west by bog. The site was a *fulacht fiadh* consisting of three inter-cutting troughs, backfilled with fire-cracked stone and charcoal and partly sealed by a truncated horseshoe-shaped mound of burnt sandstone, silt and charcoal. No trace of the site was visible prior to construction. This *fulacht fiadh* was located 10m north of Killoran 25 and 60m northeast of Killoran 22 in a cluster of similar structures. Several isolated cremations were also located at Killoran 24, 60m to the southeast.

Trough 1

The first trough, C1, was a sub-circular pit, oriented north–south and substantially truncated by a 2.5m wide machine-excavated trench; it was 2.3m x 1.8m x 0.45 m. It contained a lower fill of grey clay, 10 mm thick, which sealed 50% of the base of the trough, and an upper fill of fire-cracked stone, <0.2m in diameter, charcoal and silty clay. The upper fill was sealed by burnt mound material and peat. Charcoal from the trough produced a radiocarbon date of 2585–2195 BC.

Trough 2

The second trough, C2, was an oval cut that partially cut the northern edge of trough C1 and its fill. The cut was oriented north-northwest–south-southeast and measured 2.1m x 1.8m x 0.47m. An associated internal stake hole was cut into the northeast corner of the base of the trough. The southern corner of the cut was elongated to allow for the insertion of a clay revetment, which was packed up against the southern end of the trough, sealing the edge of the earlier trough and its fill. This course of action maintained a clean water percolation into the second trough, uncontaminated by the charcoal fill of Trough 1. It also reduced the internal dimensions of the trough to 1.7m x 1.8m x 0.47m. The trough was filled by a gritty silt in-wash from the clay revetment at the base, fire-cracked stone, charcoal and silty clay; in turn sealed by similar material containing large stones and boulders; in turn sealed by peat.

Burnt mound

The ploughed-out burnt mound was truncated by plough furrows and was roughly circular, measuring 10.8m east–west, 8m north–south and 0.15m in depth. It consisted of small fire-cracked stone, <0.15m in diameter, made up of 80% sandstone and 20% limestone, spread downslope over the troughs.

Discussion

This site represents the earliest dated *fulacht fiadh* investigated by the Lisheen Archaeological Project and is one of the earliest in Ireland. The site was located on the western break of a steeply sloping glacial ridge at the southern edge of the northwestern spit of Derryville Bog, north Killoran. This glacial ridge was the focus of activity for a number of burnt mounds and cremations, including sites Killoran 22, Killoran 24, Killoran 29 and Killoran 30.

The two troughs were both used and deliberately backfilled, in common with the other dryland sites, and had capacities of 1,000 litres (Trough 1) and 1,224 litres (Trough 2), compatible with the larger troughs of Derryville Bog and Killoran 19. The troughs may represent a time-lag between phases of use. The earlier trough appears to have the remnants of a clay lining. This use of clay as a sealant is also demonstrated in the second

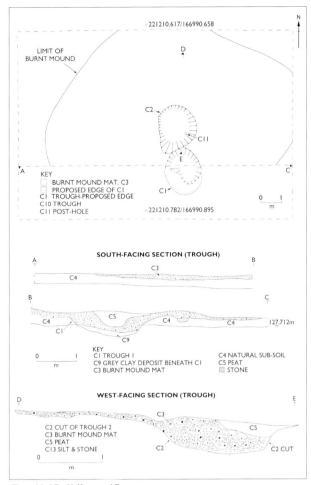

Fig. 11.17 Killoran 17.

trough. It seems that clean water was important to the builders. The use of sandstone over local limestone reinforces this view. The discovery of sheep bones at Killoran 5, Killoran 22 and Killoran 27 and the discovery of a saddle quern at Killoran 240 suggest that cooking was the function of these sites.

Killoran 19 (Fig. 11.18)

Fulacht fiadh; Bronze Age; 15m x >5m; re-cut trough and spring, sub-rectangular trough and burnt mound; glacial knoll; construction, use, reconstruction and abandonment phases; site code KIL19; licence 97E372 extension; NGR E221682 N166355; 127.04m OD (top of troughs).

Paul Stevens

This site was revealed during topsoil stripping for a car park in the area of the site owner's compound. It was located on largely flat ground, rising very gently on all sides. The site constituted a *fulacht fiadh* consisting of a re-cut trough complex and a second trough, both backfilled with fire-cracked stone and charcoal and both sealed by a truncated horseshoe-shaped mound of burnt sandstone, silt and charcoal. No trace of the site was visible prior to construction.

Trough 1

The earliest trough, C12, was a sub-rectangular pit, oriented northeast–southwest, measuring approximately 1.6m x 1.55m x 0.47m. It was partially truncated by a modern drainage ditch, 1m wide. In the base of the trough, there was a large circular pit, which appeared to be contemporary with the trough, that served as a well or sump for water percolation. The lower fill of this pit was very peaty, and there were large pieces of sandstone and some flags, possibly a disturbed stone lining. Two associated internal stake holes were cut into the north and east corners of the base of the trough, but it was not possible to establish if the other corners contained similar features. The trough contained an upper fill of fire-cracked stone, <0.5m in diameter, charcoal and peat. This was completely sealed by burnt mound material.

Later troughs

The later troughs (C2, C15 and C10) were in effect several attempts at a successful trough. They consisted of a large pit complex, truncated on both sides by modern pipe trenches, cut and re-cut by shallow oval or rectangular pits to form a large irregular trough with a flat base containing a circular post hole, C11, and an oblong depression, C9. The latter appeared to represent a spring with an associated post hole, C11, and an additional stake hole, C8. The spring had been widened to allow access,

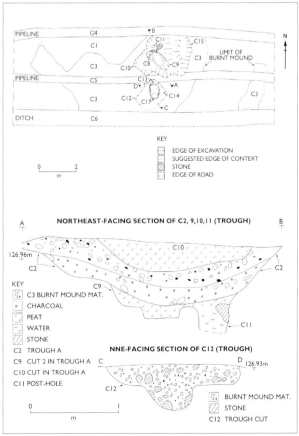

Fig. 11.18 Killoran 19.

and the posts may have supported a step, as at Killoran 27 (see below), or a winch or merely marked the location.

The sequence of construction started with the sub-rectangular cut C2, which was presumably unsuccessful. C2 was cut by C15, slightly to the west. Within C15, the spring, or rather swallow hole, was uncovered and subsequent post and stake holes, C11 and C8, were inserted. The complex was then left open for enough time for peat to accumulate in the swallow hole. It is also feasible that peat was brought in to act as a seal or plug to the spring.

The oblong depression, C9, measured 1.52m x 0.77m, was oriented northwest–southeast and had an excavated depth of 0.55m. C11 measured 0.96m in diameter and was 0.46m in depth. C8 measured 0.09m in diameter and was 0.18m in depth. C2 was approximately 2.45m north–south, 1.45m east–west and 0.4m in depth. C15 was approximately 2.5m north–south and 0.7m in depth. All features were filled with burnt mound material of occasional fire-cracked stone and charcoal, which sealed the peat fill of C9. An upper fill of more fire-cracked stone and charcoal sealed all features.

The above features were cut by a shallow northeast–southwest-oriented sub-rectangular cut, C10, probably a later trough, measuring 2m x 1.2m x 0.3 m. The cut was filled by *Sphagnum* peat formed in a hollow.

Burnt mound

The ploughed-out burnt mound was truncated by modern pipe trenches and a road drainage ditch, and was roughly circular. It measured 15m east–west, >5m north–south and 0.17m in depth. It consisted of small fire-cracked stone, <0.4m in diameter, made up of 80% sandstone and 20% limestone, spread out over the troughs.

Discussion

This site represented a multi-phase *fulacht fiadh*. Such sites were also noted at Killoran 17 and Killoran 27. The site was located on level ground subsequently overgrown by shallow bog, 120m south of the Late Bronze Age *fulacht fiadh* Killoran 26, 240m southeast of the Middle Bronze Age *fulacht fiadh* Killoran 27 and 130m west of the Middle Bronze Age cremation cemetery Killoran 10.

The troughs were all used and deliberately backfilled, in common with the other dryland sites, with the exception of C10, which may not have been a trough at all. C2, C10 and C12 had capacities of 1,421 litres, 720 litres and 1,164 litres, respectively, compatible with the larger troughs of Derryville Bog and Killoran 17. The troughs may represent a time-lag between phases of use. The earlier trough appeared to have the remnants of a stone lining, and both C12 and C11 had a considerable peat content in their fills, suggesting peat was used as a liner or plug, a feature also noted at Killoran 253. It appears that clean water was important to the builders. The use of sandstone over local limestone reinforces this idea and would suggest cooking was the function of these sites.

Killoran 21 (Fig. 11.19)

Fulacht fiadh; Bronze Age; >9.5m x 8m; circular trough and burnt mound; base of a glacial knoll; use and abandonment phases; site code KIL21; licence 97E372 extension; NGR E220529 N166807; 127.34m OD (top of trough).

Paul Stevens

This *fulacht fiadh* was revealed during topsoil stripping for the Killoran east borrow area, a sand and gravel quarry. No trace of the site was visible prior to construction. The site lay at the base of the north-facing slope of a glacial knoll, at the edge of a small peat-filled hollow. Killoran 1 was situated 90m to the northwest, on the opposite side of the hollow. The site consisted of a sub-circular trough and several associated pits backfilled with fire-cracked stone and charcoal, and partly sealed by a truncated horseshoe-shaped mound of burnt sandstone, silt, charcoal and peat. The site continued into the baulk to the north and remained unexcavated.

Trough

The trough was a sub-circular pit measuring 1.94m east–west, 1.76m north–south and approximately 0.5m in depth. It was backfilled with clayey deposits, 0.73m in depth, taking the form of a low, flat-topped mound protruding above the edges of the cut and filling the trough almost entirely. A limited amount of burnt mound material was noted within the trough at the southern side. The trough contained a total of thirteen deposits, most of which were related to the clay fill, and included large stone boulders and layers of charcoal.

A small northwest–southeast oriented pit, measuring 1.17m x 0.45m x 0.29m, was located 2m south-southeast of the trough; it contained a single fill of burnt mound material. The second feature was an east–west aligned concave linear measuring 3m x 0.75m x 0.13m. It too was filled by burnt mound material, and, although it ran directly towards the trough and slightly downslope, it stopped short of the trough by about 1.40m.

Burnt mound

The ploughed-out burnt mound covered an area of over 9.50m north–south by 8m east–west, to a maximum depth of 0.13m, however, it was very patchy, surviving mostly at its southern extent. It consisted of small fire-cracked stone, <0.15m in diameter, made up of 80% sandstone and 20% limestone, which was spread out upslope, partly concealing the trough.

Discussion

The site was located on the northern break of the slope of a glacial knoll, close to the edge of a peaty marsh. It was close to a similar site, Killoran 1, and others, not disturbed by the development, may have been located to the north and east. The trough had been deliberately backfilled, in common with other dryland sites. However, this site was backfilled with natural clay, sand and large stones, and it is possible that the trough contained a stone lining. The trough had a large capacity of 1,216 litres.

Killoran 22 Structure 1 (Fig. 11.20)

Fulacht fiadh; Bronze Age; >8m x 4m; sub-circular trough, possible circular trough and burnt mound; two *Ovis* teeth found; glacial till; construction, use, abandonment and disturbance phases; site code KIL22; licence 97E372 extension; NGR E221125 N166905; 127.45m OD (top of trough).

Paul Stevens

This heavily ploughed-out *fulacht fiadh* was revealed during topsoil stripping for a drainage channel within the magazine borrow area, a sand and gravel quarry. No trace of the site was visible prior to construction. It was situated at the base of a west-facing slope, 5.5m west of Killoran 22 Structure 2, 73m west of Killoran 25 and

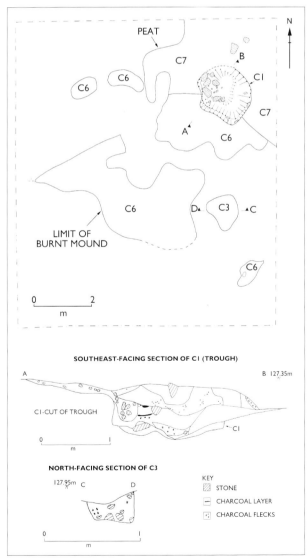

Fig. 11.19 Killoran 21.

127m southwest of Killoran 17. The terrain on all sides was undulating, except to the north, where it sloped northwards under the bog, towards Derryville Bog. The structure consisted of a sub-circular trough partly sealed by thin patches of a heavily ploughed-out spread of burnt sandstone, silt, charcoal and peat. A linear field drain cut north–south across the site and was immediately adjacent to the trough. It may have destroyed a second trough. A drainage ditch dividing the present-day field from marginal scrub forest and bog truncated the site. Part of the site continued north into the baulk, but it was not part of the development and so remained unexcavated.

Troughs

The trough was a sub-circular pit, 1.25m in diameter and 0.25m deep, containing three fills. The lower fill consisted of sand, silt and charcoal; the middle fill consisted of burnt mound material of fire-cracked stone and charcoal and the upper fill consisted of peat with gravel and large stones. Two sheep teeth, possibly molars, were recovered from the upper fill. The trough showed signs of a burnt lining of light brushwood, impressed into the sides.

A later field drain running the whole width of the site substantially truncated a second possible trough located immediately to the east. This trough was irregular in profile and had a steep, straight east side and a near vertical west side. It was 1.4m in diameter and 0.58m deep. It was filled with dark greyish-brown sandy clay silt with burnt sandstone, <0.06m in diameter. The evidence for a second trough was a curved kink in the eastern edge of the otherwise straight-sided drain and a change in the drain profile from a V-shape to a scarp. The second trough appeared to cut the first in profile and represented an earlier feature.

Burnt mound

The ploughed-out burnt mound covered an area >10m east–west by 8m north–south, to a maximum depth of 0.1m, and was very patchy due to heavy truncation by mechanical excavators during stripping. It survived mostly at its western extent and consisted of small fire-cracked stone, <0.15m in diameter, made up of 80% sandstone and 20% limestone. It spread out upslope, partly concealing the trough. The mound was cut by a large north–south aligned field drain and was peppered with tree bole disturbance.

Discussion

The site was located on the low western break of a slope of glacial till, stretching westwards and northwards to the present-day raised bog of Derryville. This land had been drained in antiquity and soil cover was thin. The site was close to similar sites at Killoran 17, Killoran 25, Killoran 29 and Killoran 30. The troughs had been deliberately backfilled in common with other dryland sites. The first trough had evidence of a burnt lining of wood; this was also revealed at Killoran 26. Examples from

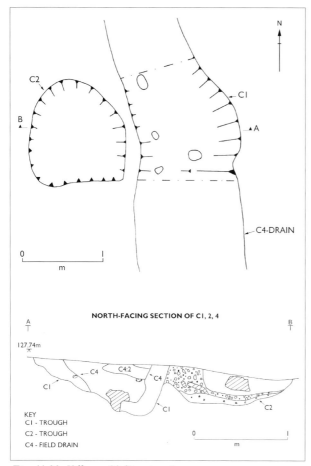

Fig. 11.20 Killoran 22 Structure 1.

Derryville Bog contained linings of wicker, planks and clay (Killoran 253, Killoran 240 and Killoran 304), and clay and stone linings were also in evidence (Killoran 17, Killoran 19 and Killoran 21). The Killoran 22 Strucutre 2 trough had a capacity of 306.7 litres, one of the smallest troughs excavated during the project. The use of sandstone rather than limestone implies that clean water was important. The discovery of sheep teeth in the trough, as occurred at Killoran 5 and Killoran 27, and the finding of a saddle quern at Killoran 240 all suggest that cooking was the function of these sites.

Killoran 22 Structure 2 (Fig. 11.21)

Fulacht fiadh; Bronze Age; >10m x >7m; sub-circular trough, linear cut and burnt mound; glacial till; construction, use and abandonment phases; site code KIL22; licence 97E372 extension; NGR E221125 N166905; 127.81m OD (top of trough).

Paul Stevens

This heavily ploughed-out *fulacht fiadh* was revealed during topsoil stripping for a drainage channel within the magazine borrow area, a sand and gravel quarry. No trace of the site was visible prior to construction. It was situated at the base of a west-facing slope, 5.5m east of Killoran 22 Structure 1, 70m west of Killoran 25 and

121m southwest of Killoran 17. The terrain on all sides was undulating, except to the north, where it sloped northwards under the bog, towards Derryville Bog. The structure consisted of a sub-circular trough partly sealed by thin patches of a heavily ploughed-out spread of burnt sandstone, silt, charcoal and peat. Part of the site continued north into the baulk but was not part of the development and so remained unexcavated.

Trough

The trough consisted of a shallow north–south aligned sub-circular pit measuring 1.7m north–south x 1.44m east–west, with a maximum depth of 0.12 m. It had steeply sloping, almost vertical, sides and a flattened base. A shallow stake hole, <0.08m in diameter and 0.04m in depth, was cut into the northwest corner of the feature. The pit contained a single fill of burnt mound material.

Burnt mound

The ploughed-out burnt mound covered an area >10m east–west by >8m north–south, to a maximum depth of 0.1m. It was very patchy due to heavy truncation by mechanical excavators during stripping. It consisted of small fire-cracked stone, <0.15m in diameter, made up of 80% sandstone and 20% limestone and spread out upslope, partly concealing the trough.

Discussion

This site was located on the low western break of a slope of glacial till, stretching westwards and northwards to the

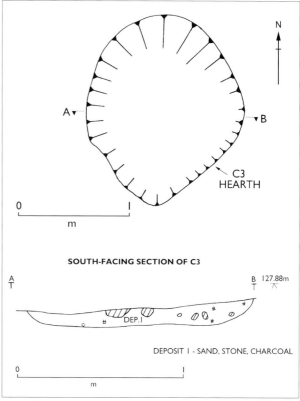

Fig. 11.21 Killoran 22 Structure 2.

present-day raised bog of Derryville. This land had been drained in antiquity, and soil cover was thin. The site was close to similar sites at Killoran 17, Killoran 25, Killoran 29 and Killoran 30. The trough had been deliberately backfilled, like other dryland sites in Killoran, and had a capacity of 195.4 litres, the smallest trough excavated during the Lisheen Archaeological Project. The use of sandstone rather than limestone implies that clean water was important.

Killoran 23 (Fig. 11.22)

Boiling pit; AD 660–880; 2.46m x 1.5m; pit and field drain; sloping glacial till; construction, use and disturbance phases; site code KIL23; licence 97E372 extension; NGR E220232 N167131; 128.32m OD (top of burnt pit).

Paul Stevens

This site was revealed during the mechanical removal of a modern farm track (a 0.30m deep layer of sub-angular stones) for the preparation of the Killoran north peat stockpile. The site was situated at the base of a west-facing sloping glacial till on the margin of a small bog and consisted of a circular pit and later linear field drain. No topsoil stripping was undertaken for the peat stockpile, and peat was deposited directly onto topsoil over the area. This caused significant disturbance to the peaty sod and underlying subsoil, and, although monitoring of the operations failed to turn up further material, the existence of further sites in the area cannot be ruled out.

Pit

The sub-circular pit was 1.80m in diameter and 0.55m in depth. It was oriented northwest–southeast. The pit had steep straight or convex sides and was filled with several layers of decayed sandstone, fire-cracked sandstone and charcoal-rich sandy clays, sealed by a layer of charcoal and silty clay loam.

Ditch

The pit was partly truncated on the western side by a north–south aligned field boundary ditch measuring over 8m in excavated length and 1.8m in width. It was 0.44m deep and was backfilled with natural clay subsoil. This ditch was noted on the first edition OS map for the area, extending 130m north–south.

Discussion

This pit was located on a localised bog also utilised by the *fulachta fiadh* Killoran 12, 13 and 14 along the southern margin. The pit closely resembled a *fulacht fiadh* in trough form and content, but the absence of a burnt mound suggested it was used for a limited period and did not represent the same tradition. The location was close to water, allowing percolation, and the use of sandstone from the glacial till rather than limestone from the surrounding bedrock suggested that the function of this pit was for boiling clean water. No other features were associated with the pit. However, activity of a similar date was located close by at Killoran 16 and Killoran 8.

Other examples of early medieval pits excavated in the area include Killoran 16, a large pit backfilled with natural subsoil, dated to AD 890–1040. Several burnt pits, probably for roasting, were excavated at Killoran 3, dated to AD 980–1270, Killoran 11 and Killoran 15. These pits were all similar in size, usually oval in plan and all showed evidence of *in situ* burning. They may be part of an Early Christian phase in the area.

Killoran 24 (Fig. 11.23)

Isolated cremations; Late Bronze Age and post-medieval; 60m x 20m; cremation pits, a post hole and pits; brow of glacial ridge; construction phase; site code KIL24; licence 97E372 extension; NGR E220292 N167126; 129.86m OD (top of cremation).

Paul Stevens

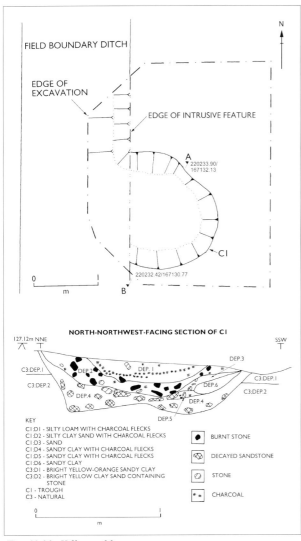

Fig. 11.22 Killoran 23.

This site was revealed during topsoil stripping for the magazine borrow area. The site consisted of two isolated cremation pits and three further pits, almost 50m apart, and later post-medieval furrows and pits. The site was located at the northern extent of a raised plateau overlooking lowlands to the east and Derryville Bog to the north. The cemetery Killoran 4 was also located on this plateau, 335m to the south, and two isolated cremation pits, Killoran 6 and Killoran 7, were located 400m and 300m to the south, respectively.

Cremations

The first cremation (C1) was a steep-sided pit measuring 0.34m x 0.24m x 0.18m. It contained an abundance of cremated bone within a particularly charcoal-rich fill, capped by re-deposited clay subsoil, similar to that in the cemetery Killoran 10.

The second cremation (C4) was a shallow concave pit measuring 0.42m in diameter and 0.16m in depth. It contained less charcoal and a few tiny fragments of burnt bone and some large fragments of fired clay.

Pits and furrows

Several other features, including a pit, a post hole and a burnt spread, were excavated within the site area, although they did not appear to have any direct association with the cremation. There were also a number of post-medieval cultivation furrows (lazy beds), similar to those recorded at Killoran 16. They were oriented northeast–southwest and set at intervals of 2m.

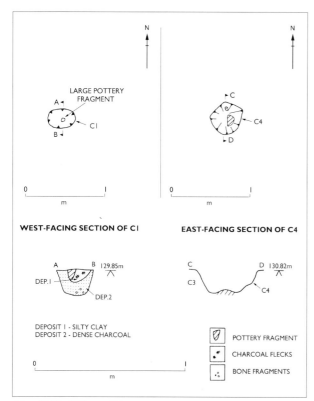

Fig. 11.23 Killoran 24.

Killoran 25

Burnt mound; Bronze Age; 30m x 23m; burnt mound; slope of glacial ridge; construction phase; site code KIL25; licence 97E372 extension; NGR E221198 N166961; 127.05m OD (base of mound).

Paul Stevens

This burnt mound was revealed during topsoil stripping for the magazine borrow area. The site was located on the north-facing break of a slope at the northern extent of a raised glacial ridge overlooking Derryville Bog to the north. It consisted of a truncated semicircular mound of burnt sandstone, silt and charcoal. A large modern drainage ditch cut the site east–west, possibly removing all trace of any trough that might have been present. The site was located 10m south of the *fulacht fiadh* Killoran 17 and 60m northeast of Killoran 22, in a cluster of similar structures. Several isolated cremations were also located at Killoran 24, 60m to the southeast.

Burnt mound

The partially ploughed-out burnt mound covered an area over 30m east–west x 23m north–south, to a maximum depth of 0.8m. Its full thickness was preserved under a field bank roughly central to the site. It consisted of two layers of burnt material sealing a layer of peat. The upper layer was 0.3m deep and composed of small fire-cracked limestone, <0.1m in diameter, in a charcoal sandy matrix. The lower layer was 0.37m deep and consisted of fire-cracked stone, <0.15m in diameter, made up of 95% sandstone and 5% limestone, with some dark, charcoal-rich sand.

Discussion

The site was classified as a burnt mound due to the absence of a trough. It is the largest example of a burnt mound, or *fulacht fiadh*, from the Lisheen Archaeological Project. Several other sites of this type were recorded, one on the dryland (Killoran 13) and two in Derryville Bog (95DER0119 in Cooleeny and Killoran 316). The site was located on the western break of a steeply sloping glacial ridge, at the southern edge of the northwestern spit of Derryville Bog. This glacial ridge was the focus of activity for a number of Bronze Age burnt mounds and cremations, including sites Killoran 17, Killoran 22, Killoran 24, Killoran 29 and Killoran 30.

Killoran 26 (Fig. 11.24)

Fulacht fiadh; 1145–795 BC; 13m x 7m; rectangular trough, burnt mound and linear ditch; glacial till; construction, destruction, reuse, abandonment and disturbance phases; site code KIL26; licence 97E372 extension; NGR E221676 N166443; 127.27m OD (top of trough).

Paul Stevens

This *fulacht fiadh* was revealed during the topsoil stripping for the administration building and car park within the plant site. No trace of the site was visible prior to construction. It was situated on flat ground within glacial till subsequently overgrown by bog. The site consisted of a sub-rectangular trough that had been re-cut by a rectangular trough and partly sealed by a heavily ploughed-out spread of burnt mound; a linear field drain that partly cut the mound north–south and a tree bole.

Trough

The trough consisted of a 0.46m deep sub-rectangular cut, 2.36m east–west by 2m north–south, through the subsoil down to the bedrock. The cut was irregular in plan and had straight, fire-reddened sides, slightly steeper to the

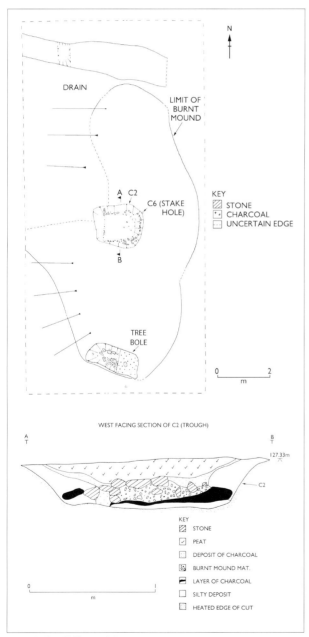

Fig. 11.24 Killoran 26.

south, and a flat base of bedrock. The cut was filled by an almost uniform layer of charcoal and charred wood, representing a burnt plank or roundwood lining, and was sealed by a hard packing of re-deposited clay subsoil up against the sides.

A second use of the trough, following the burning of the initial lining and collapse of the side lining, was represented within the smaller area delimited by the re-deposited clay packing. This second phase had vertical sides towards the base and straight (north)/convex (south) sloping sides and measured 1.44m east–west, 1.2m north–south and 0.35m in depth. It contained three fills: a lower fill of burnt mound material of fire-cracked stone and charcoal; a middle fill of large burnt stone, cobbles and occasional boulders (50% limestone and 50% sandstone); and a thick layer of charcoal, representing the last firing of the trough, subsequently abandoned and sealed by peat.

Burnt mound

The spread of burnt mound material (80% sandstone and 20% limestone) had been largely removed by machine during stripping, but it appeared to cover an area 13m north–south by 7m east–west, surviving to a depth of <10mm. The mound was cut by a large north–south field drain to the north and by a tree bole to the south.

Discussion

This site was located on level ground within a wide plateau stretching east to Derryville Bog, subsequently overgrown by shallow bog. The site lay within a cluster of similar sites that included the *fulacht fiadh* Killoran 19, 120m to the south, and the Middle Bronze Age *fulacht fiadh* Killoran 27, 150m to the northwest. The trough was partially backfilled, but was then reused before being abandoned without being backfilled. The first trough had a burnt lining of wood, as at Killoran 22 Structure 1. Examples from Derryville Bog contained linings of wicker, planks and clay (Killoran 253, 240 and 304); clay and stone linings were also evident at dryland sites (Killoran 17, 19 and 21). The earliest trough had a capacity of 1,019 litres; the second trough had a capacity of 870 litres. The low proportion of sandstone in the trough fill is unusual and may be due to a high use of limestone in the initial phases of water boiling. A more usual proportion of sandstone to limestone in the burnt mound represents the normal use of such sites in the Derryville area.

Killoran 27 (Fig. 11.25)

Fulacht fiadh; 932 BC; >10.3m x 6m; large oval pit/trough, sub-rectangular partly plank-lined trough, burnt mound and pits; *Ovis* bones and tooth found; glacial till; construction, reuse, abandonment and modern disturbance phases; site code KIL27; licence 97E372 extension; NGR E220565 N166574; 127.553m OD (top of trough).

Paul Stevens

This site was revealed during the topsoil stripping for the process building, part of the plant site. No trace of the site was visible prior to construction. It was situated on flat ground within glacial till subsequently overgrown by bog. The site was an unusual *fulacht fiadh*, consisting of a large circular pit, a later sub-rectangular trough with a wooden step and a shallow pit, all partly sealed by a heavily truncated spread of burnt mound, cut to the west by a 2m wide modern field drain.

Large pit trough

The earliest feature on the site was a large sub-oval pit representing a large trough, measuring 3.45m north–south, 2.72m east–west and 1.22m in depth. The pit had gently sloping sides at the top that became steep, near vertical sides midway down; the base was flat and uniform. The pit was filled with silt and re-deposited clay-sand subsoil, sealed by a thick deposit of sandy silt containing fire-cracked stone (30% sandstone and 70% limestone), <0.13m in diameter, up to 0.58m deep.

Sub-rectangular trough

The large pit trough was cut by a second trough consisting of a sub-rectangular cut measuring 1.64m east–west, 1.24m north–south and 0.37m in depth. The sides were near vertical and the base was flat. The trough was filled with a thick layer of slumped organic silty clay. This contained wood fragments, sheep bone and teeth and occasional fire-cracked stone, <0.12m in diameter, made up of 70% sandstone and 30% limestone. This layer slumped in from the western edge of the pit. It was partly sealed by two slump layers of sand, the lower containing heavily oxidised charcoal with some limestone cobbles. The upper level of the trough was delimited to the west by a timber step and to the north by a decayed timber. Several other fragments of timber were also recovered from the

organic fill layer. The trough was sealed by several layers of silty clay containing burnt mound material up to 0.5m thick. This in turn was sealed by a number of layers of peat and peaty clay.

Timber

A plank overlay the sandy fills of the trough. It was located at the western side of the trough and consisted of a large radially split oak plank (T5), 1.7m long, 0.25m wide and 0.07m in thickness. At each end of the plank were two square mortise holes into which two large oak uprights were slotted. The northern upright or peg, T4, was fitted flush with the mortise hole. No formal tenon had been carved, but the timber had been trimmed to a point at this end and was sunk into the lower fills of the earlier trough. The peg was a radially converted oak with a square profile, and it measured 0.87m x 0.1m x 0.08m. It produced a dendro chronologicaldate of 932±9 BC; the base plank, T5, produced a date of 939±9 BC, and both pieces are likely to be from the same tree (D. Brown, pers. comm.). The southern peg, T3, was also fitted flush with its mortise hole, and again, no formal tenon had been carved, but the timber had been trimmed to a point at this end, except this time it was slightly burned. It was sunk into the lower fills of the earlier trough and partly into the bedrock below. It was a radially split oak with a rectangular profile, slightly decayed, measuring 0.74m x 0.07m x 0.035m.

A second timber, located at right angles to the step, on the northern edge of the trough, overlay the organic slump fill. It was a large, degraded, radially split oak plank (T2), measuring 1.24m x 0.14m x 15mm. The outer face of the plank had a square mortise hole cut into the edge, but no associated timbers were found. Another timber was recovered from the organic fill and may have represented a similar feature.

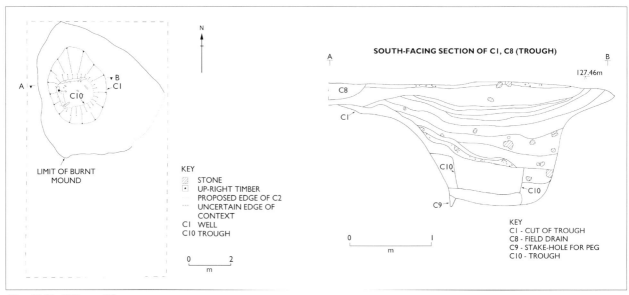

Fig. 11.25 Killoran 27.

Burnt mound

The spread of burnt mound material had been largely removed by machine prior to stripping, but appeared to cover an area 10.7m east–west by 6m north–south, surviving to a depth of 0.2m. It consisted of fire-cracked stone and occasional cobbles (50% sandstone and 50% limestone) set within charcoal-rich silty clay. The mound was truncated to the west and to the north by large (>2m wide) modern drainage trenches, aligned north–south and east–west. Associated with these drains were a number of tree boles along the line of a ditch that disturbed the mound and the upper trough fills.

Discussion

The site was located on level ground within a wide plateau that stretches east to Derryville Bog, subsequently overgrown by shallow bog. It lay within a cluster of similar sites, including the *fulacht fiadh* Killoran 19, 240m to the south, and the Late Bronze Age *fulacht fiadh* Killoran 26, 150m to the southeast. The trough was partially backfilled, but evidently was reused before being backfilled and abandoned, common to other dryland sites.

The first trough was unusually large but had similarities to Killoran 19, which appeared to be a series of re-cut troughs that created a large pit with a capacity of about 2,000 litres. The first trough at Killoran 27 had a capacity of about 1,500 litres. The later trough had a capacity of about 440 litres. The low proportion of sandstone in the trough fill is unusual and may reflect a high use of limestone in the initial phases of water boiling. However, a proportion of sandstone to limestone more consistent with other sites excavated for the Lisheen Archaeological Project, and indicative of a requirement for clean water, was found in some fill layers and in the burnt mound. The discovery of sheep bone and teeth in the trough, as occurred at Killoran 5 and Killoran 22, together with the discovery of a saddle quern at Killoran 240, suggests that cooking was the function of these sites

The presence of timber at the top of the second trough at Killoran 27 may represent a crude lining or revetment to stop slumping (which the site was prone to). Plank-lined troughs were found in *fulachta fiadh* Derryfadda 216 and Killoran 240. The timber may also have provided a step down into this deep trough which may originally have had all four sides lined at the top at least. The sand slump layers under the timber may have been inserted after the timbers, suggesting that the revetment was largely unsuccessful.

Killoran 28

Possible *fulacht fiadh*; Bronze Age; 5m x 7m; break of a sloping glacial ridge; site code KIL28; licence 97E372 extension; NGR E220979 N166453.

Paul Stevens

This ploughed-out *fulacht fiadh* or burnt mound was revealed during field walking in a ploughed field west of the plant site/Kvaerner Cemetation Ltd. contractor's compound. The site consisted of a spread of fire-cracked sandstone and some charcoal, and it was located on the west-facing break of slope of a glacial ridge, sloping westwards to a localised bog (partially reclaimed). An isolated flint flake (97E372:28:1) was also revealed during the field walking, 130m southeast of the site. The site was not developed so remained unexcavated.

Killoran 29 (Fig. 11.26)

Fulacht fiadh; Bronze Age; <12m x <11m; sub-oval trough, pit and burnt mound; glacial till; construction, use and abandonment phases; site code KIL29; licence 97E372 extension; NGR E221134 N166847; 127.94m OD (top of trough).

Paul Stevens

This structure was a heavily ploughed-out *fulacht fiadh* revealed during topsoil stripping for the magazine borrow area, a sand and gravel quarry. No trace of the site was visible prior to construction. The site was situated on level ground subsequently covered by peat and drained and ploughed in antiquity. It was 60m southeast of Killoran 22, 90m northeast of Killoran 30 and 170m southwest of Killoran 25. The terrain was undulating on all sides except the north, where it sloped northwards under the bog towards Derryville Bog. The excavation revealed a sub-oval trough, partly sealed by thin patches of a heavily ploughed-out spread of burnt sandstone, silt, charcoal and peat, and an irregular feature, possibly a hearth.

Trough

The trough was a sub-oval pit, 1.91m north–south by 1.78m east–west, with a maximum depth of 0.23m. It had rough vertical sides and a flat base. The pit contained a basal fill of burnt mound material with occasional large cobbles, sealed by a layer of charcoal-flecked clay inwash, underlying a layer of peat.

Burnt mound

The ploughed-out burnt mound was almost entirely removed during mechanical topsoil stripping. It was only apparent in patches, filling shallow hollows over an area >12m north–south x 11m east–west, to a maximum depth of 0.1m. It consisted of small fire-cracked stone, <0.15m in diameter, made up of 80% sandstone and 20% limestone.

Possible hearth

A possible hearth, irregular in plan and profile, containing burnt mound material and moderate charcoal, was excavated 1.45m north of the trough. It contained small fire-cracked stone, <0.15m in diameter, made up of 80% sandstone and 20% limestone.

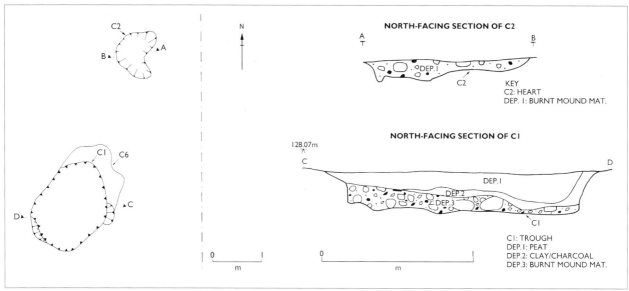

Fig. 11.26 Killoran 29.

Discussion

The site was located on a low plateau of glacial till, stretching northwards to the present-day raised bog of Derryville. This land had been drained in antiquity and the soil cover was thin. The site was close to similar sites at Killoran 17, Killoran 22, Killoran 25 and Killoran 30. The absence of a formal mound may suggest that the site was closer in form to the medieval example at Killoran 23; however, a pocket of surviving mound material suggests that it is more likely to be of the burnt mound, and therefore Bronze Age, tradition.

The trough was deliberately backfilled, in common with other dryland sites, and had a capacity of 781 litres. The use of sandstone rather than limestone implies that clean water was important. Hearths were found at sites in Derryville Bog (Killoran 304 and 240), as well as on the dryland (Killoran 5); however, these were more defined than the one at Killoran 29, which may either represent a single firing or a pre-construction shallow depression or uprooted tree.

Killoran 30 (Fig. 11.27)

Fulacht fiadh; Bronze Age; 7.6m x 6.7m; oblong trough, pit and burnt mound; glacial till; use and abandonment phases; site code KIL30; licence 97E372 extension; NGR E221117 N166808; 128m OD (top of trough).

Paul Stevens

This heavily ploughed-out *fulacht fiadh* was revealed during topsoil stripping for the magazine borrow area, a sand and gravel quarry. No trace of the site was visible prior to construction. The site was situated on level ground subsequently covered by peat and drained and ploughed in antiquity, 110m southwest of Killoran 22, 90m southwest of Killoran 29 and 190m southwest of Killoran 25. The terrain was undulating on all sides except the north, where it sloped northwards under the bog towards Derryville Bog. The site consisted of an oblong trough, partly sealed by thin patches of a heavily ploughed-out spread of burnt sandstone, charcoal and sand, and an irregular pit feature, also sealed by burnt mound material.

Trough

The trough was an oblong pit, aligned northwest–southeast and measured 1.76m x 1.04m x 0.37m. It had a rough vertical south side, an irregular concave north side and a flat base. The pit contained several fills of in-washed silt or burnt mound material. The basal fill was a sand lens overlaid by a hard oxidised layer of sand, charcoal and clay, with some sandstone. This in turn was sealed by several layers of burnt mound material consisting of silty clay and small fire-cracked stone, <0.09m in diameter, made up of 90% sandstone and 10% limestone.

Burnt mound and pit

The ploughed-out burnt mound was almost entirely non-existent, probably removed during mechanical topsoil stripping. It was only apparent in patches filling shallow hollows over an area >7.6m north–south by 6.7m east–west, to a maximum depth of 0.01m. A shallow pit feature, measuring 0.42m north–south, 0.36m east–west and 0.09m in depth, was located east of the trough. This feature, which was filled with burnt mound material, probably represented a stone socket. A second hollow to the south of the trough also contained burnt mound material.

Discussion

The site was located on a low plateau of glacial till, stretching northwards to the present-day raised bog of Derryville. This land had been drained in antiquity and the soil cover was thin. The site was close to *fulachta fiadh*

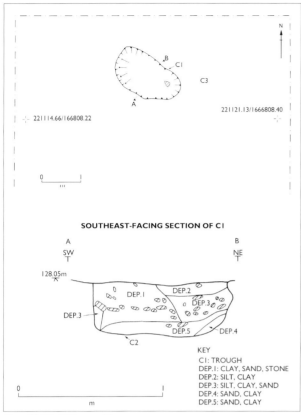

SOUTHEAST-FACING SECTION OF C1

KEY
C1: TROUGH
DEP.1: CLAY, SAND, STONE
DEP.2: SILT, CLAY
DEP.3: SILT, CLAY, SAND
DEP.4: SAND, CLAY
DEP.5: SAND, CLAY

Fig. 11.27 Killoran 30.

and a burnt mound at Killoran 17, Killoran 22, Killoran 25 and Killoran 29. The absence of a formal mound may suggest that the site was closer in form to the medieval example at Killoran 23; however, pockets of surviving mound material suggest that it is more likely to be of the burnt mound, and therefore Bronze Age, tradition.

The trough was deliberately backfilled, in common with other dryland sites, and had a capacity of 1,041 litres. The use of sandstone rather than limestone implies that clean water was important.

Killoran 31, 'Cill Ódhráin' (Fig . 11.28)

Enclosed monastic settlement; AD 450–690; 155m diameter (enclosure), 8m x 5m (cutting); furnace pits, linear gullies, post/stake holes and a linear field ditch; glacial ridge promontory; construction, iron working, delineation and abandonment phases; site code KIL31; licence 98E269; NGR E221065 N166153; 124.75m OD (top of post hole).

Paul Stevens

This site was a large, sub-circular enclosure listed in the Sites and Monuments Record (TI036-020). The site appeared as a curving circular bank and ditch preserved by the line of the modern field boundary and partly as a low raised-platform. No visible trace of the site remained to the north of the modern bisecting field boundary. The enclosure was 155m in diameter and the earthworks survived to a height of about 0.5m.

The enclosure was located at the southern end of a glacial ridge, at the southern edge of Killoran townland. It was naturally defensive, enclosed on three sides, to the south and west by bog (now reclaimed) and to the east by the Moyne Stream, which bended around the monument, 50–70m to the east.

The excavation was conducted to facilitate the installation of an electricity pole unit and took the form of a rectangular cutting located roughly central to the enclosure, in the area to be disturbed. It uncovered a cluster of pits, linear features and post/stake holes, two containing evidence of *in situ* iron working and dating to the Early Christian period. The cutting was located roughly in the centre of the enclosure up against the modern field boundary that bisected the enclosure. The features uncovered were disturbed by the modern field boundary ditch.

Iron working pits

Several associated pits (C24, C26 and C31) were clustered in the southeast area of the cutting. Two of these were rich in iron-working waste and contained iron slag and raw bloom (unworked iron). C24 was a curved ovoid in plan with a square profile and had slightly fire-reddened, vertical sides and a flat base. It was 1.32m east–west, 0.43m north–south and 0.2m in depth. It contained a single piece of slag amongst a dark, charcoal-rich matrix. C26 was located immediately north of C24, was circular in plan and measured 0.21m in diameter and 0.17m in depth. It contained a single piece of slag amongst a dark, charcoal-rich matrix. C31 was located east of C26, was oval in plan, irregular in profile and measured 0.66m east–west, 0.45m north–south and 0.15m in depth. The sides and base were heavily fire-reddened, and this feature contained a large quantity of iron slag and part of the stone base of a furnace (98E269:31:2). A tiny fragment of bone was also present. It was probably the site of a smelting or smithying furnace. Charcoal from pits C24 and C31 was identified as old wood of oak and yew branches (six years old) and was radiocarbon dated to AD 460–690.

Linear features

Two overlapping, parallel linear slot trenches (C3 and C20) bisected the cutting. These were probably associated with a cluster of post and stake holes roughly aligned to the trenches and probably post-dated the iron-working activity. A third linear feature, a gully (C32), cut almost at right angles to the line of the slot trenches and stopped short of them. The three features delineated the majority of the features on the site, which appeared to cluster against them. It is, therefore, probable that the slot trenches represented an internal partition within the large enclosure.

C3 was over 5.5m x 0.27m x 0.15m and extended into the northwestern baulk, terminating with a regular, round-end cut by a stake hole. Two other stake holes were

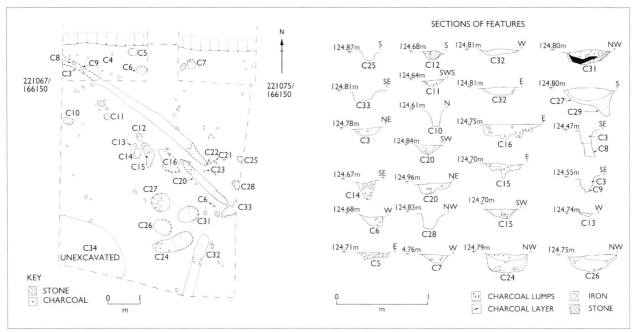

Fig. 11.28 Killoran 31.

probably associated with the linear slot trench. The trench had a square profile, vertical sides and a flat base. The fill contained tiny fragments of bone. Along the same line as C3, located 1m southeast of its terminus, was a post hole, C28, that appeared to continue the line of the slot trench. C3 was a slot trench for a wattle panel upright with associated sails and supports.

The second slot trench, C20, was located 0.2m south of C3 and the two trenches overlapped at the ends by 0.8m. It measured 2.2m x 0.23m x 0.1m and had irregular rounded ends. C20 had a square profile, steep sides and a largely flat base. The southeastern terminus was abutted by an irregular cut feature, C33, which extended northwards into the eastern baulk. C33 may represent an associated trench that continued at right angles to C20. A stake hole, located 0.2m southeast of a second stake hole, was aligned with the trench, 0.5m northwest of its western end.

The gully, C32, was oriented approximately north–south and terminated with a regular, rounded end, extending south into the baulk. This gully was located south of C20 and east of C24. It had a shallow concave profile and measured 1.9m x 0.36m x 0.1m.

Associated features

A number of other associated and unassociated features of apparent antiquity were located within the cutting. There were also several outlying features, mainly shallow post holes. One cluster of post holes around the northwestern terminus of C20 did not appear to form an alignment.

Post-medieval bank and ditch

A relatively recent east–west aligned field bank bisected the enclosure. The cutting was located against this bank, as this was the location of the development. The line of the bank was revealed to have been built on an earlier but contemporary ditch that cut the site. The bank contained various residual lumps of iron slag, post-medieval pottery sherds and two iron objects.

Discussion

This site is a large previously undated monastic settlement, measuring about 150m in diameter. It is strategically located at the southern terminus of a long glacial ridge, forming a low peninsula surrounded on three sides by boggy ground and by the Moyne Stream, and is enclosed by a single bank. The monastery was set within a marginal landscape of dry knolls rising above poor soils, marsh and vast raised bog to the east, dividing up the ecclesiastical and provincial territories. Settlement in the Bronze Age, Iron Age and early medieval periods was evidenced. Some of the early medieval activity was probably related to the monastery, such as land division and agricultural drainage.

The date of the site is comfortably within the lifetime of St Odran. It is almost certain that this monastic settlement was founded by him prior to AD 563, and it may have continued to function as a monastery for some years. The archaeological background to the area suggests a strong concentration of activity in the Early Christian period.

12. Catalogue of finds

Laureen Buckley, Sarah Cross May, Niall Gregory, Cara Murray,

John Ó Néill, Helen Roche and Paul Stevens

Introduction

Twenty-one artefacts were retrieved from Derryville Bog during the course of the project: sixteen wooden composite or broken objects, one saddle quern and four very small, fragmentary sherds of prehistoric pottery. In 1973, a 'wooden beetle' (National Museum of Ireland accession no. 1973:23) was recovered from the bog in Cooleeny at a similar depth to trackway Cooleeny 31, but the exact relationship between the find and the site is unclear.

The preservation of organic material in raised bogs is well attested (Kelly 1995, 141) and attributable to the low microbial activity and the anaerobic nature of peat. Most of the artefacts and sites uncovered in the excavations at Derryville were deposited in a fen environment. The preservation of material within fen is poorer than that in raised bog, which may explain the lack of small or fragile organic finds.

No evidence was found for metalworking, despite the obvious widespread use of a range of metal tools, as demonstrated by the woodworking evidence (see Chapter 13).

The individual assemblages of objects, such as lithics and pottery, were too small for detailed group analysis. The limited size and nature of the assemblages reflects the assumed status of the sites.

The numbering system used to record the artefacts found during the survey and excavation of Derryville Bog relate to the excavation licence numbers issued for individual sites and groups of sites excavated. Artefacts found in the pre-project assessment survey by the IAWU are recorded under licence number 94E106. The project licence numbers 96E202, 97E158 and 97E160 are used for the wetland finds; no artefacts were found under licences 96E203 and 96E298. The finds retrieved during the monitoring of enabling and construction works were covered by licence numbers 97E372, 97E439, 97E036, 97E051, 97E168, 98E066 and 98E269.

Wooden artefacts

Sarah Cross May, Cara Murray, John Ó Néill and Paul Stevens

Wood species and function

The wood identification analysis suggested a particularly high attrition of the material recovered from Derryville Bog in comparison with other bogs (I. Stuijts, pers.

comm.). This is a consequence of the particular nature of the bog at the time of, and soon after, the deposition of the objects. The wooden artefacts were made of alder, ash, cherry, yew, oak, hazel and, in one case, guelder rose.

Although it was difficult to determine the function of some of the artefacts, particularly the composite or broken pieces, the characteristics of the wood from which they were made did reveal something about their nature.

Alder is known for its durability in wet conditions. Two of the three alder artefacts were sturdy, dressed pieces of roundwood that formed part of larger composite objects, and the third was a bucket fragment that had well-preserved gouge marks.

Ash, on the other hand, is a lightweight, flexible wood. The functions of the three ash objects were unclear. One

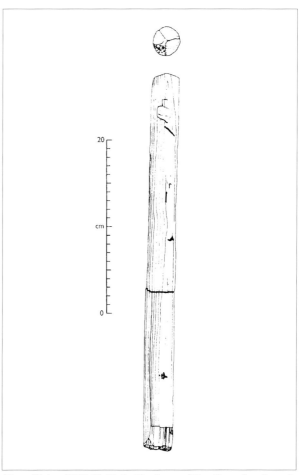

Fig. 12.1 97E160:31:1 (Cooleeny 31). Turned ash rod; c. 600 BC.

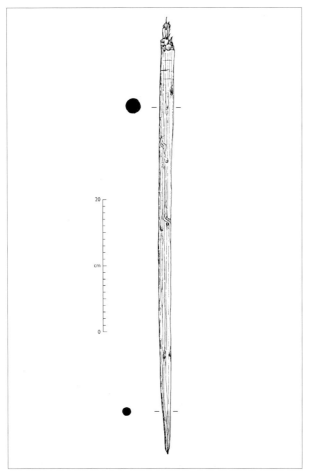

Fig. 12.2 94E106:2 (Derryfadda 13). Yew spearshaft: Late Bronze Age-early Iron Age (after IAWU 1996a).

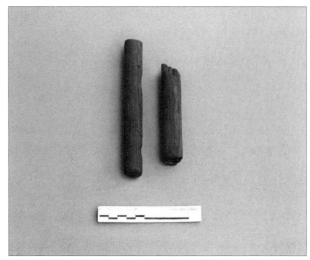

Plate 12.1 97E160:31:1 (Cooleeny 31). Turned ash rod; c. 600 BC.

Plate 12.2 97E160:237:1 (Killoran 237). Yew spear shaft in situ; Bronze Age.

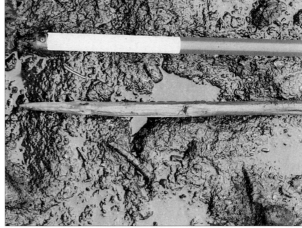

Plate 12.3 97E160:237:1 (Killoran 237). Tip of yew spear shaft; Bronze Age. Detail of Fig. 12.2.

was a lathe-turned rod (97E160:31:1; Fig. 12.1; Plate 12.1), and its association with Cooleeny 31 dated this method of woodworking to *c.* 600 BC (see Chapter 13).

Cherry wood is noted for its aesthetic appearance and is frequently used for carving. The three worked peg-like objects were all made of cherry, although the two similar pieces from Derryfadda 9 were obviously part of broken composite objects.

Two possible spear shafts (94E106:2, Fig. 12.2, and 97E160:237:1, Plates 12.2–12.3) were made of yew, which is a heavy, elastic and dense type of wood and is often used for making longbows, spears and dagger handles.

The number of artefacts retrieved was very limited. This clearly relates to the nature of the sites and the type of peat they were associated with. Of the five artefacts found during the field survey, four were found outside the study area. Three were found within the Cooleeny Complex and one was recovered during the survey of the Killoran Complex to the west of the study area. Only one artefact was found during the survey in association with a site that was later excavated during the project: Derryfadda 13.

Perforated and split lath (94E106:1)

Fraxinus; undated; max. length 368mm, width 59mm, thickness 21mm; 95DER0032 (recovered during IAWU survey); Fig. 12.3.

This four-sided, tangentially-faced ash roundwood was broken at both ends. It had two broad flat surfaces and two narrow sides, which were bevelled, creating an almost octagonal appearance. All four surfaces were abraded and there were no diagnostic tool marks. A

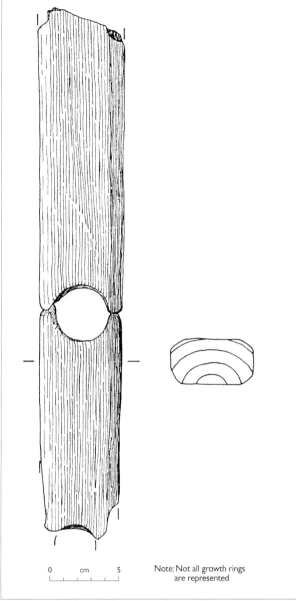

Fig. 12.3 94E106:1 (95DER0032). Perforated and split ash lath; undated (after IAWU 1996a).

dowel hole, 38mm x 36mm, perforated the central portion of the artefact. Another possible dowel hole survived at one end. Tool marks were clear and unabraded in these dowel holes.

The object was recovered on the eastern side of the Cooleeny Complex in 1994 while drain faces were being cleaned. It was not associated with any particular structure.

Spear shaft (94E106:2)

Taxus; Late Bronze Age–Iron Age; length 690mm; diameter 24mm; Derryfadda 13 (recovered during IAWU survey); Fig. 12.2.

This piece of yew brushwood was dressed on all surfaces and tapered to a point. The tip of the point was damaged,

with up to 10mm missing. The complete surface was dressed with fine, non-diagnostic tool marks. There were eleven knots in three groups representing trimmed branches along its length.

The object was recovered from among the substructural layers of trackway Derryfadda 13. It was originally deposited in wet conditions.

A second spear shaft (97E160:237:1; Plates 12.2 and 12.3) was found at the lowest level of the platform Killoran 237, on its eastern limit.

Dressed and perforated roundwood (94E106:3)

Alnus; undated; length 114mm, max. diameter 75mm; 'between' 95DER0134 and 95DER0135 (recovered during IAWU survey); Fig. 12.4.

Broken at one end, across a perforation, this object was produced from a straight-grained piece of wood with one small knot visible. The surfaces were quite smooth, and no diagnostic tool marks were visible, although all the surfaces were dressed. The largest facet was 20mm wide and 38mm in length.

The intact end was cut flat with a bevel of up to 20mm around the edge. The other end was broken across a centrally cut dowel, 26mm in maximum diameter. The hollow was made with a concave tool at least 7mm wide, possibly a gouge, and the tool marks were clear and unabraded.

This artefact was recovered close to the possible platform sites 95DER0134 and 95DER0135, within the Cooleeny Complex.

Tapered plank (94E106:4)

Quercus; undated; length 399mm, width 118mm, thickness 22mm; Cooleeny Complex (recovered during IAWU survey); Fig. 12.5.

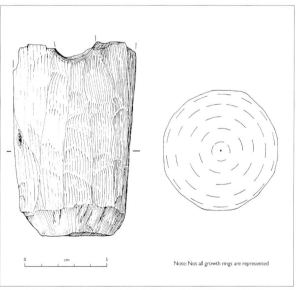

Fig. 12.4 94E106:3 (95DER0134 and 95DER0135). Dressed and perforated alder roundwood (after IAWU 1996a).

This oak plank had curved ends, and the piece tapered in thickness from 22mm at a central groove incised/worn into one surface. It was recovered in a drain face with the grooved surface uppermost. This surface and the sides were abraded, with no visible tool marks. The underside had slight traces of tool marks associated with the surface dressing. Other marks on the surface of the plank had resulted from the conditions of its deposition.

The groove had a shallow V-shaped profile, 32mm wide and 11mm in depth. It was positioned transversely, approximately midway along the length of the plank.

This object was recovered from directly on top of the silt layer between 95DER0033 and 95DER0041, within the Cooleeny Complex.

Dressed roundwood (94E106:5)

Alnus; undated; length 221mm, max. diameter 101mm; 95DER0082 (recovered during IAWU survey); Fig. 12.6.

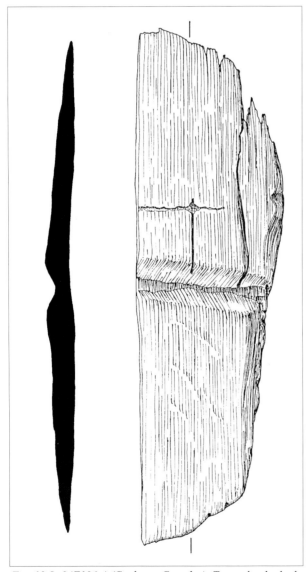

Fig. 12.5 94E106:4 (Cooleeny Complex). Tapered oak plank; undated (after IAWU 1996a).

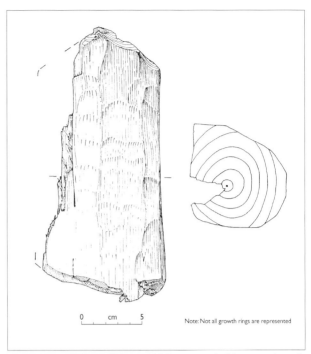

Fig. 12.6 94E106:5 (95DER0082). Dressed alder roundwood; undated (after IAWU 1996a).

The outer tangential splitting and dressing of an alder roundwood produced this artefact. Of the four remaining surfaces, one was damaged by drain cutting, one was split and the remaining two had no indications of signatures. The ends were cut with a sharp metal tool, which left faint traces of signatures. One end was cut vertically, while the other was cut tangentially.

The artefact was found in isolation and was recorded during the IAWU survey in the Killoran Complex.

Possible wedge (96E202:23:1)

Quercus; Middle Bronze Age; length 140–160mm, width 100–114mm, thickness 20–40mm; Derryfadda 23; not illustrated.

This item was split along the rays of an oak timber. There was no evidence of woodworking on either of the faces. The piece was trimmed on its outer, widest edge, and it was damaged at its narrower end, which had a slightly rounded appearance. Both sides of the piece were axe marked and were slightly concave in appearance.

The artefact was found on the northern side of the single-plank trackway Derryfadda 23 in fen peat. A large area of wood chips on the southern side of the track, indicative of woodworking, suggested that this object may have been used in the splitting of timbers during the tracks construction.

Pommel-headed baton with perforated, tapered end (broken) (96E202:9:1)

Prunus avium/padus; Iron Age; length 350mm, diameter 32mm; Derryfadda 9; Fig. 12.7.

This object of cherry wood had a multifaceted pommel- or mushroom-shaped head. The underside of the pommel was clean-cut, with no evidence of wear or damage. The head measured 42–49mm in width and 3mm in height. The facets varied in size from 7mm x 4mm to 20mm x 22mm and were flat with some evidence of signatures.

The shaft was 320mm in overall length and 32mm in diameter. The entire surface of the shaft had been whittled in a series of elongated facets. Notably, none of these facets was cut into the underside of the pommel. There was a 1mm wide compressed band in the shaft surface 55mm from the pommel, where the object appeared to have been either bound or enclosed as part of a composite artefact. The shaft tapered to an asymmetrical wedge point that had been perforated, and the object had broken at this point. The elongated perforation, which was not centrally placed, was 20mm in length, and some evidence of internal wear was evident.

This artefact was found below the northern limits of the superstructure of platform Derryfadda 9 in very wet, wood-rich fen peat on the edge of a pool within alder carr fen. It was oriented roughly north–south, with the pommel end to the south towards the structure. A second, very similar artefact, 96E202:9:2 (Fig. 12.8), was found 0.40m from it (see below). Cherry root systems were found in the area of Derryfadda 13 to the south of this site. The construction of Derryfadda 13 and the location of these artefacts nearby indicate the importance not only of the fruit from these trees but also the use of the wood for fine

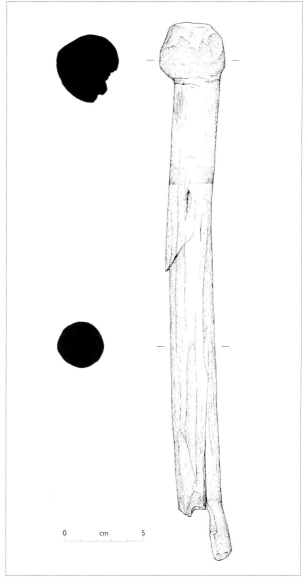

Fig. 12.8 96E202:9:2 (Derryfadda 9). Pommel-headed cherry wood baton with perforated, tapered end (broken).

wood carving. A parallel for these artefacts was found at a pool platform of the Difford's complex in the Somerset Levels (artefact no. 76.008; Coles 1989, 36, 46), which was dated to AD 235–414 (HAR-1842; Orme 1982b, 24) and AD 231–394 (HAR-1854; *ibid.*) . The exact function of this object is unknown, but the site was dated to the Iron Age, specifically 395–180 cal BC (Beta-102739), and the apparent similarity between the artefacts and the two sites would suggest a similar function.

Pommel-headed baton with perforated-tapered end (broken) (96E202:9:2)

Prunus avium/padus; Iron Age; length 345mm, diameter 31mm; Derryfadda 9; Fig. 12.8; Plate 15.2.

This piece, like 96E202:9:1, also had a pommel-shaped head, 43–46mm wide and 36mm in height. One side of the pommel was slightly crushed. The shaft was 310mm long

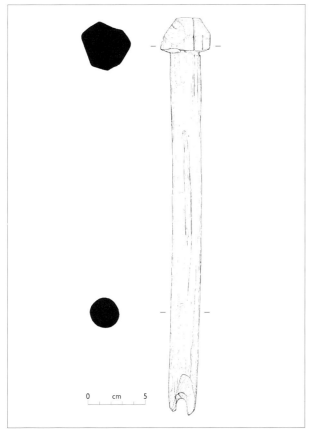

Fig. 12.7 96E202:9:1 (Derryfadda 9). Pommel-headed cherry wood baton with perforated, tapered end (broken).

and was circular in cross-section. There was a fracture line along the shaft running towards the perforated end, which gave the shaft a slightly curved profile. The face of the shaft had been whittled in a series of elongated facets. There was a 13mm wide compression 30mm from the pommel end where the object appeared to have been either bound or enclosed as part of a composite artefact, similar to 96E202:9:1. The shaft tapered to an asymmetrical wedge point that had been perforated, and the object had broken unevenly at this point. The elongated perforation was 40mm long and showed some evidence of internal wear. The perforation had remained intact for 25mm.

This artefact was found 0.40m from 96E202:9:1, at a lower level in the peat. It was oriented roughly north-south in an area where slippage of the overlying timbers of the superstructure and substructure had occurred, probably as a result of the subsequent flooding of the site. A similar artefact was found at the Difford's complex in the Somerset Levels (see above). Derryfadda 9 was dated to the Iron Age, specifically 395–180 cal BC (Beta-102739).

Possible spoke or handle (96E202:13:1)

Fraxinus; Late Bronze Age–Iron Age; length 322mm, max. diameter 49mm; Derryfadda 13; Fig. 12.9; Plate 12.4.

This cylindrical artefact had a tenon at one end and was rounded at the other, broken, end. Although broken, the object appeared to be virtually complete. Two recovered pieces were sufficient to determine that the artefact was complete when deposited. There was also an area of compression on the shaft that had occurred as a result of the overlying timber. The tenon was 39mm in length and was straight sided, with little evidence of wear or abrasion. There was no surviving evidence of tool marks, which may be the result of natural attrition during its use.

The object was located within the substructure of Derryfadda 13, which was composed of roundwoods up to 0.70m in depth. This site was dated to 767–412 cal BC (GrN-21943) and 790–395 cal BC (Beta-102756). A possible parallel for this object was found at the Glastonbury Lake Village in Somerset, where the artefact was interpreted as a billhook handle (Minnitt and Coles 1996, 29).

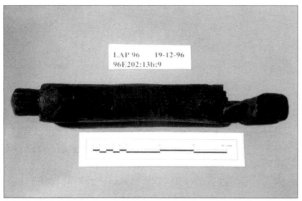

Plate 12.4 96E202:13:1 (Derryfadda 13). Possible ash spoke or handle; Late Bronze Age–Iron Age.

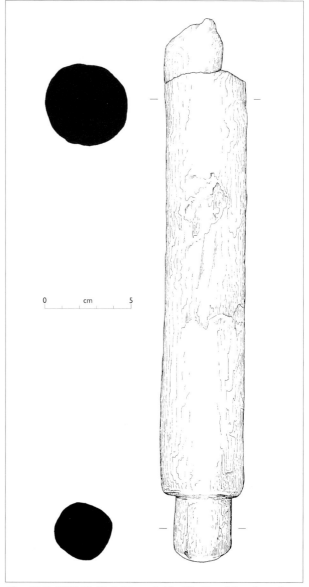

Fig. 12.9 96E202:13:1 (Derryfadda 13). Possible ash spoke or handle; Late Bronze Age–Iron Age.

Although the Derryfadda artefact is not exactly the same, it does bear a noteworthy similarity.

Bucket fragment (96E202:17:1)

Alnus; Late Bronze Age; length 139mm, max. width 85mm, thickness 8–21mm; 'near' Derryfadda 17; Fig. 12.10; Plate 15.6.

This alder bucket fragment was uncovered during the examination of the western drain face in Field 43, but could not be stratigraphically linked to Derryfadda 17. The wood was in good condition, although slightly degraded on its outer face and base. A 5mm deep croze for the insertion of a base plate was located 155mm from the base of the object. Both the area of the croze and the base of the object had been naturally degraded. The inner surface displayed narrow gouged grooves, up to 133mm wide, worked along this face. A series of small facets provided evidence of woodworking on the outer face.

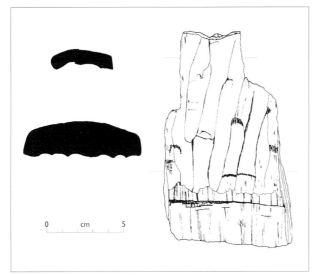

Fig. 12.10   96E202:17:1 (Derryfadda 17). Alder bucket fragment; Late Bronze Age.

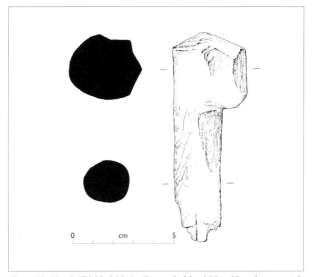

Fig. 12.11   96E202:203:1 (Derryfadda 203). Hazel peg with worked end (broken); Iron Age.

No other archaeological material was recorded in the vicinity. The object was evidently discarded, possibly during the use of the site.

The closest parallels for this piece were found in Corlea, County Longford. Of the sixteen tub or bucket fragments discovered at Corlea 1, dated to 148±9 BC, one was of willow, five were of alder and ten were of ash (Raftery 1996, 231). Eleven of the bucket fragments were on the southern landfall of the site. There is no evidence for cooperage before the first century BC; therefore, vessels were probably formed from a cylinder hollowed from a tree trunk, with a base plate inserted (*ibid.*, 263).

Turned rod (97E160:31:1)

Fraxinus; *c.* 600 BC; length 274mm, diameter 20mm; Cooleeny 31; Fig. 12.1; Plate 12.1.

This artefact was found in Context 5 of Cooleeny 31, lying horizontally amongst brushwood. It was a piece of ash roundwood that had been turned into a small rod with one carved, smooth rounded end. Marks from lathe turning were apparent under low magnification. The end that was not rounded was broken in antiquity. There was nothing in its form to indicate its function. Its position in the foundation layer of the track suggests it was deposited as scrap.

Peg with worked end (broken) (96E202:203:1)

Corylus; Iron Age; length 104mm, shaft diameter 23mm, pommel diameter 44mm, height 42mm; Derryfadda 203; Fig. 12.11.

This pommel-headed artefact was broken at one end. The head had been worked, and twenty-one separate facets had been cut into it. The shaft had been trimmed along its length, which measured 62mm.

The object was found in the central southern area of Derryfadda 203. This site, dated to 385–50 cal BC (Beta-102761) and 390–190 cal BC (Beta-102762), was an irregularly laid brushwood and roundwood trackway, and the artefact was found between two roundwoods, with the head to the north. The exact function of this artefact is unknown, but a similar object found at Corlea 1 has been suggested to have been used for securing ropes (Raftery 1990, 53). The Derryfadda object is more extensively worked than the Corlea example and is also similar in form to some of the 'handles' for socketed artefacts that have been found in the lake villages of Somerset (Bulleid 1980, 23).

Split roundwood with notches (97E160:230:1)

Viburnum opulus; undated; length 286mm, max. width 41mm, max. thickness 22mm; Killoran 230; Fig. 12.12.

This split roundwood of guelder rose had two notches cut in its edge. It was found in Context 17 of Killoran 230, lying horizontally with the rest of the foundation material, a common situation for artefacts found in tracks. It was probably deposited as scrap. There was nothing about the form of the object to indicate its function. It was probably a component of a more complex object.

Spear shaft (97E160:237:1)

Taxus; Bronze Age; length 139mm, max. width 85mm, thickness 8–21mm; Killoran 237; Plates 12.2, 12.3 and 15.1.

This artefact, a rounded and pointed stick made from yew, was found in Context 8 of Killoran 237 (a foundation layer). It was trimmed along its whole length and the tip was rounded as well as sharpened. The other end had been broken in antiquity. It was found lying horizontally in the lower layers of the context. While its appearance suggested a spear, it seemed too slight for that function and is most likely a spear shaft, as the profile of the tip matched the sockets typical of Middle Bronze Age spear heads.

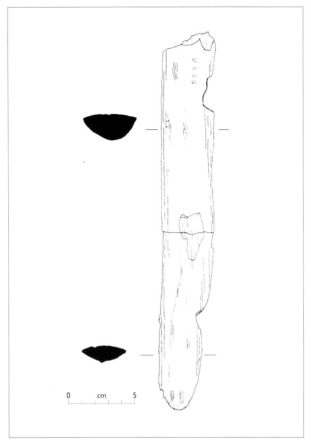

Fig. 12.12 97E160:230:1. Split guelder rose roundwood with notches; undated.

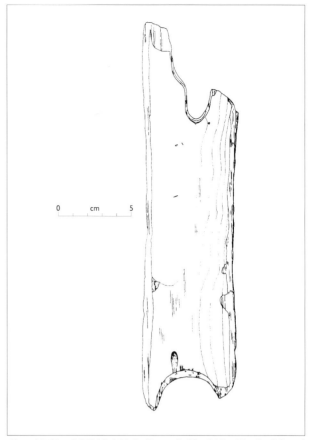

Fig. 12.13 96E202:311:1 (Derryfadda 311). Notched hazel panel: Late Bronze Age.

Notched wooden panel (96E202:311:1)

Corylus; Late Bronze Age; length 185–273mm, width 61–70mm, thickness 15–25mm; Derryfadda 311; Fig. 12.13.

This artefact was uncovered on the southern side of the site, sitting on a very thin mixed deposit of fen peat and mineral soil. It consisted of a small hazel panel broken at both ends, with a semi-circular notch at one end and a key-hole-shaped cut at the other. It was well preserved on its underside and slightly eroded on its upper face. The panel tapered slightly at the centre and was also slightly concave. It clearly formed part of a more complex artefact.

Shaft with expanded perforated terminals (97E160:325:1)

Corylus; Iron Age; length 324mm, width 53mm, thickness 32mm; Cooleeny 325; Plates 12.5 and 15.7.

Found in the brushwood at Cooleeny 325, this artefact was made from a piece of hazel that had been carved along its length to produce expanded terminals which were then perforated. Tool marks were evident along its length. One terminal was intact and one had been damaged in antiquity.

While its form was similar to a yoke, it seemed too slight for this purpose. Its position suggested deposition as scrap. It is possible that it was a handle for carrying wood; the rest of the wood at the site was also hazel.

Trough (97E158:265:1)

Quercus; undated; length 2.67m, width 0.65m, thickness 0.30m, hollowed to a depth of 0.10m; Killoran 265; Plates 12.6 and 14.1.

This partially hollowed oak trunk had been used as a trough, set in a shallow pit, at the *fulacht fiadh* Killoran 265. It had been half-split and hollowed to a maximum depth of 0.10m. The narrower end had been exposed in a BnM field drain and had been damaged when the drain had been dug. Examination by Dr Niall Gregory confirmed that it was not a reused or unfinished log boat, as it had no carrying capacity or ability to manoeuvre (full analysis below).

Plate 12.5 97E160:325:1 (Cooleeny 325). Hazel shaft with expanded perforated terminals: Iron Age.

Plate 12.6   97E158:265:1 (Derryfadda 265). Carved oak trough.

External length	2.67m
Internal length	1.5m
External width	0.7m
External height	0.32m
Internal height	0.1m
End thickness	*c.* 0.3m and *c.* 0.8m
Floor thickness	*c.* 0.15m

Table 12.1 Dimensions of carved trough 97E158:265:1.

Similar large hollowed troughs, or reused log boats, have been discovered at a number of *fulachta fiadh*, including Teeronea, County Clare (listed as No. 2a in Brindley *et al.* 1989–90); Clashroe, County Cork (Hurley 1987) and Curraghtarsna, County Tipperary (Buckley 1985).

There are a number of other recorded instances where log boats have been found in association with *fulachta fiadh*, although their use as troughs cannot be confirmed. Two reused log boats were found in Ballyglass Bog, County Mayo, at Tonregee and Cuilmore (Robinson *et al.* 1996), associated with large spreads of burnt stone and charcoal, suggesting the find sites were originally *fulachta fiadh*. Two alder log boats were found below burnt mounds at Derrybrusk, County Fermanagh (listed as 18a and 19a in Lanting and Brindley 1996), but could not be directly associated with the formation of the mounds.

Analysis of artefact 97E158:265:1

Niall Gregory

Introduction

When this artefact was recovered from Derryfadda 265, there was an initial suspicion that it was a dugout boat that had been reused as a carved trough. However, an examination of the artefact on site in 1997 suggested that it was not, nor was it ever intended to be, a dugout boat. The following report is based on extensive research and experimental work on dugout boats. Although comparative analysis of a wide range of boats indicates some similarity to this artefact, it also satisfactorily demonstrates that it was not a boat.

General description

The trough is a parallel-sided oak timber that has been worked both internally and externally. In cross-section, its exterior is rounded and follows the natural shapes of the original trunk. Most, if not all, of its softer sapwood has been removed. Internally, 1.5m of its length has a

constant cross section, while the remaining 1.17m includes very thick ends, from which the remains of a branch protrude from the thicker end. The angle of the branch indicates that the other end was towards the root of the original tree. There is no indication of any external continuation of either end. This suggests that they originally had vertical faces.

Comparative analysis of dugout boats

If this carved trough was a dugout boat, it would be considered to be a box-type boat. Of 405 recorded Irish dugout boat finds, there are two such boats (0.5%) which have the same general external shape but are flat-bottomed to some extent. As in the case of this artefact, neither has any record of boat-associated features such as ribs, tholepin hole mounts, thickness gauges, false keels, mooring holes etc.

However, unlike the box-shaped and other dugout types, the Lisheen artefact has a very thick floor and ends, and the sides are unusually low for a boat. This cannot be explained by decay or erosion, as it was found in circumstances conducive to its preservation from the period of its original deposition.

Evidence from dugout boats and experimental work shows the sequence in which they were constructed. The outside hull was always completed prior to hollowing the interior. It is plausible that the carved trough found in Derryville Bog was intended to be a boat but that it was unfinished and was subsequently reused in the circumstances from which it was excavated. This could account for the low sides, which might have become separated from the boat as a result of the wood splitting as it dried out. However, the over-rounded cross-section and the branch discount this possibility. These show that it was not externally finished prior to hollowing.

Irish dugout boat lengths vary from 1.83m to 16.9m. Those in the range 2–3m account for 11% of the boats. Widths of between 60cm and 70cm account for 15% of the boats. The dimensions of the Lisheen artefact are perfectly acceptable for a dugout boat. However, the estimated floor thickness of 15cm would be excessive. Archaeological evidence shows that the nine boats with floor thickness between 11cm and 18cm are all unfinished dugouts. The end thickness of 30cm is at the top of the boat range, but a thickness of even two-thirds of that of the other end is unheard of.

Boat manufacture is a compromise between providing a sufficient displacement to carry the intended load, speed, stability, manoeuvring and handling characteristics and seaworthiness. If the Lisheen artefact was intended to be a boat, its external shape indicates that the builders were not searching for good speed or manoeuvring qualities but for a means of conveying people or cargo over relatively short distances. However, the very thick floor and excessively thick end discounts even this purpose. The greater end thickness would not be necessary to maintain the structural integrity of the hull; it occupies space that could otherwise have been better utilised to hold cargo or crew. Both the thick floor and this end produce extra dead weight without which the hull would ride higher in the water and provide extra load-bearing capacity.

Performance capabilities

Analysis of the artefacts performance capabilities based on its measurements and a computer programme designed for dugout boats further indicates that it was not a boat. The programme results are based on a range of densities of $989 \pm 84 \text{kg/m}^3$ for fresh to seasoned oak. This figure was derived through tests for the programme that have been confirmed on life-sized replica dugout boats. The variables ($\pm$) in Table 12.2 result from using the above density. The maximum draught cited assumes a working freeboard of 10cm (10cm between the top of the sides of the boat and the waterline, after which it would be unsafe to increase the load). As the load with a 10cm freeboard would be negative, i.e. it could not carry anything, two other load conditions are considered, leaving 0cm and 2cm freeboard, respectively.

It is clear that only when a freeboard of 0cm is considered, would the artefact be able to carry one person. However, the artefacts stability when afloat would be very poor: -3mm. The righting arm is a naval architectural measure of stability. It is the horizontal distance between the centres of buoyancy and gravity when a boat heels by 10°. A weight of only 12.9kg placed on the gunwale or top edge of the side of the carved trough would lead to an angle of heel of 10°. If 32.8kg were similarly positioned, then the top edge would be at the water.

The artefact, if afloat, would have been able to carry one adult without sinking. However, the small righting arm means that the vessels stability would have been so

Unladen draught	28 ± 2cm
Max. load (10cm freeboard)	-115 ± 44 kg
Max. load (2cm freeboard)	35 ± 9 kg
Max. load (0cm freeboard)	72.5 ± 8.5 kg
Stability, righting arm	3mm
Weight on gunwale to provide 10° lateral heel	12.9kg
Weight on gunwale to provide lateral heel to water	32.8kg

Table12.2   Results of computer analysis of performance capabilities of carved trough 97E158:265:1.

poor at that draught that even in the calmest of weather conditions, the crafts occupant (or cargo) would have been continually in danger of being swamped. If the occupant so much as breathed, he got very wet.

Conclusion

The artefact was not a boat, nor was it intended to be one. Its shape, combined with aspects of its dimensions and evidence of dugout boat construction techniques, indicate that it would not have been feasible to remain in the boat and propel it in any manner without swamping.

It is possible that it was used as some form of floatation device, in which goods were placed to keep them dry while the owner waded alongside the vessel. Whether it was intended to be used on the water, it was certainly designed as a vessel to hold something.

Lithics

John Ó Néill and Paul Stevens

Introduction

Just ten tools and waste pieces of flint and chert were recovered from the excavated and surveyed sites. These sites were Killoran 1 (a chert scraper), Killoran 2 (a flint flake), Killoran 3 (two pieces of struck chert), Killoran 8 (a flint scraper, two flint flakes and two chert flakes) and Killoran 28 (a flint scraper).

There was very little good quality chert available in the area, and even less flint, so the paucity of lithics is not surprising. The dated contexts of these finds were mostly Bronze Age, such as Killoran 1 (a *fulacht fiadh*) and Killoran 8 (a Middle Bronze Age settlement). Killoran 2 is probably Bronze Age or earlier, as it appears to pre-date peat formation in the area in the Bronze Age. Killoran 3 produced dates in the early medieval period. Killoran 28 was noted during field walking but was never excavated or dated.

The poor lithic resources in the area probably led to a high level of husbandry of the available flint and chert. Although the assemblage is very small, some 40% of the flint finds and 20% of the chert appeared to be functional pieces. The largest piece of flint was a 45mm in length, while the largest piece of chert was a mere 20mm in length. Cores were also absent from the lithic assemblage recovered from the study area.

Each individual find is described below in the order of the site names.

Chert scraper (97E372:1:1)

Length 20mm, width 20mm, thickness 10mm; Killoran 1.

This dull, black chert thumbnail scraper was recovered from the burnt mound of the *fulacht fiadh* Killoran 1, on the break of the slope of a glacial ridge. It was a tertiary flake, with the striking platform present and a bulbar scar visible. There was some retouching on the cutting edge.

Flint waste flake (97E372:2:1)

Length 37mm, width 10mm, thickness 3mm; Killoran 2.

This broken, pebble flint waste flake had a buff-coloured patination. No striking platform or bulbar scar was present. Flaking ripples were present on both the dorsal and ventral surfaces. The artefact was recovered from a burnt spread, close to undated agricultural ditches at Killoran 2, on glacial till close to the modern western limit of Derryville Bog.

Chert flake (97E036:3:3)

Length 16mm, width 14mm, thickness 7mm; Killoran 3.

This piece of debitage was a small flake of black chert with a prominent bulbar scar and an irregular ventral surface. It was probably too small to reuse. It was recovered from C11 of Killoran 3.

Chert flake (97E036:3:4)

Length 18mm, width 10mm, thickness 4mm; Killoran 3.

This small piece of debitage was of good quality black chert and had no obvious bulbar scar or striking platform. It was probably detached from a larger flake during striking. It was recovered from C11 of Killoran 3.

Flint flake (97E439:8:5)

Length 14mm, width 9mm, thickness 5mm; Killoran 8.

This small tertiary flake of glacial flint was covered in caramel-coloured patination. It may have been a fragment of a larger piece, as there was damage to one side. There were no obvious signs of working or alteration, and it is most likely natural. It was recovered from the fill of C3, an internal post pit of House A at Killoran 8.

Flint scraper (97E439:8:14)

Length 30mm, width 24mm, thickness 9mm; Killoran 8.

This scraper was made from a secondary flake of grey-white pebble flint and was partially covered in a creamy-white patination, mainly on the ventral surface. Some cortex remained beside the striking platform, where there was a prominent bulbar scar. The ventral surface was partially retouched at the butt and heavily retouched at the distal end. The reinforced distal end appeared to have been damaged from repeated use. It was recovered from disturbed ground associated with a modern field boundary northwest of House B at Killoran 8.

Chert flake (97E439:8:15)

Length 18mm, width 14mm, thickness 5mm; Killoran 8.

This secondary flake of black chert had a number of flaking scars on the ventral surface; the striking platform appeared to have been removed. There were also a number of possible flaking scars on the dorsal surface. The distal end seemed to be broken. Some damage around the butt suggested that this flake may be have been used. It was recovered from the fill of C11, a pit associated with Structure E at Killoran 8.

Flint debitage (97E439:8:21)

Length 17mm, width 16mm, thickness 8mm; Killoran 8.

This was a piece of flaking debitage of poor quality pebble flint that had been heated, but not directly exposed, to a very high temperature. In two places, the surface appeared to have been created during flaking, and the piece may have become detached during flaking or heating. It was found on the surface south of House B at Killoran 8.

Chert debitage (97E439:8:26)

Length 16mm, width 6mm, thickness 3mm; Killoran 8.

This small flake of poor quality black chert had no obvious striking platform, suggesting it was a piece of flaking debitage. There were a number of flaking scars and some damage to one of the edges, suggesting this tertiary flake came from a larger artefact. It was recovered from the surface at the western side of House B at Killoran 8.

Flint scraper (97E372:28:1)

Length 45mm, width 30mm, thickness 10mm; Killoran 28.

This artefact was of grey pebble flint with a buff-coloured patination on the ventral surface. It was a secondary flake with some cortex on the ventral side, some flaking scars and some retouching. The dorsal side was not patinated. A striking platform and a bulbar scar were clearly visible, with some retouching on the cutting edge. The artefact was recovered from the same field as Killoran 28, during field walking close to the brow of a glacial ridge.

Saddle querns

John Ó Néill

Introduction

Three saddle querns were recovered from two, possibly contemporary, Middle Bronze Age sites: Killoran 8 (a settlement site) and Killoran 240 (a *fulacht fiadh*). This brings the total number of saddle quern finds from County Tipperary to six. Only two other sites in County Tipperary have produced saddle querns: Ballyveelish 3 and Lurgoe (Connolly 1994).

A number of *fulachta fiadh* have produced quern stones, although, as is the case with Killoran 240, some cannot be definitely linked to the use of the site. Three quern stones are recorded as coming from *fulachta fiadh* at Adrivale, County Cork (Cherry 1990, 50), and another was found covering a *fulacht fiadh* drain at Drombeg, County Cork (Fahy 1960). Saddles querns were also recovered, along with coarse ware and struck flint, from two *fulachta fiadh* excavated at Ballydown, County Antrim (Crothers 1997). Such quern stones could relate to a function of the troughs or may be an element of lower status ritual deposition in wetland margins, the normal location for this type of site.

In general, saddle querns are well known from Bronze Age domestic contexts (Connolly 1994, 31–2) and, more specifically, on settlement sites similar to Killoran 8, both in County Tipperary, e.g. Ballyveelish 3 (Doody 1987), and elsewhere, such as at Carrigillihy, County Cork (O'Kelly 1951) and Clonfinlough, County Offaly (Moloney *et al.* 1993b).

Saddle quern fragments (97E439:8:22)

Length 145mm, width 78mm, max. thickness 52mm (fragment 1); Length 119mm, width 86mm, thickness 32mm (fragment 2); Killoran 8.

The fragments of one or possibly two saddle querns were recovered beside each other. The larger fragment was in two pieces and was made of micaceous sandstone. It tapered to less than 11mm in thickness, possibly the reason why the stone broke. Together, the two pieces measured 145mm x 78mm, with a maximum thickness of 52mm. The smaller fragment was of the same type of stone and also tapered, in this instance to 18mm in thickness. The two fragments could not be easily matched together, but they might come from a larger quern stone that fractured around the centre, explaining why the surviving fragments are so thin on one side. They were found as packing within the wall slot (C1) of House A at Killoran 8.

Saddle quern fragment (97E439:8:25)

Length 182mm, width 129mm, thickness 49mm; Killoran 8.

This micaceous sandstone fragment was the middle section of a broken saddle quern that may originally have been up to 0.30m in length, although the two ends had broken off. The working surface was slightly concave and appeared to occupy the whole of the upper face of the quern stone. It was recovered from the fill of C3, an internal post pit of House A at Killoran 8.

Saddle quern (97E158:240:1)

Length 235mm, width 150mm, thickness 55mm; Killoran 240.

This micaceous sandstone saddle quern was recovered, base upwards, in fen peat east of the *fulacht fiadh* Killoran 240, at a level that suggested it may be contemporary with the use of the site. Otherwise, and given it was deposited upside down, it may be a ritual offering representative of the low status nature of the surrounding settlement. The quern stone was trapezoidal, with a smooth, sub-rectangular concave worked area measuring 183mm x 14mm.

Miscellaneous stone artefacts

John Ó Néill and Paul Stevens

Introduction

In all, some nine miscellaneous stone artefacts were recovered from excavations within the study area. Seven were made of sandstone (the local micaceous sandstone),

one was made of mudstone and one was possibly a piece of struck quartz. Seven of these finds were recovered from the Middle Bronze Age settlement site Killoran 8 and include hammerstones and rubbing stones, a possible whetstone and the piece of struck quartz. The two other finds were from early medieval contexts—a possible spud stone from Killoran 3 (a possible hut site) and a stone furnace bottom from Killoran 31 (the ecclesiastical enclosure *Cill Ódhrain*).

Possible spud stone (97E036:3:1)

Length 75mm, width 70mm, thickness 30mm; Killoran 3.

This micaceous sandstone artefact was recovered from the fill of a circular post hole (C5) located at the northwest end of an alignment of post holes, at an apparent right angle to a pair of large, parallel drainage ditches dated to AD 415–630 (Beta-117555). A second pit feature, also on an alignment with C5, was dated to AD 980–1270 (Beta-117556). The feature was part of a concentration of features that constituted the ephemeral site Killoran 3, interpreted as Early Christian, and later residual medieval activity, plus a phase of post-medieval drainage. The post hole was presumably part of a phase of the Early Christian activity. However, in the absence of stratigraphic layers connecting the feature with other datable features, it is impossible to relate the feature to a datable phase.

The artefact was roughly square, with a circular depression off-centre measuring 50mm in diameter and 10mm in depth. The depression was off-set to one side and the artefact appeared to have been crudely manufactured, with three flat sides and a clearly truncated fourth side.

The function of the artefact was probably as a spud to take an upright door post or gate. It appeared to have been discarded but showed some signs of use.

Rubbing stone (97E439:8:6)

Length 67mm, width 66mm, thickness 60mm; Killoran 8.

This rubbing stone was made from a grey-white sandstone pebble. It was almost cuboid in shape and wear marks on the some of the faces had left some pronounced edges. Use had also left a sheen on one face and a more abraded surface on another. The worn faces would suggest that this artefact was used as a rubbing or polishing stone. It was recovered from the fill of C12, an internal post pit of House A at Killoran 8.

Hammerstone (97E439:8:7)

Length 100mm, max. diameter 50mm; Killoran 8.

This hammerstone was made from a pink-grey cylindrical micaceous sandstone pebble that tapered slightly at both ends. There was evidence of abrasion on both ends, with a 20mm diameter area worked at an angle on one end and a 12mm diameter area worked on the other. At both ends, the damage was consistent with use as a hammerstone or pounder. It was recovered from the fill of C12, an internal post pit of House A at Killoran 8.

Whetstone (97E439:8:8)

Length 121mm, width 43mm, thickness 30mm; Killoran 8.

This possible whetstone was made of mudstone and was rectangular in shape. The upper surface appeared to have been slightly smoothed. This, along with its shape, was suggestive of a whetstone or hone stone. There was some trace metal on the upper surface, where it appeared to have been struck by a machine before excavation. It was recovered from the surface at the top of the wall slot (C13) of House B at Killoran 8.

Possible struck quartz (97E439:8:17)

Length 23mm, width 20mm, thickness 14mm; Killoran 8.

This object was a piece of white quartz that had been broken or struck. There was one rounded surface, which appeared to be original, and five fresh angular surfaces, although none of the edges appeared to be functional. As quartz does not break easily, it is possible that this piece had been deliberately altered. It was recovered from C1, the wall slot of House A at Killoran 8.

Rubbing stone (97E439:8:19)

Length 48mm, width 42mm, thickness 31mm; Killoran 8.

This sub-circular micaceous sandstone pebble had one worn and seemingly burnished surface with a large chip removed. The upper surface was rounded; the lower surface appeared to have been worn flat and had been slightly polished. This surface was heavily stained by charcoal and may actually have been heated. The large chip was removed from one side after the burnishing took place. The exact function of the object is uncertain. It was recovered from C28, an internal post pit in House B at Killoran 8.

Rubbing stone (97E439:8:20)

Length 87mm, width 85mm, thickness 37mm; Killoran 8.

This broken micaceous sandstone rubber was originally oval in shape, although just under one-half had broken off and was not recovered from the site. The original dimensions would have been approximately 120mm x 87mm. The upper surface was coarse and rounded; the lower surface was smoother and showed signs of wear. The lower surface was convex rather than flat. The object appeared to be a rubber that would have been used with a quern stone. It was recovered from the surface in the area of Structure D at Killoran 8.

Rubbing stone fragment (97E439:8:27)

Length 90mm, width 44mm, thickness 22mm; Killoran 8.

This object was a fragment of a micaceous sandstone rubbing stone with two flat working faces. The original size could not be reconstructed. Both flat faces were surrounded by an abraded edge. The faces were too smooth to have been used with a quern stone and may have been used to smooth wood or other surfaces. It was found in topsoil at Killoran 8.

Stone furnace bowl fragment (98E269:31:2)

Length 210mm, max. width 150mm, thickness 70mm; Killoran 31.

This micaceous sandstone furnace base was recovered in an upright position from the lower fill of a furnace pit (C31). The pit was within a small excavation cutting in the large ecclesiastical enclosure Killoran 31, located on the brow of a glacial ridge. Charcoal from this feature was dated to AD 450–690 (Beta-120521).

The stone base had a flat surface containing a sub-circular depression with vitrified crystallised slag residue and a flat base with fired clay and stone adhesions. The artefact was clearly broken through use and was from a square slab; one side had a right-angled corner. The circular depression had been carved into the stone and measured approximately 170mm in diameter and 40mm in depth.

Ferrous artefacts

Paul Stevens

Introduction

A single iron find (a nail) was recorded from the study area. It was found at Killoran 11, in the upper fill of a roasting pit. While the site was undated, two sherds of coarse pottery suggested an early medieval date, as did the context of the find. The artefact itself was fairly nondescript.

Iron nail point (97E372:11:2)

Length 25mm, width 9mm, thickness 4mm; Killoran 11.

This iron artefact was recovered from the upper fill of a roasting pit at Killoran 11, located on the brow of a sandy, glacial esker. Similar features were noted from the dryland excavations, one of which was dated to AD 980–1270 (Beta-117555). The artefact was very badly corroded. It was elongated and had a rough point.

Prehistoric pottery (pre-conservation assessment)

Helen Roche

The assemblages

A pre-conservation examination of the pottery assemblages from six sites excavated during the Lisheen Archaeological Project revealed a total of 196 sherds, consisting of two rim sherds, nine base-angle sherds, twenty-one base fragments, fifty-nine body sherds and 105 fragments. It is estimated that a minimum of six vessels were represented. It is envisaged that this number will change with the conservation of sherds 96E202:218:1 and 96E202:218:2. The estimated figure is based on the featured sherds and on the nature of the fabric. Vessel 5 was represented by recognisable rim sherds, whereas Vessels 1–4 and 6 were identified by their distinctive fabrics.

It is suggested at this preliminary stage that three pottery types from three distinctive periods of prehistory are represented in the assemblages: an early Neolithic round-based carinated bowl (Vessel 1), a late Neolithic Beaker pot (Vessel 6) and Middle/Late Bronze Age coarse ware (Vessels 2, 3, 4, and 5).

The majority of the sherds from the different assemblages represent coarse flat-based, probably barrel-shaped pots, dating to the Middle/Late Bronze Age (Vessel 2 from Killoran 4, Vessel 4 from Killoran 8 and Vessel 5 from Killoran 10). Although they were of the same vessel type, there were differences in the quality of the fabric and the execution of the vessels. Vessel 2 had very thick walls, but an attempt had been made to smooth the exterior surface. Although Vessel 5 was fairly thin-walled, the surfaces were very uneven and rough in texture. Vessel 3 (Killoran 8) will be included as representing a possible coarse pot, as the sherd was too small to make a positive identification.

The sherd representing Vessel 1 (Derryfadda 218) was also very small and, therefore, positive identification was again difficult. It is tempting to suggest that it also represents coarse ware, but the fabric was actually reminiscent of early Neolithic round-based carinated bowls. It will be necessary to clarify the identity of this sherd.

The sherds representing Vessel 6 (Killoran 15) were also problematic because of their small size. The fabric and surface finish were very different from those of the sherds from the other assemblages. They are, however, unlikely to represent coarse ware and were more consistent with that of the late Neolithic Beaker pottery. It would again be necessary to examine any other associated material from Killoran 15 that might clarify its identification.

Some of the sherds from Derryfadda 218 were too wet and fragile to be examined before conservation, and therefore identification was not possible.

Sherd (96E202:218:2)

Vessel 1; possible early Neolithic round-based carinated bowl; thickness 10.7mm; Derryfadda 218.

This single fragment had a hard, slightly crumbly fabric and a low to moderate grit content. Large cavities were present on both surfaces. The exterior and interior surfaces were brown; the core was black. It was possibly from an early Neolithic round-based carinated bowl. It was found at the platform Derryfadda 218 (2290–1935 BC).

Sherd (96E202:218:1)

Derryfadda 218.

This single sherd was too wet and fragile to be examined before conservation.

Multiple fragments of coarse ware (97E051:4:1)

Vessel 2; possible Middle/Late Bronze Age; thickness 9.5–12.1mm, base up to 19mm thick; Killoran 4.

Vessel 2, possible Middle/Late Bronze Age coarse ware, was represented by seven base-angle sherds, twenty fragments from the base, twenty-seven body sherds and sixty-three fragments. It had a hard, crumbly, thick-walled fabric, with a high content of mainly large calcite grits. The exterior surface was relatively smooth. The body appeared to splay out from the base in a gently rounded fashion, indicating that the vessel was probably barrel-shaped. It was orange-grey throughout. It was found at the Bronze Age cremation cemetery Killoran 4.

Multiple fragments of coarse ware (97E439:8:15)

Vessel 3; possible Middle/Late Bronze Age; thickness 11.4–14.1mm; Killoran 8.

This possible Middle/Late Bronze Age coarse ware vessel was represented by two body sherds and four fragments of hard, slightly chalky fabric with a low to moderate grit content. The vessel was thick-walled, but the fabric was compact. The exterior was orange, but the colour became grey towards the interior surface. It was found at the settlement site Killoran 8 (1860–1845 BC; 1775–1430 BC; AD 685–985).

Multiple fragments of coarse ware (97E439:8:24)

Vessel 4; possible Middle/Late Bronze Age; thickness 9.7–10.2mm; Killoran 8.

Vessel 4, possible Middle/Late Bronze Age coarse ware, was represented by two body sherds of hard fabric, slightly chalky in texture with a moderate to high grit content. The exterior surface was relatively smooth. Carbonised matter was present on the interior surface. The exterior was orange-brown, the core grey and the interior surface black. It was found at the settlement site Killoran 8.

Multiple fragments of coarse ware (97E168:10:1)

Vessel 5; possible Middle/Late Bronze Age; thickness 9.4–12.5mm; Killoran 10.

This vessel, possible Middle/Late Bronze Age, was represented by two rim sherds, two base-angle sherds, one fragment from the base, twenty-two body sherds and thirty-four fragments. It had an unexpanded, slightly in-turned rim and a hard, coarse-textured fabric with a high grit content. The vessel had been crudely executed with uneven, rough surfaces. It was orange throughout. It was found at the cremation cemetery Killoran 10 (1435–1215 BC).

Three fragments (97E372:15:1)

Vessel 6; possible late Neolithic Beaker pot; surviving thickness 4.1–5.2mm; Killoran 15.

Vessel 6, a possible late Neolithic Beaker pot, was represented by three small fragments consisting of hard, compact fabric with a moderate grit content. The exterior surface was very hard and smooth; the interior surface was missing. The exterior was brown in colour, the core black. The fragments were found at the early medieval burnt pit site Killoran 15.

Medieval pottery

Paul Stevens

Introduction

Two early medieval sites, a roasting pit (Killoran 11) and the ecclesiastical enclosure Killoran 31, or *Cill Ódhrain*, produced a small number of sherds of coarse pottery. Pottery of this type is well known from contemporary sites and is variously called Leinster cooking ware or souterrain ware, although the fabric of the sherds recovered from the study area differed from that of typical Leinster cooking ware or souterrain ware. These sherds probably represent an equivalent regional variant of the domestic coarse pottery tradition from the early medieval period.

Two coarse ware pot sherds (97E372:11:1)

20mm x 10mm x 30mm (sherd 1); 10mm x 10mm (sherd 2); Killoran 11.

Two sherds of pottery were recovered from the roasting pit at Killoran 11. They were from the same sherd and were broken in antiquity. The fabric was orange in colour and poorly fired, containing temper of grass and sand. The pottery was very thin and laminated.

Coarse ware pot sherd (98E269:31:1)

Killoran 31.

This very small and fragmentary sherd of pottery was bright orange in colour with a darker orange oxidised exterior. The sherd was broken in antiquity and only part of the thickness was present. The fabric was very sandy and poorly fired. It was recovered from a post hole at Killoran 31, an ecclesiastical enclosure dated to AD 450–690 (Beta-120521).

Burnt daub

John Ó Néill

Introduction

A number of fragments of burnt clay that were not pieces of pottery or clay moulds were recovered from contexts at Killoran 8, within the Middle Bronze Age houses. Some had the impressions of pieces of wood, suggesting that the clay had originally been attached to woven panels of wood, such as wicker walls or a fire basket. As the slot of the house appeared to have contained wattle walling, these clay fragments were interpreted as pieces of burnt daub. Their condition varied from well fired to poorly fired, and experimentation suggested that the clay was sourced locally.

Burnt daub fragment (97E439:8:1)

Length 79mm, width 75mm, thickness 49mm; Killoran 8.

This was a lump of well-fired clay that bore the impression of a 25mm diameter brushwood rod, suggesting it had been used as daub on a structure that had been burnt. The surface varied in colour from an oxidised black, through red to a creamy orange colour. There were some mica inclusions in the clay. Experiments with the boulder clay sealed between the glacial till and the limestone bedrock below the site suggested that the source was local. It was recovered from the fill of C3, an internal post pit in House A at Killoran 8.

Burnt daub fragment (97E439:8:3)

Length 30mm, width 27mm, thickness 18mm; Killoran 8.

This was a lump of well-fired clay, creamy orange to grey-brown in colour, derived from a local source. It did not appear to be a fragment of pottery and was more likely to be a fragment of burnt daub or a clay mould. It was recovered from the fill of C1, the wall slot of House A at Killoran 8.

Burnt daub fragment (97E439:8:4)

Length 23mm, width 20mm, thickness 13mm; Killoran 8.

This irregularly shaped fragment of fired clay contained mica and some larger calcite inclusions. The surface was a creamy orange colour, stained grey-black in places, resembling the naturally sourced clay found elsewhere on the site. The fragment was similar to the other burnt daub found on the site, but it could equally be part of a clay mould; identification was uncertain. It was recovered from the fill of C3, an internal post pit of House A at Killoran 8.

Burnt daub fragments (97E439:8:9)

Average size 30mm x 25mm x 25mm; Killoran 8.

A large number of fragments of fired clay, red-black in colour and made of locally sourced clay. Most of the pieces were too bulky to be pottery. The purpose and origin of these clay fragments are unknown. They were recovered from C12, an internal post pit of House A at Killoran 8.

Burnt daub fragments (97E439:8:10)

Largest fragment 25mm x 22mm x 12mm; Killoran 8.

Four pieces of fired clay were recovered from the fill of C14, an internal post pit of House A at Killoran 8. They were creamy orange in colour, with some blacker, burnished areas. There were some particles of mica present in the fabric, suggesting the source was the local boulder clay. The pieces were irregular in shape and may be more fragments of burnt daub.

Burnt daub fragment (97E439:8:16)

Length 31mm, width 21mm, thickness 17mm; Killoran 8.

This irregularly shaped fragment of fired clay was found on the land surface inside House A at Killoran 8. It was creamy orange in colour. The fabric contained mica and appeared to have been sourced locally. It may be a fragment of burnt daub, but this is uncertain.

Burnt daub fragment (97E439:8:18)

Length 24mm, width 23mm, thickness 14mm; Killoran 8.

This fragment of fired clay varied from grey-brown to orange in colour. The clay contained mica and was probably from a local source. The origin of the fragment is uncertain. While it may be a small rim sherd of coarse ware, it is equally possible that it is a fragment of burnt daub or a clay mould. It was recovered from the wall slot, C1, of House A at Killoran 8, near the western doorpost.

Faunal remains

Paul Stevens

Introduction

A number of sites produced fragmented, crushed or burnt faunal remains. In all identifiable cases, the finds were ovicaprid teeth or bone. No formal analysis of the material was possible due to the small sample size and the fragmented condition of the bone. These sites were all located on the dryland west of Derryville Bog.

Tooth

Ovis/Capra; Killoran 5.

Several fragments of a single, crushed sheep/goat premolar (mandible) were recovered from the trough backfill, C6, of the *fulacht fiadh* Killoran 5 (1750–1410 BC).

Long bone and possible skull fragment

Small mammal and possibly *Ovis/Capra*; Killoran 16.

A fragment of long bone from a small mammal and a fragment of skull, possibly sheep or goat, were recovered from C21 and C10 at the archaeological complex Killoran 16 (180 BC–AD 425/AD 890–1040).

Teeth

Ovis/Capra; Killoran 22 Structure 1.

Two sheep/goat pre-molar crushed fragments (mandible?) were recovered from the upper fill of trough C2:1 at the Bronze Age *fulacht fiadh* Killoran 22.

Carpal fragment, humerus fragment, tooth fragment and two ribs

Medium-sized mammal, *Ovis/Capra* and a small mammal; Killoran 27.

Several bones were recovered from the later fill, C1:5, of the original circular trough, C1, at the *fulacht fiadh* Killoran 27 (932 BC). They were a sheep/goat third molar (maxilla); a carpal or tarsal fragment from a medium-sized mammal, also possibly sheep/goat, and a humerus longbone shaft gnawed at both ends, also from sheep/goat.

Two further bones were recovered from the organic slump fill of the later sub-rectangular trough C10. These consisted of two rib bones; one was from a sheep or goat, and the other was an unfused fragment from an unidentifiable small mammal.

Burnt rib bone fragments, burnt long-bone fragments, burnt skull fragment and burnt carpal fragment.

All possibly *Ovis/Capra*; Killoran 31.

Several burnt bones were recovered from a sub-circular pit, C27, used for iron smelting at the enclosed ecclesiastical settlement Killoran 31 (AD 450–690). These were most likely sheep or goat but were heavily burned and fragmented. They consisted of three fragments of ribs, seven fragments of long bones, a carpal or tarsal fragment and a skull fragment.

Human remains

Laureen Buckley

Introduction

Deposits of cremated bone were recovered from four sites at Killoran as part of the Lisheen Archaeological Project. All the sites dated to the Middle Bronze Age. One site consisted of two cremation pits containing token deposits, one associated with a crushed pot. Two other sites consisted of isolated cremation deposits with only a small amount of bone present. The fourth site was the largest flat cemetery discovered in Ireland at present and contained twenty-five deposits of cremated remains. The deposits ranged in size from a few fragments to over 400g of bone. They were mostly crushed token deposits. Although there was no more than one individual per deposit, they were far from complete with the skull and femurs being selected at the expense of the spine and smaller bones. Two of the sites contained a bone fragment that appeared to have been cut with a sharp instrument prior to cremation.

Methods

Examination of cremated remains involves a description of colour and texture of the bone as this helps determine the efficiency of the cremation. Well-cremated bone is white in colour and usually has a chalky texture. Less well-cremated bone, where the temperature of the pyre was not high enough or where oxygen flow was restricted has a blue or a blue/black colour. The bone fragments are then graded by size in order to determine the degree of fragmentation. Although fragmentation of the bone occurs continuously post-deposition, due to compression pressures and also during the disturbance of excavation and processing, it is still possible to assess whether or not the bones were deliberately crushed as part of the cremation ritual. A high proportion of relatively large fragments would suggest that the bones were not deliberately crushed after the cremation whereas a small deposit of relatively small fragments would indicate a ritually crushed token deposit.

Each fragment of bone is then examined and identified if possible. The degree of identification of fragments is generally dependent on fragment size. Larger fragments are usually easier to identify although phalanges are often found intact among the smaller fragments. Successful identification depends on the number of distinguishing features present on the bone fragments as well as knowledge of the thickness and expected cross section of particular bones. However, bones shrink and warp during the cremation process and sometimes it is not possible to specifically identify long bone fragments.

The minimum number of individuals present can then be determined by the number of specific skeletal elements. It is possible to distinguish juveniles from adults from the thickness of the bone fragments, the presence or not of unfused epiphyses and by the fragmentation of the teeth. Adult teeth crowns tend to shatter during the cremation process but unerupted juvenile teeth tend to survive intact as they are protected by the jaw bones. It is usually possible to age juveniles if enough teeth are present.

Killoran 4

There were two samples of cremated bone recovered from this site. The larger sample from C2 was not very highly crushed and it was possible to identify 23% of the bone. It appeared to be the remains of one adult but there was insufficient information to determine the sex. The skull and lower limbs had been well collected from the pyre but the axial skeleton (ribs, vertebrae and pelvis) had barely been collected at all. Perhaps they were too difficult to collect from the bottom of the pyre or perhaps they were not considered characteristic of the individual and therefore of less importance.

The smaller deposit from C7 did not contain the 1600–3600g of bone expected from a normal cremation but nevertheless 190g was a large sample compared to most of the samples recovered from the sites at Killoran. The cremation was highly crushed and appeared to have been washed prior to burial (P. Stevens, pers. comm.). This explains why the bone had an eroded appearance which made identification difficult. All these factors

point to it being a token ritual deposit and the presence of an upturned fragmented pot forced into the cremation pit is a further indicator of this.

The remains appeared to be from one male adult. Most of the identified bone came from the leg bones with very little skull or arm bones remaining and no axial skeleton. This could be due to the fact that only the heavy obvious bones were selected for deposition.

Summary

The two cremations from this site are the remains of two adults one of which is a male. The male burial consisted of a highly fragmented token deposit associated with an upturned pot.

Killoran 6

Only one bone sample, from C3, was provided from this site. It consisted of a small highly fragmented token deposit which appeared to be the remains of one adult only. A small proportion of skull bone was identified, with the rest of the sample being undifferentiated long bone.

Killoran 10

This was the largest flat cemetery recovered to date in Ireland and is associated with the nearby settlement site also dated to the Middle Bronze Age. It contained sixty-seven pits and thirty of these contained evidence for burials. Some of the pits were sealed by a capping of clay or sand (P. Stevens, pers. comm.).

A total of twenty-five bone samples were received for analysis. Sixteen of these were from capped pits and nine were from uncapped pits.

A summary of the findings is given in Tables 12.3 and 12.4 for both capped and uncapped pits. Where a very small amount of bone was present, no result is given for either Weight or Percentage Identified.

The sample sizes ranged from a few fragments of bone to a maximum of 414g, with an average of 221g among the significant samples. This is clearly far short of a full cremation.

It can be seen from the summary tables that although most of the larger samples came from the capped pits, three were from uncapped pits and eight of the tiny insignificant samples came from the capped pits. Thus the presence or absence of a cap did not indicate the size of the sample within.

Most of the fragments were highly fragmented which, coupled with the low weight of bone recovered, indicates that they were probably token deposits. The degree of fragmentation reflects the percentage of bone that could be identified. Where the samples were less crushed, then a higher proportion of bone was identified. The average proportion of identified bone from significant samples on this site was 21%, one sample being significantly higher at 52%.

Context	Weight (g)	Degree of Fragmentation	Percentage identified	Min. no. of individuals
C10/C18	347	Moderate	25	1 adult
C15	n/a	6 small frags	n/a	Uncertain
C21	301	Moderate	28	1 adult
C24	73	Highly	23	1 adult
C25	1	Highly	n/a	1 juvenile
C26	15	Highly	n/a	Uncertain
C30	414	Moderate	21	1 adult
C31	231	Moderate	23	1 adult
C34	n/a	9 tiny frags	n/a	Uncertain
C35	n/a	5 tiny frags	n/a	1 juvenile
C41	166	Low	52	1 adult
C42	166	High	10	1 adult
C45	6	High	n/a	Uncertain
C48	3	High	n/a	Uncertain
C54	110	High	4	1 adult
C55	8	High	n/a	1 juvenile

Table 12.3 Results of bone sample analysis from capped pits in the Killoran 10 cemetery.

Context	Weight (g)	Degree of Fragmentation	Percentage identified	Min. no. of individuals
C7	n/a	1 frag.	n/a	Uncertain
C8	1	High	n/a	Uncertain
C9	191	High	8	1 juvenile
C11	8	High	n/a	Uncertain
C13	n/a	4 frags	n/a	Uncertain
C28	218	Moderate	30	1 adult
C32	218	High	12	1 adult
C37	1	1 frag.	n/a	Uncertain
c38	n/a	7 frags	n/a	Uncertain

Table 12.4 Results of bone sample analysis from uncapped pits in the Killoran 10 cemetery.

Also, as a consequence of the small fragment size, not enough prominent skeletal features were present to help determine the sex of the individuals. The most that could be done was to divide the samples into adults and juveniles, ignoring the smaller insignificant samples. A total of ten adults and four juveniles were present among the samples of this site, with no more than one individual per sample.

The difference between this Middle Bronze Age site and the Early Bronze Age cemeteries at Keenoge, Strawhall, Graney West and Brownstown (Buckley 1997; 1998) is startling. At the Early Bronze Age sites most of the samples, which had not been disturbed, were over 1000g in weight with one sample at Brownstown weighing over 2000g. Furthermore, the samples were not deliberately crushed and it was possible to identify 87% of one sample from Grave 8 at Keenoge. Also, most of these early cremation deposits contained every possible skeletal element including vertebrae, ribs, metacarpals, metatarsals and phalanges.

At Killoran 10 most of the samples contained very little or none of the axial skeleton (i.e. ribs, vertebrae and pelvis), and there were only a few fragments of metatarsals and phalanges in some of the deposits. It seems it was easier or perhaps more significant to the population to collect the skull and heavier long bones, such as the femur, to leave in their ritual token deposits. The smaller bones and vertebrae may have been left in the pyre or removed while the bones were sorted. A few of the smaller samples consisted exclusively of long bones but there was only one of the larger samples which contained no skull. Therefore it seems that the deposits were collected from one skeleton with no attempt to bury the long bones in a separate location from the skull.

The only pathology noted was a small cut from a fine instrument on a fragment possibly from a humerus. It appeared to be have been made antemortem as the bone would have shattered if the bone was cut after cremation.

One fragment of bone from another deposit had a green stain which is usually caused by contact with copper metal.

Summary

This flat cemetery contained twenty-five deposits of cremated bone ranging from a few fragments of bone to around 400g. A total of ten adults and four juveniles were present with not more than one individual in each deposit. Some of the cremation pits were capped with clay but no significant difference could be found among the type of deposits in the capped and uncapped pits. The deposits appeared to be mainly highly crushed token deposits the sorting of which concentrated on collecting the major bones of the skeleton such as skull and long bones and ignoring the smaller bones and the flat bones.

Killoran 24

This consisted of two isolated cremation pits. C1 contained only 1g of highly crushed and weathered long bone and C2 contained only a few tiny fragments of bone.

13. Worked wood

John Ó Néill

Introduction

During the course of the surveys and excavations of Derryville Bog, many examples of humanly altered wood were recorded and sampled for further examination. The background to current worked wood analysis emerged from excavations in the Somerset Levels (Orme 1982a; Orme and Coles 1983) and the raised bogs of the Irish Midlands (O'Sullivan 1996). Recent work (Brunning and O'Sullivan 1997; Sands 1997) has also addressed individual issues and reflects, what is now, a steadily growing corpus of comparable published work. A dedicated synthesis of worked wood studies has yet to appear, but the publication of further papers may help establish an agreed methodology and suggest further areas for research.

Admittedly, there was no pre-existing worked wood research strategy at the outset of the project, but any cursory review of the papers listed above demonstrates the vagaries of the surviving evidence on and within individual waterlogged structures. The direction of this study followed on from the construction of chronological and environmental sequences for Derryville Bog, which created the context in which the assemblage needs to be understood.

Once the archaeological background was clear, a number of potential issues were apparent. The dated sequence, from the Neolithic through to the medieval period, provided an opportunity to examine change and continuity within the surviving evidence. This heavily influenced the approach taken to the study, which also utilised the 8,000-plus detailed wood identifications of Dr Ingelise Stuijts (Chapter 7) and the observations of the individual excavation directors.

Ultimately, the identification of the introduction of iron axes became the focal issue in this study, along with the creation of a methodology to differentiate bronze and iron axes. Other issues, prominent in the work published by Bryony Orme, John Coles and Aidan O'Sullivan, are discussed briefly, where appropriate.

The woodworking assemblage recorded from the excavations in Derryville Bog amounts to over 700 roundwoods and 200 (mostly degraded) cleft- or split-wood samples from which some information could be recorded. The majority of the recorded examples were from trackways, causeways, stake/post rows and platforms dating from the Neolithic through to the ninth and tenth centuries AD. Preliminary analysis carried out at the end of the 1996 field season identified several biases in this evidence that would dictate the direction of the final report. These biases were not redressed by the sites excavated during the 1997 season but were, in fact, reinforced and are discussed below.

The quantification of the worked wood resource provided the necessary context for this analysis, which in the case of the Derryville worked wood assemblage, encompassed a range of evidence of varied quality, scattered over 3,000 to 4,000 years. This provided an opportunity to examine ways of extracting information from the assemblage and addressing issues that would normally seem to be beyond the scope of a project producing such a paucity of small finds.

The data generated by the samples were either recorded in the field or sampled for recording post-excavation. In an attempt to obtain sets of data that would allow comparisons to be made across the range of dates and locations of sites, some structures had information recorded for all the worked wood present, while others had a representative sample recorded or retained. The general guide for this was the actual condition of the site on excavation, with some sites producing few surviving worked ends and others producing large numbers.

The pieces of wood that had been worked or altered in some shape or form numbered in the thousands. Many of these simply had degraded worked ends or signs of branch-trimming. On many structures, all or most of the wood showed some form of alteration, and this information was recorded in the field where samples were not required. Generally, the evidence was in the form of roundwoods with worked ends and a lesser number of samples of cleft wood ('split' timber).

The picture from the wood identifications made by Dr Stuijts suggests that the majority of the wood used in Derryville Bog came from the fringes of the bog. This fitted with the interpretation of the wetland structures as signifying a 'vernacular' landscape (see Chapter 15). This interpretation implies that tool identifications would relate to objects that could be considered to be in everyday use. This would have none of the implications of prestige that must be considered when dealing with hoards and ritual deposition. The range of tools used could then be shown to be in circulation among the lowest level of society, a group that is often overshadowed by the more traditional archaeological focus on high status landscapes and monuments.

The level of survival of tool marks varies according to the post-depositional hydrological stresses acting on the peat containing each structure. Extremely high levels of preservation were found on sites built when the water table was rising, such as Derryfadda 6 and Derryfadda 9, constructed when the water table had recovered from the seventh-century BC bog burst. On other structures, prone to re-exposure, dehydration and compression, very poor survival conditions led to the deterioration of much of the wood. This was the main factor in the degradation of early structures, which were subjected to a number of swift drops in the water table.

Whenever hydrological or subsequent biochemical processes allow deterioration to take place, the decay first attacks the cut and exposed sapwood and pith of worked ends. On some structures, this could result in a level of degradation as high as 95%. This decay is more apparent on sites located in minerotrophic and some less-humified ombrotrophic peats. The periodic variations in sub-surface conditions were responsible for the deterioration of the worked ends on many Derryville structures down to around 300 BC.

Other biases are more apparent. The Neolithic is very sparsely represented among the Derryville Bog material, in contrast to the Bronze Age and Iron Age. The wetland excavations also did not reveal a single site from AD 50 to 640, although a few sites dating from the seventh century AD, or later, were investigated.

The constraints imposed by the nature of the development, as outlined in the introduction to the chronology of the excavated structures, were responsible for the sparsity of Neolithic evidence. Modern peat extraction in the bog removed the upper levels, which may have contained later, medieval, structures.

Apart from Killoran 18, many of the larger structures, such as Derryfadda 218 and Derryfadda 13, produced a lower percentage of identifiable worked ends (7.6% and 11.3% respectively) than many of the smaller structures. Some of the latter sites, such as Derryfadda 9, produced much higher percentages (59.9%). This figure does not take into account the quality of the surviving examples, which, in the case of Derryfadda 218, was very poor, and in the case of Derryfadda 9, was excellent.

The documentation of the hydrological history of the bog systems uncovered several episodes of drastic fluctuations in the water table as a result of bog bursts, the last of which occurred around AD 300. The higher level of survival on some structures built after 400–300 BC, when the water table had recovered from the 820 BC and 600 BC bog bursts, is in contrast to the poor condition of many sites where construction took place before that date. In any event, the earlier sites were also mainly constructed in nutrient-rich fen and marginal forest peats, and, as such, were more susceptible to decay, as discussed above. The relationship between the bog bursts and the surviving condition of structures is illustrated in

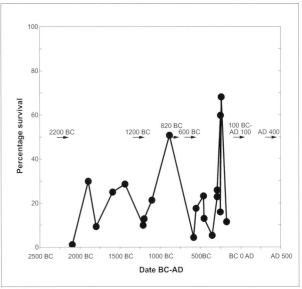

Fig. 13.1 Percentage survival of worked ends plotted against date. Bog bursts are indicated by arrows.

Fig. 13.1, where the measure of surviving condition was taken as a percentage of pieces of wood with recordable worked ends. The range of levels of survival in the Iron Age is a reflection of the greater proximity of structures to the modern bog surface.

Overall, enough samples of worked roundwood ends were recovered from a number of periods to allow comparisons to be made. Although the quality of the examples varied by site, enough were found on most sites to give an indication that the typical tool marks of the axe-type in use is represented, at least for the excavated structures. A glossary of terms is provided at the end of the chapter.

Aims and methodology

A number of woodworking studies establishing the criteria on which individual tool marks can be examined and identifying the techniques used in certain periods have been published (Orme 1982a; Orme and Coles 1983; O'Sullivan 1996). A review of the basis of these studies shows how user variability is a major factor in the performance of most tools. Data collection followed that used by O'Sullivan (*ibid.*) to allow for the creation of a woodworking database. This database could be used for inter- and intra-site comparisons. Given that statistical criteria for differentiating the tool marks of individual axe-types or for mapping chronological periods have yet to be established, the database was used to identify which information may provide an aid to the analysis of isolated sites.

The study included an examination of the techniques of felling, woodworking and carpentry, as evidenced amongst the Derryville assemblage; the dating of axe-types and chronological periods, through analysis of worked roundwoods; and the analysis of axe signatures and blade widths.

Several specific areas of study were selected on the basis of the available evidence. The use of a morphological distinction between the marks made by stone and metal tools was employed to date the early use of metal in the Mountdillon Bogs, County Longford (*ibid.*), and it was apparent from the range of dates produced that a bronze-iron series may be produced from the Derryville assemblage. Thus, actual parameters that would give an end-date for the use of bronze axes and a start-date for the use of iron axes could be identified. This was attempted by sequencing the changes in the transverse profile of the facets left by individual axes and then comparing the properties of the axes placed within each part of the series.

A further area of study concerned the use of chisels, specifically unhafted Early Bronze Age flat axes, for cutting mortise holes in timbers. This use was considered against the backdrop of 'socketed' technology, typified by a relative gap of 500 years between the evolution of socketed spearheads and axe heads.

Both of these studies were carried out alongside a consideration of the woodworking assemblage recovered from the excavated structures in and around Derryville Bog. The basic techniques of prehistoric and medieval felling, woodworking and carpentry have been examined by a number of authors (McGrail 1982; O'Sullivan 1996) and, as such, are well documented. The Derryville assemblage was examined to see how it conformed to the techniques known from other sites.

Another area that was studied was the traces (signatures) left by the individual blemishes on an axes cutting edge. These can be used to identify worked ends cut with the same axe and, under certain conditions, information regarding working patterns or internal site chronologies (Sands 1997). When used on a woodworking assemblage spread over a long period of time and across a wide area, the potential margin for error is often so great that no benefit can be gained (Brunning and O'Sullivan 1997). This caveat meant that no firm conclusions could be drawn from much of the Derryville material.

Blade widths, either from surviving jam curves or as minimum estimates from tool marks, were recorded for a number of sites. The vagaries of the information regarding jam curves on the Derryville assemblage, with some sites producing clear examples and others failing to produce any, does not allow for a detailed discussion of the numbers of tools used etc. However, an overview of the actual blade width of the axes used may provide some insight into the availability of raw materials and husbandry of resources within a tighter chronology than is normally available.

Woodworking techniques

Introduction

The bulk of the wood from the Derryville assemblage was derived from the marginal forests and deciduous woodlands of the surrounding area. It was generally poorly grown. The techniques of tree-felling and modification have been detailed for the prehistoric period in Ireland by O'Sullivan (1996). At Derryville Bog, the same basic techniques were in use. Occasionally, evidence of other activities or points of interest were identified, and these are discussed below.

Tree-felling

Some of the evidence from Derryville Bog sheds further light on tree-felling practices in prehistoric Ireland. A trunk found on Derryfadda 207, dating to the Early Bronze Age, had fallen naturally and showed evidence of a discontinued attempt at felling 0.3m above ground level. Similarly, a 0.3m high axed oak stump was found beside the Iron Age trackway Derryfadda 203. In an experiment at Flag Fen, tree-felling was carried out at 'well above ground level', though the height was not quantified (Pryor 1991, 63). The correlation between the felling heights is interesting, given that the trunk at Derryfadda 207 had a diameter of 20cm, while the trunk at Derryfadda 203 had a diameter of 10cm, suggesting continuity in felling practices. Neither of the trunks showed signs of bark stripping or ring burning before felling.

The practice of branch-trimming felled trees was evident on structures of all periods, as was the practice of using kerfs—asymmetric wedge-ends for controlling the direction in which a tree would fall. The controlled felling of trees in dense woodland, as evidenced by the kerfs, and the stripping of felled trees into more transportable bundles is unsurprising given the difficult nature of the ground surface in the fens, marginal forests and raised bogs of the area.

The felling and use of complete trees in the construction of trackways was evidenced by the chopped ends of timbers at Derryfadda 9, where the top wood had been removed from an alder trunk and a willow trunk. Similar tool marks were present on a number of other sites, such as Derryfadda 13, Cooleeny 31 and Derryfadda 208.

Charred wood and charcoal were uncovered in a number of areas around Derryfadda 6 and Derryfadda 17, and on the land surface near the end of Killoran 18 and Killoran 223, suggesting scrub clearance, even within the marginal forest. These charcoal deposits have been identified in horizons dating from the Early Bronze Age through to the post-medieval period.

The range of diameters and the quality of the wood used on the structures are discussed elsewhere (see Chapter 7). The species present on the wetland structures—alder, ash, birch, blackthorn, crab apple, elm, hazel, holly, oak, rowan, wild cherry, willow and yew—are also known as firewood from the *fulachta fiadh*. It is likely that they were also used in the construction of the settlement sites around the edges of the bog (Killoran 8 and Killoran 16). Some other species, such as Scots pine, alder buckthorn, water elder, poplar and field maple, occurred once or very rarely, but did show evidence of having been exploited. Notable absences were hawthorn, juniper, elder and buckthorn, indicating that certain species were simply not present or not selected for use on the structures. In general, a wide range of species was selected for use and a number of managed woodlands were also maintained in the area. Future studies may identify a link between certain woodworking techniques and managed woodlands.

Cleft wood

In Derryville, evidence for the splitting or cleaving of trunks into planks was found on all the major site types: causeways, trackways, platforms and stake rows. Examples were recorded on a range of structures dating from the Neolithic through to the medieval period. Structures that were built mainly with cleft wood include the *fulachta fiadh* troughs causeway Killoran 18, trackway Cooleeny 22 and trackway Derryfadda 23.

The vast majority of these cleft wood pieces were oak or ash. The wood was split into the basic conversions produced when it is cleaved along its natural splitting planes (Orme and Coles 1983). Other species split were alder, hazel, birch, crab apple and willow. Although wooden wedges would have been employed to produce planks, no definite examples were found during the excavations nor were the tracks of the wedges recorded on the faces of any of the split timbers. However, a possible oak wedge (96E202:23:1) was recovered on the northern side of the single-plank trackway Derryfadda 23 (see Chapter 12).

Much of the cleft wood was in poor condition and a number of sites contained naturally split wood, mainly deriving from the boles of bog oaks that had rotted and broken up into fragments. Half-split brushwood and roundwoods were occasionally found. On Killoran 54, half-split brushwood was used as uprights, possibly as an economising measure.

The larger pieces of cleft wood were mainly found on sites dating to the Middle and Late Bronze Age. The relative scarcity after this period suggests that the best quality trees were being cut down; the surviving oak woods consisted of the progeny of the poorer quality trees that were not selected for felling and splitting. This would have reduced the quality of timber that could be produced from those woods.

In the Middle and Late Bronze Age, timbers up to 7m in length were split with great skill into tangential and radial splits, suggesting that there was little difficulty in controlling the splitting along the planes of weakness of the wood, which would be one benefit of good quality timber. By the Iron Age, the cleft wood did not take the form of large oak and ash planks but was more often smaller half- and quarter-split trunks.

Hewing and bucking

A number of planks, notably those from the relined trough of Killoran 240, had been hewn across one or both faces. The practise of hewing or dressing timbers was not very well represented among the Derryville assemblage. A number of the planks used in the construction of the

trough of Derryfadda 216 and some of the timbers with mortise holes had been hewn along their edges.

At Derryville Bog, the tool used for hewing was generally the adze, as was most apparent on planks from Killoran 240. Here, the direction of the blows was along the main axis, suggesting that the plank was worked lying flat.

On some sites, where tooling survived on the ends of timbers, it did not appear to represent felling marks or later surface dressing. In a number of cases timbers appeared to have been worked prior to being split. This may indicate the bucking of timbers into lengths for splitting. The fragmentary nature or hewing of most of the cleft wood had modified the original ends and removed the worked area, indicative of bucking.

Carpentry joints

A small range of carpentry joints were recorded amongst the Derryville assemblage. Outside of the *fulachta fiadh* troughs, few of the Derryville structures would have relied on planks and jointing to maintain their form. It is presumed that a greater range of carpentry joints would have been employed on the roundhouses of Killoran 8 and Killoran 16, dryland sites that did not produce waterlogged finds.

Vertical lap joints were present on the side planks of the trough of Killoran 240, holding the relining in place. Barefaced tenons were also used on timbers in the original lining of the same trough. These had been inserted into holes in the side of the trough pit. A plank used as part of Killoran 226 contained a series of circular holes. This plank may have originally been a base plate for a wattled wall.

Timbers with mortise holes were identified on a range of sites, from trackways such as Cooleeny 22 and Derryfadda 23, as well as the *fulacht fiadh* Killoran 240, all dating to around 1600–1500 BC. These mortise holes were mostly cut from both faces to produce an hourglass profile. There were no clear examples of V-shaped or parallel-sided mortise holes on the Derryville timbers. A number of hourglass-shaped mortise holes (e.g. T23 at Derryfadda 23 and T24 at Killoran 18) had one half cut in a V-shape with one straight and one angled side, the straight side being the one nearer the end of the timber. The reason for this woodwork is unknown but it may have been an attempt to avoid splintering the plank around the mortise. On two structures, Cooleeny 22 and Derryfadda 23, pegs were found within mortise holes. The tracks of pegs were found aligned with the mortise holes on timbers used to reline the trough of Killoran 240.

The purpose of many of the mortise holes is unclear, although some obviously housed pegs. It is likely that many of these timbers were reused from settlements around the edges of the bog and that the presence of the mortise holes is unrelated to their secondary use. At Derryville the use of mortised planks centres on the period between 1600 BC and 1450 BC. The unenclosed settlement of Killoran 8 appears to have been in use around the same time and may be the source of some of the timbers used in these structures.

End shapes

There are three main categories of worked ends on round-woods used on wetland structures: chisel (or single-faced ends), wedge (bi-facial) and pencil (three or more faces). Torn ends were also present in some numbers but could not always be clearly distinguished due to preservation conditions. There appears to be a simple correlation between the three types and diameter (Fig. 13.2).

The range of relationships between diameters and end shapes noted on the Early Bronze Age material (Fig. 13.3) was similar to that from Ross Island, County Kerry (O'Brien 1994, 140), where the increase in diameter was reflected in an increase in the number of sides tooled. The range of end shapes found also matched that at Annaghbeg 1, where chisel-ends predominated, but not Corlea 6, where the majority were wedge-ends (O'Sullivan 1996, 307).

In the Middle Bronze Age, there was a predominance of pencil-ends (Fig. 13.4) due to the sharpening of stakes for the earliest phase of Killoran 18. The post rows at Clonfinlough, County Offaly (OF-CFL 0011 and OF-CFL 0015) showed a preference for wedge-ends, as did a trackway at Kilmacshane, County Galway (GA-KME 0008), although a majority of chisel-ends were found on a nearby trackway (GA-KME 0009; Moloney *et al.* 1995, 25). Roughly equal proportions of chisel and wedge-ends were recovered from Derryoghil 10, County Longford (O'Sullivan 1996, 317). The proportion of wedge-ends increased with diameter for the Derryville material.

In the Late Bronze Age, pencil-ends predominated on Runnymede Bridge (Heal 1991), where uprights were the major part of the structure. At Caldicot, the preference for wedge-ends on wood with larger diameters was noted (Brunning and O'Sullivan 1997, 179). The pattern from Derryville was also an increase in wedge-ends with increasing diameter (Fig. 13.5). A similar pattern was found at Derryoghil 2 (O'Sullivan 1996, 319) but not at Derryoghil 1, where chisels predominated.

The same correlation between wedge-end predominance and diameter also persisted for material from the Late Bronze Age-Iron Age transition (Fig. 13.6) and Iron Age material (Fig. 13.7) at Derryville. In Corlea and Derraghan More, chisel-ends were most prominent, with a greater number of pencil-ends than wedge-ends (*ibid.*, 332). Most of the worked ends of the medieval material recovered in Derryville were of the chisel-end category (Fig. 13.8).

Generally, where a pattern is known, it is of wedge-ends displacing chisel-ends as the predominant type, with a concurrent increase in diameters. Pencil-ends would then displace wedge-ends for larger diameters, as demonstrated amongst the firewood present at Ross Island. This is a natural progression in tree-felling, as more faces or sides are worked to control the felling direction or to cut a larger trunk. The correlation between end shape and woodland source may also be an indicator of the management strategies and woodworking techniques employed.

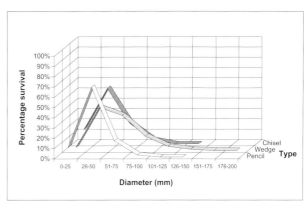

Fig. 13.2 Percentage survival of roundwood versus diameter by end shape.

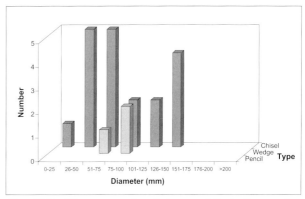

Fig. 13.3 Diameters and end shapes of Early Bronze Age worked ends.

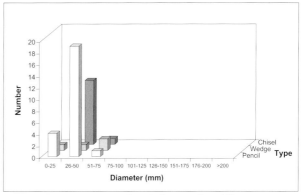

Fig. 13.4 Diameters and end shapes of Middle Bronze Age worked ends.

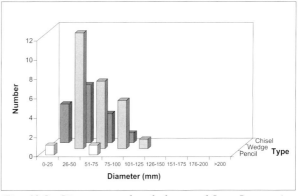

Fig. 13.5 Diameters and end shapes of Late Bronze Age worked ends.

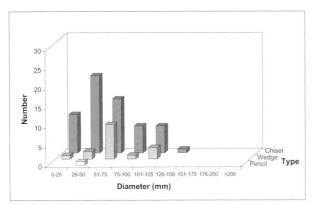

Fig. 13.6 Diameters and end shapes of Late Bronze Age-Iron Age transition worked ends.

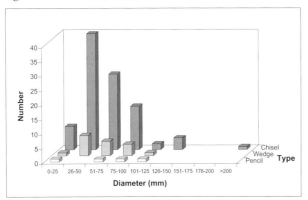

Fig. 13.7 Diameters and end shapes of Iron Age worked ends.

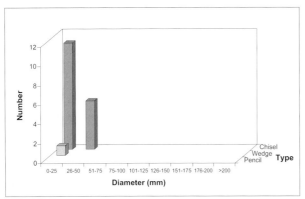

Fig. 13.8 Diameters and end shapes of medieval worked ends.

Woodcarving and fine woodworking

A limited range of wooden artefacts were found during the Derryville excavations (see Chapter 12). They give a glimpse of the range of portable wooden artefacts that were made by people dependent on, and hence practised in, the day-to-day production of wooden tools and utensils.

Many were basic implements, such as the yew spear shafts from Derryfadda 13 (94E106:2) and Killoran 237 (97E160:237:1). These tools had been whittled to a point with a knife, which must have been a very common practice. The same can be suggested for other artefacts, such as the notched peg from Derryfadda 203 (96E202:203:1) and the possible spoke or handle from Derryfadda 13 (96E202:13:1).

A number of the artefacts showed signs of having been perforated or drilled, including the two pommel-headed batons from Derryfadda 9 (96E202:9:1 and 96E202:9:2); the shaft with expanded and perforated terminals from Derryville 325 (97E160:325:1); the split roundwood with notches from Killoran 230 (97E160:230:1); the notched wooden panel from Derryfadda 311 (96E202:311:1) and the perforated and split lath from Cooleeny 32 (94E106:1).

The bucket fragment from close to Derryfadda 17 (96E202:17:1) had gouge marks on its inner surface. Socketed gouges are well known from the Late Bronze Age onwards.

Evidence of turning was found on the rod recovered from Cooleeny 31 (97E160:31:1). Under low magnification, lathe marks could be seen along the shaft where the rod from the outer rings of an ash trunk, had been turned. This piece of wood hints at a high level of competence in fine woodworking, which is generally not represented amongst the material recovered from Derryville Bog.

In most instances, these artefacts appear to have been lost or discarded on the bog surface or during the construction of some of the structures. They were made from a range of wood types, including cherry, ash, alder, yew, oak and guelder rose. In one or two cases, they may have been deliberately deposited in the bog, but as a group they merely hint at the range of wooden artefacts that would have been commonplace in antiquity.

Mortise holes and the use of unhafted axes in the Early Bronze Age

In his account of a hut uncovered in the Bog of Drumkelin, Inver, County Donegal, Wood-Martin (1886, 39) relates that a Captain W. Mudge reported 'chisel' marks on the mortises and 'axe' marks on the hewn ends of the timbers. More recent analysis has concluded that axes were used to cut the mortise holes in timbers (O'Sullivan 1996, 325). Unfortunately, the tooled areas around mortise holes on timbers from the Derryville assemblage rarely retained jam curves and other indicators of the cutting edge.

Where evidence could be collated for a timber recovered from Flag Fen (B1421), at least three different cutting edges were identified, and it could not be demonstrated that the hewn ends and the mortise holes were worked with the same blade (Taylor 1992, 490). There was no indication of whether the blades used to cut the mortises were unhafted.

Although this appears to be a semantic issue, the variability possible with an unsocketed axe blade may explain both the absence of identifiable chisels in the early phases of metal use and the late appearance of socketed axes, well after socketed technology was available.

In considering the use of axes to cut mortise holes, it was noted that not one single example was found of a stray blow missing the worked area around the hole. Experimental work on mortise holes produced a different picture, with a series of cut marks around the hole show-ing the inaccurate nature of axes (Orme and Coles 1983). Even given a high degree of skill, it is unlikely that the prehistoric woodworkers would have been so infallibly accurate. The absence of stray cut marks is not a survival issue, as the underside of some of the base planks of the trough of Killoran 240 had cut marks where it had been used as a chopping board, probably during the manufac-ture of pegs, but none around the hole.

As hourglass-shaped mortise holes are the predomi-nant type amongst the Derryville assemblage, stray cut marks should have been evident on the better-preserved undersides, but this was not the case. It may be unlikely then that hafted axe heads were used for this purpose and that chisels or unhafted axe heads were used instead, allowing more control over the removal of the wood inside the mortise hole.

Cutting tools identified as chisels do not appear in the Irish archaeological record until the Late Bronze Age. It must then be assumed that the unsocketed axes of the ear-lier half of the Bronze Age could be utilised as chisels. The potential versatility of Early Bronze Age axes has been noted elsewhere (Ramsey 1995, 15; O'Sullivan 1996, 321) and there is external evidence to support the hypothesis.

A Ballyvalley-type axe from Brockagh, County Kildare, was found in a leather sheath, and a number of axes from the Early Bronze Age series had surface deco-ration that would have been covered by a haft (Harbison 1969). The most logical explanation for this is that it was intended for use without a haft, i.e. as a chisel. The butt of the Brockagh axe is no different in profile to the butt of chisels found in Late Bronze Age hoards. We have no knowledge of the use-history of the Brockagh axe, nor do we know if short vertical hafts were used to transform flat axes into chisels.

In an attempt to understand the relationship between flat axes, decoration and hafts, the published corpus of Early Bronze Age flat axes was examined to compare these elements. The expected correlation if the axes were used unhafted would be that axes with decoration up to the butt would have a flat or hammered butt, from their use as chisels. Examination of the published flat axes (*ibid.*) produces a notable pattern.

The Early Bronze Age axes can be categorised into three groups, displaying (1) no decoration, (2) decoration confined to the portion visible after hafting and (3) deco-ration extending to the butt, which would be covered if hafted. Taking the subtypes chronologically, the Lough Ravel-type and Ballybeg subtypes are all undecorated, with a high percentage (72%) of Lough Ravel-type axes having flat butts compared with a mere 2% of Ballybeg axes. Between 11% and 17% of the undecorated Ingot, Killaha, Ballyvalley- and Derryniggin-types have flat butts. The two recorded Killaha-type axes with decora-tion confined to the blade have pointed butts, but 15% of the Ballyvalley-type and 14% of the Derryniggin subtype

Type	Undecorated	Partial decoration	Decoration at hilt
Lough Ravel	72%	0%	0%
Ballybeg	0%	0%	0%
Ingot	11%	0%	0%
Killaha	17%	0%	0%
Ballyvalley	16%	15%	34%
Derryniggin	16%	14%	50%

Table 13.1 Percentage of each type of axe with a 'flat' rather than 'pointed' butt. 'Partial decoration' does not include those with decoration confined to the hilt.

have flat butts, similar to the undecorated axes. A differ-ent pattern emerges from axes having decoration that would be covered by a haft. All of the Killaha subtype axes have pointed butts, with 34% of the Ballyvalley sub-type and 50% of the Derryniggin subtype having flat butts. These percentages are given in Table 13.1.

In summary, the pattern emerging from the Lough Ravel and Ballybeg subtypes is markedly different from the homogenous pattern for the Ingot, Killaha, Ballyvalley and Derryniggin subtypes. The pattern extends to blade-decorated Ballyvalley and Derryniggin axes but not to axes of these two types where decoration would be covered by a haft. This latter group has a markedly higher percentage of axes with a flat butt. This demonstrates some correlation between decoration on the hilt and a flat butt. While there may be no direct cor-relation between butt type and function, it is not an unlikely relationship.

Conclusions

It is not until the beginning of the Late Bronze Age that dedicated chisels emerge as a separate tool type. Their evolution is most likely a response to the appearance of socketed axe heads, which could no longer be effectively utilised in different capacities, as could flat axe heads. Both socketed axes and chisels occur in association with gouges in a series of hoards from Crevillyvalley, County Antrim (Raftery 1942); Enagh East, County Clare (O'Carroll and Ryan 1992); Callanagh, County Cavan; Bishopsland, County Kildare; Knockmaon, County Waterford and Ross, County Tipperary (Eogan 1983, hoards 50, 16, 144 and 137, respectively). Oddly, there were no chisels recorded in the 200 items of the great Dowris hoard (Crawford 1924).

The emergence of these 'woodworkers' hoards in the Late Bronze Age may chart the rise of a market in skilled woodcarving (Eogan 1964) or simply a greater function-al distinction between individual tools. The use of sock-eted technology with spearheads occurs as early as the Early Bronze Age (Ramsey 1995, 52), with a relatively late appearance of the socketed axe at the end of the Middle Bronze Age.

At Derryville Bog, the evidence for socketed axes and gouges comes from the Late Bronze Age sites. Derryfadda 17 produced a bucket fragment with internal gouge marks and axe marks consistent with those produced on other Late Bronze Age sites that appear to have been produced by socketed axes. The yew spear shaft dating to the Middle Bronze Age found at Killoran 237, again shows the presence of the technology in the area at least 300 years before its use for axes. The reason for this may need some explanation.

It is highly unlikely that socketed axes slowly evolved through various forms of flanged axes and palstaves when the technology had been available for some 500 years. This would suggest that the adoption of socketed technology was resisted, perhaps on functional grounds, as flat axe heads were more versatile. This would also explain the increasing innovation in the hafting techniques and the avoidance of sockets. The alternative is that the use of single- and bi-valve moulds was part of domestic technology, whereas the use of clay moulds and cire perdue was more a prestige or martial technology that did not percolate down through society until the later stages of the Bronze Age.

Data analysis

Introduction

In order to create a comparable approach to woodworking analysis, a series of analogues were compared to produce a recognisable sequence. This analysis was conducted mainly to distinguish bronze and iron axes, although the internal Bronze and Iron Age chronologies could also be examined.

If the available evidence could generate information regarding the date of various axe types and, by extension, the use of various raw materials, then a remarkable number of issues could be addressed. One unequivocal factor underlying this analysis is that the tool marks on the worked ends come from axes being used in their most basic, domestic function and that they survive by accident, not design.

Firstly, the chronological sequence of the dates was used to compare the woodworking from individual sites on the basis of the character of the marks left by the tools. A number of assumptions were made before examining the worked ends:

1. Different axe types will produce distinct sets of tool marks due to changes in the weight of the axe, the degree of sharpness of the blade, the longitudinal profile of the axe head and the hafting mechanism.

2. The resulting axe head trajectory, entry angle and friction co-efficient will leave a track determined by the distinct properties of the axe type. The axe head trajectory will be dictated by the weight and hafting of the axe, the entry angle by the sharpness of the blade and the friction co-efficient by the shape of the axe.

3. As axes are classified typologically, in a chronological sequence, on the basis of these same properties, a similar sequence of changes should be visible by the marks left by those axes.

4. Thus, stone axes, with a relatively low weight, fairly sharp blade and high friction co-efficient (due to the thickness of the body) are best used in short stabbing motions that do not cut deeply into the wood and leave short concave facets or chopping marks. Iron axes, with a higher weight, sharp durable blade and thin body, can be used to cut deep into heartwood to leave long flat facets.

5. Even allowing for variations in the quality of the timber worked and the skill level of the woodworker, the character of the facets of the worked ends produced by each type of axe should oscillate within identifiable parameters. If necessary, certain specifications could be used as a control, such as Alnus with diameters of 50–70mm and 40–60 growth rings. This would allow the chronological subdivision of the individual site assemblages into a dated framework.

6. Rather than matching individual tool marks to certain tools, the sequence can then be matched against a proposed typological series, for example: (i) stone axe, (ii) flat axe, (iii) flanged axe, (iv) palstave, (v) socketed axe, (vi) shafthole axe. In practice, some of these stages may be absent or overlap.

In this way, different stages where a change in tool type or hafting may have taken place could be identified. Ideally, this would allow an independent chronological framework to be created, and this framework could be compared with the conventional date ranges of each period. Given the sequence of dates from the tenth to first century BC, the actual change in character of the tool marks could be plotted more accurately across the transition from the Late Bronze Age to the Iron Age than for the period from 2000 BC to 1000 BC.

Dating the Bronze–Iron transition

In an attempt to analyse the worked ends by period, the approach adopted was to compare the individual facets left by the axes and then identify a basis on which they could be subdivided. In this way, it was hoped that any typological changes in the axes or hafting would be matched by a similar change in the character of the individual facets.

When the transverse profiles of the facets were examined, it became apparent that a series of changes in the shape, varying from concave to flat, were synchronous and reflected some form of artefactual change. The reality of this was apparent when one structure, Killoran 248, which was initially found to be 'out of sequence,' was re-dated and found to be later than first believed. The date for Killoran 248 had been obtained from a fallen oak below the structure, but a direct date from wood used in the track itself was considerably later. It also fitted back

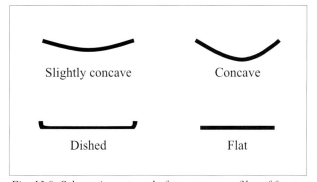

Fig. 13.9 Schematic portrayal of transverse profiles of facets.

Slightly concave   Concave

Dished   Flat

into the sequence again. On this basis, a series showing the changes that occurred through the Bronze and Iron Ages was prepared. Using this series, the following three designations could be made (see Fig. 13.9):

1. If there was an immediately noticeable curvature of the profile, it was designated as 'concave' (with the modifier 'slightly' also included).

2. If the profile had to be examined by viewing along the facet and a slight curvature could be distinguished, it was designated as 'dished' if slightly curved at its sides or 'slightly concave' if the curvature was gradual.

3. If the profile showed no signs of curvature at all, it was designated 'flat.'

The correlation of transverse profile designation and date

In the Early and Middle Bronze Age, the profiles were concave, although the actual depth of the concavity varied. The profile always formed an unbroken curve. In the Late Bronze Age, the profiles were also concave, although facets were generally smaller and the concavity more pronounced. The latest group of sites identified within this group include Killoran 229, Derryfadda 64, Derryfadda 209 and Killoran 69, whose dates span 1420–770 BC.

On structures dating from the end of the seventh century BC, such as Cooleeny 31, the profile changed to being 'dished,' with a wide, flat base and very short, curved sides. Sites falling within this group include Derryfadda 206 (which predates Derryfadda 215), Derryfadda 13 (lower level), Derryfadda 210 (510–365 BC) and Cooleeny 31 (*c.* 600 BC), whose dates span 790–365 BC.

On structures dating from around 460–450 BC and later, the profiles were very flat. Derryfadda 215, dating to 457±9 BC, is the earliest site with this type. Others include Derryfadda 6 (directly overlying Derryfadda 215), Derryfadda 9, Derryfadda 13 (upper level), Killoran 75, Derryfadda 201, Derryfadda 203, Derryfadda 208, Derryfadda 214, Killoran 248, Killoran 301, Killoran 229, Killoran 314 and Killoran 315. The medieval structures were also identified with this last group.

Comparison of analogues

As the information could be grouped into a sequence, a number of analogues were created to facilitate comparison. This was done by extracting measurements that could demonstrate the performance of the axe (or the woodworker) and give information about the axes in relative terms. Thus, the measurements of the longest surviving facet, minimum cutting angle and wood density (diameter/year rings) were logged and the inter-relationships plotted for the subdivisions of the sequence. The pre-seventh century BC material was divided further into the conventional periods of Early Bronze Age (2350–1700 BC), Middle Bronze Age (1700–1200 BC) and Late Bronze Age (1200–*c.* 650 BC). The period between *c.* 650 BC and 460 BC was treated as a transitional period, identified with the second designation of the transverse profiles as identified above. Structures after the hiatus up to AD 500 were grouped as 'medieval.'

The plots for maximum facet length against cutting angle (Figs 13.10–13.14) and cutting angle against wood density (Figs 13.15–13.19) are shown for each period from the Middle Bronze Age onwards. Comparisons of the plots for the Late Bronze Age, 'transitional' and Iron Age sequences showed that the transitional group followed the same trends as the Iron Age group, implying a similar performance. On this basis, it would be proposed that from *c.* 630 BC iron axes were in use in the vicinity of Derryville Bog, with a possible change in type, or improvement in quality after 460–450 BC (perhaps indicated by the appearance of completely flat facets).

In summary, the three subjective groupings were an unbroken curve (concave), a disrupted profile (dished) and flat. Defining what this represents in terms of the type of axe in use is more difficult. As the changes are diachronic, they appear to reflect a common property of the axes in use at any given time. Why this should be reflected in the transverse profile of the facets is uncertain, although it may be a product of the flexibility of wood. When an axe head cuts through wood, the fibres are stretched in proportion to the width of the axe, which would be likely to produce a variation in the profiles of the facets.

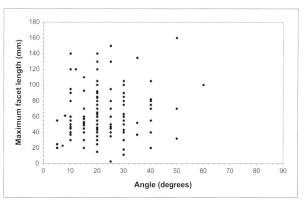

Fig. 13.10 Cutting angle versus maximum facet length for Middle Bronze Age samples.

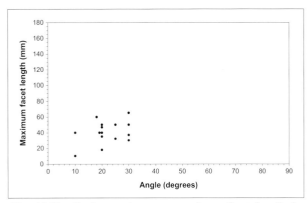

Fig. 13.11 Cutting angle versus maximum facet length for Bronze Age samples.

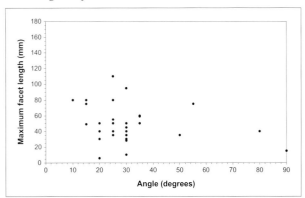

Fig. 13.12 Cutting angle versus maximum facet length for samples dating to the Late Bronze Age-Iron Age transition.

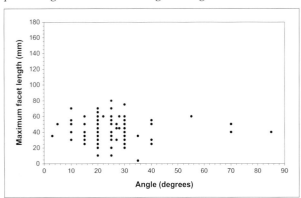

Fig. 13.13 Cutting angle versus maximum facet length for Iron Age samples.

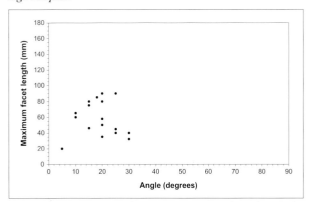

Fig. 13.14 Cutting angle versus maximum facet length for medieval samples.

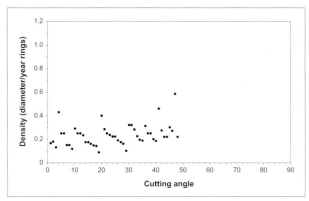

Fig. 13.15 Density ratio (diameter:year rings) versus cutting angle for Middle Bronze Age samples.

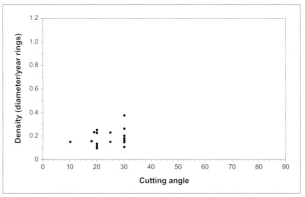

Fig. 13.16 Density ratio (diameter:year rings) versus cutting angle for Late Bronze Age samples.

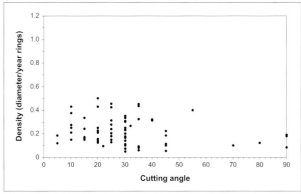

Fig. 13.17 Density ratio (diameter:year rings) versus cutting angle for Late Bronze Age-Iron Age transition.

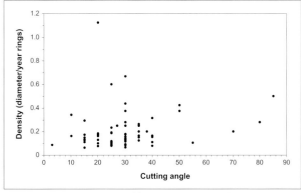

Fig. 13.18 Density ratio (diameter:year rings) versus cutting angle for Late Bronze Age samples.

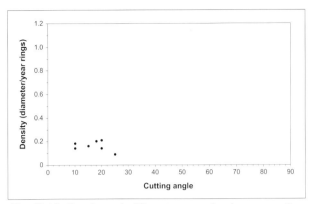

Fig. 13.19 Density ratio (diameter:year rings) versus cutting angle for medieval samples.

The flat facets appear to have been produced by iron axes, which would be heavier and thin in cross section. Socketed axes had a broader cross section, but earlier metal axe types, while not as thin as iron, are also generally flat and less broad than socketed axes. The shape of the cutting edge would not appear to influence the transverse profile, as not all iron axes have flat-blade edges and not all bronze axes have curved-blade edges. At Derryfadda 203, an axe with a curved blade produced flat profiles.

Analysis of signatures

Few of the sites in Derryville Bog had enough surviving evidence of signatures to make detailed comparison possible, such as at Derryfadda 9, where three sets of axe signatures were present, or Derryfadda 6, where one set was present. The exact meaning of this information is equivocal (Brunning and O'Sullivan 1997, 164), the only definite statement that can be made being that identical sets of signatures were produced by the same axe. This allows us to say that Killoran 18 and Killoran 305 are contemporary because the same axe signatures are present. How often the axe might have been re-sharpened (thus removing the signatures) is a matter of conjecture. Another complicating factor is that some wood species carry signatures better than others (Coles and Orme 1985).

On the Iron Age platform Derryfadda 9, three pieces of wood showed the same set of axe signatures but had been felled in spring, midsummer and late summer, respectively. As they had been used as horizontal elements of the site, there is no reason to believe they were reworked. This suggests that signatures could survive on iron axes at least from spring to late summer, a period of around six months. Further damage or blemishes do not appear to have been added to the blade over this time.

On larger structures, the use of signature comparisons has more potential. In Derryville Bog, Killoran 18 was the only structure large enough to be considered as likely to produce any intelligible patterns. The distribution of signatures was examined across the two rows of upright stakes that marked the first phase of construction. A

number of scenarios that might conform to particular patterns were created:

1. An irregular distribution could suggest stockpiling of sharpened stakes for use as pegs.

2. Where individual sets of signatures could be shown to have accumulated during the sharpening or felling of stakes, the distribution of these could identify the series in which they were added to the site (i.e. east–west or west–east). If the stakes had been stockpiled, there would be interruptions to this pattern in the reverse direction.

3. Where a wide range of signatures could be recorded in no discernible pattern, the input of a large number of people in the laying down of the staked phase of the track could be suggested.

The number of sets of signatures present at Killoran 18 (Phase 1) was as follows: Field 23 (0), Field 39 (9), Field 41 (5), Field 43 (0), Field 45 (3) and Field 46 (1). The huge variation in the level of preservation of worked ends across Killoran 18 meant that, in practice, observations would be confined to small numbers of stakes on certain fields. The signatures on Field 39, for example, were found on a range of species (*Alnus*, *Fraxinus*, *Salix* and *Malus*), diameters (31–53mm) and cutting angles (10–40°). Ideally, the analysis of signature patterns should be carried out on an element of a large structure that utilises a restricted range of species/diameter/cutting angle profiles. These conditions were not present amongst the material excavated from Derryville Bog.

Tool types and dating

The use of certain individual tool types has been recognised on a number of sites. A list of the tools likely to have been used to build the structures in Derryville Bog is presented below, along with any evidence concerning the date of their use, where relevant. One obvious absentee is the saw, which does not appear to have been introduced into Ireland until late in the historic period.

Adzes

The use of adzes in the Early Bronze Age was identified on timbers from the trough of Killoran 240. Evidence of adze use was also found on the carved oak trough from Derryville 265.

Awls/borers

A number of artefacts recovered from Iron Age contexts had been drilled or bored to a diameter unlikely to have been produced by a gouge. These include the pommel-headed batons recovered from Derryfadda 9. No evidence of their use has been identified before this date, though it should probably be assumed that they were in use.

Axes

Although it may be self-evident, it is important to bear in mind that in all periods there was free access to the use of axes, which in the case of the dated structures appear to be metal. In the case of structures built after *c.* 630 BC, axes seem to be made of iron.

Chisels

As discussed in an earlier section, the use of chisels on timbers to produce holes or mortises is known from the Early Bronze Age. The emergence of specialist narrow-bladed, flat and trunnion chisels appears to be a Late Bronze Age phenomenon. The evolution of socketed axes did not favour the use of an unhafted blade as a chisel in the same way as flat axes or palstaves may have been used. This would explain the flat axes found with decoration from the blade to the butt and also the axes found in sheaths.

Gouges

The use of a gouge is known from the bucket fragment from Derryfadda 17 and the base plate for a wattle screen from Killoran 226, dating to the Iron Age. Both bronze and iron gouges have been identified within the archaeological record.

Knives

Although it is likely that some form of flint or metal knives were in use in all periods, evidence of their use can only really be inferred from artefacts such as spear shafts, the earliest of which dates to the Middle Bronze Age.

Mallets

Wooden mallets were probably used along with seasoned oak wedges to split trunks into planks from the Middle Bronze Age onwards. Again, it should be assumed that they were probably in use in all periods.

Glossary of terms

Bucking	Dividing wood into shorter lengths before splitting.
Brushwood	Generally roundwood below 6cm in diameter.
Chisel-end	Worked roundwood end with one cut face.
Cleaving	The splitting of wood along the natural planes of weakness; types of split include half, quarter, radial and tangential.
End shape	The type of worked end produced by felling or sharpening.
Facet	The individual mark left by each cutting action of the tool used.
Kerf	An asymmetric wedge-end indicative of control over the direction of felling.
Hewing	Working or dressing the face of a timber.
Pencil-end	Worked roundwood with three or more cut faces.
Roundwood	Unmodified wood, including branches or trunks.
Signatures	The series of striations on a facet that are produced by nicks in the blade edge of the tool used to cut wood.
Timber	Split or cleft wood, as opposed to unmodified wood (roundwood).
Wedge-end	Worked roundwood end with two opposing cut faces.

14. Terrain sensitivity

Sarah Cross May, Cara Murray, John Ó Néill and Paul Stevens

Introduction

All landscapes can be defined as the interface between humans and the environment they inhabit. Terrain is the non-human component of the landscape. In Derryville Bog the inter-relationship of these two elements can be studied against the backdrop of the evidence provided by the broad -based environmental research carried out during the project. A level of awareness can be detected in the siting of the structures and in the reaction of their local environments. An assessment of this interplay between structure and environment can also be made on regional and broader levels.

Derryville Bog is part of a chain of bogs that were continually developing throughout the period in which most of the sites were constructed, that is, mainly the Bronze Age and Iron Age. It would appear from the analysis of these structures that two patterns are present. The principle one represents the vast bulk of the sites and demonstrates that the building of structures in the bog was a response to continual peat formation. The second pattern relates to several dramatic episodes. The structures, inserted into an active ecosystem, generated a range of positive and negative impacts, but most appeared to be constructed with a significant knowledge of the microenvironment.

Cooleeny 31 marks a major departure from this general pattern. It displays none of the awareness of micro- or macrotopography shown by the other structures (cf. Casparie, page 39), and it is therefore argued here that the impetus and details of its construction came from a non-local agency. It contrasts with the vernacular nature of the majority of the structures in Derryville Bog and indicates a possible regional motivation behind the bridging of the chain of bogs.

The detailed study of the peat morphology, palynology and related studies, wood identifications and microfaunal analysis combine to form a detailed reconstruction of past environments in Derryville Bog. This chapter considers this set of changing environments in a regional context and discusses the siting of structures in three groupings: those sited to take advantage of local conditions (the majority of *fulachta fiadh* and many trackways and platforms), those constructed in response to dramatic episodes of change (some platforms, trackways and causeway Killoran 18) and those imposed upon the environment (causeway Cooleeny 31). Examining these groupings leads to an understanding of how terrain sensitivity functions in Derryville Bog.

A changing environment

The geographical location, topography and climate of Ireland were conducive to peat formation throughout the Neolithic and later periods. Of course, there were times when the peat-forming conditions were more favourable than others, but in Derryville Bog the water supply was never the limiting factor to peat formation (see Chapter 3).

When considering the human response to bog development, a number of points need to be borne in mind. The rate of fen peat accumulation in Derryville Bog was not more than an average of 70mm per century, or 20mm every twenty-five years. Just as the long-term marine transgressions in the Assendelver Polders (Abbink 1986) did not impact on local consciousness or influence their actions directly, it is unlikely that people around Derryville Bog would have noticed, or reacted to, a formation rate that was barely visible across four generations. The palaeohydrological evidence suggests that the water table fluctuated from as little as 1cm below the surface to 15cm below. This may be a measure of the variability that people regarded as the normal behaviour of a bog.

As these variations could happen within short time spans, folk tradition could retain them as part of the intimate experience and knowledge of the immediate environment. A high tolerance of the bog and a culture in which it was seen as an exploitable environment, rather than a wet wasteland, could develop. It is noticeable that the pattern of exploitation of the bog changes only when the fen is acidified through the expansion of raised bog.

The impact of climatic trends on the use of a bog is reduced if the intemperance of its surface is an accepted everyday feature of its use. Thus, gradual increases and decreases in the measurable parameters, such as rainfall, peat formation rate or water table depth, would not impinge upon human consciousness, as the scale of change would be imperceptible. The continuing bog development was the permanent backdrop to any activity in the fen or raised bog. More dramatic changes to the surface of the bog would have conditioned the responses that saw the construction of individual sites, such as Killoran 18 or the structures constructed after bog bursts.

The nature of bogs as boundaries and crossing points

The range of archaeological evidence now known from wetland contexts in Ireland shows the exploitation of the estuarine, lake edge, fen and raised bog resources that were available to the local population. The construction of causeways across these wetlands represents a distinct component of this activity, reflecting a desire to bypass rather than access the wetter terrain. The type of passage afforded by large causeways was rarely available, even after a sustained period of dryness considerably lowered water tables. The focusing of attention on a single corridor across a bog represents a more complex form of activity than a local response to maintaining a route during a climatic downturn. Passage across a bog system is rarely possible without some form of structure built for that purpose.

Features such as mountain ranges, rivers and lakes often form the basis of political, social or economic boundaries as they act as physical barriers. In much the same way, expanses of bog interrupt lines of communication and, as in the case of the tract of bog that includes Derryville Bog, they can become territorial boundaries. Unlike many other topographical features, which can be crossed with conventional land or water transport, it is often impossible to cross a bog by foot. It is in this context that the causeways that were built to allow direct communication across bogs should be viewed.

The provincial, ecclesiastical and county boundary that runs along the northeastern side of the study area follows the line of the bog chain to the north and south. It is tempting to look at the narrowest sections of this bog chain and see Derryville, in particular, as an obvious choice for crossing this divide. The archaeological evidence shows that this choice seemed less obvious to prehistoric groups, with the larger Bronze Age causeways constructed across the bogs directly to the north and south. The environmental evidence gives us some insight into why this was the case.

The identification of the narrowest sections is not all that needs to be considered when choosing a bridging point. There are two important reasons why Derryville Bog may not have been chosen as a crossing point in the Late Bronze Age. Firstly, for a midland bog, Derryville Bog remained a fen for much longer in prehistory than is normally expected. Due to the greater variations in the humidity and stability of the surface, it was more difficult to construct substantial structures across this fen than a raised bog. Secondly, the five recorded bog bursts indicate that the study area may have been a particularly unstable sector of the bog chain.

As the peat development of other bogs in this chain has not been thoroughly studied, it is not clear why the nearby options were more suitable for causeway construction and associated activity. The archaeology of these bogs has undergone preliminary assessment (IAWU 1996b), and the patterning of their lower-order structures suggests similarities to Derryville Bog. The evidence from the dated causeways in Templetouhy Bog and Longfordpass, County Tipperary, suggests that, in the Late Bronze Age at least, the more substantial structures were sited elsewhere.

At a more local level, Derryville shows another attribute of boundaries. Marginal zones or buffers between different communities are often the focus of intense activity, with or without exchange between groups (Kowalewski *et al.* 1983). Once again, the environmental research helps us understand the position of Derryville, as fen has a wider variety of resources than raised bog. The structures revealed by this study show the bog as a focus for more than one group to whom it was not an obstacle to contact and movement but a place with value of its own. The degrees of knowledge of, and sensitivity to, this ecosystem are displayed in the studies of groups of sites below.

Our understanding of the role of any terrain feature in a landscape can be measured by the level of knowledge of both the environments present and the responses made to them. The degree of sensitivity shown to the environmental context is directly related to the input of local knowledge during siting and construction.

The regional position of Derryville Bog

The chain of bogs containing Derryville Bog stretches for 40km from the Slieve Blooms and runs south along the foot of the Slieve Ardagh Hills. The area is part of the midlands plateau of Carboniferous Limestone under glacial drift ridges forming patches of higher fertile ground. The bogs lie on the eastern and northern parts of the plain containing the River Suir, bordered by the Devil's Bit mountain range to the west and the Comeragh Mountains to the south. Through these valleys, the Suir winds its way towards the sea, and a series of passes through the hills on either side allow access to the Shannon network in the west and the Nore valley in the east. During the prehistoric period, much of the land now reclaimed for farming would have been dense forest, flood plain or bog.

Derryville Bog is located at the centre of this chain of bogs, with Templetouhy and Carrick Hill Bogs immediately to the north and Baunmore and Inchirourke Bogs to the south. Similar concentrations of archaeological material to the study area have been recorded in surveys of the surrounding bogs (IAWU 1996b). Although the precise nature and date of many of the structures recorded by the IAWU is uncertain, they indicate that a similar range of activities was occurring in other bogs along the chain. A number of dated and undated causeways are present and give some indication of the areas where bridging this chain of bogs was attempted.

Archaeological structures in the bogs north and south of the study area

In addition to the material within the study area, Derryville Bog contained two unexcavated clusters of material: the Cooleeny and Killoran Complexes. A further forty-four structures were recorded by the IAWU to the north in Killoran and Derrygreenagh townlands. Twenty-three structures are present in Baunmore and Inchirourke, southeast of the study area, and an archaeological complex and seventeen structures were recorded in Leigh and Longfordpass South, to the south. In the townland of Ballybeg, further south, a discrete complex of archaeological material was recorded. More structures have been noted around Derrynaflan, at the southernmost part of the chain of bogs.

In all of these areas, trackways, stake rows and other deposits of archaeological wood have been identified. The survey information from these bogs may suggest a similar pattern of activity as occurred in Derryville Bog. Thus, just as at Derryoghil, County Longford (Moloney *et al.* 1993a; Raftery 1996) and other excavated and surveyed areas of bog, the range of structures present at prehistoric levels in Irish raised bogs and fens appears to be fairly homogenous. As this activity is a product of local use of the fen or raised bog as a resource, it is not surprising that a range of comparable structures have been found.

Bridging the bog chain

North of Derryville Bog, Late Bronze Age causeways have been recorded at Dromard More (Raftery 1996, 223) and Baunaghra (NMI files; Raftery 1996, 223). Two further undated causeways have been recorded in the northern part of this chain, near Roscrea, at Monaincha (Halpin 1984, 144) and Timoney (Lucas 1975).

Southeast of the study area, an undated causeway runs from Inchirourke to Baunmore (IAWU data). In the bog to the south, a Late Bronze Age causeway has been recorded in Leigh and Longfordpass North (Rynne 1965, Togher A). A further, undated, causeway was recorded in isolation in Leigh townland *(ibid.*, Togher B), with another recorded slightly further south in Longfordpass South *(ibid.*, Togher C). In the southernmost part of the bog chain, another undated causeway (RMP ref. TI054-074) was recorded near Derrynaflan.

Longfordpass and the area around Derryville Bog are notable as narrow tracts of bog that were passable using relatively short lengths of trackway. Significantly, in Britain (Coles and Coles 1986) and Continental Europe (Casparie 1987; Hayen 1957; 1987) long lengths of trackway were constructed to cross areas much wider than the Derryville stretch of bog, which is only 5km at

its broadest and 1km at its narrowest. The majority of the recorded causeways crossing this chain of bogs occur in a fairly restricted 10km-long sector in the centre. This is where the chain is at its narrowest, around Derryville and Longfordpass. This also happens to coincide with the most direct route between the passes through the hills on either side.

Terrain sensitivity in Derryville Bog

General

The sites have been grouped into three main categories: structures that took advantage of favourable conditions, structures that were built as a response to a change in those conditions and structures that appear to have been built with little knowledge or consideration of the environmental context in which they were placed. All the structures were intended to modify the bog for human purposes, with an underlying desire to extend the potential area of use. As all the structures were placed in a developing ecosystem, each had some impact on the environment, both in the immediate aftermath of its construction and for a period of time afterwards. However, the scale of the negative impacts of the structures in the first two categories is negligible compared with those in the third.

Structures taking advantage of local conditions

Many of the *fulachta fiadh*, platforms and trackways built in Derryville Bog were opportunistically sited to exploit available resources in a variety of shapes and forms. Strictly speaking, these structures were responding to very gradual changes in the environment, such as the gradual overspill of peat from the upland Killoran Bog system into the Derryville Bog system, which led to the construction of a series of structures in the northwest of the study area. The rate of these changes was imperceptible, as outlined above. In a sense, many of these structures rectify hazardous terrain features easing access to areas of the bog.

The most obvious attempt to exploit the available resources is the selection of sites where a trough could be dug to collect percolating water. These structures, with a mound of burnt stones derived from heating the water in the trough, are known as *fulachta fiadh* in Ireland (Plate 14.1).

In general, the locations of the twenty-eight *fulachta fiadh* and burnt mounds examined in the study area can be broken down into groups. Two were constructed on the surface of Derryville Bog, twelve around contemporary margins of the bog, five on the edges of areas of surviving bog in Killoran townland and nine in areas that were formerly bog but have since been drained or reclaimed. The locations of these sites match the existing pattern in the Irish midlands (Feehan and O'Donovan 1996, 204).

Plate. 14.1   Killoran 265. Fulacht fiadh *dating to 1425–1120 BC.*

The availability of a water table that was relatively close to the surface appears to be the dominant factor in the siting of the *fulachta fiadh* examined in the study area. In all but one case, Killoran 253, there was evidence that the builders had considerable success in selecting areas in which a trough that would collect percolating water could be dug. The flaws in the construction of Killoran 253 are discussed below.

The relative success in the selection and reuse of trough sites has implications for our perception of the *fulachta fiadh* as monuments. Three unsuccessful attempts and one successful attempt by project staff to select and dig an experimental trough that would satisfactorily fill with fresh water suggests that a fair degree of skill was involved in choosing a location that would provide the right amount of percolation. This reinforces the interpretation of the *fulachta fiadh* as part of the settlement pattern of the area, rather than transient cooking sites. This is further enhanced by the deliberate backfilling of a number of dryland *fulachta fiadh* troughs, an indicator of landscape maintenance, which is more likely to be part of a permanent, rather than transient, settlement pattern.

Often, when the *fulachta fiadh* had become overgrown with peat, they became the focus of later activity in the fen and marginal forest areas in which they had been constructed. Derryfadda 216, Killoran 240, Killoran 265, Killoran 304 and Killoran 400 all survived as areas with a stable bog surface, reducing the amount of effort required to access the resources present in the fen. In this way, they became incidental, but advantageous landscape features, that enticed people to build over later structures in their immediate environment, as happened in ten instances in Derryville Bog.

These ten structures were just some of the thirty-four trackways and six platforms that were constructed within the study area. The trackways were built across expanses of fen and raised bog to allow access to those areas. In doing so, they generally took the line of least resistance to reduce the amount of effort required in construction. In this way, they utilised existing dry areas, such as the *fulachta fiadh* and root systems, as can be demonstrated for a number of structures.

Within raised bogs, there are hummocks and hollows, which act as alternating dry and wet areas. The hummocks are generally oval and measure 2–4m in diameter, while the hollows vary from 2–5m in diameter and are up to 30cm in depth. The single-plank trackways Cooleeny 22 (Plate 14.2) and Derryfadda 23 were built to access the raised bog, with the line of the planks on the hummocks identifying the direction of the trackway through the more treacherous hollows. Many of the other trackways, such as Derryfadda 13, Derryfadda 203 and Killoran 314, were built in the fen, using the marginal forest root systems to stabilise the walking surface. It is possible that if the length of the trackways represents the width of the band of marginal forest around the fen, then a number were built where the woodland was at its narrowest.

Some of the platforms, such as Derryfadda 214 and Derryfadda 218, also incorporated root systems as part of the extended area of their structure. The siting of all the platforms within the fen or marginal forest appears to reflect their intended function, and each utilises some

Plate 14.2   Cooleeny 22. Single-plank trackway dating to 1517 BC.

Plate 14.3 Killoran 312. Second-century BC hurdle trackway.

existing terrain feature as part of the structure, be that a dry area, root system or a subsurface *fulacht fiadh*. The mounds of the *fulachta fiadh* provided a sound base for later tree growth, even when the mound was completely covered with peat.

The arc of alder carr in the northwest formed over several *fulachta fiadh*. This area of marginal woodland and the pool that it surrounded were the main focus of the Late Bronze Age and Iron Age activity in that area. At this point, the mineral-rich run-off from the dryland combined with the acid discharge from Derryville Bog. A pool of bog bean peat (*Menyanthes trifoliata*) developed amongst the alder carr and was bordered to the east by a larger open water pool dominated by rannock rush (*Scheuchzeria palustris*). Trackways Killoran 75 and Killoran 312 (Plate 14.3) stretched out into this pool and used hurdles to form a stable surface in the very wet conditions. This was not simply a passive response but was a need to access the pool and its rushes. It would have been considerably simpler to approach the pool from the alder carr. In fact, Killoran 312 may have done so originally. The choice to build Killoran 75, in the place and manner it was constructed, combined knowledge of the environment with social concerns. The form of that environment was a similar combination of natural process and earlier human action.

Structures responding to localised change

In a number of instances, the siting and construction of certain structures can be seen as a considered response to a local change in the environmental conditions. As described in the preceding section, all the structures are, strictly speaking, responding to change, but some were built after a visible change, such as a bog burst or a prolonged dry period. This can be demonstrated for a range of structures, again including trackways and a *fulacht fiadh*, and this time also including stake rows and a causeway, Killoran 18. These structures also demonstrate that, at local level, the bog surface was used with a reasonably intimate knowledge of the environment.

The first structure to be built that took advantage of an environmental change was causeway Killoran 18 (Plate 14.4). This causeway took advantage of an environmental change but itself brought about a new change and thus caused its own demise. Prior to the construction of Killoran 18, Derryville Bog had two developing mires of ombrotrophic *Sphagnum* growth in the southern and northern areas of the study area. These were separated by a swath of mineratrophic fen growth approximately 120–150m wide, stretching from east to west, and marginal brook forest on the dryland margins. The wetness (approximately 90% water) of this largely reed and sedge fen would have prohibited passage across it; however, a spell of not more than five years of dry summers or a change in the drainage of the bog resulted in a drying out of the fen surface to 80%. This allowed passage across the fen, and the use of a clearly staked out route created an area of disturbed amorphous peat under the causeway, which was further compressed later by traffic over heavy stones.

The use of relatively large stones in the east and centre of the structure was in contrast to the west, where small scattered cobbles and occasional large flags were used. This would suggest a drier situation in the west, although the intermittent use of a brushwood foundation throughout showed uniformity. It is more likely that the stone gathered from the west was less extensively quarried. A discharge channel running north–south at the western end of the causeway presented a wet obstacle. To bridge this channel, a thick, irregular structure with several branching abortive excursions was constructed using wood collected from the western landfall. The change in morphology of the structure, in tandem with the change in the underlying terrain, suggests that the builders were well aware that the stones would quite probably sink in the discharge channel and that wood would provide a functional walking surface.

The construction of Killoran 18 had other unforeseen consequences. The western end of the structure acted as a dam in the discharge channel and altered the hydrology of the bog in this area. Eventually, the channel breached the dam, removing part of the structure and rendering the track unusable. The effect of the heavy stone structure, probably used to weigh down a matting of reed that

Plate 14.4 Killoran 18. Causeway dating to 1700–1200 BC.

formed a walking surface, lowered the peat surface along the line of this track, which allowed alder shrubs to root on its surface and edges. The bog system, unable to drain to the north due to the damming of the discharge channel, flooded east along the lower track surface and created ideal conditions for rapid ombrotrophic peat formation. The net result was that the causeway was flooded and later breached. Numerous deposits of the humic colloid dopplerite (see Chapter 3 in Feehan and O'Donovan 1996) highlighted a sharp transition from relatively dry conditions to wetter ones. Although attempts were made to repair wetter or sunken areas, the track could not have been in use for more than twenty years and in itself brought about the rapid spread of raised bog across the remaining central fen.

While these events may appear dramatic in hindsight, the timescale for the integration of the northern and southern raised bog formation was several hundred years, and these events should not detract from the level of knowledge demonstrated by the ability to exploit a dry period in which a causeway could be constructed across Derryville Bog for the first time.

Four stake rows were recorded in the study area. The earliest of these, Derryfadda 209 (Plate 14.5), was constructed in the Late Bronze Age, prior to bog burst C of 820 BC. With the increasing water table, areas of the raised bog not only became inaccessible but also dangerous due to the rate of discharge along the steeply sloping ridge at the base of the peat in the southeastern area. In an attempt to demarcate this dangerous area, the margin of the slope was marked with a 38m long line of stakes that separated the edge of the marginal forest from the fen environment.

This type of site was not in use again until the medieval period, when raised bog had encroached over

the entire area. Killoran 54, a dense stake row 406m in length, was constructed along the line where the peat overlay a ridge in the mineral soil. Similarly, Derryfadda 19 continued the line of a smaller sub-peat

Plate 14.5 Derryfadda 209. Stake row dating to before the 820 BC bog burst.

ridge along the 123m contour. These stakes rows would have facilitated continued use of the bog by defining the unusable areas and demonstrate a considered response to increasingly adverse local conditions.

After bog burst B of 1250 BC, which probably occurred in the Cooleeny area, the water level within the system dropped by about 1m. The pre-bog burst water levels would have taken up to two centuries to be re-established, during which time the vegetation cover would have changed, with drier species invading this newly extended zone. The effect of the bog burst would have been greatest south of the study area, in Cooleeny. Rates of fen development and raised bog growth decrease in the aftermath of a drop in water table, such as after a bog burst. During the two centuries in which water levels rose, from *c.* 1250 to 1050 BC, new areas of bog that had previously been too wet became accessible and archaeological constructions were extended into these areas.

Patches of raised bog were dry enough to cross with only limited constructions, as seen at Killoran 20. On the western margin, a larger band of fen could be used (Cooleeny 64), while on the eastern side the marginal forest and areas of the raised bog were accessible with the aid of the stone trackways Derryfadda 17 and Derryfadda 311. The use of stone would not have been an effective construction method in a wetter environment. Within the Killoran Bog system, the *fulacht fiadh* Killoran 253 had to be constructed further downslope in the marginal forest, in order to reach the level of the water table.

Although the water table would have taken up to two centuries to recover from the 1250 BC bog burst, initial increases, within the first ten to twenty years, would have been quite sharp. After this, the rise would have been more gradual. At two of the longest sites, Cooleeny 64 and Derryfadda 17, attempts were made to extend their use, in both instances by supplementing the main body of the site with additional wood. At Derryfadda 17, this was also achieved by marking the line of the track and the stone deposits within the pools of the raised bog with marker stakes.

Later bog bursts were to have similar effects on trackway construction. Killoran 69 (Plate 14.6) was constructed in the aftermath of bog burst C. The subsequent recovery of the water table, particularly the western discharge channel, led to the eventual destruction of part of the structure. Prior to construction, the western bog margin at this point was defined by a large discharge channel draining northwards and separating the western marginal forest zone from ombrotrophic raised bog forming to the east. A bog burst in the Cooleeny Bog area brought about a lowering of the water table across the whole bog. This had the effect of exposing a buried oak lying on peat and glacial till within the marginal forest drowned three hundred years earlier. The western terminus of trackway Killoran 69 utilised this oak as a dry surface on which to build from. The resulting fast recovery in the bog hydrology and water

Plate 14.6 Killoran 69. Trackway dating to after the 820 BC bog burst.

table, with further flooding, gave rise to erosion gullies, one of which smashed into the track, breaking up the timbers and removing the eastern extent.

A second bog burst (bog burst D) in the southern part of Cooleeny around 600 BC brought about another lowering of the water table. At the same time, the rapid increase of eastward discharge from the Killoran Bog system resulted in a reversal of water in the discharge channel southwards. The drop in the level of the surface of the southern portion of the bog also contributed to this reversal and could be observed as a layer of alluvium partly sealed by the trackway built along its line. The buried oak utilised by Killoran 69 was re-exposed after the bog burst. Similarly, trackway Killoran 306 utilised this oak as a dry surface and, with the addition of some stone and wood, continued into the marginal forest. Again, intimate knowledge of the local topography assisted in the construction of this trackway.

In the Iron Age, an area in the western part of the study area was subject to prolonged localised erosion and flooding, resulting from the interaction between the Derryville and Killoran Bog systems. The product of localised erosion and flooding was the removal of peat in this area and mud deposition. A number of trackways had already been constructed and re-exposed in this area (see above). One trackway, Killoran 226 Structure 2, was constructed in the mud following the erosion along the line of an earlier trackway, Killoran 226 Structure 1, largely utilising the drier surface it created to cross a wet hollow in the retreating marginal forest. Despite the wetness of this area, a number of other structures, such as Killoran 248 and Killoran 301, were also constructed to bridge the hollow.

The upland Killoran Bog heavily influenced the development of the fen in the northwest of the study area. The sites in this area responded to peat development in two different ways and contributed to the shape of the local environment as well. This pattern shows the ways in which people responded to the environmental changes. In turn, it shows how those choices conditioned later choices, just as, further south, the trackways constructed into and around the discharge channel impacted on the direction and rate of the flow.

As the slow discharge from Killoran Bog caused fen to develop amongst pockets in the sloping marginal forest, trackways and platforms were initially built as a response to the beginning of this fen growth. The *fulacht fiadh* Killoran 240 was built before Killoran Bog spilled over into Derryville Bog, indicating that the slope was already a focus for human activity. It seems likely that this area was already being used for a broader range of purposes and that the later tracks and platforms are simply evidence of the continuation of that use. As the fen developed on the slope and the water table in Derryville Bog rose, the sites running from the fen edge to the small ridge of drier peat to the east became more substantial. The entire complex of trackways, *fulachta fiadh*, platforms and casually deposited wood probably spanned a hundred to a hundred and fifty years. This is the pattern of structures taking advantage of topographic features that maintained access to the resources available in the area.

The later impact of these structures influenced the pattern of growth of the marginal forest, so that there were two 'arms' of alder carr stretching out down the slope of the fen, coincident with the positions of the earlier *fulachta fiadh*. The relationship between the *fulacht*

Plate 14.7 Killoran 230. Trackway dating to 1500–1195 BC.

fiadh Killoran 265 and the trackway Killoran 230 (Plate 14.7) shows that *fulachta fiadh*, with their high mineral content in a mineral-poor environment, encouraged the rapid growth of dense vegetation. Within as little as twenty years, the *fulacht fiadh* had become so overgrown that it was no longer visible by the time of the construction of the trackway, which entirely bypassed the area of the mound.

Structures imposed on their environment

At its simplest, a number of instances can be cited to show that some of the structures built in Derryville Bog were not constructed with the intimate local knowledge exhibited by other structures. Although the majority of the *fulachta fiadh* were perfectly functional, and some did have later unintentional impacts on their environment, only one site, from twenty-four, appeared to have been badly chosen. The boiling pit Killoran 23, on the dryland, is considered to be a short-lived *fulacht fiadh* and either did not attract enough percolating water or attracted too much and was quickly abandoned. Even the site of Killoran 253, where the trough collapsed as it was constructed within the fen, was carefully selected to provide access to a very low water table.

There were unforeseen consequences arising from the positioning of many of the structures, but most added to the cumulative impact human actions have on any fragile ecosystem. In the case of Cooleeny 31, the scale of this impact was far greater and demonstrated an acute ignorance of the properties of the Cooleeny Bog system (cf. Casparie, page 39) and the effect a heavy wooden causeway would have on it.

What this causeway shows is that people sometimes disregard environmental constraints in favour of social constraints. It also shows how imperfect knowledge can defeat the initial purpose of the construction. Bog burst B had affected this area before the construction of the causeway, destabilising the surface. The unstable drainage of this area and the uneven nature of the raised bog required a very firm structure to make crossing possible, even at this very narrow point. The builders chose a massive structure, including sand to create a stable surface. They also chose to run the causeway off the southern flank of the western underlying glacial ridge, a long crossing, rather than off the eastern point, which would have made a shorter crossing. The ridge itself would have provided a break in the drainage so that the causeway did not meet the main flow until it left the raised bog on the eastern side.

This re-entry to fen was the weakest point of construction, as the change in surface caused a structural weakness. More importantly, fen continued to grow in this area because of a strong water flow from the rest of the bog. The massive structure of the causeway dammed the outflow of water created by the earlier bog burst and caused another, more massive bog burst (bog burst D).

This would have occurred within a year of the construction of the track and damaged it beyond repair. The effects of the bog burst can be seen in the line of the track as it crosses the fen peat bordering the southwest ridge. The majority of the surveyed activity in this area is on the western margin, which would have been inundated with the outflow from the bog burst. The human cost of the error in construction, is as such, unquantified.

Other structures in the study area responded more closely to the terrain, suggesting the builders may have had better knowledge or were operating under less pressing social constraints. There were very few structures that attempted to cross this bog. Those that did were less invasive. The single-plank trackway Derryfadda 23 would have had little effect on the growth of the bog. The more substantial causeway Killoran 18 did affect the development of the bog (see above), but its initial construction exploited opportunities rather than imposing a preset design. Cooleeny 31 represents a departure from the normal terrain response for the communities using Derryville Bog. This site was constructed with imperfect knowledge of, or respect for, the local environment.

Conclusions

This chapter has examined the interaction of archaeological sites with the environment at different scales and from different perspectives. Most of the structures appear to have succeeded in modifying the local environment in the short-term, making human activity in the bog and its margins possible. In the long-term, their impacts range from negligible, as at Derryfadda 211, to disastrous, as at Cooleeny 31. There are also sites with long-term effects that cannot be seen as positive or negative but are simply another element in the developing bog. *Fulachta fiadh* affected the form of the marginal forest. Killoran 18 precipitated the spread of raised bog across the basin, but its immediate impact would not have been noticeable. The distinction between the two levels of impact appears to reflect a division between local and regional requirements.

Cooleeny 31 is not unique in Ireland in displaying this lack of understanding of the nature of the bog surface on which it was constructed. The construction of the large Iron Age causeway at Corlea, County Longford, across a bog that was unable to support its weight (Raftery 1996, 421–2) may also be a result of central rather than local desire for passage across the bog.

While the impact each structure had was significant to some degree, local topography and hydrology, in combination with larger climatic trends, always had the determining role in the formation of Derryville Bog. No pre-

historic or early historic sites intentionally changed the bog for longer than thirty to forty years, and serious transformation of this bog to suit human purposes did not occur until the early part of the twentieth century.

People building structures in the bog responded to their knowledge of the environment in three basic ways. The first reflects the continued use of the fen or bog as a resource. The second reflects the opportunistic exploitation of changes or circumstances to use the bog in new ways. Killoran 18, Killoran 69, Killoran 253, Killoran 306 and Derryfadda 17 can be seen in this light, as can the stake rows. These sites identify the position of this resource within a local context, where hazards had to be marked out or removed from a familiar landscape to prolong its use. The third response sees a departure from the passive acceptance of the limitations of a bog environment, as demonstrated by the construction of causeway Cooleeny 31 to cross a part of the bog chain that was previously unsuitable for such passage. The ambition of this last project did not match the technical skill of the builders.

The knowledge assumed in the characterisation of responses is local knowledge. Current research can show the connection between local patterns and national trends, but people respond to their immediate, rather than regional, environment. There is a similarity in the patterning of the archaeological material from the region, but that may be due to the same reaction to a comparable environment rather than a coherent regional response. Sites indicating a regional view of the environment, with the bog as a boundary with crossing points, are rare. This is probably more a product of social circumstance than environmental context.

National and international patterns of environmental change seem to have had little discernible effect on the archaeology of Derryville Bog. The well-defined climatic crises of later prehistory do not signify important changes here. While the pollen record shows broadly similar zones and changes to the rest of the country, these boundaries do not mark cultural departures.

The archaeological record of this project represents local responses to immediate environmental conditions and changes. The long-term prehistoric trend was for increasingly intense use of the fen and fen margin. This is in opposition to the environmental trend of increased raised bog growth. Archaeological sites either take advantage of environmental changes and circumstances or are constructed to continue and extend use in the face of adverse conditions. Understanding the interplay of response and impact requires detailed knowledge of how the bog and the human use of it developed and changed over time.

15. Landscape context

Sarah Cross May, Cara Murray, John Ó Néill and Paul Stevens

Introduction

The archaeology of Derryville Bog forms the wetland component of an early vernacular landscape—a landscape created by the daily activities of the people who lived in it. There is no central focus on a high-status site: it is a local, domestic pattern. In periods with proposed settlement hierarchies (Grogan *et al.* 1996), it would have formed the lowest order. Far from diminishing the importance of the archaeological material found, the vernacular position of the bog allows us to examine questions that are more often addressed in relation to high-status artefacts, sites and landscapes. Vernacular material, which constitutes a significant element of the archaeological record, has not always been the focus of archaeological analysis.

The use of the term vernacular is more than an explanation of the absence of high-status material in Derryville Bog. The nature of the material, its pattern of use within the bog and its relationship to the excavated dryland sites all support an argument for the use of the term vernacular in the interpretation of the sixty-four excavated sites that do not cross the bog. This also applies to the pattern of material investigated on the Killoran headland. In addition, there is a broader, but still vernacular, context for the causeway Killoran 18. The causeway Cooleeny 31 is unique, however, and is best understood as part of a regional social system. It provides a 'bench mark' for the interpretation of the rest of the material.

The first part of this chapter presents the argument for viewing Derryville Bog as a vernacular archaeological landscape. This vernacular interpretation is then used to explore the economic and social patterns in the region's prehistoric past. The archaeology revealed provides information on material culture, subsistence systems, patterns of work and land use and broader social relations. The picture formed by the material should be seen as complementary to the existing record and the interpretation of high-status sites or 'ritual landscapes'. While much of the evidence can and should be debated, we have attempted a broad assessment of its importance.

The vernacular argument

Absence of high status material

There is general agreement that bogs were suitable places for votive deposits (Coles and Coles 1989, 164; Raftery 1994, 184–5). Some of the discomfort that people have with this notion is the disjunction between the modern use of wetland and that of prehistoric groups (Hill 1989). Seeing bogs as sacred places does not fit with the common modern image of bog as wasteland. Some of the discomfort also stems from the fact that there are regional and chronological complexities to this practice (Bradley 1991; Cooney and Grogan 1994). Not all bogs are suitable for votive deposits, and not all objects in bogs are there as a result of ritual.

The evidence for votive deposition in Derryville Bog is tenuous at best. No metalwork is recorded from the bog, and in the course of the current project, just five objects that could possibly be the result of ritual activity were recovered. The spear shaft (97E160:237:1; Plate 12.2, 12.3 and 15.1) found on the eastern edge of Killoran 237 could have been a deliberate deposit. Spearheads are a common deposit (Bourke 1996, 11; Bradley 1991; Cooney and Grogan 1994, 197), but they are rarely found with shafts, even when found in bogs where the shafts would survive. The separate deposition of shafts may therefore be significant. Another spear shaft (94E106:2; Fig. 12.2) was recorded during the IAWU survey at Derryfadda 13 (94E106:2). The two peg-like cherry wood (*Prunus avium*) objects with bulbous heads and perforated tips (96E202:9:1; Fig. 12.7 and 96E202:9:2; Fig. 12.8; Plate 15.2) found at Derryfadda 9 are even less clearly ritual. They do not seem to have served any function associated with the track, nor do they seem to have been deposited as scrap. Nonetheless, they cannot be associated with any ritual activity, and they appear to have been broken during use. In 1973 a wooden beetle was recovered in

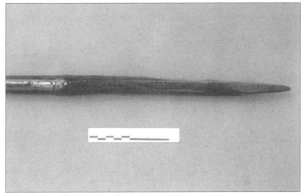

Plate 15.1 Yew spear shaft from Killoran 237.

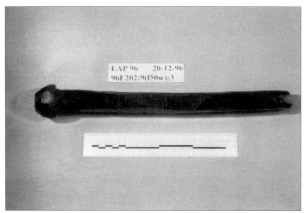

Plate 15.2 Cherrywood artefact from Derryfadda 9.

Cooleeny very close to the location of Cooleeny 31. It is described in the NMI file as likely to be medieval or post-medieval in date. As it was found at a similar depth to the causeway, it remains possible that it is, in fact, prehistoric. If so, its deposition could be a votive deposit relating to harvest, after the fashion of beehive querns (Raftery 1994, 124). The final object with possible ritual associations is the saddle quern (97E158:240:1) found at the *fulacht fiadh* Killoran 240. While this object could have been deposited as part of the use of the *fulacht fiadh*, it could also mirror the votive deposit of beehive querns.

None of these objects can be considered votive with any certainty, but their deposition in all but the last case is ambiguous. It remains possible that ritual deposits were made in Derryville Bog but were not recovered by the project. Although the scope of investigation extended beyond archaeological structures, such objects are small, and they can lie some distance from the structures with which they are associated. The Killymoon bracelet and dress fastener are the only gold objects to have been recovered in a modern wetland excavation in Ireland (Hurl 1996).

Nonetheless, this fen may not have been an important focus for votive deposits. In southeast Clare, it has been shown that deposits of metalwork are associated with high-status sites such as hillforts and that the zone of deposition is surrounded by a zone with less detectable activity (Grogan *et al.* 1996, 41). Killymoon is located beside a hillfort (now on the grounds of Killymoon Golf Course). If we describe the landscape surrounding Derryville Bog as vernacular, perhaps we should not expect high-status deposits. In southeast Limerick and north Cork, the distribution of metalwork is complementary with the distribution of burial sites, with metalwork deposits on the edges of the main concentrations of burials (Fig. 7.4 in Cooney and Grogan 1994). The unenclosed pit cemetery cremation pits, representing a phase of burial on the dryland, may point to a different type of ritual connected with fen in the Derryville landscape.

Nature of material

Of the sixty-six excavated sites in the bog, only two were causeways. Both these structures indicated a degree of prior planning (discussed in more detailed below). Six *fulachta fiadh* were excavated. The remaining fifty-eight sites were comparatively unsophisticated, degraded structures. The variation in their construction is discussed in Chapter 9. The condition of the wood suggested that the sites had been exposed for a long period of time. Taken as a whole, the wood sampled during the project had suffered extensive decay, and it was difficult to distinguish differences in patterns of use between sites on this basis, as the decay levels were all so high. While preservation is often poor in fen deposits, the sites were also degraded as a result of prolonged use. Fourteen structures were extended, seven structures were repaired and fifteen structures were modified.

Pattern within the bog

Most of the prehistoric trackways and platforms recorded approached or lay at the edge of the fen or marginal woodland. Even in the Iron Age, when raised bog virtually completely filled the basin, the archaeological focus remained in the fen. Trackways and platforms provided access to the fen and through marginal woodland. Almost all of the sites exhibited a high degree of sensitivity to the local environment and changes to that environment (see Chapter 14). This implies a familiarity with and knowledge of the bog and its local conditions.

In other bogs, tracks that did not cross the bog are considered to be markers of routes that did. Raftery (1996, 227) has suggested that the cluster of tracks from Derryoghil, County Longford, could have bridged the wettest portion of the bog and stopped as they reached raised bog, which could be crossed without tracks. This is not the case in Derryville Bog. Only one of our sites, Derryfadda 17, crossed the fen to meet the edge of the raised bog. The rest of the sites stopped well short of it, and many of them deliberately stopped at particularly wet parts of the bog (Derryfadda 203, 210, 218 and 311 and Killoran 69, 75, 305, 306 and 314). They appeared to be providing access to the fen.

Tracks constructed to access fen are not unique to this project. Bohelenweg XLII, Wittemoor, Germany, crossed from dryland to a potentially navigable stream (Coles and Coles 1989, 168). This site and the Emmerschans hurdles have been interpreted as accessing bog iron deposits (Casparie 1986, 204–6; Coles and Coles 1989, 169; Raftery 1996, 227). The Neolithic Garvins tracks in the Somerset Levels in England (Coles and Orme 1977a) accessed especially wet bog, possibly for fishing or hunting (Casparie 1982, 161). The Neolithic track at Bourtanger Moor XXI crossed to a terminal in the wettest part of the bog, though its function remains unclear (*ibid.*, 163). It is therefore possible that the Derryoghil tracks should be seen as deliberately accessing wetter sections of the bog.

However, as the peat morphology of that bog has not been studied (Raftery 1996, 227), this suggestion cannot be analysed with the same clarity as the Derryville sites.

The medieval sites represent an entirely distinct use of what had become a different landscape. At this point, the entire basin was covered with raised bog peat. It is difficult to reconstruct the use of this landscape because commercial milling has removed the peat into which the sites were set. None of the recorded sites were causeways. Killoran 66 may have been a hut site, and Killoran 54 and Derryfadda 19 were stake rows used for delimiting areas within or paths across the raised bog. The rest of the recorded sites were tracks covering short stretches of wetness. These point to activity within the raised bog; tracks may not have been needed to cross it.

Nature of the margins

The prehistoric sites mark the fen and fen margins as places; the nature of these zones needs to be considered. While the fen in Derryville Bog changed over time, there were enduring features. The fen would have been a mosaic of beds of reeds, rushes and sedges, with pools of open water and small streams or discharge channels . It would have been a suitable habitat for wildfowl, though too acid for fish. The high proportion of bog bean and reed indicates that it was a particularly wet environment, and large reed fields bordered a stretch of fen in the discharge channel identified at the north and northwest of the study area.

The fen was fringed by mixed scrub woodland, which stretched out at some points forming peninsulas and islands. No attempt was made to clear this woodland in the prehistoric period, although it was both exploited and managed in an informal way (see Chapter 7). The woodland would therefore have formed a dense boundary around the bog, rather than the fairly open forest presented elsewhere (Roberts 1998). This marginal woodland is referred to as such rather than as carr because it formed a continuous zone and there was some evident rooting in mineral soil.

Forty-seven structures occurred in this environment. The tracks and platforms were necessary to provide access through the very dense undergrowth and across pools of fen between root systems. Shrubs such as hazel, cherry, juniper and bramble would have colonised the fen edges of the marginal woodland where more light was available. Over time, the fen would have retreated from being the dominant environment of the Derryville basin in the Neolithic to being a strip of fen around the edge of a dome of developing raised bog by the Iron Age. This change and the nature of the fen and marginal woodland are discussed in Chapter 3.

It is easy to detect clusters in the general distribution of sites that are persistent over time. Certain areas retain their importance throughout the prehistoric period, suggesting continuity in nearby settlement patterns. The fen and fen margin formed part of the wider, occupied landscape from the Early Bronze Age through to the Iron Age.

Possible functions of tracks and platforms

Structures in the fen and marginal woodland would have provided access to a range of resources beyond those provided by the dryland settlement areas. These resources remained attractive to people into the medieval period, as indicated by ecclesiastical complaints about their lack of control over the resources and the benefits accruing from them. Coles and Coles refer to a sixteenth-century quote describing fen as "The fatness of the earth gathered together at the time of Noah's flood" (1989, 153). Another reference, however, describes raised bog "as a wild expanse of reed in 'an atmosphere pregnant with pestilence and death'" (*ibid.*).

There is little surviving evidence, in the form of detritus or debitage, to indicate the specific function of the small tracks and platform sites excavated in Derryville Bog. This can be attributed, in part, to the nature of the peat, which would not have preserved delicate items such as basketry, although the preservation of all other materials, apart from bone, would have been quite good. Nevertheless, their position and nature indicate a function facilitating seasonal gathering and hunting.

The location of fruit-bearing trees such as elder, crab apple, plum and cherry at the margins of these areas may explain the construction of some sites at the outer fringes of the marginal woodland. These trees would all have produced their edible fruits in the autumn, when most of the trackways were constructed. Such trees and shrubs would have occurred in greater profusion and been easier to access on the outer edge of the woodland, where the vegetation was less dense. This was seen at Derryfadda 13, where the trackway was laid around root systems of alder (*Alnus*) and cherry (*P. avium*). The need to cross the woodland for resources explains the construction of trackways that crossed from the dryland to the outer edge of the marginal woodland.

The presence of twenty sites terminating at the western discharge channel indicates that reeds and rushes, rather than sedges, were the primary resource in the fen. Sedges and other fen grasses can be used for seasonal

Plate 15.3 Killoran 75. Second-century BC hurdle trackway during excavation.

Plate 15.4 Killoran 8. Houses dating to 1860–1845 BC or 1775–1430 BC.

grazing (Feehan and O'Donovan 1996, 37). The high water levels within this fen, however, suggest that grazing may only have been possible for a very short period in very dry conditions. There is little evidence from any of the archaeological sites that they were used as droveways. The sites on the west side of the bog had greater numbers of dung beetles than those on the east. The site with the most, Killoran 75 (Plate 15.3), was the least appropriate for droving due to its light structure. Derryfadda 13 seemed to be a more likely structure for droving but had very few dung beetles and led to a sharp slope into wetter fen. The overall beetle assemblage at this site suggested that it had been washed in rather than developed *in situ*. If the fen was used for pasturing, the trackways were not an important part of the process. They have a more apparent function if the reed beds were harvested for fodder: to simply provide access to the beds.

The most direct parallel for many of the Lisheen sites is the Baker site in the Somerset Levels, England. This platform, "placed at the junction of raised bog and island, would have served as a blind or shoot for wild fowl, a fishing flat, a landing stage for water craft, a collecting centre for reeds and other plants, an entry and exit from the marshland on the trackways, and a bridge (psychological and practical) between two contemporary economic zones for early farmers and gatherers. The archaeological and environmental evidence supports some of these suggestions; the others are conjectural" (Coles *et al.* 1980, 23).

Dryland settlement pattern

The settlement pattern in the landscape around Derryville Bog was under-represented by monuments and obvious sites, especially in the prehistoric period, prior to earth-

moving work carried out during the project. Historic settlement in the study area in the Early Christian period is represented by ringforts and two possible ecclesiastical enclosures located some distance from the bog. However, the pollen record suggested several concerted episodes of forest clearance in the locality dating from the Bronze Age, with established settlement reflected in a pattern of land management evidence during the first millennium BC and again in the late first millennium AD.

Archaeological investigation over the 200ha area of development west of Derryville Bog, across Killoran townland, as well as areas in Derryfadda, Barnalisheen and Cooleeny, revealed substantial evidence for archaeological activity. This activity dated from the Bronze Age, Iron Age and early medieval period, with material evidence for activity mainly focused on raised glacial ridges and knolls within the marginal land and later on the blanket bog in central and east Killoran. Four clusters of sites, as well as isolated *fulachta fiadh* around wet depressions, represented the most abundant evidence for settlement throughout the Bronze Age. Associated with these *fulachta fiadh* were the house sites at Killoran 8 (Plate 15.4), the cremation cemeteries at Killoran 4 and 10 and isolated burials. The material in itself appeared to represent relatively small-scale and short-lived settlement in the area, but a pattern that was nevertheless persistent throughout the period.

The evidence for prehistoric settlement on the Killoran headland was in contrast to that on the more apparently fertile area of the northern Derryfadda peninsula, which did not reveal any settlement evidence. However, the existence of a bog-side or adjacent dryland settlement cannot be ruled out in view of the wealth of unexcavated material in the Cooleeny Complex (field walking in the field adjacent to the complex revealed evidence of a burnt mound outside the study area). It should also be noted that the project investigated areas subject to development rather than areas of archaeological potential, which were preserved.

Iron Age settlement on the adjacent dryland was considerably under-represented when compared with the large body of wetland material from this period. An isolated Iron Age house was excavated in the central area of the Killoran dryland, but this was possibly associated with wetland material located in bog north of the study area.

In later periods, settlement was less apparent and activity appeared to reflect the changing use of the landscape evidenced by the wetland material. Dispersed medieval activity was represented by two main date clusters, with largely sporadic activity associated with settlement, land division and agriculture. Secular settlement concentrated in ringforts was largely restricted to areas further from the bog and away from the marginal land. However, evidence of sixth to seventh century AD activity in the northwest and southwest of Derryville Bog (including the hut site Killoran 66, trackways and stake rows), may be associated with the house site Killoran 8. However, it is equally possible that this was part of a wider ecclesiastically controlled landscape associated with the enclosures at Killoran 31 and in Derryfadda (RMP ref. TI036-034).

The area was virtually devoid of activity between the thirteenth and seventeenth centuries. Only in the post-medieval period did settlement fully encroach onto the bog again with the reclamation of land and the building of estate houses in Killoran and Cooleeny close to the western margins of the bog.

Overall, the sites provide evidence for the concerted use of the headland. The wide range of archaeological material on the dryland suggests an integrated settlement pattern. The periodic use of the dryland is reflected in similar periodic activity in the bog. The wetland and dryland sites can, therefore, be seen as two aspects of one landscape possessing a range of sites relating to small, sparse community structures.

The *fulachta fiadh* and burnt mounds, in common with many other types of sites, could have been put to different uses by different groups. It is most likely that even individual sites did not have single functions. This may be one of the reasons for continuing debate on their function (Buckley 1990a; Roberts 1997). The possible use of a *fulacht fiadh* for hoard deposition at Mooghaun North, County Clare, points to possible ritual associations in high-status landscapes (Condit 1996). Possible use as saunas has met with criticism (Barfield and Hodder 1987; Ó Drisceoil 1988). Most discussions centre on more domestic functions. The *fulachta fiadh*, in this context, point to the more everyday uses of the fen and their social importance. Once again, these sites demonstrate a high degree of local knowledge and should be considered as part of an established pattern of low-density vernacular settlement.

Nothing was recorded from the dryland to indicate high-status settlement. The houses revealed were simple structures, and neither Killoran 8 nor Killoran 16 showed any evidence of an accompanying enclosure. Both the large unenclosed cemetery at Killoran 10 and the isolated and unenclosed cremation pits among the other clusters of sites can be seen as local phenomena. The relatively large number of burials at Killoran 10 need not reflect a large or powerful community. Fen edge burial in other places seems to form part of a local settlement pattern (Roberts 1998). This is in contrast to more monumental burial rites, which may have served a larger group (Chapman 1981).

Killoran 18: community cohesion

The straight line of Killoran 18 stretching across the bog seems, at first glance, like an indicator of a larger scale pattern with more central control. A more detailed assessment of the site based on excavation detail shows that it is not a high-status site and can represent a part of the vernacular landscape. It simply possesses attributes not shown on the other wetland sites. The causeway required planning, and its construction would have involved quite a large group of people, but it still served a local purpose. This site is an important indicator of community cohesion.

Killoran 18 was constructed between the northwestern tip of the Derryfadda peninsula and the centre of the Killoran headland. As one of a limited number of direct crossings in the south midlands bog chain, it could have served a wider community. Location is therefore crucial to an interpretation of the causeway as a regional access route. Nonetheless, local environment and topography may have contributed more to its nature and its siting. The site took advantage of a short-lived localised dryness to cross a stretch of fen between two wetter areas of raised bog. This type of terrain sensitivity fits in with the general pattern of the rest of the material studied.

While the location of the causeway would have been useful to a broader area, its particular siting probably followed the knowledge and needs of a local group. Traces of Neolithic activity below the eastern end of the causeway and the Early Bronze Age *fulacht fiadh* just beyond the western landfall indicate that this area of the bog was already a focus of local activity. The single-plank tracks Cooleeny 22 and Derryfadda 23, built in raised bog at the centre of the basin, also show that a need to cross areas of the bog was being met prior to the construction of Killoran 18.

Resources from both sides of the bog were used in the construction of the causeway, and it is possible that it was built from both sides at different points during construction. If the two headlands were occupied by different communities, this may imply co-operation between groups. The narrow surface, however, suggests that the causeway was used mostly for single-file, pedestrian traffic. This is indicative of a local community undertaking,

Plate 15.5 Cooleeny 31. Causeway dating to 790–380 BC during excavation.

as it could not have been used for the large-scale transport of goods, livestock or people. If this structure formed part of a regional network, a wider causeway would be more likely.

Cooleeny 31: intrusive regional activity

In order to identify the main landscape pattern represented in Derryville Bog, it is important to recognise elements that do not fit the pattern. The identification of sites with regional importance offers a control point to the discussion. The structure and location of Cooleeny 31, in conjunction with its palaeoenvironmental context, indicate that it formed part of a significant route between two regional power bases, which in turn formed links in a route from the Shannon into the southeast. Linking particular causeways with regional and national routes is always tempting (see Lucas 1985; Raftery 1994, 104). In this case, detailed consideration of both the local and regional contexts suggests that, given the present evidence, such a role is the best explanation for the structure and its position.

Cooleeny 31 (Plate 15.5) was a very substantial structure built of transverse roundwood and planks with a complex substructure including sand and brushwood. The construction included a phase of marking and preparing the straight route across the south end of the study area. It resembles 'corduroy roads' (see Chapter 9) but showed no sign of being used for wheeled transport. These structural elements make it very different from other sites in Derryville Bog.

The site was more massive than any other structure investigated in the study area. The size of the structure in most of the sites studied was partly determined by available material. At Cooleeny 31, two lengths of roundwood

overlapped to span its width. The deliberate use of sand and gravel in the substructure was unique. The complexity and regularity displayed in the constructional sequence marked it out from other tracks in the bog. Its closest parallel in the archaeological record is Togher B (Fig. 15.1) from nearby Littleton Bog (Rynne 1965). Its difference from other local sites and its similarities to another complex site in the region provide the first indication that the causeway had a more regional than local significance.

The causeway also differed from other structures in the bog in the damage it caused to the bog system. It was a massive structure laid across the least stable section of the bog system. Its weight and particular location caused a damming of water flow in an essential channel of the basins discharge system. The resulting build-up of water pressure 'upstream' of this channel caused a significant bog burst that effectively swept the structure away. Most of the sites in Derryville Bog showed a high degree of environmental knowledge and reacted to this in a fairly precise manner. Many sites took advantage of changes in bog hydrology in order to exploit previously inaccessible zones within the fen. The other causeway, Killoran 18, used the nature of the reed fen to cross it with the least effort. Cooleeny 31 was constructed in the most delicate part of the bog and used a massive structure to overcome environmental instability. This suggests that the causeway was planned and built without the detailed understanding of the bog evident in the other sites investigated. It seems, therefore, that regional location was more important than local considerations.

The finished surface of Cooleeny 31 was more than 4.5m wide, but this surface was neither used for wheeled transport nor herding. The lack of wheel rutting, the rough surface and the lack of appropriate dung beetles all

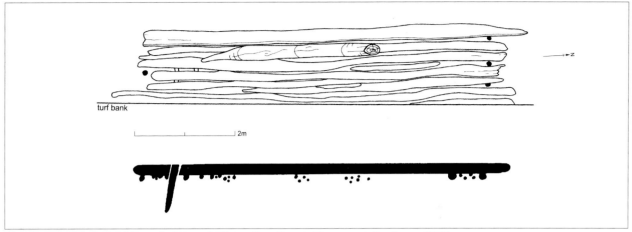

Fig. 15.1 Togher B, Leigh/Longford Pass, Littleton Bog, County Tipperary (after Rynne 1965).

indicate that the causeway was not used for wheeled vehicles. The microfaunal analysis found a lower number of dung beetles than on many other, less substantial tracks, and those that were evident were indicative of large wild mammals such as deer and horses. The massive structure may, therefore, have been intended to impress. Considerable time and resources were devoted to its construction. The width may have also facilitated the passage of large groups of people at a single time. These groups could have even been military in nature. The construction of bog roads for a military purpose is well attested (Lucas 1985), and the period the site dates to is noted for increased coercive power (Cooney and Grogan 1994). Even if user groups were involved in the transport of goods, this still shows centralisation. Long-range transport is an aspect of infrastructure that relates to group surplus. It is only needed when the economic group is larger than the local group. The transport of goods implies redistribution (Dalton 1967, 69).

Viewing Cooleeny 31 in its local context within the bog leads to looking at its regional context. Derryville Bog straddles the drainage catchments of the Suir and Nore rivers. It sits in a basin between the Silvermine Mountains and the Slieve Ardagh Hills. Both of these ranges have foothills to the north divided by passes. To the west, the valley of the Nenagh River, which runs into Lough Derg, divides Knockanora and the Devil's Bit Mountain from the Silvermines. To the east, the Nuenna River, which is a tributary of the Nore, separates two lower hills from the Slieve Ardagh Hills. These are the only passes crossing Ireland through its midpoint. Cooleeny 31 is on a direct line between them. Going south, there is no break in the Silvermines until the southern edge of the Slieve Felim. To cross would require going further south to Slievenamon or working up through the bog chain to the Nuenna River pass. Going north, the main obstacle is the wide expanse of the main bulk of midland bog. The study area is a narrow tongue in the bog chain that would be a minor obstacle in comparison with the rest of the midlands.

The known archaeology of this period reflects the importance of this natural routeway. Three hillforts overlook the Nenagh River pass. Knockadigeen and Ballincurra mark the western end of the pass (Grogan *et al.* 1996, 35; Raftery 1994, 42), and Garrangrena marks the eastern end of the pass (Raftery 1968). The pairing of hillforts is a common occurrence, and the placement of these sites could indicate their contemporaneity (Condit 1992; Grogan *et al.* 1996, 32). Knockadigeen is one of the largest sites in the country and forms part of a line of large multivallate sites running along the southern edge of the midlands (*ibid.* 33, Fig. 19). This line emphasises the importance of the natural routeway described above. A hillfort at Clomantagh (Condit and Gibbons 1988) marks the Nuenna River pass. While this site is less massive than Knockadigeen, it contains a cairn that is either a Linkardstown burial or a passage tomb (*ibid.*; Manning 1985, 53). The presence of an earlier cairn is a feature of many important hillforts throughout Ireland (Raftery 1994, 41).

Hillforts are regularly associated with the defence and control of passes (Grogan and Condit 1995; Raftery 1994, 41). According to Condit and Gibbons, "The site at Clomantagh, in common with many other Irish hillforts, is in an obvious strategic position. It commands the western approaches to north Kilkenny and may have been a focal point for the agricultural community which would have farmed the rich lands around it" (1988, 51). These sites would have had an important role beyond defence: they are centres in regional systems. We are not suggesting that these hillforts provided the impetus to build Cooleeny 31, but simply that their presence marks the importance of these passes in the period during which it was built.

The complex at Rathlogan (Mitchell and Ryan 1997) may be part of the activity marking the eastern pass. It occupies the western end of a smaller valley to the north of Clomantagh hillfort, just west of the village of Gattabaun, whose name translates as white gate. A multivallate enclosure takes advantage of a small promontory projecting into the valley. Its scale, position and scarped construction bring to mind parallels with Clenagh, County Clare (Grogan and Daly 1994) and Uisneach, County Westmeath (Donaghy and Grogan 1997). Both these sites have been suggested to have a similar date range to hillforts. A large group of substantial *fulachta fiadh* cluster along the stream in the valley, pointing to its importance in the Bronze Age. At the western end of the complex, a linear earthwork was recorded during the Lisheen project running against the line of the pass. This earthwork was similar to those found at Tara, County Meath (Condit 1993), Rathcroghan, County Roscommon (Herity 1991) and to examples with greater preserved lengths such as the Dorsey (Tempest 1930) and the Cliadh Dubh (Doody 1996). All of these monuments, and even the name of the nearby village (Gattabaun), emphasise control and closure of this pass. While any complex like this is almost certainly multi-period, it was clearly a significant place in later prehistory.

The approach to the Nuenna River pass from Derryville Bog is also marked by a monument of possible Late Bronze Age/early Iron Age date in Foulkscourt, County Kilkenny. This site, a large barrow constructed on very wet ground, perhaps fen in prehistory, was recorded by the project in spring 1996. The barrow was 32m in diameter and approximately 2m in height. A low bank survived on the northern side of the monument. Barrows of this size and in this landscape position have been noted in southeast Limerick and may be associated with larger multivallate ceremonial enclosures, with a construction date range in the transition between the Late Bronze Age and the Iron Age (Cross 1995).

The position and nature of Cooleeny 31 indicate that it could have been a link in the route between the western pass and the eastern pass. Even using the modern road network, which no longer relies on passes, it is possible to walk from one pass to the other in one day. The very similar track, Togher B, in nearby Littleton Bog (Rynne 1965) has been milled away, so it is no longer possible to check if it dates from the same period. Nonetheless, it seems to have been at a similar level, so contemporaneity remains a possibility. If it was built at the same time, it may have been part of the same phenomenon. It could have formed a replacement for the destroyed causeway Cooleeny 31. Alternatively, its orientation could suggest that it marked the route to the south of the Slieve Felim, running into the Golden Vale.

Causeway Cooleeny 31 is the only site in Derryville Bog that can be argued to have formed part of a regional communication network. The lack of environmental knowledge shown in its construction led to its demise within a year. Other than its environmental effects, the causeway seems to have had little impact on the archaeology of the bog. Local trackways and platforms accessing fen resources were built both before and after its construction.

It is possible that the preserved Cooleeny Complex largely post-dates the structure, in which case, the causeway may be associated with a new focus of local activity close to or in the bog. At a general level, however, the causeway stands outside the main pattern for the bog, which remains vernacular.

Implications

Material culture

Although most of the visible aspects of material culture were absent from the features excavated at Derryville, the surviving evidence demonstrates that a range of metal tools was in use and a high level of basic carpentry and finer woodworking were present (see Chapter 13). While it would be foolish to judge woodworking skills on the basis of casual tracks, the results of the woodworking analysis give us insight into the tools used to construct these sites. This has also been shown elsewhere (O'Sullivan 1996). There is a wide range of metal tools in evidence from the earliest periods, suggesting a broad availability of such tools. This is an integrated aspect of all social orders, not just high-status groups. In particular, the dissemination of iron tools at the early end of the transition between the Late Bronze Age and the Iron Age shows the rapid transmission of this technological change.

The few small finds that were recovered show that finished woodworking was a highly developed aspect of material culture. The turned rod 97E160:31:1 (Plate 12.1) is the earliest known evidence of lathe turning in Ireland. The bucket fragment 96E202:17:1 (Fig. 12.10; Plate 15.6) and the notched panel 96E202:311:1 (Fig. 12.13) show skill and control in carving. All of these objects are likely to have had domestic functions. The presence of finely finished domestic artefacts in a vernacular landscape underlines the fact that craft working would be important in all social spheres and was not only a product of specialised groups for competitive display.

Subsistence systems

Early indicators from the pollen record within Derryville suggest pastoral activity, including woodland farming, although small-scale cultivation would be masked within the overall pollen record (see Chapter 6). The extent of clearance and cultivation increases with recognisable cereal production by the middle Bronze Age. This may be evidenced by the ditches and a flint flake recorded at Killoran 2 and is supported by evidence from the *fulachta fiadh*: the saddle quern found at Killoran 240. A possible sheep/goat tooth was recovered from the trough at Killoran 5, and cattle and sheep bones were recovered form Killoran 27 and Killoran 22. Wild resources would have been used to provide dietary additions to agricultural production.

LAP 97
DER 17
96E202:14

Plate 15.6 Alder bucket fragment from Derryfadda 17.

Evidence for this has been found in Neolithic lake villages in Switzerland (Coles and Coles 1991, 151) and in sites in Latvia and Lithuania, such as the Šventoji sites in northwestern Lithuania (Dolukhanov 1991, 93). Wild resources also played a part in the pastoral economy, with elm and holly being used for fodder (Lucas 1963, 46). A full consideration of the role of wild resources in an agricultural economy is time consuming but rewarding (Louwe-Kooijmans 1993). When considering changes in economy, wetland sites provide essential information on the ways in which wild resources were used (Coles and Coles 1991, 151). The resources from the marginal woodland and the fen environment formed an integral part of the subsistence economy rather than an extension to it. An examination of the known uses of these resources will help explain their function as an element of a mixed agricultural subsistence strategy.

Common reeds, rushes and grasses were some of the main plants of the fen environment. The rushes and reeds from the fen and discharge channel would have provided a substantial resource, the use of which has been recorded in the early Irish law texts. Reeds were used for floor matting and thatch, and the inner pith from rushes was used in candle making (Kelly 1998, 55, 384). Such candles and lights were made by dipping a stripped rush in tallow. On the western side of the study area, rannock rush, which is rarely found in Ireland, was associated with the discharge channel. Rushes may have been used

as bedding material for cattle, for rush candles and for thatching (Feehan and O'Donovan 1996, 235), although they made poorer thatch than reeds. No direct evidence was found for the cutting of reeds. Such evidence has been found elsewhere, as at the Meare Heath track in the Somerset Levels, England (Coles and Orme 1976, 20). In Ireland, it is generally assumed that sickles, as found in the Lisnacrogher hoard, were used for cereal cultivation (Raftery 1996, 123).

Many species of birds are known to inhabit both the fen and raised bog (Feehan and O'Donovan 1996, 344). Among these are the whooper swan, greylag goose, snipe, wild duck, red grouse and a number of migratory birds, such as the white-fronted goose, which would have arrived from October onwards to overwinter. These birds are mostly ground nesters, increasing their vulnerability to hunting (Foss and O'Connell 1997, 185). Such birds would have proved a valuable source of meat, eggs and down. Evidence from Meare Lake village, England, where bones have survived, suggests that fowling was very significant even within this highly established community (Orme *et al.* 1979, 11).

In relation to the uses of wood in the distant past, it is clear that the properties of each wood species were clearly understood, and specific wood species were used for specific functions dependent on the requirements (Kelly 1976; Stuijts, this volume). The existence in the law tracts of a tree list indicates the specific importance of this material as a resource (Kelly 1998, 380). For example, alder is particularly durable in wet conditions, which explains its use in many of the trackways. In addition, the foliage, fruit, bark and resin of different trees could have been used as fodder, as well as in tanning and as pitch. An indication of some of these additional uses is recorded in the later law tracts. It is not feasible to extrapolate such usage into prehistory, but the tracts do provide an indication of the economic value of some of the resources.

There is evidence that the marginal woodlands were used for additional pasture. Within this context, the importance of holly and ivy in the tree list are attributed to their use in supplementing winter feed for animal fodder. At Cooleeny 325, a hazel artefact (97E160:325:1; Plate 15.7) that may have been used to carry such material was found among the deposit of holly brushwood. Later law texts refer to *in crann na fulachta fiannsa*, the tree of the open air cooking pit, which is thought to be holly, as the wood is very durable and suitable for using as cooking spits (*ibid.*, 382).

Trees were useful for their fruits as well as their wood. The regeneration of woodland on the fen margin would have occurred naturally and as a result of woodland pasturing. This would have resulted in the resurgence of species that are unable to persist in a closed wood, such as cherry and hazel and some of the understorey plants such as blackberry and strawberry (Kelly and Kirby 1982, 187). Blackberry and strawberry were identified at

Plate 15.7 Hazel artefact from Cooleeny 325.

the platform/togher site 95DER0178 in Cooleeny (IAWU 1996a), and the woodland evidence from this indicates the small-scale management of wild apple by the Middle Bronze Age. Hazel was harvested and possibly planted and, in this way, 'managed' to some degree. Apple and hazel were valuable food sources and highly valued in the eight-century texts (Kelly 1998, 306). Hazelnuts and apples, which can be stored for up to a year, would have provided useful supplements to the winter diet. A definite autumn bias is seen in the construction of sites in the fen and marginal woodland. This bias appears to relate to the fruit available at this time of year, which would have been collected to supplement the winter diet.

If the sites excavated at Derryville represent the local exploitation of wild resources as a complement to an agricultural subsistence strategy, what can be said about the nature of that exploitation? The structures are a gauge of the importance of wild resources in the subsistence pattern. The effort of construction was merited by the value of the resource. The structures also show that the process of gathering and hunting was structured to a certain extent, with some areas being marked out. The clustering of sites into distinct zones in the bog suggests a regular pattern of use. It is even possible that clusters represent the control of resources by different local groups.

The persistence of construction through all of the prehistoric periods shows that the use of wild resources was a factor in all of the economic systems that existed. On the other hand, the sporadic pattern of construction shows that it was not central to any subsistence strategy. The increasing intensity of construction through time may indicate either increased density of population or a change in the balance of subsistence practice, with wild resources playing a more important role. There is no evidence from the dryland to support the first suggestion, though there are large unexcavated areas on both headlands that, if excavated, could produce evidence to contradict this picture. The persistence of the clusters of structures in the bog also weighs against this suggestion, as increased density would be likely to lead to a more extensive use of the bog.

The balance between arable, pastoral and wild resources has been portrayed as fluctuating throughout prehistory, and the change in bog use may be best explained by this pattern (Mitchell and Ryan 1997, 233–4). This could also have a bearing on discrepancies between clearance evidence from pollen analysis and construction episodes in the bog. If structures represent an increase in the use of wild resources, they would appear in a different pollen zone than large-scale clearance.

The locations of the trackways and platforms seem to facilitate access to a wide range of resources. This suggests structures were sited to be useful in a variety of gathering and hunting activities. Accordingly, this means that aspects of subsistence, hunting, food gathering and the collection of building materials may have been integrated rather than distinct sets of activity. In any case, there is no basis for distinguishing between hunting sites and gathering sites.

Patterns of work

The use of the fen for economic purposes also provides us with some possible information about the social structure of groups living on Killoran headland. The organisation of work is a basic element of social interaction. Many societies involve large groups in harvesting, and the work expresses the relationships of the people involved (Dalton 1967, 69). None of the sites in the study area could have accommodated working groups of more than twenty people, and many of the tracks were designed for a single person to use. Larger platforms like Derryfadda 9 could have held five to ten people as they organised gathering or lay in wait while hunting. The platforms on the *fulachta fiadh* at Killoran 253 and Killoran 265 would have only been comfortable working spaces for two to three people. The broad surface provided by Killoran 75 and Killoran 312 together made space for more people than either track could hold at one time, but this would still be a small group. Groups not larger than a family unit did the work in Derryville Bog—gathering, hunting or heating water with hot stones.

Roberts (1998, 196) has suggested that burnt mounds studied in East Anglia were used for tanning by small groups of women and children. The composition of these work groups is not based on archaeological evidence, but it underlines the social importance of our interpretation of these sites.

This small-scale organisation represents the structure of a community at the low end of a settlement hierarchy. The only real examples we have of larger groups being organised for work is in the construction of sites like Killoran 18 and Cooleeny 31. Both these sites have more than a local significance. In the case of Cooleeny 31, the impulse for the organisation of large groups of labourers may have come from outside the local community. In the case of Killoran 18, the community may have organised itself for a particular event. In both cases, the size of groups is clearly a departure from day-to-day work practice.

Broader social relations

Many discussions on the social aspects of wetland archaeology consider trackways to be devices for overcoming the bog as an obstacle to communication. Raftery refers to "the challenge of crossing the soggy, bogland wastes" (1996, 226). Coles and Coles see tracks as providing "those essential links if early settlements were not to be cut off from their contemporaries across the bogs" (1989, 154). Sites crossing the entire basin are rare and significant in Derryville Bog, as discussed above. The tracks and platforms studied here suggest another aspect of wetlands: their attraction as places for settlement (*ibid.*, 152; Roberts 1998).

As the evidence described here suggests a vernacular landscape, the social patterns and practices of this type of group may be more difficult to distinguish archaeologically. There are, however, indicators of how the fen fits into social practices (*ibid.*). Burial and small-scale deposition point to ritual activity. Less obvious, but perhaps more important, social patterns were embedded in the wider use of the fen.

The strongest evidence we have for ritual practices that give the fen social importance is from the dryland. The flat cremation cemetery at Killoran 10 (Plate 15.8) and the smaller groups of cremation pits at Killoran 4, 6, 7 and 24 display a pattern of fen edge burial that is beginning to be recognised as important elsewhere. In East Anglia, surveys of Borough Fen and Catswater Fen have shown large clusters of barrows on the fen edge. Hall *et al.* explain this as the "intentional locating of ritual foci at the approximate boundary between dry and wet land" (1987, 191). They go on to state that "as these ritual structures all seem to relate to mortuary activities, we could suggest that the dead were being placed, or alternatively honoured, at the boundary between the 'two worlds'" (*ibid.*).

Roberts discusses a complex of fen edge inhumations and burnt mounds as indicating a vernacular use of a fen edge in East Anglia. She sees people being buried in the fen edge as "a sign of respect to both them and their envi-

Plate 15.8 Killoran 10. Flat cremation cemetery dating to 1435–1215 BC.

ronment" (1998, 196). The same pattern is visible in southeast Limerick, where very large cemeteries of ring barrows cluster along the edge of what were small lakes ringed with fen (Cross 2001).

The lack of evidence for ritual deposition from the study area has been discussed above. Consideration of deposition from the rest of this bog chain suggests a pattern that is distinct from high-status hoards. There are very few artefacts from the chain of bogs through this area. Nearby Littleton Bog has been the most prolific, producing two Late Bronze Age swords (Rynne 1965; Raftery 1958, 122), a Late Bronze Age spearhead with a portion of the shaft (Rynne 1965, 140), a decorated Early Bronze Age flat axe (Raftery 1961, 74) and an early Iron Age leather-covered wooden shield (Rynne 1962, 152). Nearby at Bawnreagh, a small bronze socketed spearhead was recovered during BnM milling. This group of artefacts shows that there was deposition in the chain of bogs, but it does not point to a very intense pattern. Small-scale deposition of single artefacts, with less emphasis on metalwork, may be part of the same concept as large-scale hoard deposition, but occurring at a local level.

Further research on votive deposition in bogs is needed. Bogs as a landscape entity need to be broken down into more concrete and useful concepts. First, a division must be made between deposition in fen and deposition in raised bog. Are fens a version of lakes, or do fens, raised bogs and lakes all play different roles in the landscape of ritual practice? Next, the nature of particular bogs must be studied. Is there more deposition in small bogs, large open tracts or mosaics of bogs with small 'islands' of dryland? Position in relation to other terrain features also needs to be considered. The large hoards at Mooghaun North, County Clare and the Bog of Cullen, County Tipperary, are both overlooked by hills (Cooney and Grogan 1994, Fig. 8.16). The combination of the two elements may have been important in the siting of such activities. These sites are also near contemporary high-status sites, and this combination needs to be studied more fully. The material discussed here suggests that there may be another less intense variant of ritual deposition associated with what we term vernacular landscapes.

With this in mind, we should perhaps view the construction of the causeways at Derryville in their chronological contexts. The fact that these two structures date to the Middle Bronze Age (Killoran 18) and the transition between the Late Bronze Age and the Iron Age (Cooleeny 31) points to their role in the development of social contacts and the increasing scale of social interaction. Both of these periods are times of significant social change in Ireland (*ibid.*). It is at these times that the local pattern of the study area becomes significant in the regional system. The landscape of this area remained vernacular throughout prehistory, but the scale of its interactions with the wider landscape and social structures fluctuated through time.

The fen as wild space

The relationship between the wild and the domestic is an enduring theme in discussions of prehistory from the Neolithic onwards (Hodder 1990, 11). Yet notions of 'nature' and the 'wild' are notoriously poorly explored and used, even in current debates (Wilson 1990, 13). Dealing with prehistory is even more difficult, but we must consider the importance of the fen in the landscape of this area. Some researchers have relied on early texts to catch glimpses of earlier visions of the natural world (Ross 1995). Others have included archaeological information (Chapman and Gearey 2000). Current novelists and thinkers can also broaden our visions of what wild wetlands represented in the past. Here, we consider the fen and marginal woodland of Derryville Bog in a general fashion, with the simple intention of pointing out the range of importance that it constituted.

Most wetlands in modern Europe have been substantially altered for human purposes. In addition to the economic benefits of these alterations, the taming of the wild space of the bog was important to gain tighter control of society. Graeme Swift, in his novel *Waterland*, presents a view of the fen as a place where socially marginal activity takes place. For the poet Séamus Heaney, they are deeply spiritual places. It is no accident that the American philosopher Thoreau sets much of his discussion of political dissent in wetland. The importance of bogs to rebels and resistance fighters has long been recognised (Evans 1973, 35), and the building of bog roads for military purposes (Lucas 1985, 50–1) shows that, in more tightly controlled societies, the wild space of the bog is more fully tamed.

In this context, the lack of clearance and reclamation in the fen and fen margin of prehistoric Derryville Bog is significant. The drainage of wetland is not unknown in prehistory (Adams 1966, 48; Pryor 1991, 52), and the clearance of considerably larger tracts of woodland is one of the major landscape changes through prehistory. Derryville Bog was probably left wild and slightly inaccessible as a choice of the local population or simply because it sustained the sparse local population in its near natural state.

Clearly, such a choice resulted in economic benefits, as discussed above. Understanding possible social benefits is more difficult. Once again, the difference between our perspective and that of the population we are studying is standing in the way (Hill 1989). Remembering the pervasive nature of spirituality suggested for Iron Age groups (Raftery 1994, 178–9), the presence of a wild space may have had religious importance (Newman 1996). The high proportion of apple wood used in the construction of the Derryville sites may have had subtle significance, given the value placed on the apple as a tree of the otherworld. In addition, the presence of a wild space close to settlement leaves room for solitude and even courting. The marginal woodland surrounding the bog offered a buffer between the fen and the settled headland. We cannot prove what social benefits accrued from this buffer, but we should regard it as a significant part of the landscape.

Conclusions

The evidence from this study of Derryville Bog suggests that it represents the wetland component of a vernacular landscape. This social model allows for the integration of the results of the various aspects of the study. The variation shown in site morphology, the continuity over time and the terrain sensitivity all lead to the main argument presented here, which is supported by the nature and positioning of the archaeological sites and the absence of high-status material, either in the bog or in associated dryland sites. Most of the structures in the bog seem to have facilitated gathering and hunting in the fen and marginal woodland. The sensitivity displayed to this local environment is striking. This feature has helped us to identify sites that may have been constructed outside of the local pattern.

Such 'everyday' landscapes have not received as much research attention as high-status or ritual landscapes. Viewing our material in this light, has led to some important suggestions. Firstly, there is a higher degree of continuity throughout prehistory than may have been expected. This is especially apparent in the transition between the Late Bronze Age and the Iron Age. This continuity does not suggest a conservative trend in material culture. New tool types seem to have been widely adopted relatively quickly. The pollen record indicates broad changes in the subsistence practice of clearance and regeneration. The tracks and platforms reaching out into the fen and its marginal woodland show that the use of wild resources was consistent but also varied over time.

Secondly, the results highlight the fact that social relations in the study area may have gone beyond competition, display and redistribution. The prehistoric communities of this area usually worked in small groups, but came together for important tasks. The relatively large cemetery on the dryland at Killoran 10 shows long-term group cohesiveness. The regional picture may suggest a low-level of votive deposition, complementing the hoards from higher status landscapes.

Thirdly, the study highlights an important difference between the modern and prehistoric landscapes. The combination of fen and marginal woodland, which forms the focus of this study, is all but absent from the present Irish landscape. The recognition that this was a desirable location for everyday domestic activity in the prehistoric period is perhaps the most important result of the project. The area contained established communities that made regular use of this varied environment. The bog itself had an important role in the structure and use of the landscape.

16. Archaeological conclusions

Sarah Cross May, Cara Murray, John Ó Néill and Paul Stevens

Introduction

The Lisheen Archaeological Project was the first opportunity to investigate a complete prehistoric wetland landscape in Ireland. While the majority of sites dated to the Bronze Age, the full prehistoric sequence is represented in our understanding of the bog as a landscape. In addition to the sixty-six sites within the bog, thirty-three sites were excavated on the Killoran headland to the west of the bog. This allowed us to consider dryland/wetland interactions in an integrated fashion. The theme of integration underlay the multidisciplinary methodology. The active participation of environmental specialists throughout the project established a sense of the bog as a dynamic environment. Derryville Bog emerges from our analysis as a vernacular landscape. Neither an obstacle to be overcome nor a place of ceremony, it was part of the daily reality for local groups.

Chronological structure

The combination of a long series of absolute radiocarbon and dendrochronological dates, their contextual environment and an analysis of tool use on the wood facilitated the construction of a very detailed chronology of the excavated wetland material. It also provided a framework for more conventional artefact-based typologies, affording an opportunity to sequence material and sites within the first millennium BC, which is usually obscured by conventional radiocarbon dating problems. We have also been able to link archaeological chronologies with environmental sequences. Throughout the prehistoric period, the site types remain similar, with a predominance of casual structures. Generally, the sites were constructed in marginal areas of the fen and raised bog. The clusters of dates within periods appear to relate to changing environments and changing uses of the bog.

There is residual evidence for Neolithic wetland activity, but the bog and surrounding woodland remained virtually uninhabited until the Early Bronze Age. From this point through to the Late Bronze Age, the margins were persistently exploited by short trackways, platforms and *fulachta fiadh*. A bog causeway was attempted in the Middle Bronze Age, but after this the bog became impassable. The transition from the Bronze Age to the Iron Age is marked by the disappearance of *fulachta fiadh*, but no other change is apparent in site typology or function. During this transition period, an impressive attempt was made to cross the bog again. There is a noticeable absence of human interference in the early first millennium AD before further occupation recommenced in the early historic period.

Comparisons of tool marks across the prehistoric material showed a striking continuity from one period to the next. The exception to the pattern is the clear introduction of iron for domestic tools during the seventh century BC. This marks the transition from the Bronze Age to the Iron Age in an otherwise continuum of typological and functional wetland sites.

In contrast, the dryland settlement matches with well-defined archaeological chronological patterns of typology and morphology. This material was evidenced in the Bronze Age by round houses, numerous *fulachta fiadh*, agricultural reclamation and burial sites, including a substantial flat cremation cemetery. Iron Age settlement was characteristically sparse, but was represented by a round house. Early Christian settlement was also revealed, with ecclesiastical enclosure, ringfort settlement, sporadic burnt pits and agricultural reclamation all represented. However, medieval and post-medieval occupation was almost totally absent, with reclamation of the area only taking place during the enclosure acts of the eighteenth century.

Chronologically, we see continuity throughout the first millennium BC. The overall impression created is that the end of the Bronze Age may not represent a collapse into a dark age, but rather a continuation into a period of slow transition. For an area particularly lacking in high-status monuments, the suggestion that iron was introduced as a basic domestic tool at this early date, and continued in use in perpetuity, reinforces the idea of gradual transition rather than dramatic social collapse.

A dynamic environment

The project provided an opportunity for a detailed examination of the interaction of the human and physical components of a marginal landscape over time and against varying scales. While the fen changed over time, some of the enduring features were a mosaic of reed beds, rushes and sedges with pools of open water and discharge channels running through them. This environment is a suitable habitat for wild fowl, but is too acidic for fish. The fen was fringed by mixed scrub woodland, which extended out at some points to form peninsulas or islands. No attempt was made to clear this woodland in the prehistoric period, although it was exploited and managed in an informal way. This woodland formed a dense boundary on the western side of the bog and a less dense one on the eastern side.

The proximity of arable land, woodlands and fen provided a biodiverse environment capable of sustaining a community through much of the prehistoric period. The siting of trackways, wooden platforms and burnt mounds maximised the positive aspects of the terrain, particularly in the period 2500–100 BC. This suggests knowledge of the immediate vicinity that would only be held by people living in the area.

Some of these structures were modified during their lifetime as a response to localised changes. Killoran 18 was constructed in a narrow window of time when a localised dry line existed in the bog. Unfortunately, the drainage channel at the western side of the bog could only be crossed with difficulty. In time, the northerly flow of this channel was reversed as Killoran 18 effectively blocked its course. Paradoxically, the weight of structural components, particularly the stone, compressed the surface of the bog. This allowed discharge from the ombrotrophic raised bog system to the south, to flood the remaining fen to the north.

The people who built the various structures in and around Derryville Bog could be viewed as taking advantage of local conditions and adapting to subtle changes in the nature of their environment. Some structures could also be viewed as being imposed, with little insight, on that environment. Adapting to and taking advantage of conditions seems to reflect a local impetus. The appearance of non-local, or regional, motivation for bridging the bog can be seen with Cooleeny 31, which was very much imposed upon its environment.

The scale of Cooleeny 31 is of a different order than much of the archaeology at Derryville Bog. This structure, and Killoran 18, represent the only two attempts to provide a continuous route across the bog. As an engineering solution, its use of sand and underlying supports reveals an organised, considered attempt to carry out this task. This same structure, however, dammed the southern discharge of the bog and caused a bog burst within a year of its construction—*c.* 600–595 BC. Later, communities saw no need to bridge the chain of bogs at this location.

At a regional level, the bog clearly acted as a considerable obstacle to east–west progress, emerging in the early historic period as the boundary between the ancient provinces of Munster and Leinster. Thus, the same terrain was a local resource, a regional obstacle and a provincial boundary marker.

Vernacular landscape

The continuity over time and the terrain sensitivity lead to the proposition that the material excavated at Derryville represents part of a vernacular landscape. This is supported by the nature and positioning of the archaeological sites and the absence of high-status sites, both within the bog and on the associated dryland areas. Most of the structures within the bog seem to have facilitated gathering and hunting in the fen and marginal woodland.

The sensitivity displayed to this local environment is clearly evident. This factor has aided in the identification of sites that may have been constructed outside of this local pattern. The structure and location of Cooleeny 31, in conjunction with its palaeoenvironmental context, suggest that it formed part of a significant regional route. This indicates that prehistoric landscapes were integrated across scales and that the results of the Lisheen project should not be considered in isolation. Wetland landscapes, and vernacular landscapes in general, are a core element of Ireland's prehistory. An exclusive focus on the centre of later prehistoric polities keeps us from understanding many important questions.

Vernacular landscapes have not received as much research attention as high-status or ritual landscapes. Further projects focusing on this social scale are to be hoped for. The analysis of the Lisheen data has led to a number of suggestions. There was a higher degree of continuity evident throughout prehistory than might have been expected. In particular, the dissemination of iron tools at the early end of the transition between the Late Bronze Age and the Iron Age shows the rapid transmission of technological change. The pollen record indicates broad changes in the subsistence practice of clearance and regeneration, which are not reflected in the patterning of archaeological material. The use of wild resources was consistent over time but still has some variation, which may benefit from further study.

It appears that the people living in the Derryville area usually worked in small groups of under ten but came together for important tasks. The relatively large cemetery on the dryland at Killoran 10 shows long-term community cohesiveness. A wide range of metal tools was in use, indicating that access to metal was not restricted to the upper levels of the social hierarchy. The few small finds, such as the bucket fragment, the turned rod and the notched panel, show that finished woodworking was a highly developed aspect of material culture.

The results also show that wetland sites do not demonstrate morphological or functional variation within a chronological pattern. Instead, continuity, both in usage and location within the bog, is one of the strongest patterns to emerge, with a populace showing an intimate knowledge of the bog exploiting the same places over time.

Conclusions

The analysis presented here is far from exhaustive. It represents a combined vision from four different perspectives, working on four different sets of material. Working alone, each of us would have pursued slightly different paths, and hopefully these will be explored in the future. All the same, the co-authoring of analysis has had substantial benefits. The division of excavated material was based on practical considerations, but working in different places within the bog allowed us to look at very different life histories for the bog. If we had looked at only one, the story would have been very different.

The Lisheen project presents a chronological picture of persistence. Enduring architectural forms, patterns of behaviour and spatial patterning draw threads between the traditional period boundaries. In contrast, the bog is dynamic. It changed in size, position, hydrological regime and ecological profile. Sometimes dramatic changes were triggered by human use of the bog. Sometimes human use took advantage of dramatic change. Persistence and adaptability are common features of vernacular landscapes. They tell subtle stories in contrast to the dramatic narratives of revolution, invasion and collapse.

This local perspective allows us to consider a common question posed of bog trackways: where are they going? While large causeways could be part of regional networks, most of the structures in Derryville Bog formed links in a web of daily life. The complex of sites in the northwest of the bog study area may have been used by the inhabitants of the nearby round houses at Killoran 8. There may have been houses near to other clusters of wetland sites. The causeway Killoran 18 may have allowed access to the cremation cemetery Killoran 10. But all of the sites formed an integral part of this inhabited landscape. Rather than searching for the end of the road, we can recognise that wetland archaeology is settlement evidence.

17. Inventory of radiocarbon dates

The radiocarbon dates listed in Table 17.1 are presented by site name, survey designation, laboratory code, date BP (plus standard error) and calibrated range (Stuiver and Pearson 1993; Pearson and Stuiver 1986). For sources of the dating samples see the individual entries in Chapters 10 and 11. A list of dates from environmental samples is presented in Table 17.2. Dated sites that were recorded during the initial IAWU survey, but not subsequently excavated, are denoted by an asterisk.

Site name	Survey designation	Lab. code	Date BP	Calib. range (2σ intercepts)
Cooleeny 31	DER31h	GrN-21822	2475±25	778–423 BC
Cooleeny 31	DER31	Beta-111367	2440±70	790–380 BC
Cooleeny 62*	DER62	GrN-21945	2500±25	792–526 BC
Cooleeny 64	DER64	Beta-111374	3020±70	1420–1020 BC
Cooleeny 100*	DER100	GrN-21820	2150±20	351–120 BC
Cooleeny 141*	DER141	GrN-21949	2100±30	197–47 B
Cooleeny 143*	DER143	GrN-21824	2240±40	771–401 BC
Cooleeny 169*	DER169	GrN-21950	2235±20	388–207 BC
Cooleeny 178*	DER178	GrN-21817	2200±20	372–194 BC
Derryfadda 6	DER6	Beta-102737	2160±70	380–5 BC
Derryfadda 9	DER9	Beta-102739	2250±50	395–180 BC
Derryfadda 13	DER13a	GrN-21943	2460±20	767–412 BC
Derryfadda 13	DER13a	Beta-102736	2200±50	380–100 BC
Derryfadda 13	DER13b	Beta-102756	2460±60	790–395 BC
Derryfadda 17	DER17	Beta-111272	2950±60	1315–980 BC
Derryfadda 19	DER19	Beta-102760	1290±60	AD 640–890
Derryfadda 201	DER21	Beta-102738	2250±40	390–190 BC
Derryfadda 203	DER203	Beta-102761	2190±60	385–50 BC
Derryfadda 203	DER203	Beta-102762	2250±60	400–165 BC
Derryfadda 204	DER204	Beta-102764	3590±60	2120–2080 or 2050–1755 BC
Derryfadda 206	DER206	Beta-102755	2420±70	785–375 BC
Derryfadda 207	DER207	Beta-102575	3660±70	2205–1875 or 1805–1795 BC
Derryfadda 208	DER208	Beta-102746	2130±60	365 BC–AD 5
Derryfadda 209	DER209	Beta-102749	3680±80	990–770 BC
Derryfadda 210	DER210	Beta-102740	2350±50	515–365 BC
Derryfadda 211	DER211	Beta-102753	2900±60	1265–910 BC
Derryfadda 213	DER213	Beta-102758	2930±70	1315–915 BC
Derryfadda 214	DER214	Beta-102741	2210±60	390–75 BC
Derryfadda 215	DER215	Beta-102754	2380±60	760–635 or 560–370 BC
Derryfadda 216	DER216/T3	Beta-102305	2980±70	1400–990 BC
Derryfadda 218	DER218	Beta-102759	3720±60	2290–1935 BC
Derryfadda 311	DER311	Beta-111375	3050±80	1450–1030 BC
Killoran 3	KIL03:Linear	Beta-117555	1550±50	AD 415–630
Killoran 3	KIL03:Pit	Beta-117556	930±80	AD 980–1270
Killoran 5	KIL05:Trough	Beta-117545	3300±80	1750–1410 BC
Killoran 6	KIL06:Cremation	Beta-117548	2860±50	1145–900 BC
Killoran 8	KIL08:House	Beta-117553	3340±80	1860–1845 or 1775–1430 BC
Killoran 8	KIL08:House	Beta-117554	1200±60	AD 685–985
Killoran 10	KIL10:Cremation	Beta-117546	3090±50	1435–1215 BC

*Table 17.1 Radiocarbon dates from archaeological structures and sites (continued overpage). *Unexcavated sites.*

Site name	Survey designation	Lab. code	Date BP	Calib. range (2σ intercepts)
Killoran 16	KIL16:House	Beta-117551	1890±130	180 BC–AD 425
Killoran 16	KIL16:Pit	Beta-117552	1050±50	AD 890–1040
Killoran 17	KIL17:Trough	Beta-117547	3930±70	2585–2195 BC
Killoran 18	DER18/C13	Beta-102750	2990±70	1405–1000 BC
Killoran 18	DER18/C15	Beta-102752	3290±80	1745–1405 BC
Killoran 18	DER18/C19	Beta-102751	3180±70	1605–1285 BC
Killoran 18	DER18/C27	UB-4082	4447±45	3339–2924 BC
Killoran 18	DER18/C37	UB-4097	4224±68	3020–2613 BC
Killoran 18	DER18/C39	UB-4083	3210±49	1620–1410 BC
Killoran 18	DER18/C41	UB-4095	3503±99	2133–1548 BC
Killoran 20	DER20	Beta-111373	2940±60	1305–940 BC
Killoran 23	KIL23:Pit	Beta-117550	1280±50	AD 660–880
Killoran 26	KIL26:Trough	Beta-117549	2780±90	1145–795 BC
Killoran 31	KIL31:Furnace pit	Beta-120521	1450±70	AD 450–690
Killoran 48*	DER48	GrN-21823	1260±30	AD 672–853
Killoran 54	DER54c	GrN-21944	1250±40	AD 668–884
Killoran 66	DER66	GrN-21945	1200±20	AD 775–887
Killoran 69	DER69	UB-4180	2646±24	838–799 BC
Killoran 75	DER75	GrN-21947	2190±20	368–190 BC
Killoran 75	DER75/C3	Beta-102766	2170±60	380–40 BC
Killoran 75	DER75/C4	Beta-102763	2190±60	385–50 BC
Killoran 75	DER75/C5	Beta-102765	2020±70	185 BC–AD 130
Killoran 98*	DER98	GrN-21948	935±20	AD 1024–1162
Killoran 226	DER226	Beta-102748	2260±60	405–175 BC
Killoran 226	DER226	UB-4181	2053±30	165 BC–AD 8
Killoran 226	DER226	UB-4182	2352±51	754–370 BC
Killoran 229	DER229	Beta-102747	2950±60	1385–930 BC
Killoran 230	DER230	Beta-111368	3090±70	1500–1195 BC
Killoran 234	DER234	Beta-111369	2470±60	795–395 BC
Killoran 237	DER237	Beta-111370	3260±70	1685–1400 BC
Killoran 248	DER248	UB-4183	2239±33	394–199 BC
Killoran 253	DER253	Beta-111378	2940±60	1305–940 BC
Killoran 265	DER265	Beta-111377	3050±60	1425–1120 BC
Killoran 301	DER301	UB-4184	2221±30	367–195 BC
Killoran 301	DER301	UB-4185	2160±35	367–195 BC
Killoran 304	DER04	UB-4186	3651±34	2138–1935 BC
Killoran 305	DER305	UB-4187	3134±53	1520–1310 BC
Killoran 305	DER305	UB-4188	3083±39	1436–1264 BC
Killoran 306	DER306	UB-4189	2394±30	756–407 BC
Killoran 306	DER306	UB-4190	2447±22	756–407 BC
Killoran 314	DER314	Beta-111376	2140±60	370–5 BC
Killoran 315	DER315	Beta-111371	2270±60	405–180 BC

*Table 17.1 (Continued) Radiocarbon dates from archaeological structures and sites. *Unexcavated sites.*

Sample name	Survey designation	Lab. code	Date BP	Calib. range (2σ intercepts)
Scots pine:TII	D21	GrN-21821	7150±20	6078–5965 BC
Peat below silt		GrN-21819	2160±30	363–114 BC
Peat above silt		GrN-21818	1980±30	72 BC–AD 79
Peat/Field 40	S667	Beta-102745	2430±70	780–385 BC
Peat/Field 41		UB-4089	2217±49	400–129 BC
Peat/Field 41		UB-4096	5908±54	4938–4687 BC
Peat/Field 46	S664	Beta-102742	2200±60	390–60 BC
Peat/Field 46	S665	Beta-102743	2440±70	790–380 BC
Peat/Field 46	S666	Beta-102744	2630±70	905–760 or 670–550 BC
Peat/Field 48		UB-4085	3611±44	2134–1883 BC
Peat/Field 48		UB-4086	2801±43	1057–844 BC
Peat/Field 48		UB-4087	2106±65	369 BC–AD 20
Peat/Field 48		UB-4088	1427±43	AD 550–667
Pollen	DER18i	Beta-100942	3820±70	2465–2140 BC (1σ)
Pollen	DER18ii/iii	Beta-100941	3050±70	1400–1200 BC (1σ)
Pollen	DER18iii/iv	Beta-100940	2969±70	1275–1030 BC (1σ)
Pollen	DER18v	Beta-100939	2360±60	415–380 BC (1σ)
Pollen	DER18vi	Beta-100938	2180±70	365–115 BC (1σ)
Pollen	DER18vii	Beta-100937	1740±60	AD 240–395 (1σ)
Pollen	DER23v	Beta-100946	3070±60	1405–1260 BC (1σ)
Pollen	DER23vi	Beta-100945	2850±60	1065–915 BC (1σ)
Pollen	DER23vii	Beta-100943	2150±80	355–290 or 230–50 BC (1σ)
Pollen	DER23vii	Beta-100944	2490±60	780–485 or 465–425 BC (1σ)
Pollen	DER75	Beta-11084	1510±60	AD 530–630 (1σ)
Pollen	DER75	Beta-11085	1740±60	AD 240–395 (1σ)
Pollen	DER75	Beta-11086	1930±60	AD 25–135 (1σ)
Pollen	DERSILT	Beta-11087	2080±70	60 BC–AD 5 (1σ)
Pollen	DERSILT	Beta-11088	2150±80	355–290 or 230–50 BC (1σ)

Table 17.2 Radiocarbon dates from environmental samples.

18. Inventory of dendrochronology dates

The dendrochronology dates listed in Table 18.1 are presented by site name, survey designation, laboratory code, date of last year ring (plus standard error or comment). For sources of the dating samples see the individual entries in Chapters 10 and 11. A list of dates for environmental samples is presented in Table 18.2.

Site name	Survey designation	Lab. code	Date of last year ring
Cooleeny 22	DER22:Plank/phase 1	Q9544	1521±9 BC
Cooleeny 22	DER22:Plank/phase 1	Q9547	1517±9 BC
Cooleeny 22	DER22:Plank/phase 1	Q9548	1526±9 BC
Cooleeny 22	DER22:Plank/phase 1	Q9549	1517±9 BC
Derryfadda 23	DER23:Plank	Q9369	1606±9 BC or later
Derryfadda 23	DER23:Plank	Q9370	1590±9 BC or later
Derryfadda 215	DER215:Plank	Q9400	457±9 BC
Derryfadda 218	DER218:Bog oak/TAQ	Q9540	3368±9 BC
Killoran 18	DER18f	Q9188	1542±9 BC
Killoran 18	DER18:Timber/phase 2	Q9470	1534±9 BC or later
Killoran 18	DER18:Plank/phase 3	Q9349	1440±9 BC
Killoran 27	KIL27:Trough timber	Q9697	939±9 BC
Killoran 27	KIL27:Trough timber	Q9698	932 BC
Killoran 226	DER226:Structure1 timber/phase 1	Q9476M	450±5 BC
Killoran 226	DER226:Structure1 timber/phase 1	Q9480	475±9 BC
Killoran 226	DER226:Structure1 timber/phase 1	Q9479M	541±9 BC or later
Killoran 226	DER226:Structure1 timber/phase 2	Q9478M	460±9 BC
Killoran 226	DER226:Structure1 timber/phase 2	Q9477M	542±9 BC or later
Killoran 226	DER226:Structure2 timber	Q9475M	161±9 BC
Killoran 235	DER235:Timber	Q9541	1212±9 BC
Killoran 241	DER241:Roundwood	Q9542	1547±9 BC
Killoran 243	DER243:Roundwood	Q9543	979±9 BC
Killoran 248	DER248:Roundwood	Q9483	641±9 BC or later
Killoran 301	DER301:Timber	Q9488	332±9 BC or later
Killoran 301	DER301:Timber	Q9486	322±9 BC or later
Killoran 301	DER301:Timber	Q9487	462±9 BC or later
Killoran 301	DER301:Combined mean felling date		324±9 BC or later
Killoran 306	DER306:Sunken oak/TPQ	Q9489M	1148±9 BC

Table 18.1 Dendrochronology dates for archaeological samples.

Sample name	Lab. code	Date of last year ring
E1:Bog oak/NE ridge	Q9119	422±9 BC
E5:Bog oak/NE ridge	Q9122	448±9 BC
E6:Bog oak/NE ridge	Q9124	403±9 BC
E7:Bog oak/NE ridge	Q9125	443±9 BC
E8:Bog oak/NE ridge	Q9126	473±9 BC
E9:Bog oak/NE ridge	Q9127	282±9 BC
E225:Bog oak/W ridge	Q9402	174±9 BC or later
E238:Bog oak/W ridge	Q9403	358±9 BC or later
E243:Roundwood/Field 7	Q9543	979±9 BC or later
DER90:Bog oak/TPQ	Q9401	2718±9 BC

Table 18.2 Dendrochronology dates for environmental samples.

Bibliography

Aaby, B. 1976 Cyclic climatic variations in climate over the last 5,500 years reflected in raised bogs. *Nature* **263**, 281–4.

Aaby, B. and Tauber, H. 1975 Rates of peat formation in relation to degree of humification and local environment, as shown by studies of a raised bog in Denmark. *Boreas* **4**, 1–17.

Aalen, F.H.A., Whelan, K. and Stout, M. (eds) 1997 *Atlas of the Irish rural landscape*. Cork. Cork University Press.

Abbink, A.A. 1986 Structured allocation and cultural strategies. In R.W. Brandt, S.E. van der Leeuw and M.J.A.N. Kooijman (eds), *Gedacht over Assendelft*, 23–32. IPP Working Paper **6**. Amsterdam.

Adams, R.McC. 1966 *The evolution of urban society*. Chicago. Aldine.

Alexander, K.N.A. 1994 *An annotated checklist of British lignicolous and saproxylic invertebrates* (Draft). Cirencester. National Trust Estates Advisors' Office.

Anderson, R., Nash, R. and O'Connor, J.P. 1997 *Irish Coleoptera: a revised and annotated list.* Irish Naturalists' Journal, Special Entomological Supplement.

Anon. 1989 Somerset Levels project catalogue of finds. *Somerset Levels Papers* **15**, 33–61.

Archer, J.B., Sleeman, A.G. and Smith, D.A. 1996 A *geological description of Tipperary and adjoining parts of Laois, Offaly, Clare and Limerick, to accompany the Bedrock Geology 1:1000,000 Scale Map Series, Sheet 18, Tipperary, with contributions by K. Claringbold, G. Stanley (Mineral Resources) and G. Wright (Groundwater Resources).* Dublin. Geological Survey of Ireland.

Baillie, M.G.L. and Brown, D.M. 1996 Dendrochronology of Irish bog trackways. In B. Raftery, *Trackway excavations in the Mountdillon Bogs, Co. Longford, 1985–1991*, 395–402. Transactions of the Irish Archaeological Wetland Unit **3**. Dublin. Crannóg Publication.

Baillie, M.G.L. and Munro, M.A.R. 1988 Irish tree rings, Santorini and volcanic dust veils. *Nature* **332**, 344–6.

Barber, K.E. 1981 *Peat stratigraphy and climate change: a palaeoecological test of the theory of cyclic peat bog regeneration.* Rotterdam. A.A. Balkema.

Barber, K.E. 1982 Peat-bog stratigraphy as a proxy climatic record. In A.F. Harding (ed.), *Climatic change in later prehistory*, 103–13. Edinburgh. Edinburgh University Press.

Barber, K.E. 1994 Deriving Holocene palaeoclimates from peat stratigraphy: some misconceptions regarding the sensitivity and continuity of the record. *Quaternary Newsletter* **72**, 1–10.

Barber, K.E., Chambers, F.M., Maddy, D., Stoneman, R. and Brew, J.S. 1994 A sensitive high-resolution record of late Holocene climatic change from a raised bog in northern Britain. *The Holocene* **4** (2), 198–206.

Barfield, L. and Hodder, M. 1987 Burnt mounds as saunas, and the prehistory of bathing. *Antiquity* **61**, 370–9.

Belyea, L.R. 1996 Separating the effects of litter quality and microenvironment on decomposition rates in a patterned peatland. *Oikos* **77**, 529–39.

Bermingham, N. 1997 Leabeg, Castlearmstrong, Cornafurrish and Corrabeg. In I. Bennett (ed.), *Excavations 1996*, 93. Bray. Wordwell.

Blackford, J.J. 1993 Peat bogs as sources of proxy climatic data: past approaches and future research. In F.M. Chambers (ed.), *Climate change and human impact on the landscape*, 225–36. London. Chapman and Hall.

Blackford, J.J. and Chambers, F.M. 1991 Proxy records of climate from blanket mires: evidence for a Dark Age (1400 BP) climatic deterioration in the British Isles. *The Holocene* **1** (1), 63–7.

Blackford, J.J. and Chambers, F.M. 1994 Determining the degree of peat decomposition for peat-based palaeoclimatic studies. *International Peat Journal* **5**, 7–24.

Blackford, J.J., Edwards, K.J., Dugmore, A.D., Cook, G.T. and Buckland, P.C. 1992 Icelandic volcanic ash and the mid-Holocene Scots pine (*Pinus sylvestris*) decline in northern Scotland. *The Holocene* **2**, 260–5.

Bloetjes, O.A.J. and van der Meer, J.J.M. 1992 *A preliminary stratigraphical description of peat development on Clara Bog*. Amsterdam. Fysisch-Geografisch en Bodemkundig Laboratorium, University of Amsterdam.

Boatman, D.J. 1983 The Silver Flow National Nature Reserve, Galloway, Scotland. *Journal of Biogeography* **10**B, 163–274.

Bourke, L. 1996 A watery end—prehistoric metalwork in the Shannon. *Archaeology Ireland* **10** (4), 9–11.

Bowman, S. 1990. *Radiocarbon dating*. London. British Museum Publications.

Bradley, J. 1991 Excavations at Moynagh Lough, County Meath. *Journal of the Royal Society of Antiquaries of Ireland* **121**, 5–26.

Breen, T.C. 1987 'The Giant's Road', Cloncraff or Bloomhill (Offaly), Ballynahownwood (Westmeath). In C. Cotter (ed.), *Excavations 1986*, 32. Dublin. Wordwell.

Breen, T.C. 1988 Excavation of a roadway at Bloomhill bog, County Offaly. *Proceedings of the Royal Irish Academy* **88C**, 321–9.

Brindley, A.L. 1995 Radiocarbon, chronology and the Bronze Age. In J. Waddell and E. Shee Twohig (eds), *Ireland in the Bronze Age: proceedings of the Dublin conference, April 1995*, 4–13. Dublin. Stationery Office.

Brindley, A. and Lanting, J. 1990 The dating of fulachta fiadh. In V. Buckley (comp.), *Burnt offerings: international contributions to burnt mound archaeology*, 55–6. Dublin. Wordwell.

Brindley, A.L., Lanting, J.N., and Mook, W.G. 1989–90 Radiocarbon dates from Irish fulachta fiadh and other burnt mounds. *Journal of Irish Archaeology* **5**, 25–33.

Brunning, R. and O'Sullivan, A. 1997 Wood species selection and woodworking techniques. In N. Nayling and A. Caseldine (eds), *Excavations at Caldicot, Gwent: Bronze Age palaeochannels in the lower Nedern valley*, 163–87. York. Council for British Archaeology.

Buckland, P.C. 1979 *Thorne Moors: a palaeoecological study of a Bronze Age site*. Department of Geography, University of Birmingham. Occasional Publication No. **8**. Birmingham. University of Birmingham.

Buckland, P.C., Buckland, P.I., Yuan Zhou, Don and Sadler, P.J. 1996 BUGS-A N.A.B.O. Project [CD-Rom]. New York. North Atlantic Biological Organisation.

Buckley, L. 1997 Skeletal report. In C. Mount, 'Adolf Mahr's excavations of an Early Bronze Age cemetery at Keenoge, County Meath'. *Proceedings of the Royal Irish Academy* **97C**, 44–57.

Buckley, L. 1998 Human skeletal report. In C. Mount, 'Five Early Bronze Age cemeteries at Brownstown, Graney West, Oldtown and Ploopluck, County Kildare, and Strawhall, County Carlow'. *Proceedings of the Royal Irish Academy* **98C**, 69–97.

Buckley, V.M. 1985 Curraghtarsna. *Current Archaeology* **98**, 70–1.

Buckley, V.M. 1990a Experiments using a reconstructed fulacht with a variety of rock types: implications for the petromorphology of fulachta fiadh. In V. Buckley (comp.), *Burnt offerings: international contributions to burnt mound archaeology*, 170–2. Dublin. Wordwell.

Buckley, V. (comp.) 1990b *Burnt offerings: international contributions to burnt mound archaeology*. Dublin. Wordwell.

Buckley, V.M. and Lawless, C. 1987 Prehistoric cooking in County Mayo. *Cathair na Mart* **7**, 32–6.

Bulleid, A. 1980 *The Lake Villages of Somerset* (7th edn). Taunton. Glastonbury Antiquarian Society.

Bulleid, A. and Gray, H.S.G. 1948 *The Meare Lake Village, vol. 1: a full description of the excavations and relics from the eastern half of the west village, 1910–1933*. Taunton. Privately printed at Taunton Castle.

Buttler, A., Warner, B.G., Grosvernier, P. and Matthey, Y. 1996 Vertical patterns of testate amoebae (Protozoa: Rhizopoda) and peat forming vegetation on cutover bogs in the Jura, Switzerland. *New Phytologist* **134**, 371–82.

Caseldine, C.J., Baker, A., Charman, D. and Hendon, D. 2000 A comparative study of luminescence and transmission properties of NaOH peat extracts: palaeoenvironmental implications for 'humification' studies. *The Holocene* **10** (5), 649–58.

Caseldine, C., Gearey, B., Hatton, J., Reilly, E., Stuijts, I. and Casparie, W. 2001 From the Wet to the Dry: palaeoecological studies at Derryville, Co. Tipperary, Ireland. In B. Raftery and J. Hickey (eds), *Recent developments in wetland research*, 99–113. WARP Occasional Paper 14, Monograph Series 2. Dept. of Archaeology, University College Dublin.

Caseldine, C.J. and Hatton, J.A. 1996 Early land clearance and wooden trackway construction in the third and fourth millennia BC at Corlea, Co. Longford. *Proceedings of the Royal Irish Academy* **95B**, 1–9.

Caseldine, C., Hatton, J. and Caseldine, A. 1996 Palaeoecological studies at Corlea (1988–1992). In B. Raftery, *Trackway excavations in the Mountdillon Bogs, Co. Longford, 1985–1991*, 379–92. Transactions of the Irish Archaeological Wetland Unit **3**. Dublin. Crannóg Publication.

Caseldine, C., Hatton, J., Huber, U., Chiverrell, R. and Woolley, N. 1996 Palaeoecological work at Corlea (1992–1995). In B. Raftery, *Trackway excavations in the Mountdillon Bogs, Co. Longford, 1985–1991*, 393–4. Transactions of the Irish Archaeological Wetland Unit **3**. Dublin. Crannóg Publication.

Casparie, W.A. 1972. Bog development in southeastern Drenthe (the Netherlands). *Vegetatio* **25**, 1–272.

Casparie, W.A. 1982 The Neolithic wooden trackway XXI (Bou) in the raised bog at Nieuw-Dordrecht (the Netherlands). *Palaeohistoria* **24**, 115–64.

Casparie, W.A. 1984 The three Bronze Age footpaths XVI (Bou), XVII (Bou) and XVIII (Bou) in the raised bog of southeast Drenthe (the Netherlands). *Palaeohistoria* **26,** 41–94.

Casparie, W.A. 1986 The two Iron Age trackways XIV (Bou) and XV (Bou) in the raised bog of southeast Drenthe (the Netherlands). *Palaeohistoria* **28**, 169–210.

Casparie, W.A. 1987 Bog trackways in the Netherlands. *Palaeohistoria* **29,** 35–65.

Casparie, W.A. 1992 Neolithic deforestation in the region of Emmen (the Netherlands). *European Palaeoclimate and Man* **3**, 115–28.

Casparie, W.A. 1993 The Bourtanger Moor: endurance and vulnerability of a raised bog system. *Hydrobiologia* **265**, 203–15.

Casparie, W. 1997 Peat development and morphology. In Lisheen Archaeological Project—Preliminary Report, 6–32. Unpublished report commissioned by Minorco Lisheen Ltd.

Casparie, W.A., Coert, G.A., de Leeuw, G., Smits, F. and Veen, H.D. 1983 De middeleeuwse keienweg van Bronneger, Gem. Borger. *Nieuwe Drentse Volksalmanak* **100**, 147–201.

Casparie, W.A. and Molema, J. 1990 Het middeleeuwse veenontginningslandschap bij Scheemda. *Palaeohistoria* **32**, 271–89.

Casparie, W.A. and Moloney, A. 1994 Neolithic wooden trackways and bog hydrology. *Journal of Palaeolimnology* **12**, 49–64.

Casparie, W.A. and Moloney, A. 1996 Corlea 1. Palaeo-environmental aspects of the trackway. In B. Raftery, *Trackway excavations in the Mountdillon Bogs, Co. Longford, 1985–1991*, 367–77 Transactions of the Irish Archaeological Wetland Unit **3**. Dublin. Crannóg Publication.

Casparie, W.A. and Stevens, P. 2001 Bronze Age stone-built way through an Irish bog: site Killoran 18. In W.H. Metz, B.L. van Beek and H. Steegstra (eds), *Patina: essays presented to Jay Jordan Butler on the occasion of his 80th birthday*, 195–206. Groningen/Amsterdam. Metz, van Beek and Steegstra.

Casparie, W.A. and Streefkerk, J.G. 1992 Climatological, stratigraphic and palaeo-ecological aspects of mire development. In J.T.A. Verhoeven (ed.), *Fens and bogs in the Netherlands: vegetation, history, nutrient dynamics and conservation*, 81–129. Dordrecht. Kluwer Academic Press.

Casparie, W.A., Streefkerk, J.G. and Zandstra, R.J. 1992 De neolithische veenweg van Nieuw-Dordrecht (Dr.). Een archeologisch monument op de helling. *Paleo-aktueel* **3**, 55–60.

Chambers, F.M. 1984 *Studies on the initiation, growth rate and humification of blanket peats in South Wales.* Department of Geography, University of Keele. Occasional Paper No.**9**. Keele. University of Keele.

Chambers, F.M. 1993 Late Quaternary climatic change and human impact: commentary and conclusions. In F.M. Chambers (ed.), *Climate change and human impact on the landscape*, 247–58. London. Chapman and Hall.

Champion, T.C. 1989 From Bronze Age to Iron Age in Ireland. In M. Stig Sørensen and R. Thomas (eds), *The Bronze Age-Iron Age transition in Europe*, 287–303. British Archaeological Reports, International series 483. Oxford.

Chapman, H.P. and Gearey, B.R. 2000 Palaeoecology and the perception of prehistoric landscapes: some comments on visual approaches to phenomenology. *Antiquity* **74**, 316–19.

Chapman, R. 1981 The emergence of formal disposal areas and the "problem" of megalithic tombs in prehistoric Europe. In R. Chapman, I. Kinnes and K. Randsborg (eds), *The archaeology of death*, 77–81. Cambridge. Cambridge University Press.

Charman, D.J. 1997 Modelling hydrological relationships of testate amoebae (Protozoa: Rhizopoda) on New Zealand peatlands. *Journal of the Royal Society of New Zealand* **27** (4), 465–83.

Charman, D.J., Caseldine, C.J., Baker, A., Gearey, B., Hatton, J. and Proctor, C. 2001 Palaeohydrological records from peat profiles and speleothems in Sutherland, Northwest Scotland. *Quaternary Research* **55**, 223–34.

Charman, D.J., Hendon, D. and Woodland, W.A. 2000 *The identification of testate amoebae (Protozoa:Rhizopoda) in peats.* Quaternary Research Association Technical Guide **9**. Cambridge. Quaternary Research Association.

Charman, D.J. and Warner, B.G. 1992 Relationship between testate amoebae (Protozoa: Rhizopoda) and microenvironmental parameters on a forested peatland in northeastern Ontario. *Canadian Journal of Zoology* **70**, 2474–82.

Cherry, S. 1990 The finds from fulachta fiadh. In V. Buckley (comp.), *Burnt offerings: international contributions to burnt mound archaeology*, 49–54. Dublin. Wordwell.

Coles, B.J. and Coles, J.M. 1986 *Sweet Track to Glastonbury: the Somerset Levels in prehistory.* London. Thames and Hudson.

Coles, B. and Coles, J. 1989 *People of the wetlands: bogs, bodies and lake-dwellers.* London. Thames and Hudson.

Coles, B. and Coles, J. 1991 The wetland revolution; a natural event. In B. Coles (ed.), *The wetland revolution in prehistory*, 147–53. WARP Occasional Paper **6**. Exeter. The Prehistoric Society and Wetland Archaeological Research Project.

Coles, J.M. (ed.) 1975–89 *Somerset Levels Papers* **1–15**.

Coles, J.M. 1987 Meare Village East. The excavations of A. Bulleid and H. St George Gray, 1932–1956. *Somerset Levels Papers* **13**.

Coles, J.M., Caseldine, A.E. and Morgan R.A. 1982 The Eclipse track 1980. *Somerset Levels Papers* **8**, 26–38.

Coles, J.M. and Darrah, R.J. 1977 Experimental investigations in hurdle-making. *Somerset Levels Papers* **3**, 32–8.

Coles, J.M., Fleming, A.M. and Orme, B.J. 1980 The Baker site: a Neolithic platform. *Somerset Levels Papers* **6**, 6–23.

Coles, J.M., Heal, S.V.E. and Orme, B.J. 1978 The use and character of wood in prehistoric Britain and Ireland. *Proceedings of the Prehistoric Society* **44**, 1–45.

Coles, J.M. and Minnitt, S. 1995 *Industrious and fairly civilised: the Glastonbury Lake Village.* [Somerset]. Somerset Levels Project and Somerset County Council Museum Service.

Coles, J.M. and Orme, B.J. 1976 The Meare Heath trackway: excavation of a Bronze Age structure in the Somerset Levels. *Proceedings of the Prehistoric Society* **42**, 293–318.

Coles, J.M. and Orme, B.J. 1977a Garvin's tracks. *Somerset Levels Papers* **3**, 73–81.

Coles, J.M. and Orme, B.J. 1977b Rowland's hurdle trackway. *Somerset Levels Papers* **3**, 39–51.

Coles, J.M. and Orme, B.J. 1977c Neolithic hurdles from Walton Heath, Somerset. *Somerset Levels Papers* **3**, 6–29.

Coles, J.M. and Orme, B.J. 1985 Prehistoric woodworking from the Somerset Levels: 3. Roundwood. *Somerset Levels Papers* **11**, 25–50.

Condit, T. 1992 Ireland's hillfort capital—Baltinglass, Co.Wicklow. *Archaeology Ireland* **6** (3), 16–20.

Condit, T. 1993 Travelling earthwork arrives at Tara. *Archaeology Ireland* **7** (4), 10–14.

Condit, T. 1996 Gold and *fulachta fiadh*—the Mooghaun find, 1854. *Archaeology Ireland* **10** (4), 20–3.

Condit, T. and Gibbons, M. 1988 Linear earthworks in County Waterford. *Decies* **39**, 19–28.

Connolly, A. 1994 Saddle querns in Ireland. *Ulster Journal of Archaeology* **57**, 26–36.

Conway, V.M. 1954 Stratigraphy and pollen analysis of southern Pennine peats. *Journal of Ecology* **42**, 117–47.

Cooney, G. and Grogan, E. 1994 *Irish prehistory: a social perspective.* Bray. Wordwell.

Coope, G.R. and Osborne, P.J. 1967 Report on the coleopterous fauna of the Roman well at Barnsley Park, Gloucestershire. *Transactions of the Bristol and Gloucestershire Archaeological Society* **86**, 84–7.

Corbett, A. 1973 An illustrated introduction to the testate Rhizopods in *Sphagnum* with special reference to the area around Malham Tarn, Yorkshire. *Field Studies* **3**, 801–38.

Crawford, H. 1924 The Dowris Hoard. *Journal of the Royal Society of Antiquaries of Ireland* **54**, 14.

Cross, S. 1995 Coolalough. In I. Bennett (ed.), *Excavations 1994*, 55–6. Bray. Wordwell.

Cross, S. 2001 Changing places: landscape and mortuary practice in the Irish Middle Bronze Age. Unpublished PhD thesis, McMaster University (Ontario, Canada).

Crothers, N. 1997 Ballydown. In I. Bennett (ed.), *Excavations 1996*, 1. Bray. Wordwell.

Dalton, G. (ed.) 1967 *Tribal and peasant economies: readings in economic anthropology.* New York. Natural History Press.

Daniels, R.E. and Eddy, A. 1990 *Handbook of European Sphagna.* London. HMSO.

Day, A. and McWilliams, P. (eds) 1993 *Ordnance Survey memoirs of Ireland vol.21: parishes of County Antrim VII, 1832–8 South Antrim.* Belfast. Institute of Irish Studies in association with the Royal Irish Academy.

Dodson, J.R. and Bradshaw, R.H.W. 1987 A history of vegetation and fire, 6,600 B.P. to present, County Sligo, western Ireland. *Boreas* **16**, 113–23.

Dolukhanov, P.M. 1991 Evolution of lakes and prehistoric settlement in northwestern Russia. In B. Coles (ed.), *The wetland revolution in prehistory*, 93–8. WARP Occasional Paper **6**. Exeter. The Prehistoric Society and Wetland Archaeological Research Project.

Donaghy, C. and Grogan, E. 1997 Navel gazing at Uisnech, Co. Westmeath. *Archaeology Ireland* **11** (4), 24–6.

Donisthorpe, H. St. J.K. 1939 *A preliminary list of the Coleoptera of Windsor Forest.* London. N. Lloyd and Co.

Doody, M. 1987 Ballyveelish, Co. Tipperary. In R.M. Cleary, M.F. Hurley and E.A. Twohig (eds), *Archaeological excavations on the Cork-Dublin gas pipeline (1981–82)*, 8–35. Archaeological Studies **1**. Cork. Department of Archaeology, University College Cork.

Doody, M. 1996 The Ballyhoura Hills Project. Interim report. *Discovery Programme reports: 4. Project results and reports 1994*, 15–22. Dublin. Royal Irish Academy/Discovery Programme.

Duffy, C. 1996 Derry. In I. Bennett (ed.), *Excavations 1995*, 52–3. Bray. Wordwell.

Duffy, E.A.J. 1953 *Scolytidae and Platypodidae.* Handbooks for the identification of British insects **5** (15). London. Royal Entomological Society.

Dupont, L.M. 1985 Temperature and rainfall variation in a raised bog ecosystem. A palaeoecolgical and isotope-geological study. Unpublished PhD thesis, University of Amsterdam.

Dupont, L.M. 1986 Temperature and rainfall variation in the Holocene based on comparative palaeoecology and isotope geology of a hummock and a hollow (Bourtangerveen, the Netherlands). *Review of Palaeobotany and Palynology* **48**, 71–159.

Dupont, L.M. 1987 Palaeoecological reconstruction of the successive stands of vegetation leading to a raised bog in the Meerstalblok area (the Netherlands). *Review of Palaeobotany and Palynology* **51**, 271–87.

Edlin, H.L. 1973 *Woodland crafts in Britain*. Newton Abbot. David and Charles.

Edwards, K.J. 1985 The anthropogenic factor in vegetational history. In K.J. Edwards and W.A. Warren (eds), *The Quaternary History of Ireland*, 187–220. London. Academic Press.

Ellis, C. 1999 Micromorphological description and interpretation of four thin sections from part of Derryville Bog. In M. Gowen (ed.), *The Lisheen Archaeological Project, 1996–98*, vol. 1b, appendix 1C. Dublin. Margaret Gowen and Co. Ltd.

Eogan, G. 1964 The later Bronze Age in Ireland in the light of recent research. *Proceedings of the Prehistoric Society* **30**, 268–351.

Eogan, G. 1983 *Hoards of the Irish later Bronze Age*. Dublin. University College Dublin.

Evans, A.T. and Moore, P.D. 1985 Surface pollen studies of *Calluna vulgaris* (L.) hull and their relevance to the interpretation of bog and moorland pollen diagrams. *Circaea* **3**, 173–8.

Evans, E.E. 1973 *The personality of Ireland: habitat, heritage and history*. Dublin. Lilliput Press.

Fahy, E.M. 1960 A hut and cooking places at Drombeg, Co. Cork. *Journal of the Cork Historical and Archaeological Society* **65**, 1–17.

Farmer, D.H. 1987 *The Oxford dictionary of saints*. Oxford. Oxford University Press.

Feehan, J. and O'Donovan, G. 1996 *The bogs of Ireland: an introduction to the natural, cultural and industrial heritage of Irish peatlands*. Dublin. University College Dublin Environmental Institute.

Field, D. 1999 Bury the dead in a sacred landscape. *British Archaeology* **43**, 6–7.

FitzGerald, W. 1898 An ancient footway of wooden planks across the Monavullagh Bog. *Journal of the Royal Society of Antiquaries of Ireland* **28**, 417–8.

Foss, P. and O'Connell, C. 1997 Bogland: study and utilization. In J.W. Foster (ed.), *Nature in Ireland: a scientific and cultural history*, 184–98. Dublin. Lilliput Press.

Foster, G.N., Nelson, B.H., Bilton, D.T., Lott, D.D., Merrit, R., Weyl, R.S. and Eyre, M.D. 1992 A classification and evaluation of Irish water beetle assemblages. *Aquatic Conservation: Marine and Freshwater Ecosystems* **2**, 185–208.

Fredengren, C. 1998 Shroove (Lough Gara). In I. Bennett (ed.), *Excavations 1997*, 156. Bray. Wordwell.

Friday, L.E. 1988 *Key to the adults of British water beetles*. Field Studies Council **7**, 1–151.

Girling, M.A. 1979 The fossil insect assemblage from the Meare Lake Village. *Somerset Levels Papers* **5**, 25–32.

Girling, M.A. 1984 Investigations of a second insect assemblage from the Sweet Track. *Somerset Levels Papers* **10**, 79–91.

Goddard, A. 1971 Studies of the vegetational changes associated with blanket peat initiation. Unpublished PhD thesis, Queen's University of Belfast.

Godwin, H. 1975 *History of the British flora*. Cambridge. Cambridge University Press.

Göransson, H. 1986 Man and forests of nemoral broad-leaved trees during the Stone Age. *Striae* **24**, 143–52.

Gowen, M. (ed.) 1999 *The Lisheen Archaeological Project, 1996–98*. Dublin. Margaret Gowen and Co. Ltd.

Greguss, P. 1945 *The identification of central European dicotyledonous trees and shrubs based on xylotomy*. Budapest. Hungarian Museum of Natural History.

Grimm, E. 1987 *TILIA** and *TILIA*GRAPH.* Version 1.12. Illinois. Springfield.

Grogan, E. 1996 Excavations at Mooghaun South, 1994. Interim report. *Discovery Programme reports: 4. Project results and reports 1994*, 47–57. Dublin. Royal Irish Academy/Discovery Programme.

Grogan, E. and Condit, T. 1995 New hillfort gives clue to the late Bronze Age. *Archaeology Ireland* **8** (2), 7.

Grogan, E., Condit, T., O'Carroll, F., O'Sullivan, A., and Daly, A. 1996 Tracing the late prehistoric landscape in north Munster. *Discovery Programme reports: 4. Project results and reports 1994*, 26–46. Dublin. Royal Irish Academy/Discovery Programme.

Grogan, E. and Daly, A. 1994 Excavations at Clenagh, Co. Clare. Interim report. *Discovery Programme reports: 4. Project results and reports 1994*, 58–62. Dublin. Royal Irish Academy/Discovery Programme.

Grosse-Brauckmann, G. 1980 Ablagerungen der Moore. In K. Göttlich (ed.), *Moor-und Torfkunde* (2nd edn), 130–173. Stuttgart. Schweizerbart.

Hall, D., Evans, C., Hodder, I. and Pryor, F. 1987 The fenlands of East Anglia, England: survey and excavation. In J.M. Coles and A.J. Lawson (eds), *European wetlands in prehistory*, 169–201. Oxford. Clarendon Press.

Hall, V.A., Pilcher, J.R. and McCormac, F.G. 1994 Icelandic volcanic ash and the mid-Holocene Scots pine (*Pinus sylvestris*) decline in the north of Ireland: no correlation. *The Holocene* **4**, 79–83.

Halpin, A. 1984 A preliminary survey of archaeological material recovered from peatlands in the Republic of Ireland. Unpublished report commissioned by the Office of Public Works.

Hammond, R.F. 1981 *The peatlands of Ireland.* Soil Survey Bulletin **35**. Dublin. An Foras Talúntais.

Hammond, R.F., van der Krogt, G. and Osinga, T. 1990 Vegetation and water-tables on two raised bog remnants in County Kildare. In G. Doyle (ed.), *Ecology and conservation of Irish peatlands*, 121–35. Dublin. Royal Irish Academy.

Hansen, M. 1987 *The hydrophiloidea (Coleoptera) of Fennoscandia and Denmark.* Fauna Entomologica Scandinavica **18**. Leiden/Copenhagen. E.J. Brill/Scandinavian Science Press Ltd.

Harbison, P. 1969 *The axes of the Early Bronze Age in Ireland.* Prähistorische Bronzefunde Abteilung IX, Band I. Munchen. C.H. Beck'sche.

Harde, K.W. 1984 *A field guide in colour to beetles.* London. Octopus.

Hayden, A. 1994 Kilnacarriag. In I. Bennett (ed.), *Excavations 1993*, 80. Bray. Wordwell.

Hayen, H. 1957 Zur Bautechnik und typologie der vorgeschichtlichen, frühgeschichtlichen und mittelalterlichen hölzernen Moorwege und Moorstrassen. *Oldenburger Jahrbuch* **56** (2), 83–189.

Hayen, H. 1970 Der bronzezeitliche Stapfweg IV(St) im Moore bei Gross-Heins, Kr. Verden. *Neu Ausgrabungen und Forschungen in Niedersachen* **5**, 376–88.

Hayen, H. 1985 Bergung, wissenschaftliche Untersuchung und Konservierung moor-archäologischen Funde. *Moorarchäologische Untersuchungen aus Nordwestdeutschland* **8,** 1–43.

Hayen, H. 1987 Peat bog archaeology in Lower Saxony, West Germany. In J.M. Coles and A.J. Lawson (eds), *European wetlands in prehistory*, 117–36. Oxford. Clarendon Press.

Hayen, H. 1991 Randmoore zwischen Marsch und Geest: Anmerkungen zu ihrer verkehrstechnischen Funktion. *Archäologische Mitteilungen aus Nordwest-Deutschland Beiheft* **5**, 109–122. Oldenburg.

Hayen, H. 1997 Holz als Werkstoff-Hinweise aus ur-und frühgeschichtlichen Moorfunden. *Probleme der Küstenforschung im südlichen Nordseegebiet* **24**, 311–65.

Heal, O.W. 1962 The abundance and micro-distribution of testate amoebae (Rhizopoda:Testacea) in *Sphagnum. Oikos* **13**, 35–47.

Heal, V. 1991 The technology of the worked wood and bark. In S.P. Needham (ed.), *Excavation and salvage at Runnymede Bridge, 1978: the Late Bronze Age waterfront site,* 140–7. London. British Museum Press.

Hendon, D. and Charman, D.J. 1997 The preparation of testate amoebae (Protozoa: Rhizopoda) samples from peat. *The Holocene* **7** (2), 199–205.

Hendon, D., Charman, D.J. and Kent, M. 2001 Palaeohydrological records derived from testate amoebae analysis from peatlands in northern England: within-site variability, between-site comparability and palaeoclimatic implications. *The Holocene* **11**, 127–48.

Herity, M. 1991 *Rathcroghan and Carnfree: Celtic royal sites in Roscommon.* Dublin. Na Clocha Breacha.

Hill, J.D. 1989 Re-thinking the Iron Age. *Scottish Archaeological Review* **6**, 16–24.

Hill, R. 1992 The origin and dynamics of the birch wood on Clara Bog, Co. Offaly. Unpublished BA thesis, Trinity College Dublin.

Hillam, J., Groves, C.M., Brown, D.M., Baillie, M.G.L., Coles, J.M. and Coles, B.J. 1990 Dendrochronology of the English Neolithic. *Antiquity* **64**, 210–20.

Hodder, I. 1990 *The domestication of Europe: structure and contingency in Neolithic societies.* Oxford. Basil Blackwell.

Hodder, M.A. 1990 Burnt mounds in the English West Midlands. In V. Buckley (comp.), *Burnt offerings: international contributions to burnt mound archaeology*, 106–11. Dublin. Wordwell.

Hodges, H.W.M. 1958 A hunting camp at Cullyhanna Lough, near Newtown Hamilton, Co. Armagh. *Ulster Journal of Archaeology* **21**, 70–113.

Horion, A. 1951 *Verzeichnis der Käfer Mitteleuropas (Deutschland, Österreich, Tschechoslowakei): mit kurzen faunistischen Angaben.* Stuttgart. Alfred Kernen Verlag.

Horion, A. 1960 *Faunistik der Mitteleuropäischen Käefer 7. Clavicornia 1. Teil (Sphaeritidae bis Phalacridae).* Uberlingen-Bodensee. Schmidt.

Hurl, D. 1996 Killymoon Demesne. In I. Bennett (ed.), *Excavations 1995*, 84. Bray. Wordwell.

Hurley, M.F. 1987 A fulacht fiadh at Clashroe, Meelin, Co. Cork. *Journal of the Cork Archaeological and Historical Society* **92**, 95–105.

Hyman, P.S. 1992 *A review of the scarce and threatened Coleoptera of Great Britain, part 1* (Revised and updated by M.S. Parsons). Peterborough. UK Joint Nature Conservation Committee.

Hyman, P.S. 1994 *A review of the scarce and threatened Coleoptera of Great Britain, part 2* (Revised and updated by M.S. Parsons). Peterborough. UK Joint Nature Conservation Committee.

IAWU. 1995 Preliminary archaeological assessment of part of Derryville Bog 1995, Bord na Móna Littleton Works, Counties Tipperary and Kilkenny. Unpublished report commissioned by Minorco Lisheen Ltd.

IAWU. 1996a Final report on the archaeological assessment of part of Derryville Bog 1995. Unpublished report commissioned by Minorco Lisheen Ltd.

IAWU. 1996b Littleton Works, pilot survey 1995. In I. Bennett (ed.), *Excavations 1995*, 93. Bray. Wordwell.

Ivanov, K.E. 1981 *Water movements in mirelands.* London. Academic Press.

Jahn, H. 1979 *Pilze die an Holz wachsen.* Herford. Bussesche Verlagshandlung.

Jelicic, L. and O'Connell, M. 1992 History of vegetation and land use from 3200 B.P. to the present in the north-west Burren, a karstic region of western Ireland. *Vegetation History and Archaeobotany* **1**, 119–40.

Jessop, L. 1986 *Coleoptera: Scarabaeidae.* Handbooks for the identification of British insects **5** (11). London. Royal Entomological Society.

Johnson, L.C., and Damman, A.W.H. 1991 Species-controlled *Sphagnum* decay on a south Swedish raised bog. *Oikos* **61**, 234–42.

Joy, N.H. 1932 *A practical handbook of British beetles: Vols 1 & 2.* London. Witherby.

Kelly, D.L. and Kirby, E.N. 1982 Irish native woodland over limestone. In J. White (ed.), *Studies on Irish vegetation*, 181–98. Dublin. Royal Dublin Society.

Kelly, F. 1976 The old Irish tree list. *Celtica* **11**, 107–24.

Kelly, F. 1998 *Early Irish farming: a study based mainly on the law-texts of the 7th and 8th centuries AD.* Early Irish Law Series **IV**. Dublin. Dublin Institute for Advanced Studies.

Kelly, S. 1995 Artefact decomposition in peatlands. In Moloney, A., Bermingham, N., Jennings, D., Keane, M., McDermott, C. and O Carroll, E., *Blackwater survey and excavations, artefact deterioration in peatlands, Lough More, Co. Mayo*, 141–54. Transactions of the Irish Archaeological Wetland Unit **4**. Dublin. Crannóg Publication.

Kenward, H.K. 1980 A tested set of techniques for the extraction of plant and animal macrofossils from waterlogged archaeological deposits. *Science and Archaeology* **22**, 3–15.

Kenward, H.K. and Allison, E.P. 1995 Rural origins of the urban insect fauna. In A. Hall and H.K. Kenward (eds), *Urban-rural connections: perspectives from environmental archaeology*, 55–77. Oxbow Monograph **47**. Oxford. Oxbow.

Kenward, H.K. and Hall, A.R. 1995 Biological evidence from the Anglo-Scandinavian deposits at 16–22 Coppergate. *The Archaeology of York* **14** (7). London. Council for British Archaeology.

Kloet, G.S. and Hincks, W.D. 1977 *A checklist of British insects, part 3: Coleoptera and Strepsiptera* (2nd edn; revised by R.D. Pope). Handbooks for the identification of British insects **11** (3). London. Royal Entomological Society.

Kowalewski, S., Blanton, R., Feinman, G. and Finsten, L. 1983 Boundaries, scale and internal organisation. *Journal of Anthropological Archaeology* **2** (1), 32–56.

Lamb, H.H. 1977 *Climate: present, past and future. Vol. 2, climate history and the future*. London. Methuen.

Lamb, H.H. 1982 *Climate, history and the modern world*. London. Methuen.

Landin, B.O. 1961 Ecological studies on dung-beetles. *Opuscula Entomologica Supplementum* **19**, 1–228.

Lanting, J. and Brindley, A. 1996 Irish logboats and their European context. *Journal of Irish Archaeology* **7**, 85–95.

Larsson, T.B. 1990 Skärvstenshögar—the burnt mounds of Sweden. In V. Buckley (comp.), *Burnt offerings: international contributions to burnt mound archaeology*, 142–53. Dublin. Wordwell.

Lawless, C. 1992 Lough More, Bofeenaun, Co. Mayo—crannóg, fulachta fiadh, deer traps and associated archaeological sites. *Cathair na Mart* **12** (1), 12–31.

Lekander, B., Bejer-Petersen, B., Kangas, E. and Bakke, A. 1977 The distribution of bark beetles in the Nordic countries. *Acta Entomologica Fennica* **32**, 1–115.

Lindroth, C.H. 1969 The ground beetles (Carabidae, excl. Cicindelinae) of Canada and Alaska. *Opuscula Entomologica* **34**, 945–1192.

Lindroth, C.H. 1973 *Surtsey, Iceland: the development of a new fauna, 1963–1970. Terrestrial Invertebrates*. Fauna Entomologica Scandinavica **5**.

Lindroth, C.H. 1974 *Coleoptera Carabidae*. Handbooks for the identification of British insects **4** (2). London. Royal Entomological Society.

Lindroth, C.H. 1985 *The Carabidae (Coleoptera) of Fennoscandia and Denmark*. Fauna Entomologica Scandinavica **15** (1) Leiden/Copenhagen. E.J. Brill/Scandinavian Science Press Ltd.

Lindsay, J.M. 1974 The use of woodland in Argyllhire and Perthshire between 1650 and 1850. Unpublished PhD thesis, University of Edinburgh.

Louwe-Kooijmans, L. 1993 Wetland exploitation and upland relations of prehistoric communities in the Netherlands. In J. Gardiner (ed.), *Flatlands and wetlands: current themes in East Anglian archaeology*, 71–116. East Anglian Archaeological Reports **50**. Norwich. Scole Archaeological Committee.

Lucas, A.T. 1963 The sacred trees of Ireland. *Journal of the Cork Historical and Archaeological Society* **68**, 16–54.

Lucas, A.T. 1965 Washing and bathing in ancient Ireland. *Journal of the Royal Society of Antiquaries of Ireland* **95**, 65–114.

Lucas, A.T. 1975 A stone-laid trackway and wooden troughs, Timoney, Co. Tipperary. *North Munster Antiquarian Journal* **17**, 13–20.

Lucas, A.T. 1977 A stone-laid trackway and wooden troughs, Timoney, Co. Tipperary: a further note. *North Munster Antiquarian Journal* **19**, 69.

Lucas, A.T. 1985 Toghers or causeways: some evidence from archaeological, literary, historical and placename sources. *Proceedings of the Royal Irish Academy* **85C**, 37–60.

Macalister, R.A.S. 1932 An ancient road in the Bog of Allen. *Journal of the Royal Society of Antiquaries of Ireland* **62**, 137–41.

Makila, M. 1997 Holocene lateral expansion, peat growth and carbon accumulation on Haukkasuo, a raised bog in southeastern Finland. *Boreas* **26**, 1–14.

Mallory, J.P. and McNeill, T.E. 1991 *The archaeology of Ulster from colonization to plantation*. Belfast. The Institute of Irish Studies, Queen's University of Belfast.

Manning, C. 1985 A Neolithic burial mound at Ashleypark, Co. Tipperary. *Proceedings of the Royal Irish Academy* **85C**, 61–100.

Manning, C. 1997 Daire Mór identified. *Peritia* **11**, 359–69.

Maynard, G.J. 1994 Annual Coleoptera report—1993. *Journal of the Derbyshire Entomological Society* **1994**, 10–15.

McGrail, S. (ed.) 1982 *Woodworking techniques before AD 1500*. British Archaeological Reports, International series 129. Oxford.

Meddens, F.M. 1996 Sites from the Thames estuary wetlands, England, and their Bronze Age use. *Antiquity* **70**, 325–34.

Meiserfeld, R.R. 1977 Die horizontale und vertikale Verteilung der Testaceen (Rhizopoden, Testacea) in *Sphagnum*. *Archiv fur Protistenkunde* **121**, 270–307.

Milner, E. 1992. *The tree book*. London. Collins and Brown.

Minnitt, S. and Coles, J. 1996 *The Lake Villages of Somerset*. [Taunton]. Glastonbury Antiquarian Society/Somerset Levels Project/Somerset County Council Museums Service.

Mitchell, G.F. 1965 Littleton Bog, Tipperary: an Irish agricultural record. *Journal of the Royal Society of Antiquaries of Ireland* **95**, 121–32.

Mitchell, G.F. 1989 *Man and environment in Valentia Island*. Dublin. Royal Irish Academy.

Mitchell, G.F. and Ryan, M. 1997 *Reading the Irish landscape*. Dublin. Town House.

Molloy, K. and O'Connell, M. 1993 Early land use and vegetation at Derryinver Hill, Renvyle Peninsula, Co. Galway, Ireland. In F.M. Chambers (ed.), *Climate change and human impact on the landscape*, 185–99. London. Chapman and Hall.

Moloney, A., Bermingham, N., Jennings, D., Keane, M., McDermott, C. and O Carroll, E. 1995 *Blackwater survey and excavations, aretfact deterioration in peatlands, Lough More, Co. Mayo*. Transactions of the Irish Archaeological Wetland Unit **4**. Dublin. Crannóg Publication.

Moloney, A., Jennings, D., Keane, M. and McDermott, C. 1993a *Survey of the raised bogs of County Longford*. Transactions of the Irish Archaeological Wetland Unit **1**. Dublin. Crannóg Publication.

Moloney, A., Jennings, D., Keane, M. and McDermott, C. 1993b *Excavations at Clonfinlough, County Offaly*. Transactions of the Irish Archaeological Wetland Unit **2**. Dublin. Crannóg Publication.

Moore, P.D., Webb, J.A. and Collinson, M.E. 1991 *Pollen analysis* (2nd edn). Oxford. Blackwell.

Morgan, R. 1988 *Tree-ring studies of wood used in Neolithic and Bronze Age trackways from the Somerset Levels*. British Archaeological Reports, British series 184. Oxford.

Morris, M.G. 1990 *Coleoptera: Curculionoidea: Apionidae, Attelabidae*. Handbooks for the identification of British insects **5** (16). London. Royal Entomological Society.

Munro, R. 1882 *Ancient Scottish lake-dwellings or crannogs: with a supplementary chapter on remains of lake-dwellings in England*. Edinburgh. David Douglas.

Needham, S. 1991 *Excavation and salvage at Runnymede Bridge, 1978: a Late Bronze Age waterfront site*. London. British Museum Press.

Newman, C. 1996 Woods, metamorphoses and mazes—the otherness of timber circles. *Archaeology Ireland* **10** (3), 34–7.

O'Brien, W. 1994 *Mount Gabriel*. Galway. Galway University Press.

O'Carroll, E. 1997 Lemanaghan. In I. Bennett (ed.), *Excavations 1996*, 93–4. Bray. Wordwell.

O'Carroll, F. and Ryan, M. 1992 A late Bronze Age hoard from Enagh East, Co. Clare. *North Munster Antiquarian Journal* **34**, 3–12.

O'Connell, M. 1986 Reconstruction of local landscape development in the post-Atlantic based on palaeoecological investigations at Carrownaglogh prehistoric field system, County Mayo, Ireland. *Review of Palaeobotany and Palynology* **49**, 117–76.

O'Connell, M. 1990 Early land use in north east County Mayo—the palaeoecological evidence. *Proceedings of the Royal Irish Academy* **90C**, 259–79.

O'Connell, M., Molloy, K. and Bowler, M. 1988 Post-glacial landscape evolution in Connemara, western Ireland, with particular reference to woodland history. In H.H. Birks, H.J.B. Birks, P.E. Kaland and D. Moe (eds), *The cultural landscape—past, present and future*, 267–87. Cambridge. Cambridge University Press.

Ó Drisceoil, D.A. 1988 Burnt mounds: cooking or bathing? *Antiquity* **62**, 671–80.

Ó Drisceóil, D.A. 1990 Fulachta fiadh: the value of early Irish literature. In V. Buckley (comp.), *Burnt offerings: international contributions to burnt mound archaeology*, 157–64. Dublin. Wordwell.

Ó Drisceoil, D. 1991 Fulachta fiadh: a general statement. *North Munster Antiquarian Journal* **33**, 3–6.

O'Flanagan, Rev. M. (comp.) 1930 *Letters containing information relative to the antiquities of the County Tipperary, collected during progress of the Ordnance Survey in 1840*. Bray.

O'Kelly, M.J. 1951 An Early Bronze Age ring-fort at Carrigillihy, Co. Cork. *Journal of the Cork Historical and Archaeological Society* **56**, 69–86.

O'Kelly, M.J. 1954 Excavations and experiments in ancient Irish cooking places. *Journal of the Royal Society of Antiquaries of Ireland* **84**, 105–55.

O'Sullivan, A. 1996 Neolithic, Bronze Age and Iron Age woodworking techniques. In B. Raftery, *Trackway excavations in the Mountdillon Bogs, Co. Longford, 1985–1991*, 291–342. Transactions of the Irish Archaeological Wetland Unit **3**. Dublin. Crannóg Publication.

O'Sullivan, A. 1997 Interpreting the archaeology of later Bronze Age lake settlements. *Journal of Irish Archaeology* **8**, 115–21.

O'Sullivan, A. and Sheehan, J. 1996 *The Iveragh peninsula: an archaeological survey of south Kerry*. Cork. Cork University Press.

Opie, H. 1997 Tintagh. In I. Bennett (ed.), *Excavations 1996*, 96. Bray. Wordwell.

Orme, B. 1982a Prehistoric woodlands and woodworking in the Somerset Levels. In S. McGrail (ed.), *Woodworking techniques before AD 1500*, 79–93. British Archaeological Reports, International series 129. Oxford.

Orme, B. 1982b The use of radiocarbon dates from the Somerset Levels. *Somerset Levels Papers* **8**, 9–25.

Orme, B. and Coles, J. 1983 Prehistoric woodworking from the Somerset Levels: 1. Timber. *Somerset Levels Papers* **9**, 19–43

Orme, B.J. and Coles, J.M. 1985 Prehistoric woodworking from the Somerset levels: 2. Species selection and prehistoric woodlands. *Somerset Levels Papers* **11**, 7–24.

Orme, B., Coles, J.M. and Sturdy, C.R. 1979 Meare Lake Village West: a report on recent work. *Somerset Levels Papers* **5**, 6–17.

Overbeck, F. 1975 *Botanisch-geologische Moorkunde*. Neumünster. K. Wachholtz.

Palm, T. 1959 Die Holz und Rinden käfer der süd und mittelschwedishen Laubäume. *Opuscula Entomologica Supplementum* **16**.

Pearce, E.J. 1957 *Coleoptera Pselaphidae*. Handbooks for the identification of British insects **4** (9). London. Royal Entomological Society.

Pearson, G.W. and Stuiver, M. 1986 High-precision calibration of the radiocarbon time scale, 500–2500 BC. *Radiocarbon* **28**, 839–62.

Petzelberger, B.E.M., Behre, K.E. and Geyh, M.A. 1999 Beginn der Hochmoorentwicklung und Ausbreitung der Hochmoore in Nordwestdeutschland-Erste Ergebnisse eines neues Projektes. *Telma* **29**, 21–38.

Price, L. 1945 St. Broghan's Road, Clonsast. *Journal of the Royal Society of Antiquaries of Ireland* **75**, 56–9.

Pryor, F. 1991 *English Heritage book of Flag Fen: prehistoric fenland centre*. London. B.T. Batsford Ltd/English Heritage.

Pryor, F. and Taylor, M. 1992 Flag Fen, Fengate, Peterborough II: further definition, techniques and assessment (1986–90). In B. Coles (ed.), *The wetland revolution in prehistory*, 37–46. WARP Occasional Paper **6**. Exeter. The Prehistoric Society and Wetland Archaeology Research Project.

Rackham, O. 1977 Neolithic woodland management in the Somerset Levels: Garvin's, Walton Heath and Rowland's tracks. *Somerset Levels Papers* **3**, 65–71.

Rackham, O. 1980 *Ancient woodland: its history, vegetation and uses in England*. London. Arnold.

Rackham, O. 1995 *Trees and woodland in the British landscape*. London. Weidenfeld and Nicolson.

Raftery, B. 1968 A newly-discovered hillfort at Garrangrena Lower, Co. Tipperary. *North Munster Antiquarian Journal* **11**, 3–6.

Raftery, B. 1990 *Trackways through time: investigations on Irish bog roads 1985–1989*. Dublin. Headline Publishing.

Raftery, B. 1994 *Pagan Celtic Ireland*. London. Thames and Hudson.

Raftery, B. 1996 *Trackway excavations in the Mountdillon Bogs, Co. Longford, 1985–1991*. Transactions of the Irish Archaeological Wetland Unit **3**. Dublin. Crannóg Publication.

Raftery, B., Jennings, D. and Moloney, A. 1995 Annaghcorrib 1, Garryduff Bog, Co. Galway. In Moloney, A., Bermingham, N., Jennings, D., Keane, M., McDermott, C. and O Carroll, E., *Blackwater survey and excavations, artefact deterioration in peatlands, Lough More, Co. Mayo*, 39–53. Transactions of the Irish Archaeological Wetland Unit **4**. Dublin. Crannóg Publication.

Raftery, J. 1942 Finds from three Ulster counties. *Ulster Journal of Archaeology* **5**, 130–1.

Raftery, J. (ed.) 1958 National Museum of Ireland: archaeological acquisitions in the year 1957. *Journal of the Royal Society of Antiquaries of Ireland* **88**, 115–38.

Raftery, J. (ed.) 1961 National Museum of Ireland: archaeological acquisitions in the year 1959. *Journal of the Royal Society of Antiquaries of Ireland* **91**, 68–107.

Ramsey, G. 1989 Middle Bronze Age weapons in Ireland. Unpublished PhD thesis, Queen's University of Belfast.

Ramsey, G. 1995 Middle Bronze Age metalwork: are artefact studies dead and buried? In J. Waddell and E. Shee Twohig (eds), *Ireland in the Bronze Age: proceedings of Dublin conference, April 1995*, 49–62. Dublin. Stationery Office.

Reilly, E. 1996 The insect fauna (Coleoptera) from the Neolithic trackways Corlea 9 and 10: the environmental implications. In B. Raftery, *Trackway Excavations in the Mountdillon Bogs, Co. Longford, 1985–1991*, 403–9. Transactions of the Irish Archaeological Wetland Unit **3**. Dublin. Crannóg Publication.

Reilly, E. 1997 The insect remains from Back Lane, Dublin, 1996/7 Excavations. Unpublished technical report for Margaret Gowen and Co. Ltd.

Roberts, J. 1998 A contextual approach to the interpretation of the early Bronze Age skeletons of the East Anglian Fens. *Antiquity* **72**, 188–97.

Robinson, M.A. 1991 The Neolithic and Late Bronze Age insect assemblages. In S. Needham (ed.), *Excavation and salvage at Runnymede Bridge, 1978: a Late Bronze Age waterfront site*, 277–326. London. British Museum Press in association with English Heritage.

Robinson, M.A. 1992 The Coleoptera from Flag Fen. *Antiquity* **66,** 467–69.

Robinson, M., Shimwell, D. and Cribbin, G. 1996 Boating in the Bronze Age—two logboats from Co. Mayo. *Archaeology Ireland* **10** (1), 12–13.

Ross, A. 1995 Ritual and Druids. In M.J. Green (ed.), *The Celtic World*, 423–44. London and New York. Routledge.

RPS Cairns Ltd. 1995 Environmental Impact Statement: Lisheen zinc/lead project. Unpublished report for Minorco Lisheen. Ltd.

Rynne, E. (ed.) 1962 National Museum of Ireland: archaeological acquisitions in the year 1960. *Journal of the Royal Society of Antiquaries of Ireland* **92**, 139–69.

Rynne, E. 1965 Toghers in Littleton Bog, Co. Tipperary. *North Munster Antiquarian Journal* **9**, 138–44.

Sands, R. 1997 *Prehistoric woodworking: the analysis and interpretation of Bronze and Iron Age toolmarks*. Wood in Archaeology **1**. London. Institute of Archaeology.

Scannell, M.J.P. and Synnott, D.M. 1987 *Census catalogue of the flora of Ireland* (2nd edn). Dublin. Stationery Office.

Schou Jørgensen, M. 1996 Oldtidens veje I Danmark. In J. Vellev (ed.), *Braut 1. Nordiske Vejhistoriske Studier*, 37–62. København. Danmarks Vejmuseum.

Schweingruber, F.H. 1978 *Mikroskopische Holzanatomie*. Birmensdorf. Eidgenössische Forschungsanstalt WSL.

Shirt, D.B. (ed.) 1987 *British Red Data Books: 2. Insects*. Peterborough. Nature Conservancy Council.

Smith, A.G. and Goddard, I.C. 1991 A 12,500 year record of vegetational history at Sluggan Bog, Co. Antrim, N. Ireland (incorporating a pollen scheme for non-specialists). *New Phytologist* **118**, 167–87.

Speight, M.C.D. 1976 *Agonum livens, Asemum striatum* and *Xylota coerulieventris*: insects new to Ireland. *Irish Naturalists' Journal* **18**, 274–5.

Stace, C. 1997 *New flora of the British Isles* (2nd edn). Cambridge. Cambridge University Press.

Stockmarr, J. 1971 Tablets with spores used in absolute pollen analysis. *Pollen et Spores* **13**, 615–21.

Stoneman, R. 1993 Late Holocene peat stratigraphy and climate: extending and refining the model. Unpublished PhD thesis, University of Southampton.

Stoneman, R., Barber, K.E. and Maddy, D. 1993 Present and past ecology of *Sphagnum imbricatum* and its significance in raised peat climate modelling. *Quaternary Newsletter* **70**, 14–22.

Stortelder, A.H.F., Hommel, P.W.F.M. and de Waal, R.W. 1998 *Broekbossen*. Utrecht. Natuurhistorische, Bibliotheek 66, KNNV Uitgeverij.

Streefkerk, J.G. and Casparie, W.A. 1989 *The hydrology of bog ecosystems: guidelines for management*. Utrecht. Staatsbosbeheer.

Stuiver, M. and Pearson, G.W. 1993 High-precision bidecadal calibration of the radiocarbon time scale AD 1950–500 BC and 2500–6000 BC. *Radiocarbon* **35**, 1–24.

Talhouk, A.M.S. 1969 *Insects and mites injurious to crops in Middle Eastern countries*. Berlin. Paul Parry.

Tallis, J.H. 1964 Studies on the southern Pennine peats I: the behaviour of *Sphagnum*. *Journal of Ecology* **32**, 345–53.

Taylor, M. 1992 Flag Fen: the wood. *Antiquity* **66**, 476–98.

Tempest, H. 1930 The Dorsey. *County Louth Archaeological Journal* **4** (2), 187–241.

Thomas, C. and Rackham, J. (eds). 1996 Bramcote Green, Bermondsey: a Bronze Age trackway and palaeo-environmental sequence. *Proceedings of the Prehistoric Society* **62**, 221–53.

Tipping, R. 1995 Holocene evolution of a lowland Scottish landscape: Kirkpatrick Fleming. *The Holocene* **5** (1), 83–96.

Tjaden, M.E.H. 1919 *Microscopisch onderzoek van hout*. Amsterdam. L.J. Veen.

Tohall, P., de Vries, H.L. and van Zeist, W. 1955 A trackway in Corlona Bog, Co. Leitrim. *Journal of the Royal Society of Antiquaries of Ireland* **85**, 77–83.

Tolonen, K. 1986 Rhizopod analysis. In B.E. Berglund (ed.), *Handbook of Holocene palaeohydrology,* 645–66. Chichester. John Wiley.

Tolonen, K., Warner, B.G. and Vasander, H. 1992 Ecology of testaceans (Protozoa:Rhizopoda) in mires in Southern Finland: I. Autecology. *Archiv für Protistenkunde* **142**, 119–38.

Tolonen, K., Warner, B.G. and Vasander, H. 1994 Ecology of testaceans (Protozoa:Rhizopoda) in mires in southern Finland: II-multivariate analysis. *Archiv fur Protistenkunde* **144**, 97–112.

Tottenham, C.E. 1954 *Coleoptera: Staphylinidae, Section (a): Piestinae to Evaesthetinae.* Handbooks for the identification of British insects **4** (8a). London. Royal Entomological Society.

Troels-Smith, J. 1955 Characterisation of unconsolidated sediments. *Danmarks Geologiske Undersögelse* **IV/3** (10), 38–73.

Tuinzing, W.D.J. 1988 *Geriefteen en andere bindmaterialen.* [Arnhem]. Stichting Wilg and Mand.

van der Meiden, H.A. 1961 De els in populierenbeplantingen. *Nederlands Bosbouwtijdschrift* **33** (6), 168–71.

van der Molen, P.C. and Hoekstra, S.P. 1988 A palaeoecological study of a hummock-hollow complex from Engbertsdijksveen. *Review of Palaeobotany and Palynology* **56**, 213–74.

van der Molen, P.C. 1992 Hummock-hollow complexes on Irish raised bogs. A palaeo/actuo ecological approach of environmental and climatic change. Thesis 211. University of Amsterdam.

van der Sanden, W.A.B. 1996 *Vereeuwigd in het veen: de verhalen van de Noordwest-Europese veenlijken.* Amsterdam. De Bataafse Leeuw.

van Geel, B. 1976 A palaeoecological study of Holocene peat bog sections, based on the analysis of pollen, spores and macro and microscopic remains of fungi, algae, cormophytes and animals. Unpublished PhD thesis, University of Amsterdam.

van Geel, B., Bregman, R., van der Molen, P.C., Dupont, L.M. and van Driel-Murray, C. 1989 Holocene raised bog deposits in the Netherlands as geochemical archives of prehistoric aerosols. *Acta Botanica Neerlandica* **38** (4), 467–76.

van Geel, B., Buurman, J. and Waterbolk, H.T. 1996 Archaeological and palaeoecological indications of an abrupt climate change in the Netherlands, and evidence for climatological teleconnections around 2650 BP. *Journal of Quaternary Science* **11** (6), 451–60.

van Geel, B. and Middeldorp, A.A. 1988 Vegetational history of Carbury Bog (Co. Kildare, Ireland) during the last 850 years and a test of the temperature indicator of 2H/1H measurements of peat samples in relation to historical sources and meteorological data. *New Phytologist* **109**, 377–92.

von Post, L. 1924 Das genetische system der organogenen Bildungen Schwedens. In *Memoires sur la nomenclature et la classification des sols*, 287–304. Comité International de Pedologie **IV**, Commission, 22. Helsinki. Comité International de Pedologie.

Waddell, J. 1995 Celts, Celticisation and the Irish Bronze Age. In J. Waddell and E. Shee Twohig (eds), *Ireland in the Bronze Age: proceedings of the Dublin conference, April 1995*, 158–69. Dublin. Stationery Office.

Warner, B.G. and Charman, D.J. 1994 Holocene changes on a peatland in northwestern Ontario interpreted from testate amoebae analysis. *Boreas* **23**, 270–9.

Watts, W.A. 1985 Quaternary vegetation cycles. In K.J. Edwards and W.P. Warren (eds), *The Quaternary history of Ireland*, 155–85. London. Academic Press.

Weir, D.A. 1987 Palynology and the environmental history of the Navan area. *Emania,* **3**, 34–43.

Weir, D.A. 1993a Dark Ages and the pollen record. *Emania* **11**, 1–20.

Weir, D.A. 1993b A palynological study of landscape development in County Louth during the first millennium BC and the first millennium AD. Interim report. *Discovery Programme reports: 1. Project results 1992*, 104–9. Dublin. Royal Irish Academy/Discovery Programme.

Weir, D.A. 1995 A palynological study of landscape and agricultural development in County Louth from the second millennium BC to the first millennium AD. *Discovery Programme reports: 2. Project results 1993*, 77–126. Dublin. Royal Irish Academy/Discovery Programme.

Wilson, A. 1990 *The culture of nature: North American landscapes from Disney to the Exxon Valdez.* Oxford. Blackwell.

Woodland, W.A. 1996 Holocene palaeohydrology from testate amoebae analysis—developing a model for British peatlands. Unpublished PhD thesis, University of Plymouth.

Woodland, W.A., Charman, D.J. and Sims, P.C. 1998 Quantitative estimates of water tables and soil moisture in Holocene peatlands from testate amoebae. *The Holocene* **8** (3), 261–73.

Wood-Martin, W.G. 1886 *The lake dwellings of Ireland: or ancient lacustrine habitations of Erin, commonly called crannogs.* Dublin. Hodges, Figgis and Co.